Contents

One Matter in Motion

Chapter 3 ₅₃
Work and Energy

Interlude Three
The Longest Jump _{after page 78}

Two States of Matter

Chapter 4 ₈₀
Solids, Liquids, and Gases

Four   New Concepts of Matter

Interlude Seven
Science Factories   after page 265

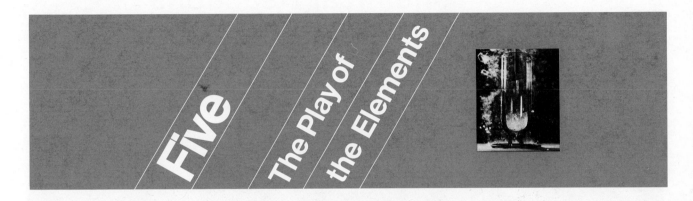

Chapter 13   266

Chemical Properties and Chemical Bonds

Chapter 14   293

Crystals, Ions, and Solutions

Six   Carbon   Chemistry

Chapter 17 363

Biochemistry

Interlude Ten
Destroying the Ozone Shield after page 385

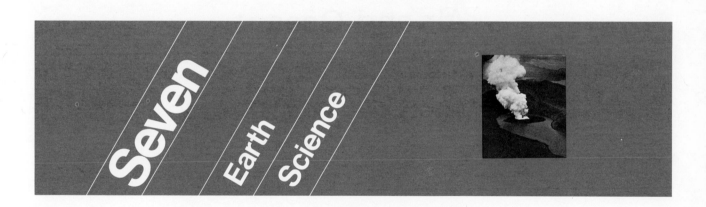

Seven Earth Science

Chapter 18 386

The Structure of the Earth

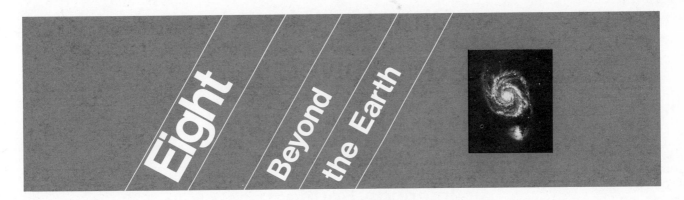

Eight

Beyond the Earth

Introduction

This book is about the whole physical world from the smallest particle to the most distant galaxy. Physical science covers four main fields: physics, chemistry, earth science, and astronomy. Each has its special branches. Physics includes the study of matter in motion (mechanics) as well as light, sound, electricity and magnetism, heat and other forms of energy, the different states of matter, the structure of atoms, and the nature of the elementary particles. Chemistry treats the changes and combinations of atoms and molecules, exploring and explaining the transformations of substances into one another. Organic chemistry focuses on the extraordinary compounds of carbon and leads to biochemistry, the basis of living matter. Earth science, or geology, investigates our home planet from its innermost part, the central core, out to the wispy edge of the atmosphere and includes the study of the oceans and continents, earthly changes from rainfall to volcanic eruptions, the earth's history, and the chemistry of the origin of life. Astronomy explores the world beyond the earth: the other planets, our moon and their moons, our sun and other stars, the gigantic assembly of stars in our own Milky Way Galaxy and the vast numbers of other galaxies far in the remote heavens. Astronomy also looks deep into the earliest age of the universe to the time when the world began, tens of billions of years ago.

Physical science is in part the story of how things work and how mankind has come to control some of the resources of the earth. It also explores some of the dangers which the application of scientific knowledge can bring. This is a book about science as well as about nature: it explains how things work—natural things like tides and man-made things like telescopes and airplanes—and it explains how we find out.

Different Kinds of Knowledge

There are different kinds of knowledge. We know some things with certainty. They simply must be so, without doubt and without possibility of future modification. The sum of 22 and 11 is 33, exactly. The area of a square with sides of 4 inches is 16 square inches, exactly. The area of a triangle is ½ (base) × (height), exactly. These examples from arithmetic and geometry are clearly true if we know the meanings of numbers, simple addition and multiplication, and words such as *square* and *triangle*. Arithmetic is the science of numbers with its laws of adding, subtracting, multiplying, and dividing.

Geometry studies the properties of squares, triangles, circles, and the rest of the plane and solid figures. But how do we know *when* to use the laws of arithmetic, geometry, algebra, calculus?

In a direct way, mathematical truths tell us nothing about the world we live in, just because they cannot say which things in the real world are triangles, squares, or spheres. Only experience can do that, in particular the experience of careful observation, experiment, and testing. Even the simplest arithmetic can only apply to things that we know add up like simple numbers. What is the sum of 6 bananas and 7 oranges? That may depend on our decision to group bananas and oranges under the joint label "fruit." Another example: What is the sum of three drops of mercury? A lot of mercury, we might suppose, but will it be in one big drop or a bunch of little droplets? This depends on *how* the addition is carried out, either slowly and carefully to join the three into one or roughly bouncing them into fragments of droplets. Or suppose you walk three miles east and then four miles north. Have you traveled seven miles or five miles? Either answer may be correct: seven miles for total walking distance or five miles distance from the starting point (since the east and north legs form the two sides of a 3–4–5 right triangle).

Mathematics as a Tool

To get knowledge of the world, then, we use mathematics as a tool for describing and analyzing. Mathematics is a precise language in which to think about the world of numbers, shapes, and the rest of the mathematical objects. When our actual world shows itself to have these numbers and shapes, then mathematics can be used. But our knowledge of the actual world is not mathematical; scientific knowledge is empirical. That is, science depends on observation and experiment to provide the materials with which to construct our pictures of what the world is and to test those pictures for their fit and their power. There is no way to know in advance what the world is like. We have to look at it with our eyes and our telescopes, microscopes, spectroscopes, and the other artificial extensions of our sense organs.

What kind of knowledge do we get when we look at the world? We can record what we observe, systematically and for many years, even for generations as in the beginnings of science in ancient astronomy. But that is never enough. Scientific explanations go beyond the reported facts to try to show how and why and when they happen. Some explanations of the facts are familiar repetitions, stories which do not genuinely explain much but keep us from realizing our puzzlement, like a recipe for a cake which succeeds but does not provide an understanding of the chemical changes involved. Such rules and recipes are not science. Science goes beyond recipes, beyond trial-and-error craftsmanship, beyond skills to give an account of nature. Scientists want to know why there are tides in the ocean and why there are two each day. A timetable of the tides will not satisfy them. How does eating bread and meat and milk produce blood, muscle, and bone? Why does white sunlight in mist become a rainbow of colors, none of them white? The list of questions is endless.

Science Unifies

But the scientific answers are not endless for science unifies the endless numbers of facts, the broad diversity of happenings, into pictures that hang together. Science groups the facts, fits them more or less together. The scientist is not so much working on a jig-saw puzzle as constructing a model. We imagine picture models of what we see and also of what we do not see. For example, we envision the air of the earth's atmosphere to be a light but mighty ocean; we imagine that an atom is designed like a tiny solar system; we guess that the forces around a magnet are like invisible elastic rubber bands; we understand the circulation of the blood to be similar to water flowing through a system of pipes. Are these pictures correct? Are they just dreams? To test the accuracy of these pictures, scientists check their predictions. The models must give the right predictions or be modified or even abandoned.

Science is inventive in another remarkable way: in science, we go beyond the facts. We see, through scientific imagination, that things which appear to be different from each other are really quite similar. Here the mathematical tools are most useful. Waves, for example, are easy to see in water, but the wave motion in a vibrating guitar string is harder to envision. When a trombone is played or a human voice heard the sound waves are invisible; light waves and radio waves are beyond observation. And yet wave motion is a unifying concept for understanding the processes of nature, for exploring and comprehending the basic properties of matter. The waves in water give a model for understanding other waving, vibrating, oscillating things, wherever they may be, however large or small, observable or imagined. The wave model, with its mathematical equations, is another scientific tool.

Some models of the world are developed from observations of everyday life. An ancient idea held that the world is fundamentally made of water. After all, we see liquid water, solid ice, steam and cloudy mist, water in solutions, water underground, water squeezed or pressed from so many substances. And early scientists knew there were fish fossils on some of the highest mountains, so they thought there had once been water covering the entire world. But their water model does not explain much, and it doesn't predict. Later, it was considered that all matter is composed of various proportions of four basic and elementary substances: earth, air, fire, and water. Today we have more elaborate models to explain the enormous variety of materials around us. Instead of 4 substances we have 106 elements. We break the elemental substances into their minimum recognizable particles, the atoms, which in turn are pictured as formed of still more basic particles, such as electrons, protons, neutrons, with forces of different sorts to hold them together. The atoms join in molecules, which then are structured into other substances. The molecular model, in its variety of pictures, has had great success in chemistry and medicine, explaining what we already knew, predicting what had not yet been observed, and offering shrewd guesses. For example, different molecular structures explain why one alcohol is a deadly poison and another alcohol is a pleasant drug.

Nature as a Mechanism

The most useful model of all has been the mechanical model, the view that nature works the way machines do. Here the key to understanding the world and all its parts is to apply Newton's mechanics, the laws of matter in motion, and nothing more. We do not need to know colors or smells or how things feel or what they want (or whether they want); whether we are studying comets in the sky or waterfalls or the eclipsing moon or ping-pong balls, we can ignore the myths and traditions about them. We need only know the masses of different bodies,

their positions in space, their present states of motion, and the forces acting upon them from outside. If we know these things, we can calculate what the bodies will do, where they will be in the future, and where they were in the past. Mass, speeds, forces—that is all. Objects act upon each other and react to each other, in accordance with a few simple mathematical laws, Newton's laws, just exactly like the gears, wheels, and spring of an enormous clockwork that was wound up at the beginning of the world. The action of the clock is completely determined, minute after minute as the hands go round, and so are all the other machines of the human and natural worlds. The world-machine, vast and complicated though it may be, would in this world-picture be as predictable as the tune played by a music box.

The mechanical model is more than a model of the universe and its parts. It is also a model for the construction of other explanations—in electricity chemistry biology even psychology Scientists look for regularities which can be summarized in laws of motion or the laws of other processes of change—melting, evaporating, exploding, ionization, forming a star from cosmic dust. If the laws are valid and the conditions fully known, then the explanations will provide predictions. And with prediction comes the opportunity for control.

With surprises eliminated, or reduced, control of our world becomes possible; in a world of surprises, of unexpected and therefore unexplained events, control is impossible. Even when the scientific laws are only partly like the clock model and prediction is incomplete, the explanation may be genuine and provide genuine power over events. A statistical law gives genuine knowledge of what happens overall but only chance predictions for each separate happening. A steel chain, for example, will support only so much weight; we do not need to know exactly which link will burst first, but it is useful to know that some one of the links will burst at a specific level of tension.

Scientific Theories as Models

Scientific theories are models. They are not really pictures of reality so much as guesses that reality for certain purposes, can be pictured that way. A scientific theory is a conjecture about the world. Like any guess, a scientific theory has to be tested against experience.

Perhaps it will be refuted and shown to be false; perhaps it will be confirmed and shown to be usable. It can be shown to be true or partly true or false, but only after it has been tested. So to be scientific, an idea or a theory must be testable.

True or false, our theoretical models are selective in what they picture. The astronomer tries to catch hold of the most important features of what he is studying: is it the color of the light from a star which gives the main clue to its nature? Scientific theories, as we shall see in every field, are models of reality but they are not so much realistic photographs as sketches and drawings, sometimes even cartoons. A cartoon is a caricature, it leaves out all the details except the most important ones. It does not tell the entire truth, in all its detail and complexity but it captures an important aspect of the truth. A scientific theory is a simplification; it claims to be no more. It will be tested and revised. And ultimately every time, it will be replaced.

Newton's mechanics is a simplified view, only a cartoon of actual things since after all nature is not just a machine: molecules are not billiard balls; the sun, moon, and planets are not parts of a solar clock; the flow of rivers and glaciers are not machine motions; digestion is not a mechanical process. But since the laws of mechanics successfully describe the ways these things move, nature is somewhat like a machine; the mechanical model is somewhat true.

These pictures are typical of science. For example, we can use the Newtonian mechanics to explain the movements and collisions of billiard balls on a flat pool table; then we can think of the balls moving in three dimensions, somewhat like extending tic-tac-toe to three dimensions. We can imagine the movement of air molecules within the walls, floor, and ceiling of a room as though the molecules were microscopic billiard balls, colliding with each other, bouncing off the walls with perfect elasticity. This model is partly successful but basically inaccurate because, after all, molecules of oxygen and nitrogen and carbon dioxide in the air are not identical to each other in the way billiard balls are, and they are neither solid nor elastic. Nevertheless, the billiard-ball model was good enough to provide a basis for the kinetic theory of gases, a beautiful and practical account of most of the properties of gases, as well as the beginning for a theory of the behavior of liquids.

The kinetic theory's picture is partly correct, and that is what counts.

Some successful models are not based on pictures, but then they call on mathematical forms to give their own reconstructions of reality. We may find a wave equation useful to describe a physical process, even though we can detect nothing waving We may use a four-dimensional geometry whereas our pictorial imagination is limited to three dimensions. Here, once again, mathematics is a tool, but this time for imagination as well as measurement and calculation.

"Free Inventions of the Human Spirit"

The scientific imagination has brought surprises from the beginning, long before Einstein's ideas about curved space. The first powerful surprise was the unity of nature. Newton's law of universal gravitation (namely that all things attract each other, just as the earth attracts an apple and makes it fall) was overwhelmingly and surprisingly successful because it described *under only one formula* the motion of the moon, the tides, the orbits of the planets, the paths of comets, the rising and falling of a pendulum, the trajectory of an artillery shell, the free fall of a stone from a tower. Newton proved that these motions, which look so different in observations, are alike in their fundamental character. In a pre-scientific picture of the world, they were as far apart as the heavens and the earth. So it is not simply observations and experiments that bring scientific progress, but thinking beyond the observed factual differences— guessing, model-making, imagining. Einstein said it this way: "Concepts and theories are the free inventions of the human spirit."

Even things that seem very stable and fixed can invite dynamic questions. Why does the map of the east coast of South America look like a rough jig-saw fit to the west bulge and indentation of Africa? Imagine great flat plates of rock, sliding and crashing over and under each other, and go on to an inventive history of the geology of the continents. Why do children look like their parents, whether animals, people or plants? Imagine that the molecules of living matter carry these inherited characteristics to new generations by some kind of electrical structure; think of the model of a cookie-cutter making cookie after cookie or of a template for a key, pressing the same form into one metal blank after another, and go on to the theory of biochemical genetics.

When the mechanical model unified different parts of the universe, the effect on humankind was tremendous. What a vast change occurred in the seventeenth century when the same physics explained heaven and earth! Our home was no longer unique, and the stars above us were no longer divine. This discovery depended partly on seeing the unity in a literal sense, seeing that, like the earth, the moon has mountains and we can estimate their heights from their shadows, as Galileo did with his pioneering little telescope.

Unifying the world depended even more on the discovery that mathematical thinking is the most powerful way into nature. But not the old mathematics of counting and measuring. It was static, useful for a purely observational science and for the skilled crafts and pre-scientific technology. Modern science needed a mathematical language of change and motion, of dynamic actions and forces, of shifting shapes and sizes, of measurements to be sure but also of errors in measurement, and of probability and statistics. How, for example, were we to express the ways that properties depend on each other as they change, especially how they change with time? Along with the new science of Galileo and Newton came a new mathematics. Logarithms were invented for rapid calculation. Algebra was joined with geometry to give equations for shapes, patterns, and orbits, with the striking result that the equation $x^2 + y^2 = r^2$ represents a circle of radius r (with center at the origin of the graph where $x = y = 0$) as fully as a drawing of that circle. The power of handling equations was combined with the insights of geometry Later the calculus was invented to deal with continuous changes and motions and with instantaneous values of changing quantities like accelerating speeds and expanding volumes.

Understanding Step by Step

Measuring too was changed, in the laboratory as well as in the new mathematical analysis. Speeds, colors, extremes of temperature, acidity of solutions, distance to the moon, to the stars, to the farthest galaxies, atomic clocks for telling ten-millionths of a second—the techniques have become marvels of sensitive accuracy far beyond what imagination might demand. And yet measurements also distort. We measure one or several properties out of a complex context, selecting what we want to know from a range of possible descriptions and properties. We neglect the remainder. Another scientist, in another specialty may go after the other properties. Science has made its way by this method of analysis and isolation, understanding step by step, bit by bit, breaking the natural world into parts, idealizing to pure cases. For example, Galileo devised a mental experiment in which he treated motion without friction and without external forces. This was an imaginative model; we are never really free from gravity on earth and only rarely approach frictionless movement, even when riding a hydrofoil on a cushion of air. When actual experiments are impossible, sometimes a physical model will do, scaled down or scaled up. We can build an artificial situation, as in the wind tunnel which simulates both the airplane and the atmospheric flow of air, showing the effect of a storm on wings, propellers, and tail, imitating but also idealizing, distorting, and falsifying. But that is the point: to see one thing, free of other complicating factors. If the size difference (scale) makes a difference of importance in the end, then that effect due to different scales will have to be identified and investigated as a new property in its turn. Scale differences may be of little importance, as in the flow of air over a wing shape, or of great importance, as in the estimation of the strength of a model bridge and a full-size bridge which are made of the same structural materials.

Our game in this book about physical science is to win an understanding of the sciences without using much mathematics, despite the undoubted fact that the history of modern science demonstrates that progress went with mathematical skill and with finely detailed calculations. We will talk *about* mathematically defined scientific ideas but not in mathematical language, for the most part. Kepler showed that the orbit of a planet as it revolves around the sun is an ellipse, not a circle; we will make that orbit, and its importance, clear without much worry about Kepler's mathematical derivations. Our mathematics will be common-sense and simple when we need it, rather than high-powered and abstract. Indeed the models of nature are almost always sensible to start with. Nature has repeatedly been found to be simple, if only a genius can be found with the proper simple idea to explain the natural events. Research is not always a search for new facts; it is even more powerful when it brings a new way of seeing old facts.

Simple and plausible as scientific explanations may be, they still contain surprises. Consider the plain old-fashioned truth of everyday understanding that time cannot go backwards, as simple as the last two lines of the rhyme about Humpty-Dumpty And yet scientific puzzles appear. First, our happily triumphant Newtonian mechanics was unable to tell the difference between forward time and reversed time. Whenever $(-t)$—which stands for time going backwards—was substituted for (t) in the physical equations the results were indistinguishable. Can you tell from a movie of a pendulum whether it is swinging forward in time, in the actual time order, or whether the film is being run backward? Some scientists said that time must be the one-way order given by cause-and-effect: the cause *defines* what is earlier, the effect being later. This seemed plausible since we never find the effect coming before its cause.

But the surprises continue. If time-order is causal-order, then events which are in no way causally related to one another might not be in the same order of time at all; they might exist in a different time-universe. Such fantasy of two disconnected worlds seems like science fiction, or an episode of *Star Trek,* but it is an aspect of Einstein's relativity theory.

Other scientists have said that the time-order is a property of our human-size slice of the world. They have shown that our ordinary experience of irreversible time sequences can be explained as the statistical result of billions upon billions of atomic-scale micro-events, each of them as reversible in its scale of size as Newton's physics appeared to claim

Seek Facts but Learn to Distrust Them

The surprises can be closer to home and still against our common expectations. We will see that light bends around corners (which is not surprising about sound!). In fact, nature is often counter-intuitive. In atomic science, we find that all matter has an innermost wave motion. In relativity physics, we see that speed of movement affects the rate at which time passes. What is the source of our common sense and of our ordinary intuitions about what is reasonable? At one time it seemed unreasonable to say that the moon falls toward the earth the way a rock falls from a cliff, but that is what Newton's

law taught us. Was Newton's theory as crazy as counter-intuitive as Chicken Little's tale about the sky falling? Earlier, people thought all motion required pushing or pulling to keep going, the stronger the push, the faster the motion. It was common sense to agree with the greatest Greek scientist, Aristotle, who wrote that "a moving body comes to a standstill when the force which pushes it along can no longer act so as to push it." But this belief depended on a lifetime of experience with motions which are slowed down by friction. When Galileo imagined motion without friction, science made progress, and common sense learned a little modesty Another example: shall we believe the earth is fixed because it feels that way or shall commonsense modesty go further and admit the counter-intuitive possibility of a fixed sun around which the earth moves? It is a fact that the earth feels solid and fixed. Science seeks facts but learns to distrust them for the facts may conceal, even while they may give clues. Indeed this is how scientists see the observations and the facts: facts are clues, and they must not be taken for the whole mystery story of nature. The scientist invents models to fit the factual clues together and to discover their causes.

But the scientist should be modest too. If we really assume that nature can be understood by analyzing it into separate parts, then we must be willing to look very carefully for explanations which will bring the parts together again. If adding distances, or adding bananas and oranges, are not obvious cases of simple arithmetic, then other bits of nature may not so simply add up to form a world either. Scientists play with models to describe the atoms and the individual properties of nature. How shall they put the parts together? A heap of bricks is not a wall, a pile of sand is not a sandcastle, and an accumulation of atoms is neither a star nor a human body. The mathematician and philosopher, Alfred North Whitehead, commented that the individuals in our world "do not blindly run, but run in accordance with the whole of which they form a part." Science must deal with wholes as well as parts.

Modesty suggests scientific investigation of the scientist, who is part of nature too. Must scientists distort what they study? Sometimes an experiment disturbs an object under investigation beyond repair. Look within an object with x-rays, to which so many substances are as transparent as plain glass is to light; then watch for the effects of the x-rays upon that object. Can we avoid such effects? We must look with *some* sort of illumination. An object observed is an object disturbed. In quantum physics we begin to understand, perhaps for the first time, that the observation, the instrument, and the observer are all part of the total situation, together with the object under study

The Potential for Good and for Evil

Modern science developed along with the modern European and American world of nations, exploration, international commerce, great cities, widespread education, religious tolerance, factory production, massive warfare, and political democracy. Science has all the marks of Western civilization: a great potential for good and also for evil and destruction. And yet, within itself, science can offer to men and women a model for understanding, modesty and independence. Newton once wrote: "If I have seen farther than others, it has been by standing on the shoulders of giants." And later, Thomas

Young: "Much as I venerate the name of Newton, I am not therefore obliged to believe he was infallible."

Science: Reason, Experiment, Imagination

Science, we have seen, is a fusion of three distinct components: first, the reasoning and logical structures of mathematics; second, experimental discoveries, the great inventive craft of instrument-making and the patient genius in their use, the endless technology of science; third, imagination, with disciplined conjectures about how to unify and explain the complexities of nature by simple and beautiful models of reality. Science should be studied carefully as a creation of the human species by which we can reach to the microcosm and to the universe, each beyond the experience of one person or one generation. A great writer, Anatole France, observed: "The wonder is not that the field of the stars is so vast, but that man has measured it."

Robert S. Cohen
Professor of Physics and Philosophy
Boston University

One of the largest and most dramatic black clouds so far known is a dark and diffuse nebulosity, a portion of which can be seen at the left side of this picture. The nebulosity is located near the middle of the constellation Cygnus (the Swan).

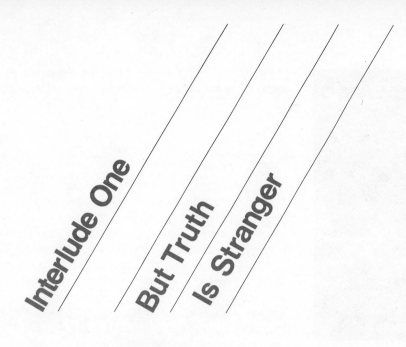

Interlude One

But Truth
Is Stranger

'Tis strange—but true; for truth is always strange,
Stranger than fiction.

Lord Byron, *Don Juan*

In science-fiction writing, human adventure, exploration, and social experiment are played out in imaginative ways against a backdrop of the attitudes,
methods, and terminology of science. Thus, science fiction provides us with
another way of looking at the scientific process. Using the same logic and
many of the same premises as the scientist (though often in an exaggerated or
prophetic form), the writer arrives at startling conclusions. Some of the novel-
ists' visions, such as Jules Verne's submarine and H. G. Wells' atomic energy,
are eventually realized by scientists, who are no less imaginative but more
restricted by the need for evidence to refute or confirm their conjectures.
    The following excerpt is from *The Black Cloud,* a science-fiction novel written

These two photographs of Pluto record its motion in 24 hours against a background of relatively fixed stars. Placed in a blinker microscope, such as the one used by Jensen in *The Black Cloud*, these two plates would be alternately visible to an observer, and the image of Pluto would appear to jump back and forth, making it easy to distinguish the planet from the background stars.

in 1957 by Fred Hoyle, the distinguished British astronomer. In this section, taken from a longer sequence in the first chapter, Hoyle relies on scientific techniques and logical argument to establish yet another menace to the earth's existence.

On this late January morning of 8th January, 1964, Jensen was down in the basement of the Observatory buildings setting up an instrument known as the 'blinker'. As its name implies, the 'blinker' was a device that enabled him to look first at one plate, then at the other, then back to the first one again, and so on in fairly rapid succession. When this was done, any star that had changed appreciably during the time interval between the taking of the two plates stood out as an oscillating or 'blinking' point of light, while on the other hand the vast majority of stars that had not changed remained quite steady. In this way it was possible to pick out with comparative ease the one star in ten thousand or so that had changed.

He finished the first of the batch. It turned out a finicking job. Once again, every one of the 'possibilities' resolved into an ordinary, known oscillator. He would be glad when the job was done. Better to be on the mountain at the end of a telescope than straining his eyes with this damned instrument, he thought, as he bent down to the eye-piece. He pressed the switch and the second pair flashed up in the field of view. An instant later Jensen was fumbling at the plates, pulling them out of their holders. He took them over to the light, examined them for a long time, then replaced them in the blinker, and switched on again. In a rich star field was a large, almost exactly circular, dark patch. But it was the ring of stars surrounding the patch that he found so astonishing. There they were, oscillating, blinking, all of them. Why? He could think of no satisfactory answer to the question, for he had never seen or heard of anything like this before.

Jensen found himself unable to continue with the job. He was too excited about this singular discovery. He felt he simply must talk to someone about it. The obvious man of course was Dr. Marlowe, one of the senior staff members.

Jensen's luck was in, for a phone call soon elicited that Marlowe was at home. When he explained that he wanted to talk to him about something queer that had turned up, Marlowe said:

"Come right over, Knut, I'll be expecting you. No, it's all right. I wasn't doing anything particular."

Marlowe was waiting.

"Well, what've you got there?" he said, nodding at the yellow box that Jensen had brought.

Somewhat sheepishly, Knut took out the first of his two pictures, one taken on 9th December, 1963, and handed it over without comment. He was soon gratified by the reaction.

"My God!" exclaimed Marlowe. "Taken with the 18-inch, I expect. Yes, I see you've got it marked on the side of the plate."

"Tell me why you're so surprised, Dr. Marlowe."

"Well, isn't this what you wanted me to look at?"

"Not by itself. It's the comparison with a second plate that I took a month later that looks so odd."

"But this one is singular enough," said Marlowe.

"I don't see why you're so surprised by this one plate."

"Well, look at this dark circular patch. It's obviously a dark cloud obscuring the light from the stars that lie beyond it. Such globules are not uncommon in the Milky Way, but usually they're tiny things. My God, look at this! It's huge, it must be the best part of two and a half degrees across!"

"But, Dr. Marlowe, there are lots of clouds bigger than this, especially in the region of Sagittarius."

"If you look carefully at what seem like very big clouds, you'll find them to be built up of lots of much smaller clouds. This thing you've got here seems, on the other hand, to be just one single spherical cloud. What really surprises me is how I could have missed anything as big as this."

Marlowe went over to the sideboard to renew the drinks. When he came back, Jensen said:

"It was this second plate that puzzled me."

Marlowe had not looked at it for ten seconds before he was back to the first plate. His experienced eye needed no 'blinker' to see that in the first plate the cloud was surrounded by a ring of stars that were either absent or nearly absent in the second plate. He continued to gaze thoughtfully at the two plates.

"There was nothing unusual about the way you took these pictures?"

"Not so far as I know."

"They certainly look all right, but you can never be quite sure."

Marlowe broke off abruptly and stood up.

"Something crazy may have happened. The best thing we can do is to get another plate shot straight away. I wonder who is on the mountain tonight."

"You mean Mount Wilson or Palomar?"

"Mount Wilson. Palomar's too far."

"Well, as far as I remember one of the visiting astronomers is using the 100-inch. I think Harvey Smith is on the 60-inch."

Marlowe drove to the Observatory offices. His first step was to get Mount Wilson on the phone and to talk to Harvey Smith.

Marlowe next put through a call to Bill Barnett of Caltech.

"Bill, this is Geoff Marlowe ringing from the offices. I wanted to tell you that there'll be a pretty important meeting here tomorrow morning at ten o'clock, I'd like you to come along and to bring a few theoreticians along."

When Bill Barnett's party of five arrived they found some dozen members of the Observatory already assembled, including Jensen,

Rogers, Emerson and Harvey Smith. A blackboard had been fitted up and a screen and lantern for showing slides. The only member of Barnett's party who had to be introduced round was Dave Weichart, a brilliant 27-year-old physicist.

"The best thing I can do," began Marlowe, "is to explain things in a chronological way, starting with the plates that Knut Jensen brought to my house last night. When I've shown them you'll see why this emergency meeting was called."

Emerson put in a slide that Marlowe had made up from Jensen's first plate.

"The center of the dark blob," went on Marlowe, "is in Right Ascension 5 hours 49 minutes, Declination minus 30 degrees 16 minutes, as near as I can judge."

"A fine example of a Bok globule," said Barnett.

"How big is it?"

"About two and half degrees across."

There were gasps from several of the astronomers.

"Now look at the next plate."

"It's fantastic," burst out Rogers, "it looks as if there's a whole ring of oscillating stars surrounding the cloud. But how could that be?"

"It can't," answered Marlowe. "That's what I saw straight away. Even if we admit the unlikely hypothesis that this cloud is surrounded by a halo of variable stars, it is surely quite inconceivable that they'd all oscillate in phase with each other, all up together as in the first slide, and all down together in the second."

"No, that's preposterous," broke in Barnett. "If we're to take it that there's been no slip-up in the photography, then surely there's only one possible explanation. The cloud is moving towards us. In the second slide it's nearer to us, and therefore it's obscuring more of the distant stars."

"Actually there's no doubt at all about it," went on Marlowe. "When I discussed things with Dr. Herrick earlier this morning he pointed out that we have a photograph taken twenty years ago of this part of the sky."

Herrick produced the photograph.

"We haven't had time to make up a slide," said he, "so you will have to hand it round. You can see the black cloud, but it's small on this picture, no more than a tiny globule. I've marked it with an arrow."

He handed the picture to Emerson who, after passing it to Harvey Smith, said:

"It's certainly grown enormously over the twenty years. I'm a bit apprehensive about what's going to happen in the next twenty."

It was then that Dave Weichart spoke up for the first time. "Is the center of the cloud staying in the same position, or does it seem to be moving against the background of the stars?"

"A very good question. The center seems, over the last twenty years,

to have moved very little relative to the star field," answered Herrick.

"Then that means the cloud is coming dead at the solar system."

Weichart was used to thinking more quickly than other people, so when he saw hesitation to accept his conclusion, he went to the blackboard.

"I can make it clear with a picture. Here's the Earth. Let's suppose first that the cloud is moving dead towards us, like this, from A to B. Then at B the cloud will look bigger but its center will be in the same direction. This is the case that apparently corresponds pretty well to the observed situation."

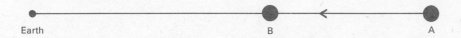

There was a general murmur of assent, so Weichart went on:

"Now let's suppose that the cloud is moving sideways, as well as towards us, and let's suppose that the motion sideways is about as fast as the motion towards us. The cloud will move about like this. Now if you consider the motion from A to B you'll see that there are two effects—the cloud will seem bigger at B than it was at A, exactly as in the previous case, but now the center will have moved. And it will move through the angle AEB which must be something on the order of thirty degrees."

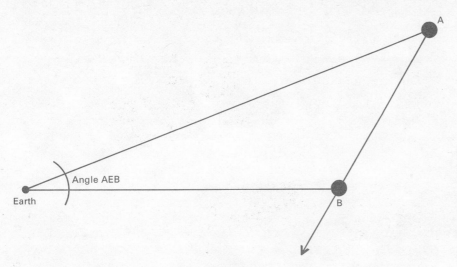

"I don't think the center has moved through an angle of more than a quarter of a degree," remarked Marlowe.

"Then the sideways motion can't be more than about one per cent of the motion towards us. It looks as though the cloud is heading towards the solar system like a bullet at a target."

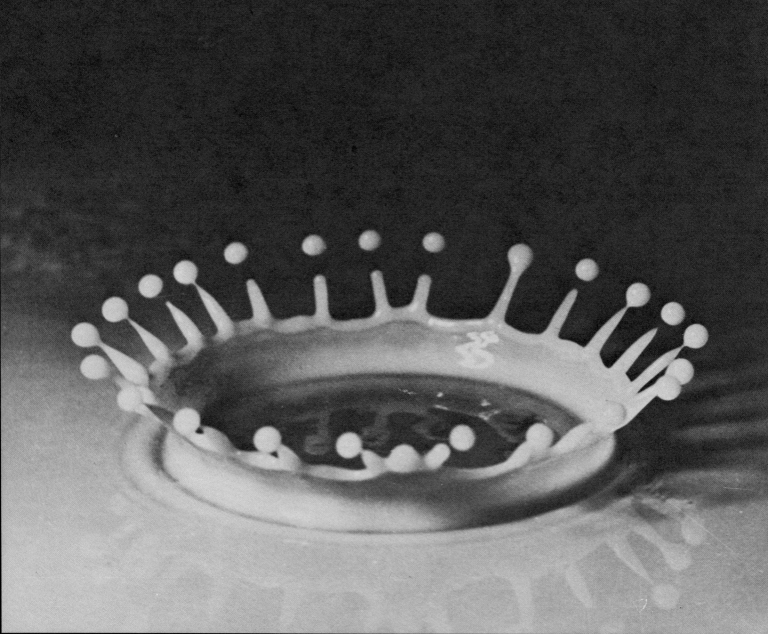

One

Matter in Motion

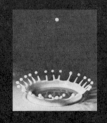

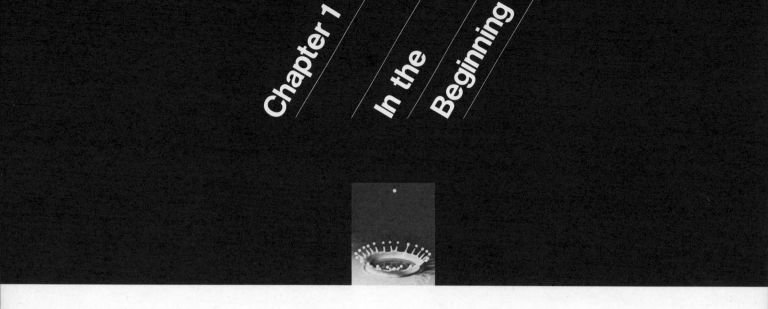

Motion is a phenomenon familiar to us all—people walk, clouds drift, rain falls. We observe it on a small scale in dust particles momentarily caught in a ray of sun, or in tiny organisms seen through a microscope. On a large scale we see it in the movement of the moon, the planets, and the stars in the heavens above us. There are even photographs recording evidence of the rapid movements of electrons and other submicroscopic particles.

Questions about motion have been the subject of thought and discussion throughout the history of mankind. Many early civilizations had theories and philosophies to describe, explain, and predict motion; but mathematical description and measurement of the many forms of motion are relatively recent accomplishments. From them theories have been developed which explain all forms of motion—from the movement of the smallest particles to the movement of galaxies—in terms of the same natural laws.

The simplest thing we can say about motion is that it involves a change in position of some object. A thrown ball, for instance, leaves the place from which it was thrown, travels over some distance, and arrives at the hand of the person catching it. A leaf, falling from a tree, breaks away from a twig many feet above the ground and drifts in an irregular zigzag path to a point on the ground somewhere below the twig. But what is it that makes the ball travel, or the leaf fall, from one place to another? What accounts for the path the leaf or the ball follows and the time it takes to move from one point to another? Is the motion of the smallest particles different in kind or principle from that of the largest bodies—the moon, the planets, the stars—in the universe around us? The search for answers to such questions about motion was the basis for the development of modern physics. In fact, the study of motion is the beginning of all modern science.

Motion always involves a change in the position of an object.

Greek Physics

Until the sixteenth century of our era, the science of motion was largely governed by a body of theory developed in ancient Greece. The establishment and unification of these ideas were for the most part due to Aristotle, a Greek philosopher-scientist of the fourth century BC, and Ptolemy, a Greek astronomer of the second century AD.

In the physics of Aristotle, every object on earth was held to be composed of one or more of four basic elements. These elements were air, fire, earth, and water; it was their presence in an object and the quantity of each that gave that object its physical properties. According to this theory the four elements were distinct (that is, different from each other), and each had its natural, or proper, place in nature, in a hierarchy ranging from below the surface of the earth to the highest heavens. Fire and air were naturally light elements; their proper place was above the surface of the earth, with fire above air. Water was an intermediate element, too heavy for the sky (rain falls to earth from the sky) but lighter than earth, which explained why it was found mainly on the surface, above a base of earth. Earth itself was the heaviest of the four elements; its natural place was not in the heavens but below the sky and air and water, making up the soil and rock of the world's surface. Combinations of these elements were many and varied. A rock was thought to be composed of earth alone, because it was found in or on the earth and returned rapidly to earth when thrown or dropped. Steam, rising from the surface of a pot of heated water, and hot air, rising from a fire or from a volcano, were composed of fire and water. Dust, found at the surface of the earth, was composed of earth and air, which accounted both for its earthy appearance and for its tendency to rise into the air when stirred up. Aristotelian physical science actually provided a crude chemistry of the forms and variety of matter as they were observed in the world.

According to Aristotelian physics, all of the objects on the earth are made up of one or more of the four basic elements: earth, water, air, and fire.

The motions of earthly objects were explained in this system in accordance with the elements that each contained. A rock, composed entirely of earth, would fall quickly, seeking its own place; whereas a leaf, containing water, air, and only a little earth, would drift some distance before it touched the ground. Of course, a rock could be hurled upward or carried on the water by a raft. Such movements contradicted the natural propensity of an object to seek its natural place, and any contradictory motion was temporary because it was unnatural. Unnatural motions required force, while natural motions did not. A rock would rise a certain distance into the sky when thrown, but it would soon fall back. When the raft tipped and the rock fell overboard, the rock would sink through the much lighter water and find its place on the lake or sea bottom.

To the four elements on or about the earth, Aristotle added a fifth element, the ether, of which all the bright heavenly bodies—planets, sun, moon, and stars—were held to be composed. These *ethereal* bodies were constant and

Figure 1-1
This sixteenth-century German engraving of Atlas supporting the universe is based on Aristotle's hierarchical world order. The earth is at the center with water on its surface; above are air and fire. Beyond are the concentric, transparent spheres of the heavenly bodies (Cristalline Firmament), whose movement is controlled by the Prime Mover (Primum Mobile). Note also the band of the zodiac.

unchanging. Substances made up of the four basic elements exhibited changing properties if mixed with each other. Iron, composed of earth, rusted when mixed with water. Water mixed with earth became muddy and developed a foul odor. But heavenly bodies were pure and incorruptible. Not only were they believed to be composed of a uniform ether, but they were thought to follow a uniform and unchanging pattern of motion about the earth. This standard elementary motion of the sky was the circle, which to the geometrically minded Greeks was the most perfect shape possible. Circles are undistorted, smooth, with no corners or irregularities. They believed that each heavenly body revolved in its own circular path, or orbit, and that the circular shape of each orbit was constant because each body was fixed within one of a set of concentric, separately moving, transparent spheres. The original motion of the heavenly bodies was ascribed by Aristotle to the action of the outermost sphere, the Prime Mover, which caused the rotation of the other spheres. And the rotation of the celestial spheres was more noble than the motion of objects on and near the earth, where some visible external force, such as the propulsion of a rock by an arm or a catapult, was often necessary to create the phenomenon of motion.

The earth, placed at the center of the Aristotelian world system, was held to be immovable, neither changing position in space nor spinning about a pole-to-pole axis. This immovable quality of the earth was ascribed to its being composed almost entirely of the element earth, the heaviest of the four mundane elements. The belief was supported by direct evidence of the senses—after all, no motion of the earth can be seen or felt by an observer standing on the earth.

Aristotle believed that the earth was stationary, immovable, and at the center of the universe.

The simple examples of motion already cited—a baseball speeding from a pitcher to catcher, a leaf falling from a tree, or the planets moving through the sky—all demonstrate the two essentials of motion: (1) that an object in motion changes its position in space relative to some other object and (2) that time passes as it changes its position.

Speed, Velocity, and Acceleration

Take the example of an automobile moving along a highway from New York to Chicago: as time passes it changes its position relative to both cities. With each hour, minute, second, and fraction of a second, it moves farther away from New York and nearer to Chicago. It is also changing its position relative to towns, cities, houses, and trees along its route. This is an important factor in the measurement of motion: *all motion is relative.* Since motion is a change of position with time, it is measured in terms of the distance covered between two points that are contained in a frame of reference such as the earth. The movement of a car traveling from New York, for example, can be measured with New York as the starting point. After a given time, the car has traveled 50 mi from New York. It has changed its position in a given time; that is, it has undergone motion.

Motion takes place in time and is measured in time.

The motion of an object on or near the surface of the earth is ordinarily measured in relation to some point or object on the earth's surface. For the purpose of measuring motion on or near the earth, we can consider the earth to be stationary. By choosing the stationary earth as our frame of reference, we avoid the complexity of calculating the effect of the rotation of the earth, the earth's orbit around the sun, and the sun's motion in the galaxy on the motion of the body we are measuring. On the other hand, if we chose the sun as our frame of reference, the earth and everything on it would be seen as moving all the time.

How can we make our concept of motion more precise? How can we measure motion, the change of place over time? In order to do this, we must become familiar with three further concepts: speed, velocity, and acceleration. Each has a very precise definition.

Speed

Simply stated, the speed of an object is how fast it moves from one place to another. That is, speed expresses the time rate of change in position.

We all have a general understanding of the concept of speed. We know that a racing car in the Indianapolis 500-mile race travels at a greater speed than an automobile on a neighborhood street, that the winner of a swimming race moves through the water with greater speed than the other competitors, that a jet flies at a greater speed than a propeller-driven airplane. In physics, **speed** is defined as the rate at which a given distance is covered by a moving object in a given time. Let's go back to the automobile traveling between New York and Chicago. If we measure out a length of highway, we can determine the car's speed, that is, its rate of travel over that distance. In order to do this, we must know how much time it takes for the car to cover that particular length of highway. Let us assume that we have chosen to measure the car's rate of speed over a distance of one mile and that the car covers this mile in 120 sec, or 2 min. Since rate is actually a ratio of distance covered to time elapsed, we can express this as follows:

$$\frac{1 \text{ mi}}{2 \text{ min}}$$

We can change this to miles per hour, thus:

$$\frac{1 \text{ mi}}{2 \text{ min}} \times \frac{60 \text{ min}}{\text{hr}} = \frac{30 \text{ mi}}{\text{hr}}$$

The term *minutes* in the denominator of the first expression is divisible into, and thus cancels, the term *minutes* in the numerator of the second expression. In an expression of this type, we treat the words for units in the same manner as algebraic symbols.

No matter what units of time and distance we use, speed is expressed as a rate, which is a ratio of distance to time. In the customary notation of physics, speed is abbreviated by the letter v, distance by d, and time by t, so that speed is conveniently expressed by the following equation:

$$v = d/t$$

Figure 1-2
To a stationary observer at a bicycle race, the riders appear to be in rapid motion with respect to the background. But in a photo, the camera freezes the action of the cyclist, and the background appears blurred, as if it is in motion—in effect, the motion is transferred to the background. Motion is never absolute; it can only be observed relative to some fixed point—here, the cyclist himself.

AVERAGE SPEED   In most instances the speed we measure is the **average speed** of an object. An automobile can seldom maintain precisely the same rate of speed during an entire journey. It slows down for a traffic light, stops for passengers to rest or eat, or travels at high speed on a stretch of open highway. If we suppose that the car travels 40 mph for 2 hr, 60 mph for 1 hr, and 25 mph for 3 hr, we can calculate its average speed by the following logic (note that to find the average speed for the trip, we need the *total* distance and the *total* time). Since $v = d/t$, $d = v \times t$. If $d_1 = 40 \times 2 = 80$ mi, $d_2 = 60 \times 1 = 60$ mi, and $d_3 = 25 \times 3 = 75$ mi, then the total distance, $d_1 + d_2 + d_3 = 215$ mi, the total time $= 6$ hr, and the average speed (or v_{av}) $= 35.8$ mph.

Under some conditions objects travel at constant speed. This may be true of electrons and other subatomic particles over the very short distances they travel within an atom or in some atom-smashing experiments; it may also be

Time (seconds)

| 0 | 60 | 120 | 180 | 240 |

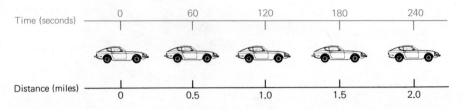

Distance (miles)

| 0 | 0.5 | 1.0 | 1.5 | 2.0 |

Figure 1-3
Speed is the rate of travel over a certain distance. In this case, a car travels 1 mile in 120 sec, or 2 min. The speed, v, equals d/t, or $\frac{1.0 \text{ mi}}{2.0 \text{ min}}$ equals 0.5 mi/min or 30 mph.

true for an automobile over a small portion of a longer journey or for ice skaters coasting across a pond. At constant speed, the distance an object travels and the time required to travel this distance are directly proportional to one another. Suppose, for example, that an arrow shot from a bow is traveling along the middle of its path to the target at a constant speed of 80 ft/sec. In terms of the equation for speed, we may write:

$$v = \frac{80 \text{ ft}}{\text{sec}}$$

If the speed of the arrow remains constant, it will have traveled 80 ft after 1 sec, 160 ft after 2 sec, 240 ft after 3 sec, and so forth. A graph of the speed of the arrow, plotting time as the x-axis and distance as the y-axis, is shown at the left. Note that, as represented in the graph, the speed is a straight line, with the slope equal to the constant rate of speed $d/t = 80$ along its entire length. Because of the straight-line graph, the equation for constant speed is called a linear equation. If we wish to determine the distance the arrow has traveled in a certain time, t, we have only to find this time on the x-axis, carry it up to the point at which it intersects the straight line, and read over horizontally from that point to the y-axis. The point at which this horizontal line intersects the y-axis will give us the distance the arrow has traveled in the time, t. Suppose that the arrow has traveled at constant speed for 4.5 sec, we can plot this from the graph and find that it has gone 360 ft. Rather than using a graph, we might also use the equation for speed, setting $t = 4.5$; then:

$$d = vt, \text{ and } d = (80 \text{ ft/sec}) \times 4.5 \text{ sec} = 360 \text{ ft}$$

The seconds in the denominator of the first part of the expression cancel the seconds in the last part.

In reality, however, the speed of an object is rarely constant. The speed of an interplanetary rocket, as it leaves the launching pad, increases from zero to its maximum speed through the first few minutes of its flight. An automobile slowing to a stop gradually decreases its speed to zero, and a stone dropped from the top of a building may fall with increasing speed until it strikes the ground. How can we describe the speed, v, of an object when speed is not constant? For an automobile whose speed is decreasing, we can measure the distance it travels during each of a series of given time intervals until it comes to a

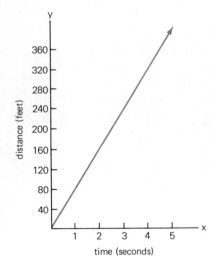

stop. The result will be a table similar to that below. The interval of time measured is $(t_2 - t_1)$, where t_2 is the time at the end of the interval and t_1 the time at the beginning; and the distance traveled is $(d_2 - d_1)$, where d_2 is the end point of the distance traveled in that time interval and d_1 is the beginning point of that distance. The beginning of each interval is reckoned as zero, even though the car has moved through some finite distance in the preceding interval.

	$t_2 - t_1$	$d_2 - d_1$
A	1 sec	150 ft
B	2 sec	90 ft
C	3 sec	40 ft
D	6 sec	10 ft

The equation for average speed for each of these intervals looks like this:

$$v_{av} = \frac{d_2 - d_1}{t_2 - t_1}$$

As used in science, the word *speed* describes only the rate at which an object is moving through a given distance; it does not describe the direction of the object's motion. If we know only that an airplane is traveling at a speed of 650 mph, our information is incomplete; in order to plot its motion we also need to know its direction. If, however, we say that the airplane is moving at 650 mph on a south-southeasterly course (that is, at an angle of 22°30′ from due south), we then have a description of what is called the airplane's **velocity**, a measure of motion consisting of two components: speed and direction.

When we speak of a *constant* velocity, we mean that both the speed and direction of an object in motion are constant. If, however, either one of these two components should change, the object is said to be *accelerated*. In case of the car stopping for a traffic light, we noted that the speed of the car gradually decreased as it came to a halt. This information is sufficient to tell us that it was undergoing acceleration. More precisely, the car was undergoing negative acceleration because its speed was decreasing. The speed of the rocket leaving its launching pad is increasing, so we say that it is undergoing positive acceleration. An object may also be accelerated if its direction is changed, even though its speed remains constant—any point on a spinning phonograph record, for instance. A train going around a curve is accelerated since its direction is changing, although its speed may remain constant. If the speed of the train changes as it rounds the curve, it will experience an acceleration with two components—a change in speed and a change in direction.

Acceleration, then, is a change in velocity. But how is this change expressed? Recall that velocity has two factors, speed and direction. For the moment, let us neglect the effect of a change in direction and consider an object moving in a

Velocity and Acceleration

Velocity means speed with the added specification of direction.

In physics an object undergoes acceleration whether it speeds up, slows down, or changes direction.

Figure 1-4
The speed of a car on a moving ferris wheel may be constant, yet it is accelerating because its direction is continually changing.

straight line. The speed of the object along that line is expressed by the equation $v = d/t$. The acceleration, a, of the object is the change in the speed of the object *with* time. This is expressed mathematically as speed/time, so

$$a = \frac{(d/t)_2 - (d/t)_1}{t_2 - t_1} = \frac{\Delta d/t}{\Delta t} \text{ or } \frac{\Delta v}{\Delta t}$$

The Greek letter delta, Δ, is used in science as a shorthand symbol to indicate a small change in the quantity named.

The numerator is the expression for a difference between two speeds; the denominator represents the elapsed time. An automobile, for instance, may be increasing its original speed of 75 ft/sec by adding 10 ft/sec for every second that it travels, thus

$$a = \frac{10 \text{ ft/sec}}{\text{sec}} \text{ or } \frac{10 \text{ ft}}{\text{sec}^2}$$

Note that the unit, the second, which is the denominator of the fraction in the upper term, and the second which is the denominator of the entire expression, are identical. Their product in the denominator is (sec × sec) or sec².

The force of gravity causes an acceleration. When we drop a stone from the top of a high cliff, it strikes the ground at a greater speed and with more force than it would if we dropped it from a lower height, because during every instant of its fall its speed is increasing. As we shall see later, gravity causes a constant acceleration of every freely falling object. If we neglect air resistance and friction, there is no difference between the acceleration of an object weighing 1 ton and one weighing 2 lb. The value of the acceleration of gravity is 32 ft per sec per sec or 32 ft/sec². That is, the speed of the object increases by 32 ft/sec (or by 980 cm/sec) during each second that the object is in free fall.

The **acceleration of gravity** is symbolized by the letter g. If we replace a in the equation for acceleration by g, we will have the following equation:

$g = v/t$ or $v = gt$

To demonstrate the use of the equation, let us suppose that we drop a rock from a cliff and that it takes the rock 3 sec to reach the ground. Using these figures in the equation, we get:

$$v = \frac{32 \text{ ft}}{\text{sec}^2} \times 3 \text{ sec or } \frac{96 \text{ ft}}{\text{sec}}$$

Thus, the final speed of the rock as it strikes the ground will be 96 ft/sec. If we dropped it from a height from which it would require 4 sec to reach the ground, its final speed would be 128 ft/sec, considerably faster than if we dropped it from a lower height.

Speed, velocity, and acceleration are all concepts essential to the branch of physics known as **mechanics**. Specifically, mechanics is the division of physics concerned with the motions of objects and with the effects of forces on those motions.

At the beginning of the sixteenth century, Aristotelian physics still provided the basic concepts used to explain natural phenomena. The earth was assumed by astronomers and other scientists to rest immobile at the center of the universe. Judaism, the Catholic Church, and the relatively new Protestant faith all officially accepted this description of the universe. It was thought to express the doctrine that man and his place of life were central in God's creation. The clergy insisted that a literal reading of the Bible supported this viewpoint.

The phenomenon of a falling object was described in Aristotelian terms. The rate at which an object fell back to the earth was thought to be determined by the amounts and proportions of the four basic elements in that object; a larger rock should thus fall to ground more rapidly than a smaller one because it contained a larger amount of the element earth. This had never been demonstrated, but it was assumed to be so by the vast majority of people on the basis of intuitively held principles. The disproof of the Aristotelian theory of mechanics was first achieved by the great Italian scientist, Galileo Galilei (1564–1642).

Galileo—The Acceleration of Gravity

Figure 1-5
If this sky diver were falling in a vacuum, he would continue to accelerate until he hit the ground. However, air resistance will eventually reduce his acceleration to zero; from that point he will fall at a constant rate of speed called the terminal speed. In free fall, his terminal speed will be about 120 mph, and when he opens his chute it will be only about 14 mph.

Galileo entered the University of Pisa as a medical student but he soon lost interest in medicine and turned instead to mathematics and physics. He suspected that there were faults in Aristotelian physics; among the theories he questioned was the one explaining the motion of falling objects. It did not seem plausible to him that a 2-lb rock would strike the ground in half the time taken by a 1-lb rock if both were dropped from the same height. In the course of his scientific work he noted that objects of very different weights, when dropped from the same height, actually reached the ground at *very nearly* the same time.

Galileo thought that the velocity of a falling object was in some regular and simple way regulated either by the distance the object fell or by the time it was in flight before striking the ground. He planned a series of physical experiments by means of which he could discover which was decisive—distance fallen or time of fall.

The Inclined Plane Experiment In order to demonstrate his ideas about the speed of falling bodies, Galileo set up an experiment in which heavy, solid balls were allowed to roll freely down a series of inclined planes. Galileo reasoned that the downhill motion of a ball under these conditions would be determined by the same gravitational force, or forces, that would draw it to earth if it was allowed to fall freely and vertically, unconstrained by a tilted plane. The ball would reach the ground more slowly over a longer path than if simply dropped through the air, giving him the advantages of greater distance and more time in which to observe it.

In the sixteenth century there were no accurate mechanical watches; the hourglass and sundial were the accepted instruments for measuring time. They worked perfectly well for marking the longer time divisions but were of no use to Galileo for measuring the very brief time between the moment of releasing the ball at the top of the plank and the moment it reached the ground. To solve this problem, he hit upon the idea of using a pendulum, whose swings he had observed to be relatively constant over a period of time. By counting the number of swings made by the pendulum in the period of time it took the ball to travel to points along the plank or to the bottom, he could measure the relationship of distance to time.

As the experiments progressed, Galileo increased the pitch, or steepness, of the angle between the planks and the earth. The balls rolled increasingly rapidly as the pitch was increased. Galileo concluded that the planks would in the end be completely vertical (i.e., effectively removed); the balls would then fall straight down to earth but would still obey the rule of motion he had already observed in his use of the inclined plane. What was this observed rule of motion? It was that the distance traveled by any ball along any plank was always proportional to the square of the time (as measured by the pendulum) that it took the ball to cover this distance. Thus, if the ball traveled over 1 unit of distance with 1 swing of the pendulum, it would cover 4 units of distance in 2 swings of the pendulum, 9 units in 3 swings, 16 units in 4 swings, and so forth. This led Galileo to formulate the equation

$d = kt^2$

With this equation, Galileo related distance directly to the square of time, through a proportionality constant, k.

From the definition of velocity, $v = d/t$, Galileo knew that the average speed at which an object falls from rest is equal to the ratio of the total distance through which the object travels to the total time required for it to travel through that distance, regardless of brief speed-up or slow-down during the movement. This he expressed in

Falling Bodies

$v_{av} = d/t$

as before. Also, from the definition of acceleration,

$a = v/t$

and by cross multiplication,

$v = at$

Finally, Galileo knew that the average speed over a given distance for an object that starts from rest and is uniformly accelerated is equal to its final speed divided by 2, expressed as

$v_{av} = v/2$

By substituting from the previous equation for v into the numerator of the right-hand term and from the meaning v_{av} in the left-hand term, Galileo derived the equation

$$\frac{d}{t} = \frac{at}{2} \text{ or } d = \frac{1}{2}at^2$$

By returning to the equation he had derived from his inclined plane experiment, $d = kt^2$, it became apparent that the constant k was equal to $a/2$, since d and t^2 were the same in both equations.

Galileo lacked a device that could time the vertical free fall of objects, so he could not verify his hypothesis directly; he used the inclined plane for indirect evidence. With better timing devices it became possible to calculate how far an object would fall vertically in specific intervals of time. As we noted earlier, the acceleration of gravity has been measured at 32 ft/sec². Since $d = \frac{1}{2}at^2$, the distance an object falls in the first second of its earthbound flight would be as follows. Since $t = 1$ sec,

The acceleration of freely falling bodies is always the same under ideal conditions. In a vacuum, a feather and a rock fall at the same rate of speed.

$$d = \frac{1}{2} \times 32 \frac{\text{ft}}{\text{sec}^2} \times (1 \text{ sec})^2$$

Cancelling,

$d = \frac{1}{2} \times 32 \times 1$, or 16 ft

In summary, what did Galileo do? First of all, he derived the formula that the distance through which an accelerating object travels (neglecting friction and air resistance) is directly proportional not only to the rate of acceleration of the object but also to the time squared. Next, with the inclined planes, he made use of one of the first rational experimental procedures to demonstrate a relationship among several factors of the physical world (in this case, distance, time, and acceleration). Modern scientific procedure, however more sophisticated, is based on these procedures of observation and experiment linked with reasoning and mathematical deduction.

Inertia

Galileo's work with inclined planes yielded other important results, as well. It had long been believed that rest, a state of no motion at all, was the natural state of earthly objects. This reasoning was based on the fact that a body at rest tends to stay at rest—that is, force is needed to set it in motion; then, if left alone, it sooner or later comes to a stop. Aristotle concluded, therefore, that a moving body must have a continually acting cause (his term for what we call force) to keep it in motion. Galileo devised a further experiment with inclined planes that demonstrated a different conclusion. He let a smooth ball roll down one smooth inclined plane and allowed it to roll up another, which he set at various angles of incline, and he found that no matter what the angle of the second plane, the ball always rolled almost as far up the second plane as it had down the first. The more polished the ball and the smoother the planes, the farther the ball rolled up the second plane, although it never quite attained the height from which it had started.

How did Galileo account for this set of observed motions? Going beyond the physical experiment to imagine a situation in which the ball was perfectly round, the planes were infinitely smooth, and the angle of the second plate was progressively reduced closer and closer to zero (that is, to the horizontal), Galileo imagined that the ball would continue moving forever, without speeding up or slowing down. The resistance to motion caused by the surfaces of objects in contact with each other, which we call **friction**, would in such a situation be eliminated. Therefore, the object would travel indefinitely, without stopping or slowing down, unless stopped or speeded up by an external force. This property of moving objects to continue in motion in a straight line Galileo

A body at rest resists any attempt to start it moving; a body in motion resists attempts to stop or change its motion.

called **inertia** (from the Latin word for "idleness"). He held that the inertia of the balls accounted for the motion he had observed in his experiment with the inclined planes and that their continued motion was limited only by the forces of friction and gravity. This finding is an almost exact reversal of the earlier Aristotelian concept. Galileo achieved it by combining his method of physical experiment with a bold conception of taking the experiment to its logical conclusion—a conclusion that cannot be demonstrated physically. His discovery revolutionized the study of motion and laid the foundation for the work, a generation later, of Issac Newton, whose own great contributions are discussed in Chapter 2.

Thus far we have studied Galileo's work with two kinds of motion: purely vertical motion under gravity and purely horizontal motion of bodies moving freely. His work with vertical motion led to the discovery and formulation of the laws governing falling bodies, and his work with horizontal motion led to the discovery of inertia.

In the real world, however, most objects in motion do not follow either the purely vertical or purely horizontal paths that Galileo stipulated in his idealized mental experiments. Objects falling through the air rarely fall directly down. For example, if a bale of hay drops from an airplane in horizontal flight to an isolated herd of deer it does not actually fall to earth in a straight line. Although it is instantly affected by the force of gravity, inertia causes it to persist along the direction of the plane's horizontal motion during the whole course of its descent. Or take as another example a bullet fired from a rifle. The explosive gases impel the bullet from the barrel and give it its initial horizontal velocity. At the moment of its leaving the barrel it begins its acceleration under gravity, and despite its horizontal velocity, which inertia continues, its path is changed to curve downward. Of course, if it were not subject to the force of gravity it would move, according to the Galilean formulation of inertia, at a constant speed (its initial horizontal velocity) in a straight line forever, or until it struck a target.

There are regular patterns in the motion of objects moving both horizontally and vertically at the same time. An object thrown upward from ground level tends to follow a path, or trajectory, that traces a symmetrical curve called a **parabola**. This is a geometric shape that was known to Greek mathematicians; Galileo's familiarity with their work was of substantial help to him in his studies of projectile motion. Some of the earlier scientists had thought that an object propelled upward reached a certain height and then dropped back to earth in a straight line. By a process of mathematical reasoning, Galileo determined that when the object reaches its greatest height its path is level momentarily before it starts downward. The curves of its ascent and descent are identical; the entire path can be divided at its high point into two equal halves. The path of the bale of hay and the path of the bullet each describe half a parabola, because their motion has been initiated at the high point of the curve.

Galileo then proceeded to investigate how the path of an object falling both vertically and horizontally is related to pure vertical motion. The stroboscopic

Projectile Motion

Figure 1-6
A graphic representation of the second part of Galileo's inclined plane experiment. If the surfaces of the ball and the planes are smooth, the ball will roll up the second plane to the height from which it started, as shown in the first three drawings. The angle of the plane does not affect the height that the ball will reach if friction is eliminated. Therefore, if the second plane were horizontal, the ball perfectly round, and the plane infinitely smooth, the ball should go on without ever stopping, as in the last drawing.

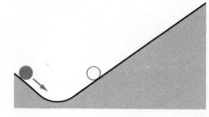
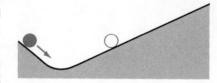

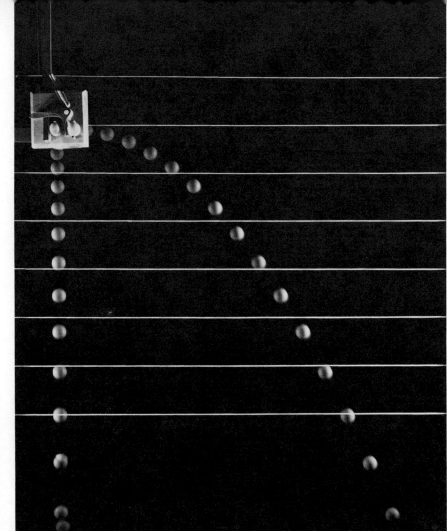

Figure 1-7
In a stroboscopic photograph, such as this one, many exposures are made on the same film at short, equal intervals of time. The stroboscope provides very brief repeated flashes of light at the rate of up to 150,000 per minute, which effectively freeze the image of a moving object. In this composite of 15 images, two balls are released at the same instant. One ball is allowed to drop straight down, while the other is projected horizontally. Regardless of the horizontal motion, both experience the same downward acceleration, and they reach the ground at the same time.

The vertical and horizontal motions of falling objects are independent of each other.

photograph (Figure 1-7) records a modern experiment that shows the vertical component of such motion is calculable on the same basis as that of bodies falling straight down. A ball dropped vertically and one projected horizontally from the same height at the same time reach the ground simultaneously. Experience has shown that the action of gravity downward is independent of any horizontal motion. Why should this be so?

Since both objects start their descent from the same height, they both become subject to the acceleration of gravity at the same instant, regardless of the combined vertical and horizontal motion of the one that is projected and the purely vertical motion of the one that is dropped. Because gravity acts in the same fashion on both — producing a downward acceleration of 32 ft/sec² — the same law of vertical motion applies to both, and it is not affected by the horizontal motion of the object.

Galileo was able to demonstrate this by his experiments and to state that the vertical and horizontal motions of falling objects are independent of each other. For purposes of calculation—that is, to determine in practical rather than theoretical terms the actual path of an object's flight—we must remember that its motion is a combination of vertical and horizontal components. It does not travel either vertically or horizontally; it follows a curve, which includes elements of the two directions—vertical and horizontal. However, the vertical and horizontal aspects of the motion must first be analyzed separately: the vertical motion in relation to the acceleration of gravity and the horizontal motion in terms of the magnitude of the initial horizontal speed (remember that because of inertia the object will maintain constant horizontal speed in the absence of air resistance.)

Suppose that you are standing on the edge of a sheer precipice with a baseball in each hand. You hold the balls in front of you at exactly the same height over the edge of the precipice. Then, you let one of the balls fall off of your hand; at the same time, you throw the other ball horizontally in front of you as hard as you can. Which ball will hit the ground first? Commonsense reasoning might lead you to believe that the one dropped straight down would have a shorter distance to travel and, therefore, reach the ground before the one thrown in front of you. But—ignoring air resistance and assuming that there is no wind—both of the balls will hit the ground at the same time because the action of gravity is independent of any motion or force perpendicular to it. The determining factor is the height from which the balls fall. If they both fall from the same height, their vertical acceleration is calculable in terms of gravity alone. Since gravity is pulling down on the two balls with equal force, they will hit the ground at the same instant.

We can use the height from which an object falls and the speed at which it was propelled horizontally to calculate the horizontal distance the object travels before it comes to rest on the earth. Suppose a bullet with a velocity of 2000 ft/sec is shot from a height of 4 ft; using the formula $d = \frac{1}{2}at^2$ in the form $t^2 = \frac{d}{\frac{1}{2}a}$,

$$t^2 = \frac{4 \text{ ft}}{\frac{1}{2} \times 32 \text{ ft/sec}^2} \text{ or } t^2 = \frac{4 \text{ ft}}{16 \text{ ft/sec}^2}$$

thus

$$t^2 = \frac{1}{4} \text{ sec}^2 \text{ or } t = \frac{1}{2} \text{ sec}$$

Using the time the bullet is in flight and its velocity, we can determine the distance it travels (we are neglecting the resistance of air friction or the possibility of wind) before it hits the ground, using the equation $d = vt$. Since $v = 2000$ ft/sec, then $d = 2000 \times \frac{1}{2}$ sec, or 1000 ft horizontally.

Figure 1-8
Galileo questioned the Aristotelian theory that heavy bodies fall faster than lighter bodies. He was able to show the error in Aristotle's theory by thinking about what would happen if he did this experiment: Take two 1 kg bricks, identical in size, and drop them from a tower. They hit the ground at the same time. Now drop the same bricks from the tower side-by-side. They still hit the ground at the same time. Put a tiny drop of glue between the bricks and drop them. Next, glue the bricks firmly together, making one 2 kg body, and drop them. Each time they take the same time to fall. From this mental experiment, Galileo concluded that *all* objects fall at the same rate. He is said to have tested his reasoning by dropping cannon balls of different weights from the Leaning Tower of Pisa, although we do not have any record of this experiment, and some claim the story is just a myth.

How then can we calculate the shape of the paths, the precise trajectories, of the two balls thrown upward? We know that, horizontally

$$d = vt$$

and that, vertically, for both,

$$d = \tfrac{1}{2}at^2$$

The component of motion that is the same for both is the time. What we seek is a relation between the two distances that can be plotted on a graph. Bringing the two equations together, using x to represent the horizontal distance each ball travels, we can substitute x/v for t (from $x = vt$) in the equation for vertical motion, so that we have

$$d = \tfrac{1}{2}a\left(\frac{x}{v}\right)^2$$

Taking the acceleration of gravity and the horizontal velocity to be constants,

$$d = \left(\frac{a}{2v^2}\right) x^2 \text{ or } d = kx^2, \text{ where } k = \frac{a}{2v^2}$$

This is the equation for a parabola; therefore any projectile fired so that it is only acted upon by gravity will follow the path of a parabola. The exact shape will depend on the velocity with which the projectile was fired.

The Mechanics of the Heavens

As an observer and experimenter in astronomy, as well as with motion on the earth, Galileo made important contributions to the new account of celestial motion that the scientists of his era were developing. This view differed from traditional ideas of heavenly motion as widely as Galileo's account of motion on the earth differed from Aristotle's.

The celestial system of the Western world had long depended on the concept of a fixed earth. Aristarchus, a Greek philosopher who lived a century after Aristotle, had suggested that the *earth* might revolve about the *sun,* but his idea did not gain wide acceptance. The fixed-earth, earth-centered (geocentric) theory was plainly supported by observed experience: standing on the surface of the earth, it is impossible to perceive that it is in motion, whereas the sun clearly is seen to rise in the east, move across the sky, and disappear in the west.

To the Greeks and other early observers, just as to us, the heavenly bodies appeared in different parts of the sky from day to day, week to week, season to season. Throughout the course of a year a planet appeared against several different fields of the stars. At different times of the month, the moon as well clearly occupied different regions of the sky. As explained by early astronomers, however, these phenomena did not contradict their theory of nested, invisible, crystalline spheres rotating at uniform rates about the earth. The difference in positions of heavenly bodies with respect to one another was accounted for by the idea that the spheres in which each heavenly body was fixed rotated independently of one another. Each sphere was fastened to a larger outer, or smaller inner, sphere by pivots at opposite ends of an axis about which that particular sphere rotated. The axes of rotation of the different spheres were set at angles to one another. Thus, although each sphere rotated at a constant, uniform rate about the earth, the observed position of a body could still change in relation to the position of bodies in other spheres.

The Ptolemaic System

Claudius Ptolemy, a Greek astronomer at Alexandria in the second century AD, was the first to develop a systematic theory of the observed motions of the heavenly bodies. Like earlier astronomers, Ptolemy believed that the heavenly bodies followed circular orbits. He did not think, however, that the earth was necessarily at the exact center of each and all of these orbits. In the Ptolemaic system, the earth might be at the center of the orbits of some of the celestial bodies, but others might rotate in circular orbits, of which the centers were not exactly at the earth (eccentric). Furthermore, Ptolemy claimed that a heavenly body might itself follow a second, small, circular orbit (epicycle) while the center of the epicycle rotated on a larger earth-centered circle (deferent). Thus in Ptolemy's system the motions of heavenly bodies follow paths that are quite complex, even though made of circles. Some even resemble three-leaf clovers. The earth, because it does not lie at the center of all of the major orbits, is not necessarily equidistant from every point along the path of each heavenly body, and so a planet may be farther from the earth at one point in its path than at another. By means of this system of complex orbital paths, Ptolemy explained the observed changing relationships of the visible planets. The path of the moon had its own shape, quite independent of the paths of other heavenly bodies. Since these bodies did not revolve about the earth at the same speed, the moon would appear against a different background of stars in the winter than it did in the spring. The Ptolemaic system resembled earlier celestial systems in its insistence that the earth is heavy, hence immovable, and in its use of the pattern of uniform and circular motions to explain astronomical observations.

Ptolemy developed the first systematic theory of celestial motion, which included complex orbits to explain how all of the other heavenly bodies rotate around the earth.

Ptolemy put forth his theory of celestial motion and the orbits of heavenly bodies in a book later called the *Almagest* (an Arabic word meaning "the greatest"), a book that for more than 1400 years was the fundamental text in astronomy. The *Almagest* contained careful calculations, based upon observations, that permitted the reader to determine the past and future positions of celestial bodies. In the light of modern knowledge about the true nature of celestial motion, the long survival of Ptolemaic astronomy is a testament to the usefulness of its system, especially for navigation.

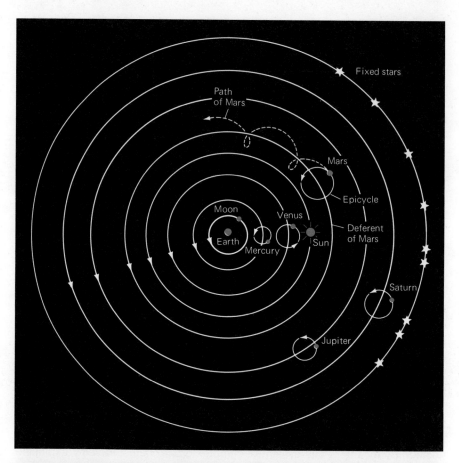

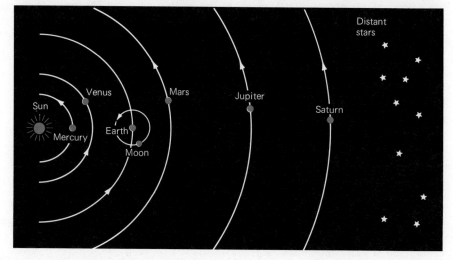

Figure 1-9
Ptolemy's theory of the universe, shown above in a simplified diagram, was geocentric. It was based on the idea of a fixed and immovable earth. The Copernican system, shown below, is heliocentric and forms the basis of the currently accepted view of the solar system.

Ptolemy argued for a spherical earth, noting that sunrise, eclipses, and other celestial phenomena were seen at different times from different places on earth. Astronomers had long wondered why the stars seem to show no **parallax**, no apparent change in position due to actual change in the position of the observer (Figure 1-11). Rather than explain this by assuming that the stars were immensely distant from the earth, as some Greek observers had suggested, Ptolemy accepted the more traditional thesis of a sphere of stars moving about the fixed earth. His system also incorporated earlier theories that explained the phases of the moon, in which the tilt given to the axes of the moon's sphere resulted in a motion for the moon that caused it to pass between earth and sun in such a way that the illumination of the moon by the sun, as observed from earth, varied in a monthly cycle.

In the early Renaissance, research in astronomy still took place within the framework of the Ptolemaic system. Despite increasing sophistication in the understanding of many aspects of nature, few new instruments had been invented to measure celestial motion, and no one had come forth with a theory that could predict the positions of the heavenly bodies with greater accuracy than Ptolemy and his followers had done. But as time passed, there were more and more discrepancies between the positions of the planets, moon, and stars as calculated and the positions actually observed. In attempting to account for the relative positions of the great numbers of heavenly objects, the Ptolemaic system became increasingly elaborate. Though Ptolemy himself had urged that scientific explanations be as simple as possible, consistent with the observations, more and more epicyles, describing more and more complex paths, were needed to explain accumulating observations.

The Copernican System

By the early sixteenth century the Ptolemaic system included more than eighty different special motion factors, with 19 epicycles for Mars alone. To simplify celestial theory, Nicholas Copernicus, a Polish astronomer and mathematician (1473–1543), developed a new system of celestial motion. He did not believe the wisdom of God could be reflected in a system that was no longer a set of simple and perfect circular motions. In his sun-centered (heliocentric) system the *sun* was fixed in position, and the planets, *of which the earth was one,* revolved around it; the sphere of the stars was motionless. There was more to Copernicus' theory; he postulated that the earth itself spins once a day about an axis passing through its center, and the moon revolves around the moving earth. The proposal was radical (even though it was essentially the same idea put forth by Aristarchus), because it contradicted Aristotelian belief and religious dogma as well.

Copernicus proposed a theory of the movement of heavenly bodies in which the sun was fixed at the center of the universe and the planets, including the earth, revolved around the sun.

The Copernican proposal solved a major problem in the Ptolemaic system, the so-called retrograde motion of the planets. Astronomers had long noted that planets did not move across the sky in only one direction. Nor did they move at constant speed. Sometimes they seemed to speed up, and then slow down. Sometimes they even reversed direction. Ptolemy had explained this phenomenon as the result of epicyclic loops in the planetary orbits, which make them appear to go first forward and then backward, as they looped around into another forward phase of their orbit. In the heliocentric theory of Copernicus, these retrograde motions of the planets were ascribed not to irregular orbital paths but to the planets' positions relative to the earth, which was also moving, but not at the same speed or the same distance from the sun. Planets whose orbits were outside that of earth moved more slowly; as the earth passed them, they *seemed* to fall back, but later in the earth's motion they were observed to have moved forward again (see Figure 1-10). The retrograde motions of the inner planets, which are traveling faster than the earth, occurred because sometimes they are moving with the earth on the same side of the sun, and at other times (because of their faster orbital speed) they are on the opposite side of the sun. In both positions they are moving forward, but at the position in their orbit where they pull away from the earth because of their faster speed, they seem to move first backward and then ahead once again.

Despite the Copernican system's simplification of the description of the universe, its predictions were no more accurate than those of the Ptolemaic

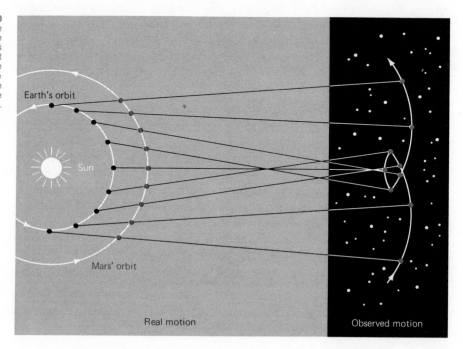

Figure 1-10
This picture shows the apparent retrograde motion of Mars. The real movement of the earth relative to the movement of Mars is depicted in simplified form on the left. But because Mars moves more slowly than the earth, it is seen from the earth to move against the stellar background in the fashion shown on the right side of the diagram by extension of the lines of sight.

Earth's orbit

Sun

Mars' orbit

Real motion

Observed motion

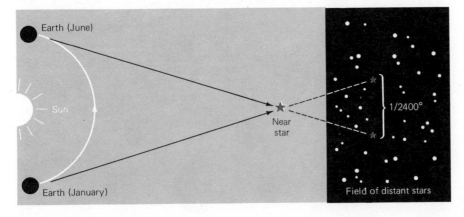

Earth (June)

Sun

Near
star

1/2400°

Earth (January)

Field of distant stars

Figure 1-11
As the earth revolves around the sun, the
observed position of a star relative
to more distant background stars will shift
with the observer's change of position. This
shift can be seen only by using powerful
telescopes. The largest observed parallax
of a star (due to the earth's motion around
the sun) has been found with refined
instruments of measurement to be 1/2400°.
Geometric calculation from this angle
indicates that the nearest star is about 25
trillion miles away. (The drawing is, of
course, not to scale.)

system. In order to maintain the circularity of the planetary orbits, Copernicus,
too, was forced to resort to some epicycles to explain discrepancies between
the observed positions of heavenly bodies and their positions as calculated
from his theory. This feature made his celestial system similar in principle to
Ptolemy's. Nor could the Copernican theory, at the time it was formulated, account any better than Ptolemy's for the lack of parallax in the fixed stars. The
Copernicans were not able to detect such a shift, even when they viewed a particular star at each end of the earth's huge orbit around the sun, and were
forced (until the development of new instruments of observation) to abandon
any attempt to calculate the distance of the stars from the earth.

The lack of observable phases for Venus also posed a problem to the acceptance of the Copernican theory of the heavens. Venus is seen in the western sky
after sunset and in the eastern sky before sunrise—we know it as the morning
and evening star. It is never seen during the night, but travels with the sun
across the high part of the sky during the day. Its close association with the sun
suggests that Venus, if it were in orbit about the sun, should exhibit phases similar to those of the moon (see Figures 1-12 and 1-13). Early observers had
suggested that Venus might exhibit phases since its brightness varies considerably, depending on its position relative to the sun, but in fact observation with
the naked eye did not reveal any phases, which supported Ptolemy's theory of
an earth-centered universe.

The major innovation of the Copernican system, then, was the theory of circular
planetary orbits around a fixed sun, rather than a fixed earth. Copernicus' theory
might have been accepted as a simplified description of the motion of heavenly
bodies, but to do so meant also accepting a new view of the earth, which
became just another planet rotating on its axis like all the other planets. The
implications of this idea were extremely hard for the learned men and religious

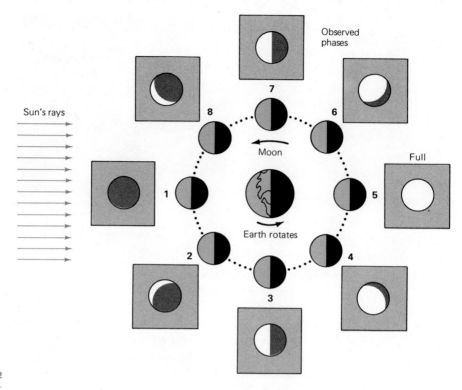

people of the time to accept. It meant that the difference between Aristotle's four earthly elements and the heavenly fifth element, the ether, disappeared. If the earth is a planet then we must face the possibility that the other planets may be solid, liquid, or gaseous, as well as fiery or ethereal. The Copernican world view overturned both Biblical and Aristotelian notions of the universe, which were based on the separate and contrasting qualities of earth and heaven.

Brahe and Kepler

At the same time Copernicus was working out his system. Tycho Brahe (1546–1601), a Danish nobleman with a passionate interest in astronomy, was making detailed observations of many of the heavenly bodies. Brahe's observations, made both with the naked eye and with instruments of his own invention, were extraordinarily accurate. They have been found to differ very little from modern calculations of the positions of heavenly bodies—and in Brahe's time the telescope had not yet been invented. Though Brahe was a meticulous direct observer and recorder of the relative positions of the planets, he drew no fundamentally new conclusions about the heavens. He did, however, devise a modification of the old geocentric system. In it the five known planets (Mercury, Venus, Mars, Jupiter, and Saturn) revolved around the sun, which itself revolved around the stationary earth, as did the moon and the fixed stars, all at their own rates. Brahe's system was a compromise. He knew the Copernican view was simpler but he could not agree that the earth moved. He felt that if it did, he should be able to observe stellar parallax from the extreme positions of the earth's orbit, which he could not do. He thought his system combined the best of Copernicus and Ptolemy.

New conclusions remained to be drawn from Brahe's data after his death by his trusted assistant, the German astronomer and mathematician, Johannes Kepler (1571–1630), who favored the Copernican explanation of motion in the universe. Kepler at first believed, like Copernicus and earlier astronomers, that the orbits of the planets were circular. However, after a long period of arduous calculations, using Brahe's observations, he found there were positions recorded in Brahe's observations that could not be reconciled with the concept of circular planetary orbits. Although the discrepancy involved only 8″ of an arc (an exceedingly small angular measure, about 1/8 of a degree), Kepler was not inclined to doubt Brahe's observations, because he knew they had been carefully done. Kepler felt obliged instead to abandon the ancient concept of circular orbits. He then tried elliptical orbits for the planets, with the sun *in the center* of each **ellipse**, but this hypothesis did not fit Brahe's data either. His next attempt was to place the sun *at one focus* of an elliptical planetary orbit. After four years of work he found, much to his delight, that this arrangement correlated with a substantial body of Brahe's data, in particular the observed position of Mars. Kepler thus established what we know to be true today: that the planets move in elliptical orbits, with the sun at one focus of a family of ellipses.

Figure 1-13
The diagram on the left shows the phases of Venus as they appear to an observer on the earth. Venus' orbit goes around the sun, as Copernicus said, and Venus exhibits phases, much like the phases of the moon, from our perspective on earth. When it is visible nearest to the earth, we see a thin crescent shape. When it is further away from the earth, we see more of the sphere illuminated. The diagram on the right shows why these phases of Venus would not be possible in the Ptolemaic system. According to Ptolemy's theory, the epicycle of Venus remains on the near side of the sun. If this were so, we would never see the phases of Venus in which more than half of her sphere is illuminated because the planet would always be between the earth and the sun. Galileo's observations of the gibbous phase of Venus, in which more than half but not all of the sphere is illuminated, were decisive evidence against the Ptolemaic theory.

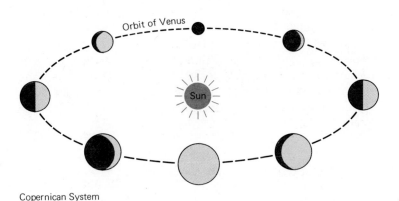

Copernican System

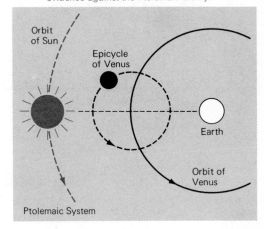

Ptolemaic System

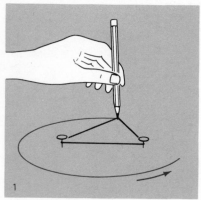

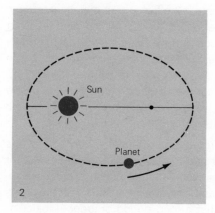

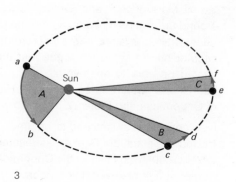

Figure 1-14
To draw an ellipse (1) and demonstrate Kepler's first law, wrap a loop of string around two tacks which are a short distance apart, being sure to keep the string taut. To draw ellipses of different shapes, simply vary the length of the string or the distance between the tacks. The focuses are the points in an ellipse which correspond to the positions of the tacks. The orbits of the planets are ellipses with the sun at one focus (drawing 2); this is Kepler's first law. Drawing 3 shows Kepler's law of equal areas. A radius vector (an imaginary line joining any planet with the sun) sweeps out equal areas in equal lengths of time. When the planet goes from a to b, it sweeps out area A. If the planet goes from c to d or from e to f in an equal amount of time, then areas A, B, and C are equal.

Kepler discovered that the planets move in elliptical orbits with the sun at one focus of a family of ellipses.

Kepler found that actual observations indicated a slightly changing speed for Mars as it moved in its orbit, a discovery that harmonized with the hypothesis of an elliptical orbit for the planets. This changing speed made it difficult to predict a planet's position. After many years of effort, Kepler reached his next conclusion, the law of equal areas, which states that the speed of a planet at one position in its orbit can be related to its speed at another position in such a way that the nearly triangular areas swept out by an imaginary radius line between the sun and the planet, as the planet moves, are equal for equal periods of time. His conclusion was based on the recognition that a planet would move faster the closer it was to the sun, while at the most distant parts of its orbit—the end of the ellipse farthest from the sun—it would move most slowly. This was an insight of considerable importance in Newton's later development of the theory of universal gravitation.

Kepler persisted further in his calculations. He was hopeful that the laws governing the solar system would be uniform for all the planets and that the orbital movements of the earth could be related to those of the other planets in that system. For this reason, he sought a general law governing a planet's speed about the sun. After ten years Kepler discovered this uniformity: the square of the time required by a planet to revolve once fully about the sun (planetary year) is proportional to the cube of its average radius (distance from the sun). This proportion is the same for all of the planets. If T is the time required for a planet to move once fully about the sun, and r the average length of its radius from the sun, Kepler's relationship is expressed mathematically, where the sign $\propto$ means "is proportional to," as

$$T^2 \propto r^3$$

or, where k is the constant of proportionality for the entire family of the sun's planets,

$$T^2 = kr^3 \text{ or } \frac{T^2}{r^3} = k$$

We can find the constant k by using the modern values of the distance between earth and sun.

$$\frac{T^2}{r^3} = \frac{(365 \text{ days})^2}{(93,000,000 \text{ mi})^3} = \frac{(365)^2}{(9.3 \times 10^7)^3} = \frac{1.33 \times 10^5 \text{ days}^2}{8.04 \times 10^{23} \text{ } mi^3}$$
$$= 1.65 \times 10^{-19} \text{ day}^2/\text{mi}^3$$

Changing units from days to years, we set $T = 1$ year and $r = 1$ astronomical unit (AU), the earth–sun distance. Then $k = 1$.

Kepler's reliance on mathematical calculations for his researches into the structure of the universe was a new departure in science. Before him, most researchers worked primarily with geometry. In fact, Kepler's own preoccupation with ideal geometrical orbits (such as the circle) had first led him astray. His eventual reliance on exact data was unprecedented, however, and provided him with a new vantage point for the problems he set himself to solve. Because of Kepler's work, search for mathematical equations replaced the assumption of geometrical perfection in a celestial model. In this way algebraic forms were added to geometric patterns in the description of motion. This new mode of work led Kepler to the conception of a mathematical relationship as the governing force in the universe, a relationship later worked out by Isaac Newton.

The New Science of Astronomical Observation

By the end of the sixteenth century the celestial system of the ancients had begun to fall apart. In 1572 Tycho Brahe discovered a supernova, which we now know to be an exploding star. The sudden appearance of this bright, glowing point of light, followed by its gradual dimming, undermined the belief that the heavenly bodies were perfect and unchanging. At the time Kepler was formulating his laws, Galileo was working at his studies of motion on the earth, but he also had a keen interest in astronomy. Hearing of the invention in Holland of a new instrument called the telescope, Galileo constructed several for himself; the largest brought objects 30 times closer than when seen with the naked eye. The first celestial object he explored with his new instrument was the moon. Up to that time the moon had been thought to be perfectly spherical, smooth, and unblemished. But Galileo discovered that the moon had a very rough terrain with high mountains and large, dark areas, which came to be known as maria, or seas. Galileo was able to estimate—very accurately—the height of the mountains by measuring their shadows on the lunar surface.

Galileo also studied the planets. With his telescope he was able to observe that Venus passed through a full set of phases. He also noticed that the disc of Venus varied in size over time. The correlation of the phases with the size variance was a virtually irrefutable argument that Venus orbits the sun, as Copernicus maintained.

Galileo was the first to demonstrate the existence of sunspots, and he noticed what he called "ears" projecting from the sides of Saturn. We now know that these were the outer regions of Saturn's rings. Another of his major discoveries was the moons of Jupiter, which, he found, revolved in orbits about Jupiter. This fact meant that there was more than one center of revolution in the known universe.

Observing the planets, Galileo found them to be large and round, but when he looked at the stars he could see only glistening points of light; they looked no larger, only somewhat brighter than when observed with the naked eye. He reasoned that if the stars appeared this way in his telescope they must be very distant indeed. This evidence of the stars' vast distance explained why no one had been able to detect parallax in the course of observing the positions of the stars during an earth-year. In fact, the first astronomer to detect stellar parallax was the German, F. W. Bessel, in 1838. His observations form the last, best evidence for the validity of the Keplerian system.

For those who accepted Galileo's telescopic observations, the truth of the new astronomy had become irrefutable. Together with the ideas of Copernicus, the observations of Brahe, and the calculations of Kepler, Galileo's findings about motion on the earth and in the heavens provided the material from which, in the next generation, Isaac Newton was to develop the magnificent synthesis that revolutionized scientific understanding of the structure and operation of the universe. Newton was born in the year Galileo died.

Glossary

The acceleration of a body is the rate of change of its velocity. An object is accelerated if its speed changes or its direction of motion changes.

The acceleration of gravity is the acceleration of bodies in free fall. It is the same for all bodies. Near the earth's surface it is 32 ft/sec².

An ellipse is a curve—the shape you see when you view a circle on a slant. It has two focuses: the further apart they are, the flatter and longer the ellipse.

Friction is the resistance to motion of an object due to its contact with other substances.

The tendency of an object to resist any change in its state of motion is called inertia. Objects left to themselves continue at constant speed in one direction.

Mechanics is a branch of physical science that deals with motion. It treats the movement of bodies and the forces that affect their movement.

Parallax is the apparent shift in position of an object relative to a background due to an actual shift in the viewer's position.

Speed is the rate at which a body covers distance: the distance a body moves in a period of time divided by that period of time.

The curve a moving object describes in space is its trajectory.

Velocity refers both to the speed and the direction of motion (Speed equals, for example, 30 mph; velocity equals, for example, 30 mph to the northwest.)

1 A group of students decide to drive nonstop from Newark to Miami, a distance of 850 miles. If the trip takes 17 hours, what was their average speed in mi/hr?

2 After the speed limit was lowered in 1974, the Greyhound Bus Company found it necessary to revise their schedule for the 214-mile New York to Boston run from 4 hr 12 min to 4 hr 40 min. Find the original average velocity of the bus and its new average velocity.

3 The speed of sound is about 1100 ft/sec. If the interval in time between a flash of lightning and a thunderclap is 7 sec, how many miles away is the lightning?

4 Galileo watched a smooth ball roll down an inclined plane from rest to a speed of 10 mi/hr in 3 sec. What was the average acceleration of the ball?

5 A test driver is rating the performance of the Fiat 124 against the Jaguar XKE. Compute the two average accelerations if the Fiat accelerates from 0 to 60 mph in 7 sec and the Jaguar accelerates from 0 to 60 mph in 5 sec.

6 How far will a race car have moved after 10 sec at a constant acceleration of 22 ft/sec²? What will be its velocity in mph? If the car begins to brake after 10 sec, with an acceleration of −11 ft/sec², how many seconds pass and what distance is covered before it reaches a state of rest?

7 A snowball dropped from the roof of a building hits the pavement 144 ft below. With what velocity is the snowball moving just before it hits the pavement? How long will it take to fall?

8 The planet Mercury completes its orbit of the sun in 88 days. How long is one Mercurial year? How many hours is the average night on Jupiter if one rotation on its axis takes 9.8 hr?

9 Why can a swordsman slice an apple in half in midair when seemingly nothing is holding the apple against the blade?

10 A passenger in an airplane drops her pen during a free-fall drop. Would the pen appear to move toward the ceiling of the cabin, move toward the floor, or remain stationary? Why?

11 The period of a simple pendulum (the time that it takes to swing from one side of the arc to the other side, *and return* to the first side) varies according to its length. What would happen if the pendulum of a grandfather clock were shortened?

12 What was Aristotle's theory of motion and rest and how was it deduced from his concept of "natural place"? How did Aristotle account for the absence of rest in the movement of heavenly bodies?

13 What is parallax, and how is it used to determine the distance to planets and nearby stars? Why were the stars thought to be stationary points for so many years?

14 Compare the theories of Ptolemy and Copernicus with regard to the motion of heavenly bodies. What observable phenomena proved troublesome in supporting each theory?

15 How did Kepler improve upon the Copernican system?

Interlude Two Keeping Time, Time, Time . . .

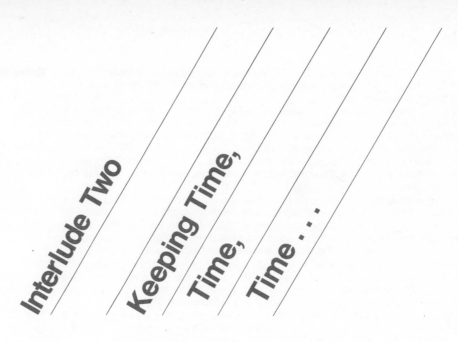

Science and timekeeping have been intertwined since their beginnings: the development of one has nearly always depended on the other. The discovery of the laws of motion, for example, required that scientists be able to measure time intervals accurately enough to determine that movement in time follows certain laws. What exactly *is* time? Poets have compared it to a river; it is the order of events, the measure of movement, and that measure is what the clock says. The clock can be any repeating cycle: a heartbeat, the phases of the moon, the opening of a marigold, sand flowing from one glass to another, the pendulum's swing, the vibrations of crystals or atoms. Accurate timekeeping is essential to our industrialized way of life, and time is also one of the central concepts by which we order our personal experience. For many people, time is money. As we grow older, we measure our accomplishments, celebrate personal and cultural anniversaries, and plan for the future within the framework of the passage of time.

Sundials were fashionable accessories in nineteenth-century Europe, and gentlemen often carried them in collapsible models in their pockets. The brass and marble French timepiece shown above sports a cannon that could be loaded with gunpowder. The gunpowder would heat up when the sun was focused on it through the magnifying glass, and the cannon would go off at noon.

In the beginning, when life was simpler, our ancestors used the sun to tell time. One of the most spectacular early monuments built, apparently, for astronomical observations and perhaps also for worship is Stonehenge (above and facing page) in Wiltshire, England. This circular setting of huge, standing stones permits an observer to mark the appearance and position of the sun and the moon against the horizon. The Heelstone, for example, marks the point on the horizon where the sun rises on the summer solstice, the longest day of the year. Stonehenge was erected in several stages between 1800 and 1400 BC.

Sundials are often incorporated into the design of contemporary public buildings. The sundial sculpture shown (below) was created by Henry Moore and stands in the forecourt of the offices of the London *Times*. The bronze sculpture is about 12 feet tall and gives the local time.

The first devices used to keep track of time are called *gnomons* (from the Greek meaning "one who knows"), shadow clocks, or sundials. The gnomon can take various forms, but it always includes a stick, pillar, or other object that casts a shadow whose length or position can be measured by marks inscribed on a field. The simplest portable gnomon is a hand dial, shown here (above center) in a rendering from an old German engraving. The stick is held by the thumb with the palm facing the person holding it at an angle appropriate to the latitude. Before noon the gnomon is held in the left hand and pointed west; after noon it is held in the right hand, pointed east. The stick casts a shadow that falls on the fingertips or joints, and the person holding the stick tells the time by learning the hours that correspond to the various shadow positions.

For nights and cloudy days, clocks that used oil, candles, water, or sand were devised. The Egyptians crafted early water clocks in the form of vessels with small holes in the bottom which were inscribed with marks for the hours on the inside. They were filled with water, and as the water flowed out, the passage of time could be noted. The cast of an Egyptian waterclock shown (above) dates from about 1400 BC.

Sand glasses were introduced during the first century AD and continue to be used today. This example from the American colonial period (left) is a ship's hour glass, inscribed "Keep this end up while running."

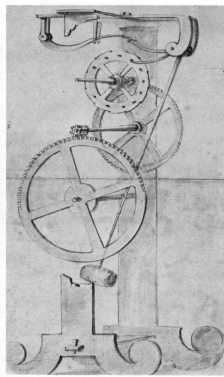

As human affairs grew more complex, so did the means of keeping time. The first mechanical clocks were large, weight-driven machines known as turret clocks because they were housed in towers. The earliest of these devices had neither hands nor dials; they were simply used to mark the hours for a bell-ringer who would then toll the tower bells by hand. The first public striking clocks appeared in the fourteenth century. The famous tower clock shown (above right) is called Big Ben and was installed in the Victorian tower of Westminster Palace in London in 1859.

The pendulum is especially useful to measure time because the full arc of a small swing is quite regular, and since the arc depends almost entirely on the length of the pendulum, it is easy to regulate. Galileo noted this timekeeping characteristic of the pendulum and incorporated it into a design for a clock mechanism, shown (above center) in a sketch drawn by his son in 1641. Weight-driven clocks with short pendulums were introduced in the late seventeenth-century. The nineteenth-century American grandfather clock (above left) is from the home of Abraham Lincoln in Springfield, Illinois.

Time measurement has become increasingly precise in modern times. In 1929 the quartz crystal was first applied to timekeeping. These crystals vibrate at a steady frequency of 100,000 cycles per second, which in quartz-crystal clocks are scaled down electrically to give impulses at frequencies of 1 cycle per second. These clocks are so accurate that their error comes to about one second every ten years. The picture (below) shows the quartz ring of a crystal clock.

Like mechanical clocks, atomic clocks use and record regular oscillations—in this case, the electric and magnetic vibrations of electrons in individual atoms. The frequencies of these oscillations are the same for all atoms of the same element. The atomic cesium clock shown (below), for example, uses Ce^{133} atoms, each of which has 54 electrons—53 filling the inner shells and 1 electron alone on the outside shell. This lone electron can undergo a quantum jump in its spin, emitting a fixed frequency quantum of radiation. In the clock, the cesium atoms are placed in an alternating magnetic field whose frequency is varied until there is a maximum resonance from that spin flip of the cesium electron. Then the alternating magnetic field has been tuned to the exact frequency of the cesium. The resulting oscillation is so precise that some cesium clocks are accurate within one second in 1000 years. Since 1967, Ce^{133} oscillations have been used to provide the standard primary definition of the second; according to International Time Bureau, a second is the duration of 9,192,631,770 cycles of cesium radiation.

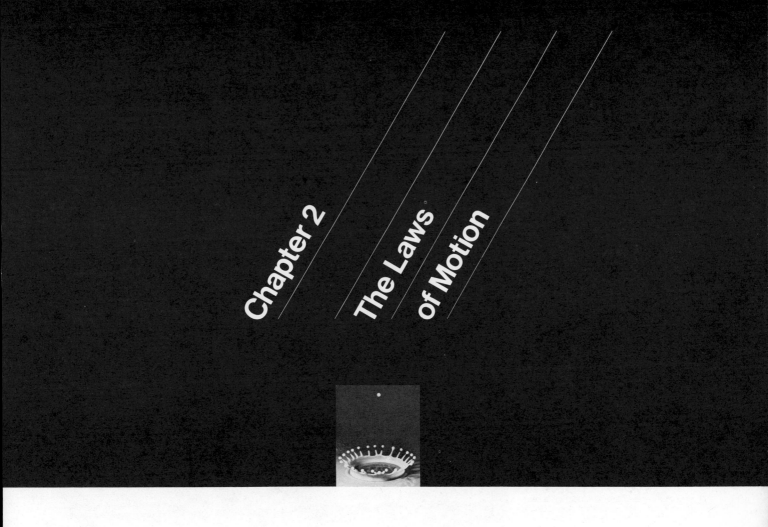

Chapter 2
The Laws
of Motion

Motion has both direction and magnitude; we can, for example, measure the direction in which a soccer ball is rolling and the speed at which it is traveling. Motion can be defined as the end result of the exertion of one or more forces on an object. The direction and magnitude of the forces in part determine the direction and magnitude of the motion — but they are not the only determinants. Equally important is the amount of matter contained in the object.

To refer to the amount of matter in an object, scientists use the term **mass**. There are two kinds of mass: **inertial mass** is a measure of an object's resistance to any change in its motion; **gravitational mass** is a measure of the mutual force of attraction between objects. **Weight** is the force with which the earth pulls downward on an object. When we study vertical motion, such as the movement of a ball tossed up in the air from the earth's surface, we must be

The term *mass* refers to the amount of matter in an object.

concerned with both the inertial and the gravitational mass of the ball. But when we study horizontal movement alone—for example, a ball rolling on a perfectly level surface—we are concerned only with inertial mass; gravity does not affect horizontal motion.

Sir Isaac Newton (1643–1727) was the principal architect of our scientific view of the world. A brilliant physicist and mathematician, he began developing his original insights into the mechanics of motion and gravitation when he was only 23. These contributions were published in 1687 in his great book, the *Principia*. One of the most important works in the history of human thought, the *Principia* sets forth Newton's three laws of motion and the law of universal gravitation.

Newton's Laws
First Law

Newton's **first law of motion**, the law of inertia, restates Galileo's concept: *Every body continues in its state of rest or of uniform motion in a straight line, unless compelled to change that state by an external force.* A force is needed to start a body moving, to change its motion, or to stop it. No force is needed to keep a body moving at constant speed in a straight line.

A body at rest or in uniform motion in a straight line will continue in that state until it is acted upon by an external force.

Is the first law contrary to everyday experience? A rolling stone eventually stops and gathers moss; a bicycle coasting on a straight road will soon come to a halt unless the cyclist pedals some more. Why do the rolling stone and the bicycle seem to disregard Newton's law? In these examples, there is a second force at work—friction. One kind of friction, called **kinetic friction** or friction of

Figure 2-1
This sled, designed to test automobile seat belts under crash conditions, provides an excellent example of the law of inertia. As the sled brakes to an instant halt, the passenger continues moving forward, except where he is held in place by the seat belt, the "external force" of Newton's first law.

motion, is the force that arises with the relative movement of two bodies whose surfaces are partly in contact; the larger and rougher the touching surface areas, the greater the friction. A second type of friction, **static friction**, exists between two *unmoving* bodies when there is a force along their mutual contact surface (a gluey surface provides just such a force). Both kinds of friction act as forces opposing motion.

In an idealized situation, such as that imagined by Galileo in the last or horizontal stage of his inclined-plane thought experiments, there is no external force to interfere with inertial motion, and constant velocity is maintained. Earth satellites, planetary moons, or the planets themselves come closest to realizing a frictionless environment. Here on earth oils, greases, and other lubricants are used to reduce friction, and ball bearings, which reduce the area of contact between two surfaces, also reduce friction to a considerable extent.

The first law of motion tells us how matter, either at rest or in motion, behaves in the absence of any external force. But as with the moving bike and the rolling stone, most motion *is* affected by other forces, such as friction, gravitational pull, or collisions. What is the effect of an external force on the inertial mass of an object? Newton answered this question in his **second law**: *The acceleration of a body is directly proportional to the force applied, and inversely proportional to the mass of the body*. The acceleration will be in the direction of the applied force. As an equation, the second law may be stated as:

$$a = \frac{F}{m}$$

or

$$F = ma$$

where m is mass, a is acceleration, and F is force.

If we apply the second law equation to a constant mass of 1 kg, and successively increase the units of force from one to two and then to three units, the acceleration will also increase from one to two and then to three units. Although the equation $F = ma$ does not sound familiar, it expresses something commonly understood and used. Everyone who has ever tried to push a stalled car off the highway knows the more people pushing, the sooner the car gets off the road.

Suppose that we hold F constant but increase m. Contrast the effect of one man pushing a Volkswagen and the same man pushing a 4-door Cadillac sedan. If we keep the force at one unit, and successively double and triple the mass, we can see from the equation that acceleration will first be cut to one-half, and then to one-third. The relationship between acceleration and mass is one of *inverse* proportion. On the other hand, acceleration and force are *directly* proportional.

Second Law

Force is equal to mass multiplied by acceleration.

Figure 2-2
The great mass of NASA's Vanguard II rocket resists the enormous thrust of the engine. As it lifts off the launch pad, its initial speed is slow.

The first law tells us that mass resists change in motion; the second law tells us how mass and acceleration are related to motion change. To produce the same acceleration in a large mass as in a small mass requires a proportionately larger force. The unit for force is defined in terms of mass and acceleration by using the second law. In the metric system it is called a **newton**, (N), or that force which, when acting on a mass of 1 kg, will produce an acceleration of 1 meter/second². In the English system, the unit of force is the pound, which is the force that will produce an acceleration of 1 foot/second². One pound is equal to 4.4 N.

Suppose we have a mass of 6 kg. What acceleration will result from successive forces of 6 N, 18 N, and 30 N? Since $a = F/m$, we get:

$a = 6/6 = 1$ m/sec²

$a = 18/6 = 3$ m/sec²

$a = 30/6 = 5$ m/sec²

As F increases, so does a.

Now suppose that a force of 60 N is applied to masses of 5 kg, 12 kg, and 24 kg. What is the acceleration in each case? $a = F/m$, so:

$a = 60/5 = 12$ m/sec²

$a = 60/12 = 5$ m/sec²

$a = 60/24 = 2.5$ m/sec²

For a constant force, then, acceleration decreases as mass increases.

Third Law

Can you apply a force to an object and remain unaffected yourself? At first glance, it seems so. Stand on a dock and pull on a rope attached to a rowboat a short distance away; you remain where you are, because your feet are pressing down on the dock. But stand in a second rowboat and pull on the rope attached to the first boat. Now as the first rowboat moves toward you, you and your rowboat move toward it.

In Newton's words, "Whatever draws or presses another is as much drawn or pressed by that other. If you press a stone with your finger, the finger is also pressed by the stone. If a horse draws a stone tied to a rope, the horse (if I may so say) will be equally drawn back towards the stone." Newton summed this up in his **third law of motion:** *To every action there is always opposed an equal reaction.* Action and reaction forces are equal in magnitude, but they act in opposite directions, each against the opposed object; they act on different bodies. If you push down on a table, the table pushes up equally on you. If an astronaut on a space walk pushes himself away from the spacecraft, the craft will drift off, since it is pushed in the opposite direction.

Forces always occur in action–reaction pairs.

Perhaps the classic demonstration of the third law is the reaction engine, of which the rocket is the best known example. In its simplest form, it is like a balloon filled with air, which, upon being released, flies off in an erratic path. The air pushed out of the balloon by the elasticity of the rubber container (action) pushes back (reaction) with an equal and opposite force, propelling the balloon forward. An only slightly more complicated version of this reaction supplied the propulsive force of the sky cycle Evel Knievel used in his attempt to jump the Snake River Canyon. Hot steam pushed out of the combustion chamber of the sky cycle (action) pushed back on the walls of the chamber (reaction), accelerating the rocket.

Examples such as that of the sky cycle are complex because they involve two forces which may have different directions and different magnitudes. How can we calculate the net effect of two or more forces acting on the same object? A **vector** is a quantity that specifies both magnitude and direction. Force is a vector, as is velocity. An arrow in the desert pointing to a town and reading: "Dry Gulch, 6 m south" can be considered a vector since it conveys information about distance (magnitude) and direction. It indicates the vector **displacement,** or the distance away from a fixed point in a given direction.

To understand the concept of a vector, imagine a block of wood resting on a table. You push on it with a force of 5 newtons directed east. An opponent exerts a force on the block of 5 newtons directed due west. These forces are balanced and effectively cancel each other out; the block remains at rest. These conditions satisfy Newton's first law. (These forces are equal and opposite, but they act on the same body, so the third law does not enter into this consideration.) The magnitudes of these vectors can simply be added because they are in the same line. If we arbitrarily assign a positive value to west, then the opposite direction (east) becomes negative. We then have:

$$(+5) + (-5) = 0$$

This equation yields no unbalanced or net force; the block is in a state of equilibrium. Now assume that you increase the eastward force to 10 newtons. Using vector addition, the equation reads:

$$(+5) + (-10) = -5$$

This represents an unbalanced force of 5 newtons eastward. By Newton's second law, the block will experience an acceleration in this direction.

These forces, in acting through the block, may be considered to act on the same point or concurrently. Since they are in a straight line, the angle between their lines of action is zero. There may, in other cases, be forces that act concurrently but at an angle to each other. Picture the two forces in our first example rearranged like the two hands of a clock: they can assume any arrangement between 12 o'clock (0° apart) and 6 o'clock (180° apart). Direction must be

Vectors

Figure 2-3
Action and reaction are vividly portrayed in the split second when a tennis ball is smashed by a moving racket. The ball is momentarily flattened in reaction to the force of the racket surface.

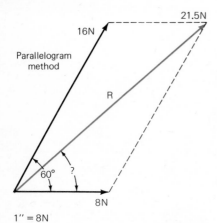

Parallelogram method

16N

21.5N

R

60°   ?

8N

1″ = 8N

Figure 2-4
The vector sum of two forces acting concurrently depends on their configuration. The resultant of two concurrent vectors may be determined geometrically by constructing a parallelogram of forces.

Both magnitude and direction are specified to describe the net effect of two or more forces that act on the same object.

Resultant forces can be determined graphically.

taken into account to find the net effect of these forces or the resultant. The **resultant** can be defined as the single force whose action is equivalent to all the forces in play.

Imagine a boat with a small outboard motor running at low throttle in a river. Its forward thrust exerts an 8-newton force due east. The river current exerts a 16-newton force directed 60 degrees north of east. By constructing a parallelogram of these forces, we can obtain their resultant both in magnitude and direction. To construct such a parallelogram of forces (See Figure 2-4):

1 With the help of a ruler and a protractor, draw vectors representing the forces in their proper scale with the given angle between them, so that they begin at the same point and their heads point in their respective directions.

2 Draw a dashed line from the head of each vector that runs parallel to the other vector, constructing a parallelogram.

3 Draw the diagonal of the parallelogram so that it represents a vector from the point of origin to the opposite corner of the parallelogram.

4 Measure the length of this vector to find its magnitude, and use the protractor to find its direction.

The resultant is $2^{11}/_{16}$ in. long, representing a force of 21.5 newtons. Its direction is 42° north of east. Although there are two forces acting on the boat, it will move as if under the action of one force. In vector addition, both the direction and the magnitude of the forces are taken into account. Changing either one will change the resultant. The magnitude of the resultant is greatest when the two components are acting in exactly the same direction. In this case, it is simply the sum of the two component forces. The greater the angle between the forces, however, the smaller the magnitude of the resultant. It reaches a minimum when the two forces are acting in diametrically opposite directions. In that case, the magnitude of the resultant is the difference between the absolute magnitudes of the component forces.

What happens when more than two forces are involved? We can go back to our original example of the block on the table, with a third opponent adding another force. You push with a 12-newton force east; your old opponent pushes with a 4-newton force north; the new opponent pushes with an 8-newton force 30° north of east. To show this graphically requires the construction of a polygon (See Figure 2-5):

1 Draw a vector diagram of the forces to scale, labeling each force according to direction and magnitude.

2 Redraw vector E.

3 Move the vectors parallel to themselves, first NE and then N, so that the tail of each is at the head of the other.

4 Finally, draw the resultant R from the tail of the first vector, E, to the head of the last, N.

5 Measure the length of R to get its magnitude, and use a protractor to get its direction.

Direction is always calculated with reference to some base line; a horizontal or vertical usually suffices. In our example, the resultant is $2^{1}/_{2}$ in. long—a magnitude of 20 N. Its direction is 23° north of east.

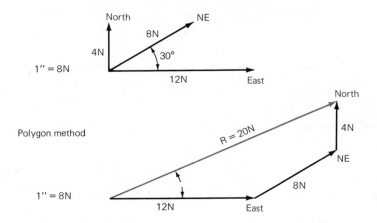

Polygon method

Figure 2-5
The resultant force of the three concurrent vectors in this diagram can be added by constructing a polygon of forces, shown here.

VELOCITY Velocity, which has direction as well as magnitude, is also a vector, and the same method can be used in its calculation. (We use **speed** to indicate the magnitude of velocity.) Imagine a tug of war, in which identical twins of equal weight and strength pull on opposite ends of a rope. Ruby pulls on the rope with a force of 50 N. Transmitted by the rope, the force (action) pulls Esmeralda. Esmeralda pulls back on the rope, and thus pulls on Ruby with an equal but opposite force (reaction). Both sisters, therefore, are pulled toward each other with a force of 50 N.

Assume, further, that there is a frictional force of 50 N between each sister and the ground. (Frictional force between an object and the surface it stands on is usually only a percentage of the object's weight. By choosing these values, we are assuming an ideal situation.)

When these forces are added together, their vector sums are zero—no net force on either sister. By the first law, both sisters remain at rest. Now imagine that Esmeralda is wearing ball-bearing roller skates that have no friction in the wheel bearings, nor between the wheels and the ground. A vector diagram shows that Esmeralda experiences an unbalanced force to the left because Ruby is pulling her; by the second law, she experiences an acceleration in the direction of the force. If Esmeralda suddenly lets go of the rope, she will no longer experience an unbalanced force. Acceleration will become zero, and she will move to the left with constant speed in a straight line, as stated in the first law.

How can we calculate the acceleration Esmeralda experiences? Suppose that her mass is 50 kg (about 110 lb), and that the force acting upon her is 220 N. From the second law, $F = ma$, we get:

$$a = \frac{F}{m} = \frac{220 \text{ N}}{50 \text{ kg}} = 4.4 \text{ m/sec}^2$$

If Esmeralda lets go of the rope after 3 seconds, what is her velocity?

$$v = at = 4.4 \times 3 = 13.2 \text{ m/sec}^2$$

Speed is the magnitude of velocity. Velocity is speed in a given direction.

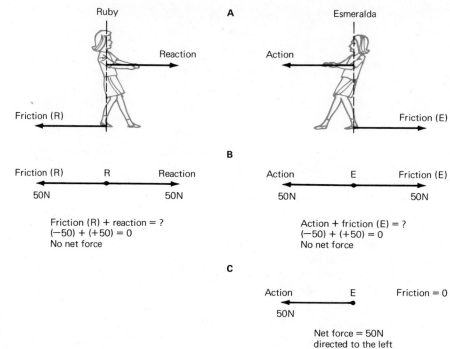

Figure 2-6
All three of Newton's laws can be
demonstrated in the tug of war between the
twins, Ruby and Esmeralda.

What distance did Esmeralda cover in 3 seconds?

$$d = 1/2 \ at^2 = 1/2 \times 4.4 \times (3)^2 = 19.8 \ m$$

Since Esmeralda's mass was initially at rest, her final velocity will have the same direction as her acceleration.

The situation is slightly different for masses already in motion. For example, a ball thrown vertically in the air is under the constant influence of the force of gravity, which produces a continuous downward acceleration opposite to the ball's velocity. The result will be a gradual loss in speed, arriving at zero speed at the moment when the ball achieves maximum height. Although speed then equals zero, the ball is not at rest and there is still acceleration. The ball will then move in the direction of the force, and the resulting acceleration will increase the ball's speed as it plummets downward. Since the ball experiences no change in its line of motion but only changes direction, its speed and velocity change together.

Scalars   Measurements that consist of magnitude only are called **scalars**. Speed is a scalar, since it provides us with magnitude but not direction. The number of bananas in a bunch, the speed of light, and humidity are all scalars. Weight, however, is a vector because it is a measure of the force with which the earth attracts an object in a particular direction—towards the center of the earth, which we call downward. From Newton's second law, $F = ma$, we can easily see that the weight of an object must be equal to the product of its gravitational mass and the acceleration of gravity:

$W = mg$

Some important physical quantities require only magnitude for complete description.

Scalars most frequently used in physics are mass, speed, and time. Since they are different in kind, like apples and oranges, vectors and scalars cannot be added or subtracted together. But vectors can be multiplied or divided by scalars; the results are vectors. For instance:

F (vector) $= m$ (scalar) a (vector)

v (vector) $= d$ (vector)$/t$ (scalar)

a (vector) $= \Delta v$ (vector)$/t$ (scalar)

But:

speed (scalar) $=$ distance (scalar)/time (scalar)

Momentum

A huge lead wrecking ball arcs through the air at the end of its cable and smashes in the side of a brick building. How fast would a brick have to move to have the same effect? A bowling ball would surely do more damage when dropped onto a plank of wood supported only by its ends than would a baseball dropped from the same height. The baseball weighs about 150 g, the bowling ball perhaps 4,000 g; yet a bullet weighing only 1 g, but moving at 1 km/sec, would bore right through the wood, causing more damage than either of the heavier objects. Two factors account for the different results achieved in these impacts: mass and velocity. The impact depends on mass, but evidently it also depends on velocity. The product of these two is **momentum**, which Newton in his *Principia* called "the quantity of motion:"

Momentum is the product of mass × velocity.

momentum $=$ mass $\times$ velocity

$= mv$

$=$ kg-m/sec

Calculating Momentum

Suppose we compute the momentum of several moving objects. A projectile with a mass of 45 kg (about 100 lb) is fired out of a cannon with a velocity of 750 m/sec. What is its momentum?

momentum $= mv$

$= 45$ kg $\times 750$ m/sec

$= 33,750$ kg-m/sec

Now let us compute the momentum of a car whose mass is 1,350 kg that is moving with a velocity of 25 m/sec.

$$\text{momentum} = mv$$
$$= 1{,}350 \text{ kg} \times 25 \text{ m/sec}$$
$$= 33{,}750 \text{ kg·m/sec}$$

To have the same momentum as the car, the projectile must have a velocity 30 times greater, because the projectile has $\frac{1}{30}$th of the car's mass.

IMPULSE   What happens to momentum when there is a variation in the period of time taken to reduce a large momentum to zero by changing the velocity? Catchers know the answer to this question; Johnny Bench of the Cincinnati Reds instinctively draws his hands back toward his body as the ball hits his glove, thereby increasing the time period of impact by a fraction of a second, enough to lessen its sting. Most drivers also know the answer because they have experienced the difference between the effect of slowing a car from 60 mph to 0 by gradual application of the brakes over a period of 1 minute, and the frightening effect of making the same stop by one swift stomp on the brakes.

The second law states:

$$F = ma$$

But we know that:

$$a = \frac{v_f - v_i}{t}$$

where v_f indicates the final velocity and v_i indicates the initial velocity. Substituting from the second law equation, we get:

$$F = m(v_f - v_i)/t$$
$$= \frac{mv_f - mv_i}{t}$$
$$Ft = mv_f - mv_i$$

The expression Ft on the left—the product of force and time—is called **impulse**, or the change in momentum. This equation tells us that change in momentum occurs as a result of force applied over a given time interval. It also tells us that any force will produce the same change in momentum as any other force when applied over a sufficiently long time interval.

Force applied over a period of time changes the momentum of a moving object.

By applying this equation for impulse and momentum, we can understand the influence of time on the magnitude of the impulsive force and hence on the effect of an impact. Suppose that a man with a mass of 100 kg (220 lbs) jumps from the window of a burning building into a safety net with a velocity of 10 m/sec. When he is caught in the net, he comes to a halt in 2 sec. What is the impulsive force exerted by the net on the man?

$F = m(v_f - v_i)/t$

$= 100 \text{ kg} \times (0 - 10 \text{ m/sec})/2 \text{ sec}$

$= -1000/2 = -500 \text{ N}$

The minus sign indicates that the force is acting against the motion.

Now suppose that the man, with the same initial momentum, misses the net and strikes the sidewalk, coming to a halt in 0.1 sec. What is the impulsive force of the sidewalk on the poor fellow in this case?

$F = m(v_f - v_i)/t$

$= 100 \times (0 - 10)/0.1$

$= -1000/0.1 = -10,000 \text{ N}$

The difference in the magnitude of the force in each case is evident. Shortening the duration of the impulse increases the amount of impulsive force; that is why a karate chop can be so lethal.

The equivalence of impulse and momentum yields one of the fundamental concepts of mechanics: the **conservation of linear momentum** (linear is used here to mean motion in a straight line, as opposed to curved motion). Imagine two billiard balls of mass m_A and m_B respectively, that are moving freely over a frictionless billiard table. Ball m_A has an initial velocity, v_{Ai}, directed to the right. Ball m_B has an initial velocity, v_{Bi}, directed to the left.

Conservation of Momentum

The balls collide and exert equal and opposite forces on each other: F_A on ball m_A (action) and F_B on ball m_B (reaction). Let us calculate their respective changes in momentum as they rebound, with velocities v_{Af} and v_{Bf}.

For mass m_A: $F_A t = m_A v_{Af} - m_A v_{Ai}$

For mass m_B: $F_B t = m_B v_{Bf} - m_B v_{Bi}$

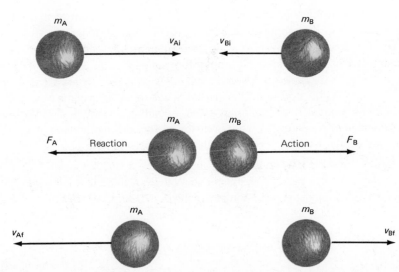

Figure 2-7
Conservation of momentum is shown here in three stages. In their collision, the two balls show equal and opposite changes in momentum.

Figure 2-8
A large ball (entering from the lower
right-hand edge of the picture) and a smaller
ball (entering from the top, right-hand edge
of the picture) collide. The large ball is
slowed down, the smaller ball is speeded
up, and their total momentum is conserved.

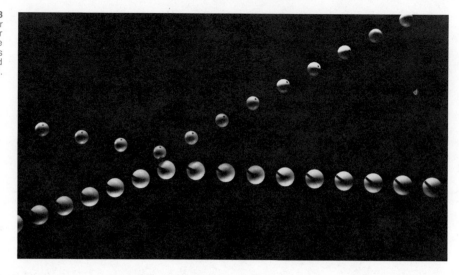

The time of the impulse, t, is the same for both balls. And by Newton's third law, $F_B = -F_A$. Therefore:

$$F_B t = -F_A t$$

The impulses are equal but opposite. This means that the changes in momenta are also equal but opposite:

$$(m_B v_{Bf} - m_B v_{Bi}) = -(m_A v_{Af} - m_A v_{Ai})$$

or

$$m_A v_{Ai} + m_B v_{Bi} = m_A v_{Af} + m_B v_{Bf}$$

or

total initial momentum = total final momentum

This means that there has been no change in the momentum of this system, regardless of the rearrangements (action–reaction pairs) generated by the two forces. The same holds true even when there are more than two objects colliding. Thus the **law of conservation of momentum** states that *the total momentum of all the bodies in an isolated system before collision is exactly the same as their total momentum after collision.* This law holds true regardless of the way in which the bodies interact—head on, at an angle, or with one body at rest. What one body gains in momentum, the other body loses. The total momentum of a system of bodies cannot be changed by internal forces between the bodies.

> Interactions between colliding bodies in an isolated system always conserve their total momentum.

To demonstrate this law, consider a gun with a mass of 2 kg which fires a 1 g bullet, with a muzzle velocity of 900 m/sec. With what velocity will the gun "kick" or recoil back against the hand and arm of the person firing it? (*b* is for bullet; *g* is for gun.)

$m_b = 0.001$ kg

$v_f = 900$ m/sec

$v_i = 0$

$m_g = 2$ kg

$v_f = ?$

$v_i = 0$

Initial momenta are zero since initial velocities are zero, and so the sum of the final momenta must also be zero:

$$(m_g v_g)^f + (m_b v_b)^f = 0$$

or, $(m_g v_g)^f = -(m_b v_b)^f$

thus

$$2v_g = -(.001)(900)$$
$$v_g = -0.45 \text{ m/sec}$$

directed to the left.

Our discussion of motion so far has focused only on motion in a straight line, or linear motion. Another concept important to Newtonian physics is motion in a curved path, or **curvilinear** motion. Circular motion is the simplest form of curvilinear motion.

Whirl a stone tied at the end of a string, with your hand fixed in position. Tension in the string is transmitted to your hand, and is experienced as a pull outwards. Your hand pulls back on the string, and this force is transmitted to the stone. The string, at any given instant, actually *is* the radius of the stone's circular path. Thus, the force exerted by your hand on the stone—and the resultant acceleration—acts radially inward toward the center of the circle. Such a force is called **centripetal**, which means center-seeking, and the acceleration is called centripetal acceleration.

At any given instant, the stone has a velocity that is tangential to the circle at the point where the stone is located. If the string were to break, the centripetal force would cease and there would no longer be any acceleration. The stone then would satisfy Newton's first law, continuing to move at constant velocity, in a straight line with the speed and direction it had at the moment the string broke. The straight line of its motion would be a tangent to the circle at the moment the string broke.

When you twirl the string, the outward pull on your hand is a reaction force exerted by the stone on your hand. This is a **centrifugal** (the word means center-fleeing) force. Since this force does not act on the stone, it in no way affects its motion.

Circular Motion

Circular motion results in a constantly changing velocity and, therefore, in a constant acceleration inward toward the center.

Figure 2-9
Velocity (v) and acceleration (a) vectors of a stone traveling in a circle at a constant speed.

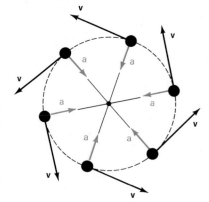

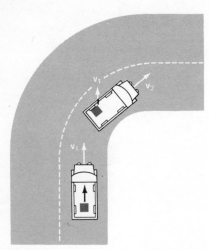

An example will help clarify centripetal and centrifugal forces. Imagine a car with a package on its back seat rounding a curve. As it enters the curve, both car and package have velocity v_1. Once into the curve, the car assumes velocity v_2, but the package, which is not held in place, tends to slide along the seat in the direction of its original velocity, v_1. This motion brings it in contact with the side of the car. The car then pushes in with a centripetal force on the package, causing it to stay on the curve. The package pushes out on the car with a centrifugal force.

The equation for centripetal force (See Figure 2-11 for its derivation) is:

$$F_c = ma_c$$
$$= m \times v^2/r$$

What centripetal force must be exerted to keep a 10 kg stone moving in a circle of radius 2 m with a velocity of 3 m/sec?

$$F_c = mv^2/r$$
$$= \frac{10 \text{ kg } 9 \text{ m}^2/\text{sec}^2}{2\text{m}}$$
$$= 45 \text{ kg-m/sec}^2$$
$$= 45 \text{ N}$$

Figure 2-10
The package tends to maintain its original velocity despite the change in the velocity of the car as it goes around the curve.

This equation tells us several things. First, a large velocity requires a large centripetal force to hold an object to a curved path; since velocity is squared it has a greater effect. Second, a sharp curve of small radius requires a large centripetal force, which is why a car has difficulty negotiating a tight curve. In this case it is friction on the tires of a car that exerts the centripetal force. The tires exert a centrifugal force by pushing out on the road.

Figure 2-11
The vector triangle was arrived at by vectorially adding $-v_1$ to v_2 to obtain Δv, which is $v_2 - v_1$. There is a second triangle made of the radii and a straight line, or chord, between x and y. Both triangles are isosceles since the radii are equal, and the magnitudes of v_1 and v_2 are equal, so the subscripts can be dropped. The vertex angles, O and P, are equal because v_1 and v_2 are perpendicular to the radii; the triangles are, therefore, equal. So, $\Delta v/v = xy/r$. Acceleration is defined as $a = (v_2 - v_1)(t_2 - t_1) = \Delta vt$. Solving this equation for Δv, we get $\Delta v = v(xy)/r$. Dividing by t gives $\Delta vt = v(xy)/rt$ or $v/r \times xy/t$. As the distance between x and y gets smaller, arc xy and chord xy become nearly equal, and $xy/t = v$, the velocity of the stone along chord xy and arc xy. Substituting into the equation: $\Delta v/t = v/r \times v = v^2 r$. If, as seen above, $a = \Delta vt$, then $a = v^2 r$, that is, the centripetal acceleration.

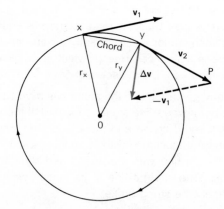

Angular momentum is equal to the product of the linear momentum of the moving body (mv) and the radius of its circular path (r). This may be stated as:

angular momentum $= mvr$

The larger the circle, the larger the momentum, for a given velocity. The distribution of mass is also important in circular motion. Each part of the turning object has its own radius to the center; the angular momentum of a solid sphere is much less than that of a hollow sphere of the same size and mass.

The **law of conservation of angular momentum** states that *if there is no external force on a rotating or revolving body, angular momentum remains constant*. The law holds for any body rotating on its axis or revolving around a point.

If the radius about the axis of rotation increases, then velocity must decrease in order for momentum to be conserved. Conversely, if the radius decreases, then velocity must increase. A living example of this principle is a spinning ice-skater, such as Janet Lynn. If Janet begins spinning with her arms out, and then draws them in, a portion of her mass is brought closer to her axis of rotation, thus reducing r, and her velocity increases. If she once more extends her arms, r increases and v decreases: she spins more slowly.

What is the centripetal force that holds the planets in their orbit, opposing the centrifugal inertia of their motion? Newton sought to answer this question by going back to the work of Kepler, verifying Kepler's laws and also providing a base from which they could be derived. We shall show how Newton's law of gravitation is proven for the simpler case of a circular planetary orbit. His full proof for Kepler's eliptical orbits is more complicated. Centripetal acceleration is expressed: $a_c = v^2/r$. Tangential velocity is written:

$$v = \frac{2\pi r}{T}$$

where $2\pi r$ is the circumference of the circular orbit, and T, the **period**, is the time for one complete revolution. Thus:

$$a_c = \frac{4\pi^2 r^2}{T^2 r} = \frac{4\pi^2 r}{T^2}$$

Kepler's third law states that $T^2 = kr^3$, therefore, we have:

$$a_c = \frac{4\pi^2 r}{kr^3}$$

$$= \left(\frac{4\pi^2}{k}\right) \cdot \frac{1}{r^2}$$

But $\left(\dfrac{4\pi^2}{k}\right)$ is a constant of proportionality. We can therefore write:

$$a_c \propto \frac{1}{r^2}$$

where r is the mean distance from a planet to the sun. By Newton's second law:

$F_c = ma_c$ and, therefore, by substitution,

$$F_c \propto \frac{1}{r^2}$$

The inverse square statement means that the centripetal force between the sun and any planet is inversely proportional to the distance between them.

By Newton's third law, if the sun attracts a planet, the planet will also attract the sun. But by the second law ($F = ma$), the sun's attraction for the planet is proportional to the planet's mass. Similarly, the planet's attraction for the sun is proportional to the sun's mass. Thus, the mutual attraction between sun and planet is proportional to both masses, or:

$$F \propto m_p m_s$$

where m_p is the planet's mass, and m_s is the sun's mass.

If we join both proportionalities for F we get the following:

The mutual attraction between the sun and a planet is proportional to the masses of the two bodies and inversely proportional to the square of the distance between them.

$$F \propto \frac{m_p m_s}{r^2}$$

or, the force of attraction between the sun and planets is directly proportional to the product of their masses and inversely proportional to the square of their distances. This proportionality can be written as an equation by introducing G, a constant of proportionality, giving us a law of planetary attraction:

$$F = \frac{G \, m_p m_s}{r^2}$$

Newton made no attempt to explain what caused gravity, but he did describe many of its properties. In contemplating the fall of an apple, he wondered whether the same force acted at a greater distance — does the earth pull on the moon? How far does gravity extend? Could it be possible that the moon, too, was a falling body under the influence of the earth's gravity?

All of the particles of matter in the universe are attracted to each other.

Newton reasoned that the centripetal force exerted on the moon — and the force exerted on the other planets by the sun — was the same force of gravity experienced here on earth. He then extended this result to everything in the universe, to formulate the **law of gravity**: *All bodies are attracted to one another by a force proportional to the product of their masses and inversely proportional to the square of the distance separating them.*

In the equation for universal gravitational force:

$$F = \frac{G\,mm_e}{r^2}$$

where m is the mass of an object at the earth's surface, there are two un-knowns — G, the gravitational constant, and m_e, the mass of the earth. Force here is the weight, acting on a known mass m, and r is the radius of the earth. If G were known, then we could determine the gravitational mass of the earth. Then, using the earth's mass and orbital radius, the masses of the sun and, finally, of the other planets could also be determined.

In 1798, Sir Henry Cavendish determed G experimentally. By precisely mea-suring the force of attraction between two known masses at a known separation, G could be determined as the only unknown in the gravitational equation. The apparatus he used — known as Cavendish balance — consists of a rigid T-shaped bar supported by a thin quartz thread. At the ends of the horizontal are

Mass of the Earth

Figure 2-12
The centripetal force that keeps these cyclists in a 6-day race at Antwerp from flying off the curve is provided by friction between their tires and the track, and by the banking of the track itself.

The Discovery of Neptune

Six planets were known in ancient times, although the ancients did not realize that they were standing on one of them. Uranus, the seventh, was discovered unexpectedly by William Herschel in the course of a telescopic sky survey. The existence of Neptune, by contrast, was predicted and its position calculated from its gravitational influence on the orbit of its neighbor, Uranus, before it was ever seen.

Uranus had been observed as early as 1690 and had been recorded as a star. There were more than 20 such sightings before Herschel recognized it as a planet in 1781. Within a few decades, Alexis Bouvard was able to compute the orbit of the new member of the solar system. By the early nineteenth century, Newtonian mechanics had been developed to the point that it could be used to calculate such orbits with great accuracy, even taking into account the perturbations produced by the gravitational forces of other planets. Bouvard had carefully included these effects in his calculations. But after a time astronomers

William Herschel

noticed that Uranus was straying from its predicted path. Bouvard's work was checked and found to contain no error, and scientists were not eager to doubt the validity of Newton's laws. The most plausible explanation was the existence of another planet beyond the orbit of Uranus.

Few astronomers, however, had the courage to attack so intricate a problem as calculating the mass and orbit of the as yet unseen planet. But in October, 1845, after computations lasting more than three years, John Couch Adams (1819–1892) sent his prediction for the planet's location to Sir George Airy, the Astronomer Royal of Britain. Airy was skeptical. He was reluctant to trust the work of so young an astronomer, one who had only recently graduated from Cambridge, and so the Astronomer Royal sent him a simple problem to test his abilities. Adams, insulted, did not respond.

Airy paid no more attention to the matter until June, 1846, when the French mathematician Urbain J. J. Leverrier (1811–1877) published a prediction that was very close to that of Adams. Airy then requested a search of the predicted area, and the new planet was actually seen; unfortunately, however, it was not recognized.

John Couch Adams

Urbain J. J. Leverrier

A month later Leverrier asked Johann Galle of the Berlin Observatory to look for the planet, and Galle, aided by new charts of that region of the sky, found it after just one hour of searching. The date was September 23, 1846.

The discovery of Neptune was a spectacular triumph for Newtonian physics and a spur to astronomers studying the motions of the heavenly bodies. It also conferred great reputations on Adams and Leverrier. In the euphoric days following the discovery of the new planet, their calculations were hailed as extraordinary examples of mathematical genius. We know now, however, that luck played a large role in their success. In order to simplify their calculations both men had assumed that the planet obeyed Bode's law (see Chapter 22). Neptune, however, is the only planet that does *not* conform to this law (see Table 22-4). This and other erroneous assumptions led them to estimates of Neptune's mass and orbital characteristics that were far from correct. Fortunately, Uranus had passed Neptune in 1822 and was still relatively close when the predictions were made. If this had not been so, the predictions would have been quite inaccurate, and Neptune would have had to await the arrival of astrophotography before taking her bow.

mounted two identical spheres of mass m. A small mirror placed on the vertical reflects a beam of light onto a scale. Identical larger spheres of mass m_1 are brought into position near the small masses as shown. The gravitational force of attraction between the large and small spheres produces a force that twists the balance through a small angle. As the balance twists, the vertical part of the T—containing the mirror—rotates. The mirror itself rotates out of its original plane. This movement deflects the original reflection of the light beam along the scale, providing the means to measure the angle of rotation. Knowing from other measurements the force necessary to twist the quartz thread through such an angle enabled Cavendish to calculate G directly:

$$G = 6.670 \times 10^{-11} \text{ N-m}^2/\text{kg}^2$$

We can now find the mass of the earth. A 1 kg mass experiences a force of 9.8 newtons at the earth's surface, which is at a distance of 6.37×10^6 m (nearly 4000 miles) from the center. Using this information, we can determine the mass of the earth by considering that:

$$m_e = \frac{Fr^2}{Gm}$$

$$= \frac{9.8 \times (6.37 \times 10^6)^2}{6.67 \times 10^{-11} \times 1}$$

$$= 5.98 \times 10^{24} \text{ kg}$$

$$= 6.6 \times 10^{21} \text{ tons}$$

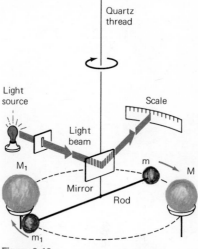

Figure 2-13
A torsion-balance apparatus such as Cavendish used to demonstrate the law of gravitation. The very small amount of twist in the wire occasioned by bringing the two large outer spheres close to the smaller spheres at the ends of the bar suspended from the wire provide a means for calculating the gravitational forces acting between bodies.

Shape of the Earth

One of the effects of gravity and centripetal acceleration is the shape of the earth. As the earth rotates on its axis, every point on its surface may be considered as a rock on a string whose center lies at some point on the rotational axis. Centripetal force reaches a maximum at the equator because tangential velocity, v_T, is zero at the poles and maximum at the equator (about 1050 mi/hr). As we approach the equator, the square of the velocity has the greatest effect on the force—despite the increase in r.

The net observed weight of a particle is the difference between centripetal force and the force of the earth's gravitational attraction:

$$w = mg - mv^2/r$$

From this equation we see that an object weighs most where centripetal force is zero (at the poles) and weighs least where the centripetal force is greatest (at the equator).

What happens in effect is that the earth is pulled toward its center more at the poles than at the equator. It is therefore somewhat flattened at the poles, and bulges at the equator. Actually, the earth is 27 miles wider at the equator than at the poles.

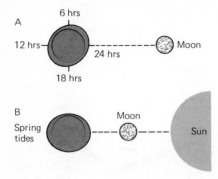

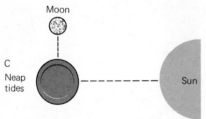

Figure 2-14
The moon is 60 earth radii (R) from the earth. The moon's gravitational pull on the nearest oceans is $1/(59R)^2$ compared with $1/(60R)^2$ for its pull on the rigid earth, and $1/(61R)^2$ for the oceans on the opposite side. The tides are caused by this difference. When the sun's gravitational pull reinforces the moon's, the result is an unusually high (spring) tide; when the two forces operate at right angles, the tide is lower (neap tide).

Oddly enough, however, it was found that the earth does not quite conform to shape dictated by the effects of gravitational and rotational forces. Through detailed measurements taken from an earth satellite orbit during the International Geophysical Year, July 1957–December 1958, it was found that the earth was slightly pear-shaped. Calculations showed that there is an excess of matter in the north end of the earth equivalent to a layer 50 feet thick and covering an area the size of the Atlantic Ocean. A comparable deficiency exists at the south end. It was also estimated that there is a 25-foot deficiency in sea levels in north temperate latitudes and a similar excess in south temperate latitudes. Hence, the pear shape—virtually unnoticeable visually.

TIDES It was known long before Newton that tides change with phases of the moon, but it remained for him to explain the connection. By the laws of mutual attraction, the moon pulls on the solid earth and its oceans. The water closest to the moon will experience the greatest pull. As a result, the water is pulled away from the earth, creating a bulge (see Figure 2-14A). But why is there a bulge on the other side of the earth? As a system of masses taken together, the earth and the moon rotate around a common center of gravity 1000 mi below the surface of the earth. The water on the side away from the moon is not attracted as strongly as the rest of the earth; hence, it fails to move along with the earth around the center. Its relatively unbalanced inertia makes it lag and bulge from the earth on the other side as does the delay in water flow due to friction with the irregular ocean floor. Any given point on the earth—and the waters there—will pass a region of high tide every 12 hours. The moon effectively holds the bulges in place while the earth rotates beneath its envelope of water. The sun exerts a similar influence, but to a lesser degree because of its distance. When the earth, sun, and moon are in a line (Figure 2-14B), all of the forces explained above are increased and the higher spring tides occur. When the three bodies form a right angle (Figure 2-14C), the sun's gravitational force diminishes the tidal effects of the moon, producing neap tides.

Glossary Curvilinear motion is motion in a curved path.

Centrifugal force refers to the apparent force on a body in curvilinear motion that seems to pull the body outward along a radius.

Centripetal force is the inward force required to move a body around in a circle. The force is directed toward the axis of rotation.

The law of **conservation of momentum** states that the total momentum of colliding bodies in an isolated system is the same before and after their collision.

The distance away from a fixed point in a given direction is called **displacement**.

Newton's law of gravity states that all of the matter in the universe is attracted to all the rest of the matter in the universe by a force that is directly proportional to the product of their masses and inversely proportional to the square of their distances from each other.

A change in momentum is called **impulse**, and is the product of the force applied over a given period of time.

Kinetic friction is the force between two moving bodies whose surfaces are in some contact. It opposes their relative motion.

Movement in a straight line is called **linear motion.**

Mass refers to the amount of matter in an object. **Inertial mass** is a measure of an object's resistance to any change in its motion. **Gravitational mass** measures the property by which all objects exert a mutual force of gravitational attraction.

The **momentum** of a moving object is the product of its mass times its velocity.

A **newton (N)** is the force which, acting on a mass of 1 kg, will produce an acceleration of 1 meter/second².

Newton's first law states that every body continues in its state of rest or uniform motion in a straight line, unless compelled to change that state by an external force. **Newton's second law** states that the acceleration of a body is directly proportional to the force applied, and inversely proportional to the mass of the body. **Newton's third law** states that to every action there is always opposed an equal reaction.

A **period** is the time of one complete revolution.

The net effect of two or more concurrent forces is a single force called the **resultant.**

A **scalar** is a measurement that consists of magnitude only.

A **vector** is a quantity, such as velocity, that has both magnitude and direction.

The **weight** of an object is equal to the product of its gravitational mass and the acceleration of gravity.

Exercises

1 A bus traveling down a country road at 40 mph suddenly brakes for a deer crossing the road. In light of Newton's first law, explain why the passengers are thrown forward as the bus slows down. If the same bus makes a sharp right turn, what effect does this have on the passengers and why?

2 A student sits on a spinning lab stool with her arms extended and a heavy book in each hand. What will happen if she lowers her arms to her sides? Why?

3 A contestant in a drag race can accelerate his racer to 80 mph in 10 sec. How would you determine (a) the car's acceleration, (b) the distance traveled in 10 seconds, and (c) the car's velocity after 6 seconds?

4 A delivery man drops a 2 lb package from a height of 5 feet in an elevator that is *descending* at the rate of 15 ft/sec². How long will it take the package to hit the floor of the elevator? If the elevator were *ascending* at the same rate, how long would it take the same package dropped from the same height to reach the floor?

5 A horse is pulling a carriage on a straight, level path through the park. According to Newton's third law, it would seem that the carriage is pulling back on the horse with the same force that the horse is exerting to pull the carriage. If this is true, can you explain how the horse is able to move the carriage?

6 Describe Cavendish's famous experiment of 1798. Using the gravitation formula, show how Cavendish was able to deduce the value of the gravitation constant from this experiment.

7 A tourist wants to drive to a national park whose exit on the highway is 24 mi north. He takes a wrong exit 12 mi north and must drive 1 mi east then 5 mi south and then 1 mi west, before getting back onto the highway heading north. 45 minutes after starting the trip he arrives at the correct exit. What is the net displacement? What is the average velocity for the entire trip? What is the average speed for the entire trip?

8 A 0.3 kg cue ball rolling across a pool table with a velocity of 3 m/sec north, collides head on with a 0.25 kg billiard ball at rest. If the cue ball remains at rest after the collision, with what velocity does the billiard ball move and what is its momentum?

9 A glass flask sitting on a tripod suddenly explodes, breaking into three pieces. Two 25g pieces fly off north and east at right angles to each other, with identical speed. The third piece flies off at half the speed. What is the direction of its momentum? What is its mass?

10 A 400g football thrown by a quarterback moves from rest to 25 m/sec in 0.25 sec. What is the average force exerted on the football?

11 A 20,000 kg jet plane is driven forward with a constant force of 10,000 N. How many seconds will it take the plane to reach a velocity of 250 m/sec from an initial velocity of 160 m/sec?

12 Newton stated that the moon is continuously accelerating toward the earth, yet the two never collide. Explain this seeming contradiction.

13 A 70 kg astronaut is in a spacecraft that is blasting off with an acceleration of 39.2 m/sec². What is the total downwards force experienced by the astronaut?

14 When American astronauts touched down on the moon, they experienced a gravitational force only 16% of that on earth. What does a 50,000 kg spacecraft weigh on the moon's surface?

15 Kepler found that if he divided the square of a planet's period by the cube of its average distance from the sun, he would arrive at the same quotient for all of the planets. What is the relationship between Kepler's third law and Newton's inverse square law of universal gravitation?

16 In reporting the lunar flight of the Apollo 18 spacecraft, the press noted that the ship had escaped the earth's gravity in the early stages of the flight. How accurate is this statement?

17 The mass of the sun is about 2×10^{30}kg. The mass of the planet Jupiter is about 19×10^{29}g. Its average orbital speed is about 7.7×10^{11}m. What is the gravitational force exerted by the sun on Jupiter? Jupiter must maintain a certain orbital speed so it won't fly into the sun or spin off into space. What is this speed?

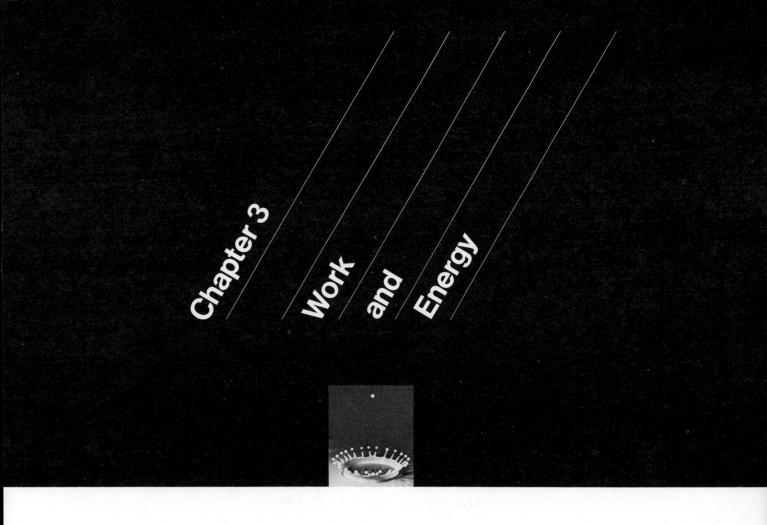

Chapter 3

Work and Energy

All the sources of energy available to man, from the breeze that turns the simplest windmill to the radioactive isotopes that fuel the most complex modern nuclear reactor, are important for one reason: their capacity to do work. The work of the industrialized world is performed largely by more and more machines that demand energy for their operation. Recent realization that we are quickly exhausting the nonrenewable energy resources of our planet has focused public attention on the relationship between energy and work.

The scientist's definition of the word *energy* is very similar to that of the layman: the capacity to do work. But what does the term *work* mean to a physicist? In this case the technical definition diverges somewhat from the ordinary sense of the word. In everyday usage, work implies some necessary, difficult, and often unpleasant activity. There may be compensation, in the form of money or gratitude or personal fulfillment, but work always demands the exercise of our strength and the expenditure of some of our energy. Necessity, unpleasantness, reward, and motivation are all social or psychological, and irrelevant to the scientific concept of work, which involves only two elements: the application of a *force* to move an object or objects, and the *distance* the object moves as a result. The *amount* of work performed depends solely on the magnitude of the force and the distance the object moves in the direction of the force:

$$W = F \times d.$$

The work of any machine may be analyzed in terms of force and distance.

Machines

We can understand the relationship between force and distance in the performance of work more clearly if we analyze a few simple machines. A **machine** is a device that transfers force from its point of origin to another point where it performs work. Rube Goldberg drew pictures of absurd machines that performed silly work by means of elaborate and bizarre but comprehensible mechanisms. The fun of the drawings is tracing the transformations of the initial force and the initial motion through all the ridiculous gears, pulleys, weights, and levers, to the final job to be done. One reason we enjoy these cartoons is that modern machines by and large frustrate our attempts to follow their operation in this way. The simple elements of work, force, and motion are seemingly lost to commonsense understanding in a maze of complex mechanical parts, or in the still more baffling intricacies of electrical circuits and chemical processes. In the solid-state components of modern electronic devices they seem to disappear altogether. Yet the most sophisticated machines perform work that can be analyzed in terms of force and distance, even though the movement may be that of electrons or molecules.

Figure 3-1
Cartoon by Rube Goldberg.

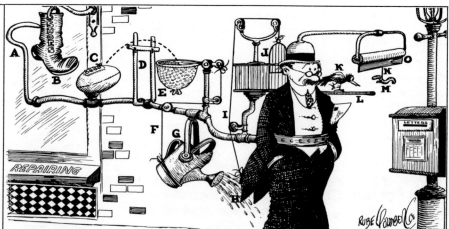

One of the interesting things about machines is that they can function not only as transmitters of force but also as multipliers of force (Figure 3-2). Consider the very simple arrangement involving ropes and a wheel called a pulley. Suppose a weight of 100 lb is hanging from the pulley, and we wish to lift this weight by pulling on the rope that passes around the wheel. Discounting friction, how much force need we apply? When all the weight is being supported by the pulley, the downward force obviously must be equally distributed between the fixed end of the rope and the free end on which we are pulling; each will transmit a force of 50 lb, as we can see from the force diagram in Figure 3-2. Apparently, then, we need only provide a force of 50 lb, and that will be sufficient to lift a 100-lb weight. How is this possible? Are we getting something for nothing? In mechanics, one never gets something for nothing; a closer look at the diagram will reveal the cost of this saving in force. The pulley multiplies the force we exert, but we must exert the force over twice as great a distance; to raise the weight 1 ft, for example, we must pull the rope 2 ft. The work put in is still equal to that produced, although the relative values of force and distance are changed. In the example:

$$W = \tfrac{1}{2}F \times 2d$$

the product of force and distance remains constant.

Mechanical Advantage

The factor by which a given machine increases the force at our disposal is called its **mechanical advantage**. The pulley we have just analyzed has an *ideal* mechanical advantage of two. The *actual* mechanical advantage can be calculated by dividing the force exerted by the machine by the force exerted on the machine, a formula that takes into account the friction involved.

A series of pulleys in combination is called a block and tackle; by increasing the number of pulleys in such a device, we can achieve any mechanical advantage we desire, but of course the distance to be pulled, in order to raise a very heavy mass, will be correspondingly very great. The large forces attainable with a block and tackle need not be exerted only against gravity; cars can be pulled out of the mud with this simple machine and sails hauled tight against the great force of a stiff wind.

For any machine in which force and distance are manipulated for mechanical advantage, one thing remains the same: the amount of work done.

If we inspect other simple machines, we will find that exactly the same principle seems to operate in all instances: we gain force, but we pay for it in distance. You probably remember that two people of equal weight can balance a seesaw if they both sit equally distant from the supporting pivot at the center. If one weighs more than the other, however, the heavier person must sit closer to the pivot to achieve equilibrium. The seesaw is actually a simple machine called a lever. The point of support is known as the fulcrum. Suppose we wish to use a lever to lift a 500-lb weight, but we have at our disposal a force of only 100 lb. We need only arrange the fulcrum so that it is five times as

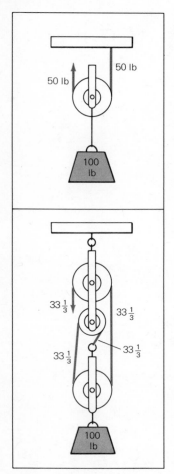

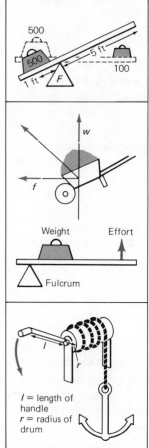

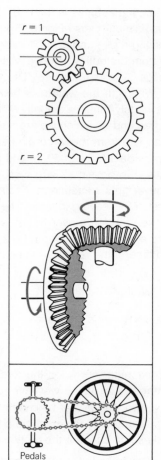

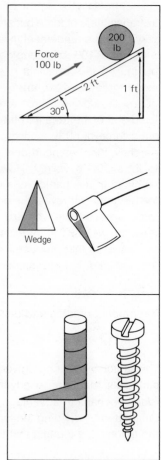

Figure 3-2
Simple Machines

The pulley (top) employs a wheel and rope to take advantage of the force–distance relationship. By increasing the number of pulleys in combination we create a block and tackle (bottom) capable of multiplying the mechanical advantage many times over, in this case by 3.

The lever (top) also takes advantage of this force–distance trade-off. The wheelbarrow (center) uses the lever principle. The force required to lift decreases as it is applied at an increasing distance from the wheel, which is the fulcrum. The load is carried forward on the wheel with a resultant force acting at a 45° angle. The windlass (bottom) is a simple lever moving in a continuous 360° arc. A 5-ft handle on a 1-ft drum gives a 5-to-1 advantage.

Gears (top) allow us to decrease force by applying it through a greater distance. A large gear turns a gear of smaller radius. The vertical gear (center) turns one in the horizontal plane. The bicycle employs the gear principle in reverse (bottom).

The inclined plane (top) distributes the expenditure of force over a greater distance. For an object on an inclined plane, the vertical downward force of gravity is divided into one part pressing perpendicularly against the plane, and a second part pushing downward but parallel to the plane. Only the parallel part along the plane must be opposed by the lifting force. The wedge (center) is two inclined planes back-to-back, and the axe is simply a moving wedge; the screw (bottom) is an inclined plane wrapped about an axis so as to transform rotary into straight-line motion.

far from the point at which we apply our force as it is from the weight to be lifted, and the force exerted on the weight will be multiplied fivefold. Indeed, we can attain any mechanical advantage we wish simply by appropriate placement of the fulcrum—a fact realized by the Greek mathematician Archimedes, who is reputed to have said that given a suitable fulcrum and a place to stand, he could move the whole world. A little geometry will show, however, that we must once again sacrifice distance for this gain in force; in the example above, of lifting a 500-lb weight, we must move our end of the lever 5 ft to lift the weight 1 ft. For Archimedes to do what he boasted, he would need an immensely long lever arm—a cosmic seesaw.

The windlass (see Figure 3-2) is still another illustration. In a sense, it is nothing but a lever that can move in a continuous 360° arc, a kind of curved lever. One revolution of the handle produces one revolution of the drum, but in the course of that revolution the handle moves through a circle of much greater radius than the drum, covering a much larger distance. Thus a small force on the handle is exerted through a large distance and a proportionally greater force exerted through a proportionally smaller distance at the drum. This device, however, is quite awkward in one respect: to increase our mechanical advantage we must increase the length of the handle. Thus for a mechanical advantage of ten, the handle must be ten times the radius of the drum. Obviously,

Figure 3-3
Ancient Chinese drawings, depicting the use of simple machines, illustrate the principle of mechanical advantage. In the first, an ox-drawn series of gears is used to raise water. In the second, as the water wheel turns, the beaters on the shaft attached to the wheel first raise and then release the hammers.

if we wish to exert a force through a distance of 50 ft, it would be more convenient to turn a 1-ft handle eight times than to turn an 8-ft handle once. For this reason most machines of this type employ toothed wheels, or gears, of different size, which enable us to alter the number of revolutions of a wheel or handle instead of increasing its size indefinitely. For a mechanical advantage of five, then, we may use two wheels of identical size, geared together so that the first makes five revolutions for every one made by the second.

An automobile transmission is basically a set of such gears connecting engine and wheels. Thus when you shift gears, you alter the mechanical advantage of the engine in relation to the wheels. A car is heavy and requires enormous force to accelerate it from rest. That is why a car is started in low gear, either by interlocking gears or through a hydraulic process—to provide the greatest mechanical advantage. To achieve this gain in force, however, the engine must make many more revolutions than the wheels; the engine runs rapidly while the car moves slowly. At highway speeds, for which little additional acceleration is needed, some of this engine force can be dispensed with. Thus in high gear the engine force does not have to be multiplied by the gear ratio. High gear provides a lower mechanical advantage, enabling the car to move faster for the same number of engine revolutions per minute; that is, for the same total distance through which the engine parts rotate. A bicycle employs the same principle applied in reverse; force is traded for distance. The gears of a bicycle are

Figure 3-4
The spinning wheel is a good example of a flywheel, which conserves momentum by having most of its mass far from the center. As the woman pumps the treadle, a wheel of large radius is set in motion, causing one of smaller radius to spin much more rapidly.

so arranged that the rider's feet on the pedals travel a small distance in order to make the wheels travel a greater distance. That is why a bicycle can sometimes move almost as fast as a car, although the rider's legs cannot come anywhere near the 4000–6000 revolutions per minute (rpm) attained by an automobile engine. Of course this entails a sacrifice of force: a large force applied to the pedals yields a small force applied to the wheels, but because of the light weight and low friction of a bicycle this lesser force is sufficient.

For a final illustration we can consider what is perhaps the most primitive machine: the inclined plane. This device is so simple that we hardly think of it as a machine at all. But an inclined plane is a device that enables us to distribute the expenditure of force over a longer distance. Think how much easier it is to walk up a gently sloping ramp than to climb a vertical ladder. If we roll a 200-lb boulder up a ramp that makes a 30° angle with the ground we will find that only 100 lb of force is needed to do the job, although we will have to push the boulder twice as far to attain the same vertical height as we would have if we simply lifted it. The principle of the inclined plane is utilized in several other important simple mechanical devices. The wedge, for example, is nothing more than two inclined planes placed back to back; the screw is an inclined plane wrapped around an axis. When we turn the head of a screw through five inches around its circumference to advance the point of the screw an inch into the wood, we are employing an inclined plane in disguise; when we use a screwdriver with a fat handle to make the work easier, we are using the windlass as well, by adding the radius of the handle as a kind of lever arm.

In all these devices force is freely exchanged for distance and distance for force. What remains constant, however, in all these exchanges is the quantity physicists call **work**, the product of the force exerted and the distance through which it is exerted. In mathematical notation, we write

Work

$W = F \times d$

A full verbal definition would be as follows: when an object is moved by a force, the amount of work done on it by that force is equal to the magnitude of the force multiplied by the distance moved in the direction of the force.

In the mks system the unit of work will thus be the newton-meter, called the joule, after the English physicist J. P. Joule (1818–1889) whose experiments are discussed later in this chapter. In the cgs system the unit of work is the dyne-centimeter, or erg. Since 1 m = 100 cm, and 1 newton equals 10^5 dynes, 1 joule is equal to 100×10^5, or 10^7 ergs. The corresponding unit of the British system is the foot-pound; 1 ft-lb = 1.36 joules. The phrase "in the direction of the force" in the definition given above is very important. If the force operates at an

The work that is done does not depend on how fast it is done.

Figure 3-5
While he holds these weights motionless above his head, this Olympic weight-lifter is doing no external physical work, since no force is operating through a distance.

No matter how much force is applied, no work is done if there is no displacement.

angle to the direction of motion, only its component in that direction performs work on the object, and thus can be included in the calculation of the work being done. The rest is not useful and does no work. Thus if there is an angle of 60° between the force and the direction of motion, the amount of work will be only one-half of what it would be if the two were operating in the same direction. If the force is operating at right angles to the direction of motion, no work at all is done, no matter what the magnitude of the force or how far the object moves. For example, pressing down vertically on a ski will not move it at all along the horizontal snow surface.

We can see, then, that the physicist and the layman do not always refer to the same thing when they speak of work. If you carry a heavy knapsack around all day you may feel that you have done a great deal of work; as long as you walk at a constant speed on a level surface, however, you are doing no work on the knapsack in the scientific sense of the term, in spite of your fatigue. Pushing against the ground, you are exerting a force equal to the weight of the knapsack to support it against the pull of gravity, but that force is applied upward, at right angles to the direction in which the knapsack is moving.

By contrast, if you were to climb a hill, a certain part of your motion would be upward, in the same direction as the force you exert on the knapsack, so you would be doing work. And if you did not always walk at constant speed, if you broke into a run, or slowed down to rest, you would have to exert a horizontal force on the knapsack to accelerate or decelerate it, in accordance with the formula $F = ma$; again, you would be doing work. Similarly, where there is no movement there can be no work, no matter how much force is applied. If you strain your muscles to the utmost trying to unscrew the lid of a peanut butter jar without budging it, you have done no work. If you apply a force of 150 lb to a heavy rock without moving it, you have done no work. Yet if you lift a pebble weighing only an ounce, you have done work. (Whether you get tired depends on the energy you use on internal bodily work processes in the muscles themselves. You do internal work, one might say, but no external physical work.)

Power Scientists and engineers naturally find it convenient to have a measure of the rate at which work is being done. An engine that can do 100,000 ft-lb of work in a minute is generally more useful than one that takes half an hour to do the same amount of work. We say that the former engine is more powerful. **Power** is the rate of work performed in a unit of time. In the British system power can be expressed in foot-pounds per second (ft-lb/sec) or foot-pounds per minute (ft-lb/min), but the most common unit is the horsepower. This unit was originally defined by the Scottish engineer James Watt (1736–1819), famous for his work in the development of the steam engine. Wishing to measure the power of his engine against that of horses doing the same work, he tested the horses and found that a typical workhorse could sustain a work output of 22,000 ft-lb/min over a long period of time. To be on the safe side he increased this figure by half, and set 1 horsepower (hp) equal to 33,000 ft-lb/min, or 550 ft-lb/sec. In the metric system the most prevalent unit of power is the joule/sec, also called the watt in honor of James Watt; 1 horsepower is equal to 746 watts. Both units are widely used in this country. The horsepower measures the power output of machines such as automobile engines. The watt is generally employed in measuring the power output of electrical and electronic devices such as light bulbs or amplifiers. Larger quantities of electrical power are often expressed in terms of kilowatts, where 1 kw = 1000 watts.

Energy

In discussing machines we have seen that the input of work is always equal to the output of work (except for losses to friction, which we shall discuss presently). But where does the work that we put into the machine come from? We know that work input is available from many sources: the burning of fossil fuels (coal, oil, gas), movement of electricity through a wire, falling water, the movement of the wind or tides, the sun's radiation, the decay of radioactive isotopes, the action of human or animal muscles. These are all energy processes. Energy can be defined as the capacity to do work. At one time energy was thought to be a unique property of living things; only what was alive could "do" something. Therefore the seventeenth century German philosopher and mathematician Leibnitz, one of the first scientific investigators of work and energy, gave the energy of motion a Latin name, *vis viva*—"living force." But he was wrong in thinking that moving things alone possess energy, and it was soon realized that there are a number of kinds of energy that can be stored in both animate and inanimate objects. Energy is the potential to do work, but this potential may be released by nature—as in an earthquake that sets off a landslide that converts potential for work in the rocks at a height—or by a human being or an ape who throws a rock.

Energy is expressed in the same units as work, because it is measured by the work it can do.

The modern term *energy* is derived from the Greek word for "at work." It was in the nineteenth century that men came to understand the transformations of the various kinds of energy, an understanding culminating in one of the most fundamental and far-reaching principles of modern science, the law of conservation of energy.

Kinetic Energy

Any object in motion possesses energy, for if it strikes another object it will certainly exert a force upon it and may cause it to move, thus performing work. A moving hammer drives a nail into a piece of wood; a gust of wind—moving air—propels a sailboat or drives a windmill; moving water turns an electric generator. These are instances of work being done by matter in motion. The energy of a moving body is called **kinetic energy,** (KE) from the Greek word for "mo-

Figure 3-6
The dips and slopes of the roller coaster illustrate the conservation of mechanical energy.

tion." We can derive a fairly simple expression for calculating the kinetic energy of a body in motion by combining the formula for work, $W = F \times d$, with some of those governing motion and acceleration, discussed in Chapters 1 and 2. Suppose a hammer of mass m moving at an initial velocity v strikes a nail, driving it a distance d into the wood. The hammer, of course, exerts a force on the nail, and by Newton's third law we know that the nail is simultaneously exerting an identical force on the hammer, thus decelerating it. By the time hammer and nail have moved the distance d, the hammer has lost most of its initial velocity, and it has given up most of its kinetic energy to the nail in the form of work done against the resistance of the wood. The hammer has performed work on the nail equal to the force it has exerted on the nail, times the distance it has driven it:

$$W = F \times d$$

But in this equation we can substitute other expressions for both F and d. Since the hammer has been decelerated by the force F, and since $F = m \times a$, we can write

$$W = ma \times d$$

And since d, the distance traveled by an accelerated (or in this case decelerated) body, is $\frac{1}{2}at^2$, the equation becomes

$$W = ma\left(\tfrac{1}{2}at^2\right) = \tfrac{1}{2}m(at)^2$$

Finally, since acceleration times time, at, equals the velocity v, we can write a formula for the work done by the hammer in terms of its own mass and the velocity at which it was traveling when it struck the nail.

$$W = \tfrac{1}{2}mv^2$$

Thus

$$KE = \tfrac{1}{2}mv^2$$

We have calculated the kinetic energy from the amount of work into which it can be converted. Using the formula for work $W = F \times d$, we would expect that the units of kinetic energy should be the same as those of work. This does not at first seem to be the case. The unit that emerges from the formula $KE = \frac{1}{2}mv^2$ in the metric system, for example, is the kg-(m/sec)². That this is identical to a joule can easily be seen if we remember that a joule is a newton-meter, and a newton is defined as a kg-m/sec². Kinetic energy, then, is expressed in joules in the mks system, ergs in the cgs system, and foot-pounds in the British system. The formula $KE = \frac{1}{2}mv^2$ calls attention, however, to the fact that the kinetic energy of a moving object is proportional to the square of its velocity. That is why speed is such an important factor in the destructiveness of automobile accidents. A car traveling at 60 mph has four times the kinetic energy that it has at 30 mph, and nine times what it possesses at 20 mph.

Motion is an important form of energy, but it is not the only form. That there must be other kinds of energy besides kinetic is suggested by a number of familiar experiences. For example, a roller coaster car, as it inches over the top of the highest rise in its track, seems to possess very little energy. But as it accelerates down the slope, it quickly acquires more and more kinetic energy. When the car is at its lowest point its velocity is greatest, and its kinetic energy is therefore at a maximum. Then, as it starts up the next incline, it begins to lose its kinetic energy, until at the top its kinetic energy is almost zero again. Where does the car's kinetic energy come from, and where does it go? If we were to lift the car from its lowest to its highest point on the track, we would have to do a lot of work—an amount equal to the weight of the car multiplied by the distance through which it was raised. Energy has been defined as the capacity to do work; therefore it seems logical to conclude that at the top of its path the car has acquired a certain energy solely by virtue of its position in the earth's gravitational field, that is, its height above the low point. We call this energy of position **potential energy** (PE). The equation for this gravitational potential energy is

$$PE = \text{weight} \times \text{height} = mgh$$

where g is the acceleration due to gravity.

We can see that the potential energy of the roller coaster is at its greatest when the car is at the highest point of its track, and least when it is at the lowest. But consider a car sitting on the ground under the tracks. Its potential energy would be lower still. Would it be zero? Then what would be the potential energy of the same car at the bottom of a pit 100 ft deep? This example suggests an important point: potential energy is not an absolute quantity, but must always be defined in relation to some point of reference. The earth's surface is often a convenient point of reference for measuring potential energy, but it is certainly not the only one. We can define the potential energy of an object with respect to any point we wish—or to a number of such points. If you place a 1-lb book on the seat of an airplane flying at 10,000 ft, for example, the book has a potential energy of zero with respect to the seat, but its potential energy with respect to the earth's surface is 10,000 ft-lb. If the seat is 18 in high, the book's potential energy with respect to the floor of the airplane is 1.5 ft-lb, and will remain so regardless of the altitude.

Very often, we are more interested in changes in potential energy than in its total amount as computed relative to some fixed reference level. For example, in some situations an engineer might be much more concerned about the total amount of potential energy gained or lost by an elevator moving from one floor to another than in its potential energy relative to the bottom of the shaft. Thus a 1-ton elevator in a building with floors 10 ft high acquires 20,000 ft-lb of potential energy for each floor it ascends. This will be true no matter what reference point is employed.

Potential Energy

Potential energy must always be measured with respect to a fixed point of reference.

It is easy to compare the amount of potential energy acquired by a mass when it is raised to a given height with the energy it acquires if accelerated to a given velocity. When we equate the formulas for kinetic and potential energy, we find that where g is the acceleration due to gravity and h the distance through which the mass is raised, then

$$\tfrac{1}{2}mv^2 = mgh$$

F in the work formula ($f \times d$) is $F = ma$, but a is identical to g because we are doing it against gravity only. And d is identical to h. Thus

$$v^2 = 2gh$$

In the British system, then,

$$v^2 = 64h$$

Consider, for example, a car traveling at 60 mph. This is a mile a minute, or 88 ft/sec. Applying the above formula, we find that $h = 121$ ft. To acquire an amount of potential energy comparable to the kinetic energy it possesses at 60 mph, then, a car would have to be raised to a height of 121 ft. If the car were to be dropped from such a height, all of this potential energy would be converted back into kinetic energy, and the car would be traveling at 88 ft/sec when it hit the ground. Conversely, kinetic energy can be transformed into potential energy, and the same formula can be used to find how high a body will rise if it is propelled upward with initial velocity v. Transposing the equation, we have

$$h = v^2/2g$$

or

$$h = v^2/64$$

in the British system. You may recall from Chapter 1 that the identical formula, $v^2 = 2gh$, can also be obtained directly from the basic equations that describe the behavior of falling bodies.

Transformations of Mechanical Energy

In a frictionless system, the sum of PE and KE is constant (i.e., the decrease in PE equals the increase in KE).

The example of the roller coaster (Figure 3-6) shows this exchange of potential energy for kinetic energy and vice versa quite clearly. If there were no losses of energy to friction, the cycle could go on indefinitely, the roller coaster always acquiring just enough speed on the downward slope to bring it back up to the same height again on the next uphill climb. The roller coaster is useful, too, in that the path of the car over the tracks creates a perfect graph of its potential energy. At the top its potential energy is at a maximum, and its kinetic energy is at a minimum; at the bottom, the reverse is true. The sum of the two, when friction is neglected, is always constant. Together, potential energy and kinetic energy are called total **mechanical energy**; we can restate the previous defini-

Figure 3-7
The potential energy of the arched bow will soon be converted into the kinetic energy of the released arrow.

tion by saying that the mechanical energy of the system is conserved. And if we add to this formulation the convertibility of energy and work, we can state this law even more broadly. Potential energy can be converted into kinetic energy; kinetic energy can be converted back into potential energy; and work converts kinetic into potential energy (lifting a weight with a pulley, for example) or potential into kinetic energy (stretching a bow preparatory to shooting an arrow, for example, as shown in Figure 3-7). In an idealized system where other forms of energy are not involved and friction and air resistance are ignored, we can say that there is complete conservation of the two types of energy, the two physical states characterized by ability to do work. Neither can be created from nothing or disappear into nothing; any gain of one means some corresponding and equal loss of the other.

Other Forms of Energy

Yet this conservation law for mechanical energy has very substantial and serious limitations. In practice, friction and air resistance can be lessened in many ways—by using oil or other lubricants, with ball bearings, through the use of an air cushion or a vacuum—but they can never be entirely eliminated. Eventually they will rob any isolated system of all its mechanical energy. A swinging pendulum will move in a smaller and smaller arc until eventually it is motionless; a roller coaster will slow down, losing kinetic energy, no longer able to return to its original height after each descent without the input of additional work. Furthermore, there are forms of energy besides those we have discussed so far. The enormous electromagnets that are used in many junkyards to pick up iron and steel use electromagnetic energy, and of course the electromagnet or any other device requiring electric current for operation works only because of the electrical potential energy which is provided either by a battery or an electric generator. Because an electric current can run a motor, electricity must also be considered a form of energy. And electrical potential energy can be stored in a battery or a capacitor.

A coiled spring in a watch also stores potential energy. You put this energy into the system by doing work on the spring when you wind the watch. It is released slowly, in the form of work done by the gears that move the hands around the dial. Similarly, you do work when you pull back the bowstring of a bow and arrow, and that work is stored as potential energy in the taut string and bent bow. It is released in an instant when you release the string, in the form of work done in accelerating the arrow to the high velocity it possesses when it leaves the bow. The uranium and plutonium used in atomic bombs and reactors are storehouses of nuclear energy, and many substances store chemical potential energy. A chemist uses various forms of energy to create an unstable explosive compound; all of this energy can be violently released in a fraction of a second if the explosive substance is detonated. The common fuels with which we are all familiar release chemical potential energy more slowly, burning rather than exploding. Even sound is a kind of energy. Because sound is a vibration in the air (as will be discussed in Chapter 8), and vibration is a form of motion, sound can be considered a species of kinetic energy. Finally, all varieties of electromagnetic radiation—visible light, ultraviolet and infrared light, radio waves, x-rays, gamma rays—represent energy.

Solar Energy
Electromagnetic radiation is most important to us, for it is by means of such radiation that we receive energy from the sun. And although we sometimes make direct use of the sun's radiation—a source of power which may become increasingly practical in the future—we make indirect use of it in a variety of ways. In fact, surprising as it may seem, almost all the energy sources on earth (except for atomic and tidal energy) that we exploit to do our work, and on which we rely for life itself, have the sun as their ultimate source. The various fossil fuels—coal, oil, natural gas—that we burn to heat our houses, run our machines, and generate our electricity, are actually chemical substances formed from vegetable and animal matter decomposing over a period of thousands of years.

Plants are living chemical factories, with sunlight as their direct energy source. The chemical process of photosynthesis, which makes it possible for most plants to grow and to manufacture the organic compounds that eventually become our fuels, would be impossible without energy from the sun. When we burn wood we are thus tapping the sun's energy. Even the most primitive

Figure 3-8
In this solar furnace near Odeillo, France, huge parabolic reflectors concentrate solar rays on a small area in order to produce high temperatures.

energy-gathering devices depend on energy from the sun. Man harnesses the wind with windmills, but it is solar radiation that heats the earth and its atmosphere unevenly in various regions, causing wind. Man harnesses falling water to turn mills and power the turbines that generate electricity. But again it is the heat of the sun that evaporates water and causes it to rise in the air, only to fall as rain on mountains and hills. This process gives rise to the swift rivers and streams that work for us. Sunlight even supplies the energy that makes possible the synthesis of vitamin D ("the sunshine vitamin," one of the few vitamins that man can synthesize) in the cells of our skin.

These examples suggest that there must be still another form of energy we have not yet considered. For it is evident that the various kinds of energy we have listed so far seem readily convertible to heat. The sun's radiation heats our air, land, and water. An electric current passing through a wire creates heat, as we can see in such common household appliances as electric toasters and broilers. Chemical and nuclear fuels release their energy largely in the form of heat. If the conservation of energy is to hold for all forms of energy, not merely mechanical, then we must conclude that heat is a form of energy too. And this hypothesis is confirmed by the fact that heat can do work. The steam engine, which made possible the Industrial Revolution and changed the world in the eighteenth and nineteenth centuries, burned coal or wood, and in some in-

A Perpetual Motion Machine

Is there anything wrong with this drawing? The man at the bottom gazing proudly up at his perpetual motion machine doesn't seem to think so. His waterwheel does much of his work, providing power for his house. He has ample leisure to enjoy nature, to tend that garden of strange plants just below the waterwheel, and to contemplate the vagaries of existence. But is the waterwheel really doing work? Water falls down onto the wheel, causing it to turn; the wheel then forces the water through its zig-zag path. From a scientific point of view, this step can definitely be called work. But it looks like no work is actually being done to move the water back to the "top" of the course. Some simple calculations can help us see what's happening here. Assuming the drawing is to scale, and the woman hanging her wash in the lower right is about 5 feet tall, we can deduce values for the other elements of the system. The water is falling a distance of about 30 feet, and the diameter of the wheel is about 8 feet. Because you can see each individual spoke of the wheel as it turns, it can't be moving very fast. Yet the wheel is pushing the water through its sharply angled path a total distance of about 90 feet to the top of the watercourse. We know that the work done on an object when lifting it against the force of gravity (the water pushed forward by the wheel) is equal to the work done by the object when it descends again (the water falling onto the wheel). It appears then, that more force comes out of the wheel than is put into it. We also know that when an object is moved by a force, the amount of work done on it is equal to the product of the force exerted and the distance through which it is exerted. Does this information tell you anything about the unusual system in the drawing?

stances corncobs and buffalo chips; the internal combustion engine that plays so large a part in our lives today burns gasoline or related fuels. Both of these engines are devices for converting heat into mechanical energy, the kinetic energy of moving wheels and pistons.

We can now plug a serious leak in our previous statement of the conservation of mechanical energy. When we first stated that principle, we found it necessary to add "neglecting friction and air resistance." Now, however, we can even account for the losses of energy to these factors. We all know that friction produces heat—enough to ignite the tip of a match or start a fire in wood shavings when two sticks are rubbed together Indian-fashion. A screw that has been driven into a piece of wood, a knife that has been sharpened on a whetstone, a pencil eraser that has been used vigorously will feel warm to the touch. Air resistance, too, is simply a form of friction. In everyday situations it seldom produces enough heat to be noticeable, but vehicles returning to earth from space at high velocities must be heavily insulated against the enormous amounts of heat produced by atmospheric friction. Meteors seldom survive their passage through the earth's atmosphere; they are vaporized by the heat of air friction. If friction produces heat, and heat is a form of energy, we need only show that the amount of energy lost from a system to friction is exactly equal to the amount of energy represented by the heat produced, and we have a complete conservation law. But to do this we must first have a way of measuring heat and determining its equivalent in work or energy.

How can heat be measured? Long before the nature of heat was known, scientists were devising means to measure it. The first step was to establish a temperature scale, to quantify our imprecise, subjective perceptions of "hot" and "cold." Temperature is a measure of the degree of heat and is important to the physicist because the various properties and behavior of nearly all substances change as their temperature changes. With rising temperature, solids become liquid; liquids boil and become gases. Even within the temperature range of a particular state, materials change their volume, usually expanding as the temperature rises and contracting as it falls, although water close to its freezing point is more dense than at the freezing point itself (see Chapter 4). This phenomenon is most marked in gases, as we shall see in the following chapter, but it is observable in liquids and solids as well. Engineers building bridges must allow a few inches of space between adjacent sections of the steel structure to permit expansion, to prevent the bridge from buckling in hot weather. You make use of expansion when you run hot water on a stuck jar lid; the heat causes the lid to expand slightly, loosening it.

The first attempts to measure temperature scientifically were based on the phenomenon of thermal expansion. Liquids are easier to use for this purpose than are solids. For one thing, they expand more; with a change in temperature, the average liquid has a change in volume ten times that of the average solid. Furthermore, it is possible to make the change in volume of a liquid much more easily visible. Medical and room thermometers, the shape of which is so familiar to all of us, are designed to do just that. When the liquid in the bulb ex-

Perpetual motion machines are impossible to construct if only because of the fact that friction can never be entirely eliminated. Friction changes the mechanical energy into heat, causing the machine to slow down and stop.

Temperature

Temperature is the measure of the degree of hotness or coldness of a solid body, a gas, a liquid, or even empty space.

Temperature Scales

Temperature is the property of a body that determines whether heat will flow into the body or out of it when it is brought in contact with another body.

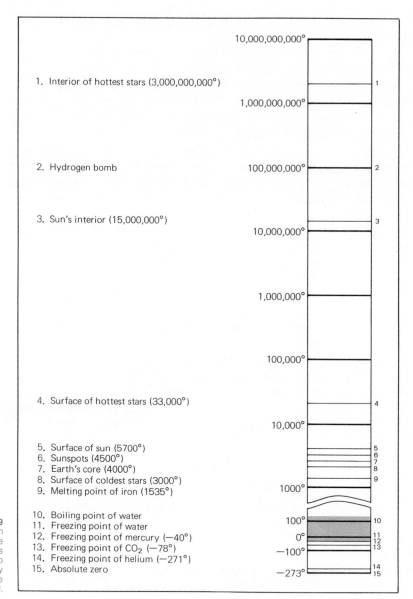

1. Interior of hottest stars (3,000,000,000°)

2. Hydrogen bomb

3. Sun's interior (15,000,000°)

4. Surface of hottest stars (33,000°)

5. Surface of sun (5700°)
6. Sunspots (4500°)
7. Earth's core (4000°)
8. Surface of coldest stars (3000°)
9. Melting point of iron (1535°)

10. Boiling point of water
11. Freezing point of water
12. Freezing point of mercury (−40°)
13. Freezing point of CO_2 (−78°)
14. Freezing point of helium (−271°)
15. Absolute zero

Figure 3-9
The temperatures of some known phenomena using the Celsius, or centigrade temperature scale. The gray bar represents the range of familiar temperatures (−40° to 120°). The temperature of the human body is 37°C. Note that two different scales are used.

pands with increasing temperature, it is forced into a long, very thin tube. The actual change in the volume of the liquid per degree of temperature is very small, but if the tube is thin enough, that small change will produce a rise in the fluid level large enough to be easily measured. In the early days of experimentation with crude thermometers, during the seventeenth century, both water and alcohol were tried as the thermometric liquid. Water, however, is useless for low temperatures because it freezes at the inconveniently high figure of 32°F; moreover, it responds to changes in temperature in a rather irregular way. Alcohol, although it stays liquid to much lower temperatures, boils at a much lower temperature than water. Neither, therefore, proved very suitable except over a very limited temperature span. In 1714 a German scientist, G. D. Fahrenheit, first employed mercury in a thermometer. Mercury proved ideal for this use, since it remains liquid over a wide range of temperatures (−39° to 357°C) and has a uniform rate of expansion over the entire range.

Fahrenheit was also one of the first to devise a practical scale of temperatures. For his zero point he used a mixture of ice and salt (the lowest temperature that could be created at that time) and set the temperature of the human body at 96°. On the scale the freezing point of water was 32°, and he then fixed the boiling point of water at 212°, to make a difference of exactly 180°. This necessitated some minor changes in the marking of the degrees, and thus body temperature on Fahrenheit's scale was adjusted to 98.6°. In 1742 the Swedish scientist Anders Celsius proposed a seemingly simpler temperature scale. On the Celsius, or centigrade, scale, the temperature at which water freezes is defined as 0° and the temperature at which it boils is defined as 100°. Both scales are still in use today. The Celsius scale is universally accepted by scientists, and it is also employed in everyday life in most countries. The Fahrenheit scale, however, continues to be used for nonscientific purposes in Britain and the United States. For the weatherman, at least, it is a convenient scale, since the range of 0° to 100°F corresponds to the common temperature range found in most parts of the world. The formula for conversion of Fahrenheit to Celsius and Celsius to Fahrenheit is a simple one:

$$C = \frac{5}{9}(F - 32)$$

$$F = \frac{9}{5}C + 32$$

Coefficients of Expansion

Different substances, however, exhibit different changes in their size for the same temperature change. For liquids and gases it is usually easiest to measure the fractional change in volume that takes place for each degree centigrade that the temperature rises or falls. This fraction is called the **coefficient of cubical expansion.** For solids, however, it is more convenient to measure the **coefficient of linear expansion,** which is simply the fractional change in any one dimension per degree centigrade. For most common substances this coefficient is the same for all directions; that is, length, height, and width change by equal fractions of their initial values. And if these fractions are small—less than 0.1% or thereabouts—then the coefficient of cubic expansion will be almost exactly three times the coefficient of linear expansion. This is, in fact, the case for all solids; steel, for example, has a coefficient of linear expansion of about 10^{-5}/°C. Thus when the temperature rises from 20° to 30°C, a steel

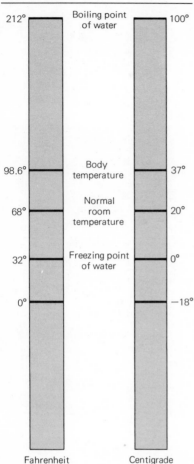

Figure 3-10

The Fahrenheit and Celsius, or centigrade, scales.

cube initially measuring 1 cm on each side increases by about a thousandth of a millimeter in length, and three-tenths of a cubic millimeter in volume. An approximate calculation of this would be

$$1 \text{ cm} \times \frac{10^{-5}}{°C} \times 10°C = 10^{-4} \text{ cm}$$

$$1 \text{ cm}^3 \times \frac{3 \times 10^{-5}}{°C} \times 10°C = 3 \times 10^{-4} \text{ cm}^3$$

Such changes in size, though they are quite small, are often quite important. Ordinary glass or china, for example, cannot withstand abrupt changes in temperature; if you pour boiling water into a glass at room temperature, it may crack. This is the result of uneven thermal expansion and low thermal conductivity. The part of the glass that first comes in contact with the hot water expands before the temperature change has had time to reach the rest of the glass. This may create enough strain to break the glass, although the actual change in size may be only a few hundredths of a millimeter. For this reason quartz or specially fabricated glass such as Pyrex, which have much lower coefficients of thermal expansion, are often employed where large changes of temperature must be withstood. Similarly, manufacturers now market "oven-proof" and "freezer-safe" dishes made of special low-expansion materials.

The thermostat (Figure 3-11), used to regulate the operation of air conditioners, refrigerators, and furnaces, is a simple device that makes use of the fact that materials vary widely in their response to temperature changes. It consists of strips of two different metals welded together. Metals are chosen that have different coefficients of thermal expansion—steel, for example, and copper. The copper strip changes its overall length more than two-and-one-half times as much as the steel strip for any change in temperature, making the bar bend one way—with the copper on the outer or convex side—if it gets warmer and in the opposite direction if it gets colder. In a thermostat one end of the bar is commonly fixed and the free end allowed to establish or break electrical contacts at certain preset points, which correspond to the maximum and minimum temperatures desired. In this way a thermostat can be set to turn a refrigerator on when the temperature inside reaches 45°F and turn if off again when it drops to 39°F, for example.

Heat Energy

Heat is a form of energy equivalent to a given amount of mechanical work.

The determination of temperature is essential for the measurement of heat, but it is not the only factor involved. This may seem confusing at first, for many people think of heat and temperature as synonymous. But this is not so. Consider, for example, this question: Which contains more heat, a cup of hot coffee or a bathtub full of warm water? Even though the coffee is hotter by some 100°F, the water in the bathtub contains more heat. It takes more energy just to warm up a tub than it takes to bring a cup of coffee water to boil. Intensity of heat is different from total energy: intensity is temperature; total energy is heat content.

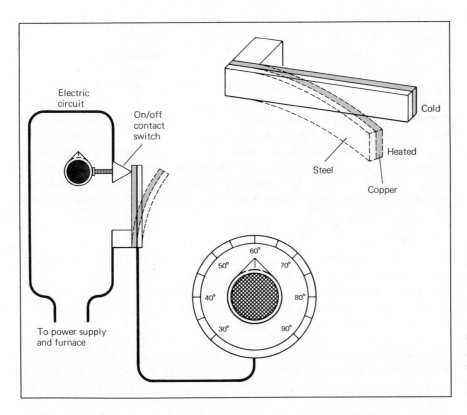

Electric circuit

On/off contact switch

Cold

Heated

Steel

Copper

60°
50° 70°
40° 80°
30° 90°

To power supply and furnace

Figure 3-11
The thermostat. Setting the temperature on the calibrated dial adjusts the size of the gap between the electrical contact and the bimetallic thermometer, thereby determining the temperature at which the furnace will be turned on or off.

We can see this easily if we try to measure the heat content of the cup and the tub in some way—by comparing how much ice each can melt, for example. A tubful of warm water might easily melt a large block of ice, whereas a cup of coffee would hardly make a hole in it. To understand why this is so, it is useful to know something about the nature of heat. As we shall see in more detail in Chapter 6, heat is actually a form of kinetic energy. The particles, or molecules, of a substance move faster—measured by an increase in temperature—and so possess more kinetic energy. The particles can show this in several ways—by vibrating, oscillating, and spinning, as well as by motion and collision in a straight line. The temperature of a substance is, in fact, nothing but a measure of the average kinetic energy of its individual molecules. Heat is a more complicated concept involving several additional factors. The heat of a body represents the total internal energy of its molecules. This will depend not only on its temperature, but also on the number of molecules it is composed of. The number of molecules in turn varies with the mass of the body and the kind of molecules. It is easy to understand the importance of the first of these factors, the mass of the substance. A drop of boiling water will contain less heat than a cupful of boiling water, a cupful less than a kettleful—simply because there is more of the water in the larger containers.

We can measure heat in terms of the temperature change it can cause in a given mass of a given substance. The amount of heat necessary to raise 1 g of water 1°C is called a calorie. We are familiar with this unit because the energy content of food is commonly expressed in calories, but this usage invites confusion, since the word "calorie" used by nutritionists and dieticians is actually the kilocalorie, equal to 1000 of the physicist's calories defined above. Thus a gumdrop, listed in nutritional tables as having 35 calories, actually releases enough heat when metabolized to raise 100 g of water by 35°C, or 35,000 g of water 1°C. To minimize this confusion the nutritionist's calorie is usually written Cal, whereas the physicist's calorie is written cal. In the British system the unit of heat is the British thermal unit, Btu. The Btu is the amount of heat needed to raise the temperature of 1 lb of water by 1°F. One Btu is equal to 252 cal.

Specific Heat

Bodies of equal mass may have different numbers of molecules, for some molecules are much heavier than others. A gram of lead contains a small number of massive molecules, whereas a gram of water contains a much larger number of lighter molecules. If the same amount of heat energy is imparted to a gram of water and a gram of lead, there will be more energy available for each molecule of lead, because they are fewer in number. The molecules of lead will acquire a higher average energy, and so the lead will attain a higher temperature than the water, even though the same amount of heat was applied to each. Similarly, we can see that it will require much less heat to raise the temperature of a gram of lead 1 degree than to raise that of a gram of water 1 degree—33 times as much, in fact. The amount of heat required to raise 1 g of a substance 1°C is called the specific heat of the substance. To calculate the amount of heat needed to produce a particular temperature change in a body, we must multiply the mass of the body and its specific heat by the number of degrees C of temperature change:

$$H = m \times \text{spec. ht.} \times (T_2 - T_1)$$

To raise the temperature of 30 g of iron (specific heat 0.11 cal/g°) from 15° to 35°C, therefore, requires

$$30 \text{ g} \times 0.11 \text{ cal/g-}° \times 20° = 66 \text{ cal}$$

The Conservation of Energy

The measurement of heat paves the way for a complete theory of conservation of energy, if it can be proved that a given quantity of heat is always equivalent to and exchangeable for a given amount of mechanical work. This was demonstrated in the 1840s by the gifted English amateur scientist, J. P. Joule. In a series of meticulous experiments, Joule measured the amount of heat produced by a variety of physical processes and showed that it always took 778 ft-lb of work or energy to generate 1 Btu of heat. In the mks system, 4.185 joules are equivalent to 1 calorie, a relationship called the **mechanical equivalent of heat.** When this is taken into account and extended to include chemical and electromagnetic energy, it is found that no energy ever appears or disappears, but only changes its form. This law has been a cornerstone of physics for more than a hundred years. Early in the twentieth century it was made even more all-inclusive by Einstein, who proved that even matter could be transformed into energy. The famous formula $E = mc^2$ expresses the equivalence of mass and energy, and explains how minute quantities of matter in a nuclear reactor or nuclear bomb can disappear, releasing enormous amounts of energy. At present the law of conservation of energy (including mass) seems inviolable in the physical world as we know it, and indispensable for understanding that world (see Chapter 12).

Energy Cycles

Knowing that heat is a form of energy enables us to trace in more detail the elaborate energy cycles, both natural and man-made, in the midst of which we live, and to see how energy, heat, and work are repeatedly and continually transformed one into the other. Let us take, for example, a pile driver. This device consists of a heavy mass that is lifted by an engine to a height of perhaps 20 ft. When the mass is dropped, it strikes the beam to be driven into the ground with tremendous force. If an internal combustion engine is used to power the pile driver, the cycle begins with the chemical potential energy stored in the gasoline used to fuel the engine. When this burns in the cylinders of the engine, it liberates large quantities of heat. The heat causes the gases in the engine cylinder to expand, moving the pistons and thus performing work. Some of this energy is lost to the cycle at this point, for some of the heat is dissipated into the air surrounding the engine, and much of the work done driving the pistons is dissipated in the form of friction. A good portion, however, is transmitted by the machine to the weight by lifting it against gravity. This work is transformed into the increased potential energy of the mass. When the mass falls, its potential energy is converted into kinetic energy, although not without some small loss due to air resistance as the mass descends. When the mass strikes the beam, the kinetic energy causes work to be done on the pile against the resistance of the earth, although some of it is again lost in the form of heat generated by the impact.

The cycle that results in hydroelectric power (Figure 3-13) is even more complex and interesting, in that it is largely natural—we only exploit one phase of it. Energy from the sun arrives in the form of radiation, which heats the oceans and lakes of our planet. Thus radiant heat energy is transformed into a form of kinetic energy. The molecules with the greatest kinetic energy break free of the water's surface in the process called evaporation, overcoming the forces of cohesion (see Chapter 4) and rising through the atmosphere, their kinetic

Figure 3-12
Joule's paddle-wheel experiment. By measuring the height through which the weight falls and the rise in the temperature of the water, the heat equivalent of a given amount of gravitational potential energy can be measured.

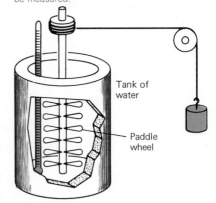

Tank of water

Paddle wheel

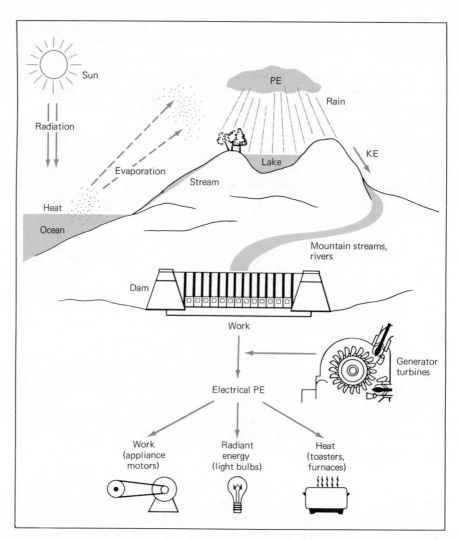

Figure 3-13
The cycle of energy for hydroelectric power.

energy transformed into potential (gravitational) energy. When the air cools, they condense into droplets (losing some of their energy to the surrounding atmosphere in the process, as we shall see in Chapter 5) and fall as rain, giving up potential energy for kinetic energy as the rain flows downhill in rapid mountain streams and rivers. This kinetic energy is tapped by man—the falling water expends some of its kinetic energy in turning generator turbines. The water performs work on the turbines, which transform that work (by a process described in Chapter 7) into electrical potential energy. This is the energy used in the various electrical devices in your home. Some of it may perform work (turning a record player or powering a refrigerator motor), some of it may be transformed into heat (in a toaster or broiler), and in a light bulb some of it may be changed back into radiant energy, the form in which it originally came from the sun.

Work is the net result of a force acting through a distance. The amount of work done is equal to the magnitude of the force multiplied by the distance moved in the direction of the force.

Energy is the capacity to do work. Energy is expressed in the same units as work because it is measured by the work it can do.

Gravitational potential energy is the energy of some body due to its position relative to some fixed reference point. There are many other forms of potential, or stored, energy, such as chemical, mechanical, electrical, magnetic, nuclear, and so on.

Kinetic energy is the energy of a moving body and depends on its mass and speed.

The **machine** is a device that transfers force from its point of origin to another point where it performs work, changing either the magnitude of the force required (to do the work) or the direction of the forces so that it is more conveniently applied.

Mechanical advantage is the factor by which a machine increases the force at our disposal.

Power is the rate at which work is being done. The unit in which it is usually expressed is the watt.

Heat is the energy of rapid, random motion of molecules.

Temperature is a measurement of the degree or intensity of heat.

1 What simple machines constitute a bicycle, and how do they translate rotational work into distance?
2 How does a thermostat utilize dissimilar expansion? Show how a thermometer uses the principle of expansion in its operation. Why is mercury used in most thermometers?
3 Cite examples that illustrate how the following forms of energy are utilized to do work: (a) heat; (b) potential; (c) chemical; (d) electrical.
4 If energy is always conserved, how do you account for the fact that a rubber ball bounces lower and lower with each succeeding cycle?
5 A mover, exerting an average force of 25 newtons, pushes a 50-kg bookcase a distance of 20 m. How much work is accomplished?
6 In another room, the mover's partner spends 10 min trying to lift a 100-kg crate but fails to raise it off the floor. If he exerts a constant force of 100 newtons and is paid at a rate of 10¢ per joule of work, how much does he earn?
7 If your father performs 300 joules of work by pushing a vacuum cleaner with a force of 25 newtons from one end of a room to another, how long is the room?

8 A weight-lifter raises a 200-lb barbell to a height of 6 ft. What is (a) the force required to lift it, and (b) the amount of work done?

9 After the weight-lifter leaves, a 90-lb science major approaches a 300-lb barbell and uses a crowbar and block of wood as a leaver and fulcrum. If the distance from the fulcrum to the weight is 1 ft, and the distance from the fulcrum to the science major is 5 ft, how much force must be exerted to lift the weight?

10 Two boys weighing 80 lb and 115 lb, respectively, are playing on a seesaw. If the lighter boy is sitting 4 ft from the fulcrum, how far must the heavier child sit to balance the seesaw in midair?

11 A maintenance person exerting 75 newtons of force takes 20 sec to push a lawnmower 15 m. What is (a) the amount of work accomplished, and (b) the rate of work?

12 A tractor pulls a 6000-lb thresher 500 yd in 5 min. Compute the horse-power of the tractor's engine, assuming 100% efficiency.

13 How many joules of energy are expended in 15 sec by a 60-watt light bulb? By a 100-watt bulb?

14 If a 20-hp engine is 80% efficient, how much real power is provided by the engine in ft-lb/min?

15 In each of the following pairs of simple machines, which has a greater mechanical advantage: (a) 5-in screw with a 20-in thread; 4-in screw with a 17-in thread; (b) 6-ft-long inclined plane, 2 ft high; 5-ft long, 3 ft high; (c) four-strand block and tackle; five strand?

16 A sculptor uses a four-strand block and tackle to raise a 300-lb marble sculpture. Neglecting the friction, (a) how much force must he exert? (b) How many feet does the rope travel if the stone is raised 6 ft?

17 Two billboard painters on a wooden platform raise themselves to the top of a 60-ft building by means of two pulleys of four strands each attached to the roof. If the combined weight of the two men, their platform, and equipment is 400 lb, what was (a) the average force exerted by each man on his rope, and (b) the work performed by each man?

18 What is the kinetic energy of a 1000-kg automobile moving with a velocity of 30 mps?

19 Compute the potential energy of a raindrop of 5×10^{-1} g mass, just before it falls from a cloud 400 m high? What is its kinetic energy when it hits the ground?

20 A 50-lb toboggan carrying five people with an average weight of 150 lb/person, coasts down a hill 100 ft high on an icy day. Ignoring friction, calculate the velocity of the toboggan as it reaches the bottom of the hill.

21 What is the kinetic energy of a pile driver with a 60-kg drop hammer just before it hits a pile with a velocity of 20mps? From what height was the hammer dropped?

22 If the pile in problem 21 is driven 30 cm into the ground, (a) with what force was the pile struck? and (b) how much work was accomplished?

23 How much energy in joules will an 800-Cal bar of chocolate provide?

Interlude Three / The / Longest / Jump

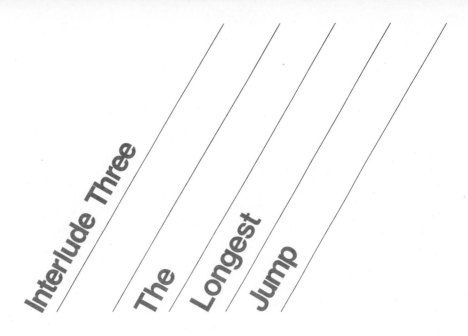

It was 1968, and the site was Mexico City. The nineteenth modern running of the Olympic Games was in progress, and Bob Beamon was standing 139 ft from the running broad jump pit. In a day when human achievements are measured in tenths of a second and fractions of an inch, he was about to make with his first jump the sort of advance usually reserved for the products of technology.

In modern industrial societies there are few places other than athletics where a person's physical abilities are really recognized. Machines have taken over most of society's work requiring brute force—the lifting, carrying, digging, pushing, and pulling once done by men and domestic animals. The physical exertions of a mineworker, a salesgirl, or a hardworking waitress during a busy lunch period are usually not publicly praised. In an earlier day, a farmer pushing his plow behind a team of oxen or a riverboat crew poling and pulling a

barge upstream could claim some credit. Those who talk about force, mass, friction, velocity, and acceleration today are more likely to discuss NASA's latest space shot, or perhaps the winning car in the Indianapolis 500.

Despite this tendency to associate the laws of motion, mass, and energy with machines, they also apply to human beings. They even applied to Bob Beamon on that day in Mexico City, although there are some who swear that he defied the laws of nature on that first jump.

The running broad jump, or the long jump, is a simple event at first glance. Like all track-and-field events (except, perhaps, the pole vault), it stems from one of early man's natural activities. Running, jumping, and throwing were all pursuits followed by every hunter and warrior for thousands of years. The long jump probably began when the first man decided to jump over a stream rather than get his feet wet or be caught by a pursuer. The Greeks included it in their Olympic games, and it has been a part of modern track meets since the mid-nineteenth century.

The object of the long jump is—just as its name implies—to travel as far as possible horizontally between the time you leap into the air and the time you come down. What factors influence how far you go before you come down? The two most important are the amount of time you spend in the air and your forward velocity while you are up there. With the right combination of time and speed in the air, you can do phenomenal things—as Bob Beamon did.

Before following him on his record-breaking flight, however, let us glance at the role played by these same physical laws in other athletic events. Consider baseball, for instance. What happened when Hank Aaron hit his record-breaking 715th home run in May of 1974? What forces were responsible for propelling that ball of cork, yarn, and horsehide 385 ft horizontally, and an undetermined distance above the ground?

It started with the pitcher. Energy in his body from the food he had eaten earlier—energy that ultimately came from the sun—moved his muscles in a well-trained rhythm. That rhythm culminated in an explosively powerful motion, releasing the ball toward the plate. At that moment the laws of gravity, friction, and aerodynamics took over.

Gravity pulled at the ball, accelerating it toward the ground at the rate of 32 ft per second^2. Then, air friction slightly slowed its forward motion, changing its

path from the smooth segment of a parabola it would have described if it had been traveling in a vacuum. The flow of air around and tangent to the rough, spinning surface of the ball applied still more forces to change its path. A professional pitcher can throw the ball about 90 mph, so that 5-oz sphere had over 100 joules of kinetic energy when it reached the plate.

At that point Hank Aaron's muscles released a bundle of energy which *they* had stored, as he swung his 33-oz bat. There was a solid *crack* as the two accelerating masses met. Some of their energy was translated into heat (did you ever play "burn out" while playing catch?), and a lot of it was translated into accelerated motion. The more massive bat (a little warmer now) traveled a little farther in the same direction that it had been traveling and fell to the ground. The smaller mass, which had absorbed an enormous amount of Hank Aaron's energy via the bat, abruptly changed direction and accelerated both horizontally and vertically, traveling in a modified parabola to a point 385 ft from the plate. Its path through the air was determined by the same factors—forward motion, gravity, and air friction—that influenced Bob Beamon during his own record-breaking flight.

O. J. Simpson, one of the greatest running backs in football history, provides another example of the physics of human effort. Simpson weighs about 215 lb and has been clocked at 9.4 sec in the 100-yard dash (Bob Beamon was

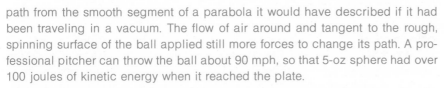

clocked at the same speed—an important factor in his long jump). When Simpson explodes through the line he is a bundle of mass and velocity that can add up to over 450 joules of kinetic energy. To bring that to a sudden stop you need a healthy amount of opposing energy. The offensive backfield has a head start in building up speed before they reach the line of scrimmage, so the defense has to find some way to compensate for its lack of motion. Since kinetic energy is the product of mass and velocity, the solution is obvious. Since the defensive line has less time to accelerate, it is composed of the most massive players in the game.

Mass and acceleration are important in every athletic event, and in none are they more important than in the shot-put. It seems obvious that a 300-lb man will have an average over a 190-pounder when it comes to heaving a 16-lb iron ball. Yet in the 1948 Olympics the first three places were taken by men who weighed no more than 190 lb, and the first man ever to put the shot over 58 ft was a tall, slim 190 lb. They competed against men who weighed up to 100 lb more than they did.

Their secret, of course, was technique—another way of saying that they knew how to combine their mass with speed to produce greater total energy. Rather than pushing it from shoulder height, they learned to start with the shot closer to the ground, giving them a greater distance over which to apply their strength before releasing it, thus building up greater velocity.

Randy Matson, the first man to put the shot over 70 ft, is a tall, 260-pounder who has developed great speed to go along with his mass and is able to capitalize on his height. This concentration on speed—explosive power—has led shot-putters to spend time working out with sprinters—as do many long jumpers.

The pole vault is another event in which velocity is important. So it is no surprise to learn that vaulters are also exceptionally fast in the 100-yd dash. The original pole vault utilized a stiff wooden pole. The vaulter applied muscular energy down against the pole at the top of the jump and by reaction was pushed even higher, above the top of the pole. With the new fiber-glass pole, the vaulter begins as he always has, with a sprint down the runway to build kinetic energy. After the pole is planted he takes another step into it, transferring his energy into the pole by bending it. The pole stores his energy as he leaps up, then it straightens and releases it to catapult him over the bar.

Technology has been used to increase our athletic achievements. Lighter track shoes, scientifically designed spikes for maximum friction, and faster track surfaces have been joined by the fiberglass pole. Despite these technological advances, track meets remain contests between men and the forces of nature. Bob Beamon's first jump at the 1968 Olympics was one of the best advances against these forces ever accomplished by a single modern athlete.

As we said earlier, the distance of a long jump is determined by the amount of time the jumper is in the air and his forward speed while he is up there. What determines how long he stays in the air? The pull of gravity, of course. And the higher you jump, the longer it takes for gravity to pull you back down. Long jumpers want to stay up in the air for as long as possible, so they aim high when they jump.

Few jumpers carry the equation for falling bodies in their heads, but world class competitors do have a solid grasp of its importance. Jesse Owens, the famous American runner, who was watching closely through binoculars when Beamon launched himself into the record books, told a reporter immediately afterward: "His body went up 5½ to 6 ft in the air, and with his speed that will do it." He did not have to see where Beamon landed to know that he would be hard to beat. Ralph Boston, the American record-holder up to the moment Beamon touched back to earth, leaned over to fellow record-holder, the Russian Igor Ter-Ovanesyan, and said: "Man, he sure was up in the air a long time!"

It is not enough, of course, to simply hang in the air a long time. You have to have forward motion, and the faster the better. What determined Bob Beamon's speed while he was in the air? Bodies in motion will remain in motion at a constant speed unless a force acts to slow them down or speed them up. The only forces acting upon him once his feet left the earth were gravity and air friction. Gravity was pulling him back to earth, vertically downward so it did nothing to slow his horizontal forward motion. Air friction at that speed can make a few inches difference. So can wind (sometimes to an athlete's advantage, depending on its direction).

Since there is no way for a jumper to change speed once he is in the air, a long jumper's air speed is determined by how fast he is going when he launches himself. It is no accident that champion long jumpers have all been exceptionally fast.

Jesse Owens, one of the all-time great broad jumpers, did little special work on that event because he usually entered—and won—four events at every track meet. For the long jump he relied on his record-breaking speed and a little practice on getting height, which resulted in a gold medal at the 1936 Munich Olympics.

Height and speed. Those were the elements that Bob Beamon knew he had to put together for a good jump. The world record when he began his 139-ft run at the pit was 27 ft, 4¾ in. He accelerated down the track, his spiked shoes giving him maximum friction with the runway and transferring his energy into ever faster forward motion. His speed peaked just before he reached the launching board. His pace was perfect, one foot pushing off from the board as he swung his opposite leg up transferring about one-third of his energy into an upward motion.

For any given starting speed, a projectile will travel a maximum distance, ignoring the effects of air friction, if it is launched at an angle of 45°. This is an impossible angle for long jumpers who approach the pit at full throttle. Calculations will show that converting all the energy of a top sprinter into upward motion would result in a high jump of over 30 ft. The human body—at least at this stage of athletic development—is not capable of that transfer. The greatest distance is achieved if you use maximum speed and settle for a lower angle, rather than slowing down so you can launch yourself at 45°. Beamon left the ground at top speed and an angle of something like 30°. He came down 29 ft, 2½ in. later. He had jumped 1 ft, 9¾ in. farther than the previous world record.

Jesse Owens had jumped 26 ft, 8¼ in. in 1936, a distance that was unmatched for 25 years. The best jumpers in the world had been able to add only 8½ in. to that record in the 33 years between Owens' triumph and Beamon's. In 1968, world-class jumpers were dreaming of 27½ ft. Twenty-eight feet loomed as a barrier as formidable as the 4-min mile before Roger Bannister ran the "impossible" in 1954. Bob Beamon skipped 28 ft entirely, adding as much to the record in one try as had been added in almost 50 years previously.

The feat becomes even more amazing when it is compared to other events. He had improved the record by 6.45 percent. A similar improvement in the 100-yd dash would result in an 8.5-sec dash. Applying that measure to the mile run gives us a time of 3:37:1. It would mean about a 6-in. improvement in the high jump, taking it to just under 8 ft.

When his teammate John Thomas translated the metric distance announced at the Olympics into feet and inches, Beamon slipped out of Thomas' arms, sank to the earth, put his forehead on the ground, and cried. He was as overwhelmed by his feat as the rest of the athletic world.

There is some controversy about the records set at the Mexico Olympics, because Mexico City is 7349 ft above sea level. The air at this altitude is very thin, which gave distance runners a real problem with their breathing. Sprinters were not bothered, however, because they rely on a short, explosive burst of energy accomplished almost entirely with oxygen already stored.

The thinner air affects the explosive events in another way, however. The lower density—76 percent of normal sea level density when Beamon jumped—offers less air resistance, and all sprinters ran about 1 percent faster than would have been expected in 1968. This decreased drag may have resulted in a jump 3 in. longer than it would have been at normal sea-level air density. That still leaves an impressive human achievement.

This higher altitude also meant that the pull of gravity was very slightly less than at sea level. The farther you are from the center of the earth, the weaker the pull of the earth's gravity on you. This difference, though infinitesimal, is enough to be measured by scientific instruments, but not enough to make a measurable difference in Beamon's jump.

An interesting historical footnote is that one other man in modern times has jumped over 29 ft, and he did it in 1854! J. Howard, an Englishman, was a professional sprinter. He jumped 29 ft, 7 in. with the help of a pair of 5-lb weights. Howard knew his physics, and practiced sprinting at full speed down the track with the weights held high. With precision timing he threw both weights to the ground at the very moment that he launched himself into the air. This well-timed action added considerably to the height he was able to reach, and therefore to his distance. He used the same principle of reaction that propelled the rocket of the Apollo astronauts to the moon—Newton's third law.

Ancient Greek records claim that an athlete named Phayllus jumped 50 ft using this method at the 18th Olympiad in 708 BC. Since the first Olympics were held on the side of the sacred Mount Olympus, it is suspected that this prodigious leap was accomplished by jumping downhill. The ancient Greeks knew a lot about physics, but it is not recorded that they could defy gravity.

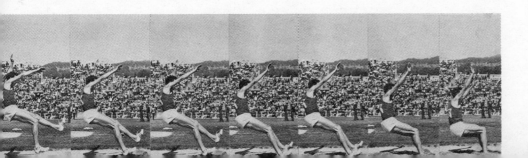

TWO

States of Matter

Chapter 4
Solids, Liquids, and Gases

We are all familiar with the three basic states of matter: solid, liquid, and gas. And if we stop to think for a moment it is easy to see how these three states differ in a few simple physical qualities. Solids are relatively rigid: their volume (the amount of space they occupy) is fixed, and they do not change shape easily. Liquids also maintain constant volumes, but they do not have fixed shapes—a liquid will flow and take the shape of its container. A gas has neither a definite shape nor a definite volume; it not only takes the shape of its container, but it will expand to fill any space it enters. Conversely, gases are highly compressible, unlike liquids and solids, which vary greatly in the degree to which they can be compressed; a lot of gas can be forced into a limited space, as anyone who has ever pumped up a bicycle tire knows. Because liquids and gases both flow, they are sometimes classed together as fluids.

The first section of this chapter examines the characteristic properties of these three states of matter more closely. The second section deals with the concept of pressure in fluids, and some of its consequences; the third develops an important set of laws that governs the behavior of gases.

Solids

We encounter hundreds of different solid substances every day—natural and man-made, light and heavy, strong and fragile. The coins in your pocket, the hair on your head, the wood of your pencil and the plastic of your pen, the natural cotton or synthetic polyester fibers of your clothing, the sugar in the sugar-bowl: all these are examples of solids. If you have ever seen a snowflake or a grain of salt under a microscope, or watched ice forming on a puddle in freezing weather, you are already familiar with the crystal, the basic structural building block of most solids. Crystals, many of which are among the most beautiful objects found in nature, always have highly regular geometric forms, with smooth planes and straight edges intersecting at sharply defined angles. The study of crystals sheds much light on the nature of matter, as we shall see in later chapters.

Elasticity

Although large crystals sometimes occur naturally, and can be artifically produced in the laboratory, most common solids are not themselves pure crystals. Their constituents, however, are usually crystalline substances. The same forces that are responsible for this crystalline composition give rise to one of the characteristic properties of the typical solid: its elasticity. **Elasticity** means the ability of a substance to return to its original size and shape after being stretched or deformed by some external stress. Solids vary greatly in their degree of elasticity; a china plate, for example, can withstand hardly any deformation before reaching its breaking point, whereas some metals are quite elastic—think of a steel spring or sword blade. But for most solids this elasticity, whether small or great, is found to obey a very simple law: the deformation of the object is directly proportional to the applied force. A metal wire, for instance, that is stretched .25 in. when it supports a weight of 100 lb will be elongated by .50 in. if the weight is increased to 200 lb. If the weight is removed, the wire will return to its original length. This principle is known as **Hooke's law**, after Robert Hooke (1635–1703), the English contemporary of Newton who formulated it, and it can be given a simple mathematical expression. If F is applied force, and s the resulting deformation of the object, then

$$F = ks$$

where k is a proportionality constant determined by the nature of the particular material of the sample under stress, as well as by its size and its shape. Steel, for example, stretches less easily than rubber, so its elastic constant is much

Figure 4-1
The smooth planes and sharp edges of the crystalline structure are illustrated in this photograph of a large quartz rock crystal. Although most solids are not themselves pure crystals, the crystalline form is the basic building block of most solid substances.

Water, Water, Everywhere . . .

Water, which is found almost everywhere on earth, is one of the few compounds that is commonly known in all three of its physical states. We are familiar with many forms: the gaseous vapor that makes a tea kettle whistle, the liquid that pours from the showerhead in the morning, the solid that clinks in a Coke. Each physical state of water is abundant in nature, although liquid water is by far the most prevalent: there are about 330 million cubic miles of liquid water on the earth's surface; from 0.1 to 4 percent of the lower atmosphere is composed of water vapor; and 23 percent of the earth is covered temporarily or permanently by snow and ice. Not only is water all around us, it is also within us: more than two-thirds of our bodies, of every living cell, is water. The properties of water have influenced all living things on earth since life began in the salty water of the seas more than three and a half billion years ago.

Chemically, water is extraordinary, a compound whose distinctive characteristics have yet to be fully explained. Its molecular formula is H_2O, so each molecule of water consists of two atoms of hydrogen and one atom of oxygen. The structure pattern of the molecule always shows the oxygen atom in the middle, flanked by two hydrogen atoms, somewhat off a straight line and therefore not quite symmetrical, so that the three atoms form an angle of about 105°. Every water molecule shows this same structure whether it is gaseous, liquid, or solid.

Because water molecules are asymmetrical in shape, they do not show the same electrical charge on all sides. Near the two hydrogen atoms, the water molecule acts as a positive charge; the other end of the molecule at the oxygen atom acts as a negative charge. This electrical-charge polarity gives each water molecule the power to attract nearby water molecules with hydrogen bonds. Many of the unique properties of water come from the property of forming extended groups of water molecules, bonded together three-dimensionally. But in their gaseous state, water molecules are largely independent of each other and relatively so far apart that they interact only on colliding, which is fairly rare. It is still puzzling how water molecules in the liquid state are arranged. Almost all liquids are normally composed of single molecules, more or less independent of each other and moving in random fashion. But water molecules in the liquid state have a structural interrelationship that seems to be somewhere between the unrelated molecular state of gaseous water and the structured state of solid water—ice. It must be the hydrogen bonds between liquid water molecules that account for such unusual physical properties as its high surface tension (in which the molecules cling together) and high boiling point (since boiling requires that the molecules be separated).

The many solid states of water are made up of crystal lattice patterns of water molecules that interact to form differing ordered structures. There are nine known forms of ice. In Ice I, the familiar form on which we ice-skate, the water molecules are linked together in a tetrahedral arrangement, one molecule in the center surrounded by another molecule on each of the four sides. This cluster of five molecules then groups with other such clusters to form a lattice of hexagonal rings. Since there is a relatively great amount of space between the molecules in such a structure, water is one of the few substances that is *less dense* in its solid state than in its liquid state. This accounts for the fact that Ice I floats in liquid water.

In addition to the liquid, rain, which is the most common precipitation, water may fall from the sky in ten different solid forms, including seven different types of snow crystals. Almost all snow crystals have a hexagonal pattern, although three- and twelve-branched forms have also been recorded. The hexagonal shape reflects the internal network of lattices formed when the water freezes.

greater. And thick objects stretch less easily than thin ones of the same material; a rubber band would have a much lower constant than a rubber tire. Other things being equal, the constant is inversely proportional to the length of the object being deformed, so that a wire two feet long stretches twice as much as a one foot wire under the same tension.

Hooke's law is not a complete description of the behavior of solid objects under stress, however. It is accurate only up to a certain degree of stress, known as the elastic limit. The elastic limit also depends on the size, shape, and material of the object; when the limit is exceeded the object is permanently bent or stretched or broken and does not return to its original configuration. Certain solids can be deformed far beyond their elastic limit without reaching their breaking point. Copper, gold, and lead, for example, possess great malleability: that is, they can be hammered or rolled into extremely thin sheets. In addition, many metals, including copper, gold, platinum, and silver, are highly ductile, that is, they can be drawn without breaking into wires so fine as to be almost invisible.

A solid that can resume its original shape after it has been deformed is elastic.

Liquids

Liquids represent an intermediate state of matter; they resemble solids in some respects, gases in others. Like solids, liquids have relatively stable volumes; they cannot be compressed to any great degree, as can gases. Like gases, though, and unlike solids, liquids can flow. But this borderline between the solids and liquid states is not always clearly defined. There are some fairly common substances that seem to be solids, but that could technically be classed as liquids because of their ability to flow, even though they flow very slowly. Ordinary tar or pitch, for example, that is used in making roads, feels hard to the touch at ordinary temperatures; but if left to stand for a long time it will gradually spread and take the shape of its container like a liquid. Among these amorphous substances is much ordinary window glass, although a pane of glass must stand undisturbed for many years before it shows any perceptible signs of flow. It is common in old houses to find window panes that are measurably thicker at the bottom than at the top.

Viscosity and Flow

Even familiar liquids vary widely in their rates of flow; some, like water or alcohol, flow very rapidly, but thicker liquids such as honey, glycerin, heavy motor oil, or the concentrated liquid detergents and shampoos that are now so common, flow much more sluggishly. This resistance to flow in a fluid is called **viscosity**; we can say that honey is more viscous than water. Most gases have viscosities far lower than that of any liquid; the viscosity of air, for example, is only $1/55$ that of water. If you have ever stored molasses or maple syrup in the refrigerator, you have probably noticed that they seem even stickier and more slow-moving when cold than at room temperature. This suggests that viscosity varies with temperature, and the viscosity of a liquid does in fact decrease as it gets warmer. One of the reasons it is hard to start a car on a cold day is that at low temperatures the oil becomes more viscous and impedes the movement of the parts of the engine until it warms up to normal operating temperatures. That is why special light oils must be employed in arctic conditions, where ordinary lubricants would be useless. Conversely, racing drivers, whose engines must perform at very great temperatures, use motor oils with an unusually high viscosity; conventional oils would "thin out" too much under such heat. In a gas, however, the relationship between temperature and viscosity is just the reverse; gases normally flow more freely at cooler temperatures than warm.

The phenomenon of viscosity arises from the way in which liquids flow. Probably you've noticed that if you tilt a cup full of sugar, the entire contents of the cup does not pour out uniformly. The top layer of sugar moves easily and rapidly, sliding over the layer beneath it; the middle layers move more sluggishly; and at the bottom of the cup there is hardly any movement at all. Liquids, which are also composed of tiny particles called molecules (although these are far smaller than grains of sugar), seem to behave in the same way, flowing in layers that slide over each other. If you've ever followed the paths of leaves or twigs carried along in the current of a brook, you will have seen the same phe-

Figure 4-2
The configurations of smoke in this wind tunnel photograph shows the phenomena of streamline flow. The dark object is a wing cross-section, and the flow of smoke shows a lifting effect.

nomenon of flow at work. The water near the edges and bottom of the stream clings to the stream bed and moves very little; the adjacent layers of water move more rapidly past these nearly stationary ones; at the center of the stream, farthest from the stream bed, the water flows most swiftly. Of course the layers of a fluid cannot slide over each other in this way without some friction, and the viscosity of a fluid is simply a measure of this internal friction.

The kind of motion we have been describing is called **streamline flow,** and it is characteristic of fluids at relatively low velocities. In streamline flow a fluid will flow smoothly around an obstacle or an irregularity without disrupting the order of the layers described above. At high velocities, streamline flow is replaced by turbulent flow, with its intricate, shifting patterns of eddies, loops, and whirlpools. If you have ever watched a canoeist trying to shoot the rapids of a fast-moving river or stream, you have a good idea of what turbulent flow is like. An object moves through a liquid or a gas more smoothly and with much less friction if streamline flow around it can be maintained. For this reason engineers take great pains to devise shapes that will cause minimum turbulence in the fluid they move through. Such shapes, with their characteristic rounded fronts and tapering rears, are called "streamlined." We can observe them in natural creatures such as seals and porpoises, which move so rapidly and smoothly through the water, and in such man-made creations as submarines and airplanes.

Among the properties of liquids are miscibility and diffusion. **Miscibility** is simply the Latin version of the word "mixability." Some liquids are miscible with each other, others are not. Alcohol and water, for example, mix easily in all proportions. All alcoholic drinks are mixtures of alcohol and water (100-proof whiskey is only about 50 percent alcohol). But oil and water, as nearly everyone knows, are immiscible. Thus you cannot use water to extinguish an oil fire; the burning oil will simply float on the water. Similarly, the dangerous oil spills from tankers at sea can sometimes be "cleaned up" before they have done too much damage; the petroleum floats on the water without mixing with it and can be absorbed by chemicals or sucked up by specially designed equipment.

Miscibility and Diffusion

Vinegar and oil, the main ingredients in many salad dressings, are also immiscible (vinegar is mostly water). If you shake salad dressing, you get an unstable emulsion: globules of one liquid suspended in the other. Left to stand, the liquids will soon separate into two distinct layers again. But it is possible to combine immiscible liquids into an emulsion in such a way that the components will remain in suspension almost indefinitely. Ordinary commercially available milk is such an emulsion; the butterfat has been broken up by homogenization into droplets so fine that they will not separate out again. There are also various chemical emulsifiers that can achieve the same effect.

Figure 4-3
Clean-up of oil spills is facilitated by the fact
that oil is immiscible with water.

Miscible fluids exhibit another interesting type of behavior: if two of them are brought gently into contact, without stirring, they will gradually interpenetrate each other until eventually a complete mixture results. This process is called **diffusion**; it is quite slow in liquids, but takes place very rapidly in gases.

Cohesion and Surface Tension

The force of attraction that holds particles of a substance together is cohesion.

If you pick up a solid object such as a ruler, the entire mass of wood moves in unison because of **cohesion**—an attraction among the particles of which it is composed that holds it together as a rigid body. If you dip your hand into a bucket of water and try to grasp the liquid, it will not move in one piece as the ruler does. Cohesion is not present in gases; it is present in liquids, but its force there is very slight compared to that of cohesion in solids. Nevertheless, it is sufficient to cause some rather unexpected phenomena. If you place a razor blade or a sewing needle carefully on the surface of a glass of water, it will float—in fact, it will not even seem to be made wet by the water—despite the fact that steel is denser than water and would ordinarily be expected to sink. But if you push these objects under the surface, they will be wet by the water and will drop to the bottom. It is as if the surface of the water were behaving like a thin film or membrane, preventing objects outside it from coming in contact with the rest of the liquid. Yet no matter how many times this surface film is broken, it restores itself immediately. This effect, known as **surface tension**, results from the cohesive forces within the liquid—the mutual attraction of its individual molecules. Throughout most of the liquid these forces operate equally in all directions, but at the surface the force will be entirely downward (or, if we are speaking of a free surface like that of a raindrop, inward). Because of this the surface of a liquid behaves as if it were under tension—as in fact it is. It is surface tension that causes droplets of liquid such as raindrops or the globules in a suspension always to assume a spherical shape—the shape with the least possible surface area.

A closely related force is **adhesion,** or the attraction between the particles of one substance—a liquid—and those of another—e.g., the surface of an immersed object, or the container of the liquid. Both the forces of adhesion and cohesion vary in strength for different substances. The cohesive forces in water, for example, are weaker than the adhesive forces between water and glass. Consequently, if a thin glass tube is partly immersed in a container of water, the water will actually rise in the tube, in apparent defiance of the law of gravity. The thinner the tube, the higher the level the liquid will attain. Seen from the side, the surface of the water in the tube has the concave shape of a crescent moon; hence it is called a **meniscus,** from the Greek word for "little moon." Adhesion causes the liquid to creep up the walls of the tube, producing this characteristic shape; cohesion and surface tension pull the rest of the column of liquid along behind. This phenomenon is known as **capillarity,** or capillary action, and it has many familiar manifestations. Why does water spread through a dry towel or sponge, or coffee through a lump of sugar, when only one corner is moistened? The liquid is spreading through the tiny channels and crannies of these porous substances by capillarity (see Figure 4-5).

The forces of cohesion in mercury, however, are stronger than the force of adhesion between mercury and glass. We can see that this is the case simply

Adhesion and Capillarity

A liquid exhibits capillary action when it rises in a narrow tube due to a combination of adhesive and cohesive forces.

Figure 4-4
Ink diffusing into water. Though slow, this process eventually results in a uniform mixture of substance. Gases diffuse in an analogous fashion.

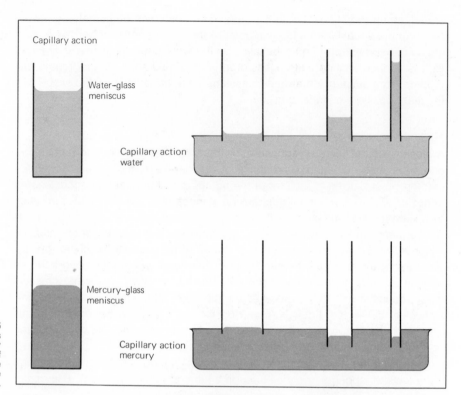

Capillary action

Water-glass
meniscus

Capillary action
water

Mercury-glass
meniscus

Capillary action
mercury

Figure 4-5
Adhesion between water and glass accounts
for capillary action in tubes of increasingly
smaller diameter. With mercury the level of
liquid in the tube actually falls because
forces of cohesion are stronger than those
of adhesion.

by dropping a little mercury and a little water on a glass plate. The water will spread out as far as possible over the surface of the plate, maximizing its area of contact with the glass in response to the strong adhesive force between the two substances. The mercury, by contrast, will assume the shape of an almost spherical globule, minimizing its contact with the glass, its strong cohesive forces prevailing over the much weaker adhesive force between mercury and glass. Consequently, if the capillarity experiment described above is repeated with mercury instead of water, the result will be just the opposite: the surface of the mercury in the tube will take on a convex shape, and its level will drop below that in the surrounding container. Cohesion pulls the mercury away from the walls of the tube, forming the convex shape, and surface tension (another manifestation of the same cohesive forces) pulls the level of the liquid down.

Gases   Since most gases are invisible, odorless, and intangible, our senses tell us very little about their nature. In spite of this, gases are the best understood of the three forms of matter, and it is through the study of gases that we have learned much about the nature of liquids and solids. Two of the principal reasons that physicists have been so successful in elucidating the properties of gases are that, unlike solids and liquids, gases are all very much alike and their behavior is easily quantifiable.

Because gases are generally so much lighter than most liquids and solids, people are often surprised to find that they do have appreciable weight: The air in a room of average size, for example, will weigh between 100 and 200 lb. A gas will expand until it fills a container of any size and shape. Conversely, gases can be greatly compressed; a relatively small metal cylinder can hold enough helium to blow up hundreds of balloons, or enough oxygen to keep a deep-sea diver or a patient undergoing surgery alive for many hours. All gases are miscible with each other in all proportions, and gases diffuse into each other much more rapidly than do liquids. If an open bottle of Chanel No. 5 or a bouquet of strong-smelling flowers is left in one corner of a room the scent, which indicates the presence of a gas given off by the perfume or the flowers, will soon spread to all parts of the room by diffusion through the air, even though the air seems absolutely still.

When a gas is fully ionized—when its component particles become electrically charged—it becomes subject to electrical and magnetic forces, and thus acquires certain unique properties. Such ionized gas is called plasma, and scientists sometimes regard it as a fourth state of matter. Plasma is rarely found on the earth's surface, outside of the laboratory, but it is the principal state of solar matter, the other stars, and much of the astronomical universe. Portions of the outer layers of the earth's atmosphere—the ionosphere—consist of plasma, and it is known on earth in lightning bolts and such phenomena as the *aurora borealis*.

Plasma, a Fourth State of Matter

The concept of density is applicable to matter in any of its states. The **density**, d, of a substance, whether solid, liquid, or gas, is simply its mass per unit volume. In the metric system density can be expressed in kilograms per cubic meter (kg/m^3) or grams per cubic centimeter (g/cm^3).

Density

Density is mass per unit volume.

$$1 \text{ kg} = 10^3 \text{ g} \quad \text{and} \quad 1 \text{ m} = 10^2 \text{ cm}$$

thus

$$\frac{1 \text{ kg}}{m^3} = \frac{10^3 \text{ g}}{10^6 \text{ cm}^3} = \frac{10^{-3} \text{ g}}{cm^3}$$

conversely

$$\frac{1 \text{ g}}{cm^3} = \frac{10^3 \text{ kg}}{m^3}$$

In the British system, weight density in pounds per cubic foot (lb/ft³) is more commonly used than mass density. Since weight is equal to mass multiplied by *g*, the gravitational constant, the weight density is simply the mass density multiplied by *g*. The density of distilled water at standard conditions of pressure and temperature is 1 g/cm³, or 62 lb/ft³; Table 4-1 lists the densities of several other substances in both systems.

Fluids: Pressure and Buoyancy

The concept of pressure is a familiar one from everyday life. We all know that fluids (liquids and gases) will flow in response to pressure, always flowing from an area of high pressure to one where the pressure is lower. The air in a toy balloon or an automobile tire, for example, is under pressure, and will rush out rapidly if allowed an opening. But what exactly is the scientific definition of pressure? The last examples suggest that pressure is closely related to force, but the two terms are not identical. **Pressure** is defined as the perpendicular force per unit area acting on a surface:

$$P = \frac{F}{A}$$

In the British system, therefore, pressure will be expressed in pounds per square foot or pounds per square inch (abbreviated psi); in the metric system the units will be newtons or dynes per square centimeter.

The inclusion of area in this definition is very important, for a force whose operation is concentrated in a very small area may produce an effect very different from that of the same force distributed over a large area. You have prob-

Table 4-1
Density of Common Substances at Standard Temperature and Pressure (STP)

Substance	Mass density (kg/m³)	Weight density (lb/ft³)
Air	1.3	8×10^{-2}
Alcohol	7.9×10^2	48
Aluminum	2.7×10^3	1.7×10^2
Carbon dioxide	2.0	0.12
Gasoline	6.8×10^2	42
Gold	1.9×10^4	1.2×10^3
Helium	0.18	1.1×10^{-2}
Hydrogen	0.09	5.4×10^{-2}
Ice	9.2×10^2	58
Iron	7.8×10^3	4.8×10^2
Mercury	1.4×10^4	8.3×10^2
Wood (oak)	7.2×10^2	45
Oxygen	1.4	9×10^{-2}
Water, pure	1.00×10^3	62
Water, sea	1.03×10^3	64
Lead	1.1×10^4	7×10^2

ably noticed that if you try to carry a heavy package by a thin cord, the cord will dig into your hand painfully; but the same package can easily be carried under your arm or in a knapsack on your back—its weight is better distributed, and does not press so heavily on a small area. Similarly, a man wearing skis or snowshoes which have a large area can walk over the surface of a snowdrift without sinking in, although this would be impossible for a person wearing ordinary shoes. A little calculation will show why. Suppose the man weighs 160 lb and wears shoes roughly 3 in. wide and 8 in. long. The total weight of his body is supported by an area of snow under his shoes measuring 48 in.^{2}, or $\frac{1}{3}$ ft^2. The pressure on the snow under his feet is thus

$$\frac{160 \text{ lb}}{48 \text{ in.}^2} = \frac{3.33 \text{ lb}}{\text{in.}^2}$$

or

$$\frac{160 \text{ lb}}{\frac{1}{3} \text{ ft}^2} = \frac{480 \text{ lb}}{\text{ft}^2}$$

If the man were to put all his weight on one foot while walking, the snow under that foot would be subjected to twice that pressure, or 960 lb/ft^2. We would not be surprised to find that snow could not support such pressures. But suppose the same man wears skis 4 in. wide and 6 ft long. The total area of the two skis is then

$$576 \text{ in.}^2 \quad \text{or} \quad 4 \text{ ft}^2$$

and the pressure under them is only

$$\frac{160 \text{ lb}}{576 \text{ in.}^2} = \frac{.28 \text{ lb}}{\text{in.}^2}$$

or

$$\frac{160 \text{ lb}}{4 \text{ ft}^2} = \frac{40 \text{ lb}}{\text{ft}^2}$$

The weight of the man is distributed over an area twelve times greater, so the pressure on the snow is only $\frac{1}{12}$ as great.

Pressure in Fluids

You have probably heard that the water pressures in the depths of the oceans are enormous, and that divers and even submarines cannot descend more than a few hundred feet without danger of being crushed. At a depth of one mile the pressure is well over a ton per square inch. Why should this be so? The cause is simply the weight of the water itself: the further down you go, the greater the pressure exerted by the weight of the water above you. This pressure is called **hydrostatic pressure,** which literally means pressure due to standing water. We can easily see that this must be so if we consider the example of a solid instead of a liquid. Suppose a block of wood weighing 1 lb and having a bottom area of 1 ft² rests on a table. The pressure on the section of tabletop under the block will be 1 lb/ft². If three more such blocks are piled on top of the first one, the pressure will then be 4 lb/ft².

But fluids differ from solids in certain important respects. In the example just mentioned all the pressure will be exerted downward on the table, whereas the table will exert an equal force upward on the block. There will be no pressure outward on the sides of the blocks. In a fluid, on the other hand, the pressure at any point is *the same in all directions,* horizontal as well as vertical.

Pascal's Principle

When pressure is applied to one part of a solid it is not transmitted to the entire body; you can squeeze the handle of a baseball bat without subjecting the other end to any added pressure. But if you squeeze the bottom of a tube of toothpaste, the pressure is transmitted throughout the entire tube. If the cap is closed, it will be subjected to precisely the same pressure as the part of the tube you are squeezing; if the cap is open, the toothpaste will flow out in response to the pressure. This is an example of **Pascal's principle**, formulated by the French mathematician Blaise Pascal (1623–1662), which states that any external pressure exerted at any point upon a confined fluid is transmitted unchanged to all parts of the fluid and perpendicularly to all surfaces of the container. Pascal's principle follows logically from the characteristics of fluid pressure we have described above. It is this property of liquids that makes them so useful as transmitters and multipliers of force. A familiar example of this is seen in a hydraulic device such as the brakes of a car (Figure 4-8).

As an example of this application, consider the apparatus diagrammed in Figure 4-7, two pistons of unequal diameter in contact with the liquid in a closed, filled vessel. If either piston moves downward, the other will be forced upward by the fluid. Suppose the larger piston has an area of 60 in.², the smaller of 5 in.². If a force of 10 lb is applied to the smaller piston, it will exert a pressure of 2 lb/in.² upon the fluid beneath it. According to Pascal's principle, exactly the same pressure will be exerted at right angles upon all the surfaces of the container, including the larger piston. If a pressure of 2 lb/in.² operates against an area of 60 in.² it will exert a total force of 120 lb. We have used a force of 2 lb to generate one 12 times that great. As with the lever, though, we pay for the increased force with a decreased distance through which it operates; we must move the small piston a foot to make the large one move an inch. The total work output is the same as the work input; where work and energy are concerned there is never any free lunch.

Since the hydrostatic pressure in a liquid at a given depth is caused by the weight of the overlying liquid, it is easy to calculate. Suppose we consider an

Figure 4-6
In any fluid, pressure increases with depth. Thus, when two holes are punched in the side of a barrel, water will flow faster from the hole nearer the bottom.

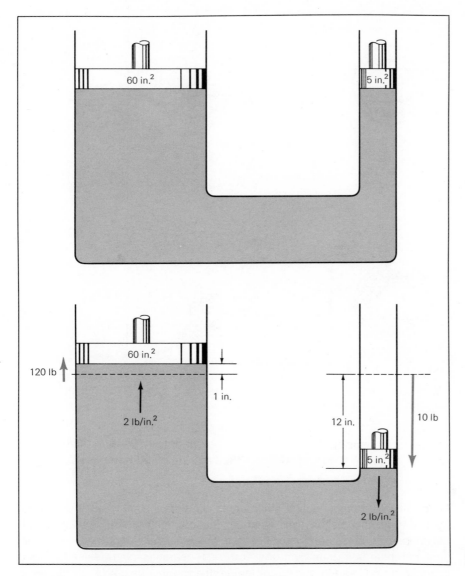

Figure 4-7
Pascal's principle

area A at a depth h below the surface of a liquid. This area is subjected to a force which is simply the weight of the liquid above it, which we can find by multiplying the weight density dg of the liquid by the volume V above our area A. This volume is simply the area A times the height of the column of water h. Thus

$$P = \frac{F}{A} = \frac{w}{A} = \frac{dgV}{A} = \frac{dghA}{A} = dgh;$$

the pressure is simply the weight density of the liquid multiplied by the depth.

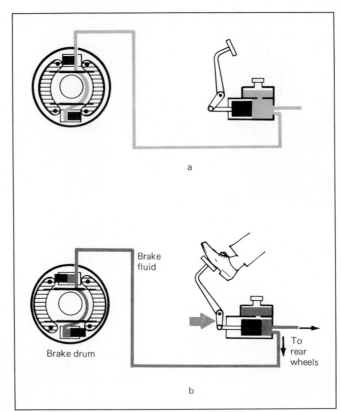

Figure 4-8
Hydraulic auto brake.
The pressure exterted at one end of a liquid
system is transmitted unchanged to the other
end. A small force at one end of a hydraulic
brake system exerted through a large
distance results in a large force exerted
through a small distance at the other end.

Brake
fluid

Brake drum

To
rear
wheels

a

b

The pressure beneath 45 ft of seawater, weight density 64 lb/ft^3, is thus 2880 lb/ft^2, or 20 lb/in.^{2}, if we consider only the pressure of the water itself.

Atmospheric Pressure

In actuality we cannot neglect other forces, for all points on the earth's surface lie under another ocean—an ocean of air, our atmosphere. All gases, air included, exert pressure because of their weight, just as liquids do. Air is very light, but there is a great deal of it above us—the atmosphere extends over a hundred miles from the surface of the earth—and the weight of all that air is enough to produce a pressure of about 14.7 lb/in.^{2} at sea level. (Thus in the problem worked out above, the actual pressure at an ocean depth of 45 ft must include the atmospheric pressure on the surface of the water, so that the total pressure is 34.7 lb/in.^{2}).

Because air is so light, because the gases that compose it are invisible, and because people are not ordinarily conscious of atmospheric pressure, it took scientists many years to become aware of this phenomenon and to measure it. The first important experiments were those of the great Italian physicist Galileo in the early seventeenth century. Galileo observed that if all the air were pumped out of a pipe and one end of it were immersed in a body of water, the water would rise in the pipe to a height of about 34 ft. The explanation of this fact was provided by his pupil Torricelli in 1644, just two years after Galileo's death. Torricelli realized that it was the pressure of the atmosphere on the body of water that forced it up the pipe; the water rose to a height at which its own weight exactly balanced the atmospheric pressure. Thus this pressure could not only be detected, but also measured; a column of water 34 ft high exerts a pressure of 14.7 lb/in.^{2}, as we can easily determine just as we did in the previous example.

$$\frac{62 \text{ lb/ft}^3 \times 34 \text{ ft}}{144 \text{ in.}^2/\text{ft}^2} = \frac{14.7 \text{ lb}}{\text{in.}^2}$$

Note that the density of fresh water is slightly less than that of salt water.

The Barometer

Since working with a column of water 34 ft high would be rather ungainly, Torricelli made his experiments more practical by substituting mercury, a liquid of much greater density (density 830 lb/ft^3). Atmospheric pressure was found to support a column of mercury about 30 in. high; Torricelli had constructed the first mercury barometer. Using this instrument, scientists soon discovered that atmospheric pressure is not constant, but varies day by day and hour by hour with changes in temperature, humidity, and wind pattern; in the eye of a hurricane the barometer may read as low as 27 in. of mercury. These fluctuations can be used to predict the weather. Atmospheric pressure is still commonly expressed in inches of mercury (Hg) as you hear on the weather report "barometer 30.05 and rising." Several others are employed as well: one sometimes encounters centimeters of mercury (cm Hg), millimeters of mercury (mm Hg, also known as Torr, or torricellis), atmospheres (at or atm), and millibars (mb). One atmosphere is simply the average sea-level atmospheric pressure of 14.7 lb/in.^{2}, which is equivalent to 30 in. Hg, 76 cm Hg, or 760 mm Hg. The millibar, used mainly by meteorologists, is defined as 100 newtons/m^2; normal atmospheric pressure is 1013 mb. This is 1.013×10^5 newtons/m^2. This crushing force—more than 11 tons/m^2—is counteracted by an equal pressure within our bodies.

Water pressure increases with depth, as we have seen; by the same principle, atmospheric pressure should decrease with height above the earth's surface, because the higher we go, the less air remains above us to press down upon us. This fact was confirmed experimentally by Pascal in 1646; he observed that the mercury in a barometer fell 3 in. when the instrument was brought to an altitude of slightly more than 3000 ft. Air differs from water in one important respect, however: water pressure increases uniformly with increasing depth, but the pressure of the atmosphere does not change uniformly with altitude. This difference occurs because water, being nearly incompressible, does not vary

As altitude increases, atmospheric pressure decreases. As depth increases, water pressure increases.

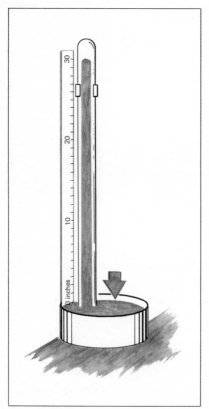

Figure 4-9
The mercury barometer

much in density as the pressure increases—but air, like all gases, is compressible, and is therefore much denser near the surface of the earth where it is under the greatest pressure from the atmosphere above it. As we ascend from the surface of the earth both the height of the atmosphere above us *and* its density decrease; the pressure, which is proportional to both, thus decreases very rapidly at first. If you have ever climbed to the top of a mountain or visited an area that is high above sea level, such as Mexico City or Denver, Colorado, you may have noticed that breathing becomes more difficult in the thin air; athletes competing in such locations must train for a long time before their bodies can adapt to the reduced oxygen supply. Yet a baseball player may find that he can hit a ball farther in a ballpark high above sea level; the thinner air offers less resistance to the ball's flight. At great heights the pressure and density of the air fall below the level needed to sustain life, and mountain climbers or aviators must use oxygen masks. Yet traces of air can still be found over a hundred miles above the surface of the earth.

Fish living in the ocean depths can withstand the vast pressures to which they are subjected because the pressure inside their bodies is maintained at exactly the same level. In the same way, we remain unaware of the atmospheric pressure of nearly 15 lb on each square inch of our bodies; there is a constant equilibrium between the internal and external pressure. But when the external pressure is suddenly changed—when we dive into a swimming pool, for example, or ascend rapidly in an express elevator—this equilibrium is momentarily disrupted. The pressure within our bodies cannot adjust so quickly; the resulting inequality of pressure is the cause of that uncomfortable popping sensation in the ears which most of us are familiar with. We can best appreciate atmospheric pressure by observing its full force unbalanced by any opposing pressure. To do this we must create a vacuum. A German engineer named von Guericke first devised a pump capable of creating a near-vacuum in 1650, and with it performed some very dramatic experiments to show the awesome power of atmospheric pressure. In one famous demonstration he joined two hollow metal hemispheres and used his pump to evacuate the interior. Unopposed by any pressure from inside, the pressure of the atmosphere on the outside of the hemispheres held them together so tightly that sixteen straining horses could not pull them apart. The common rubber suction cup works on exactly the same principle: when air is squeezed out from under the rubber, atmospheric pressure holds the cap firmly to any smooth surface. If the surface is not smooth, air will seep back under the rubber and destroy its holding power.

Buoyancy One of the important consequences of the fact that fluid pressure increases with depth is the phenomenon of **buoyancy**—the upward force that a fluid exerts on bodies submerged in it. We are all so familiar with the fact that certain bodies float that we seldom give the matter much thought. If asked, most peo-

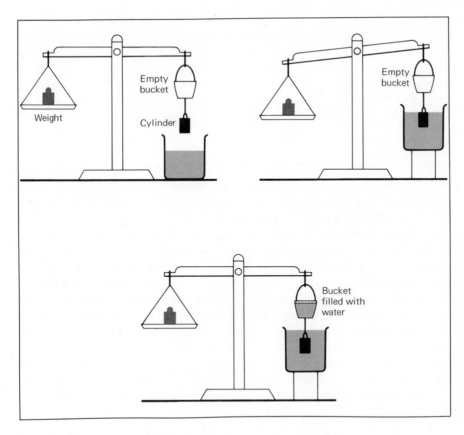

Figure 4-10
When a cylindrical solid is immersed in water, the balance dips down, indicating that the water exerts a buoyant force on the cylinder. The water level rises as the volume of the cylinder displaces a volume of water equal to its own volume. When the bucket above the cylinder is filled with water, the scales balance again. Cylinder and bucket have the same volume, so the amount (and weight) of the water in the bucket is equal to the amount of water the cylinder displaces. Since the scales balance, the buoyant force on the cylinder is equal to the weight of the water displaced.

ple would probably say that "light" objects float and "heavy" ones sink. This is obviously not right, though, for a steel battleship weighing hundreds of tons will float, whereas a pebble weighing half an ounce will sink. What's more, the upward force of buoyancy is at work even on objects that do not float; that is why heavy objects seem to weigh so much less in water. If you are lifting a rock or an anchor out of the water it seems suddenly to gain weight when it reaches the surface, and you may have noticed how your own body, too, feels heavier when you emerge after a swim. Why does this buoyant force exist, and what determines its strength?

The force that causes an object to float in a liquid is buoyancy.

One way of analyzing the phenomenon of buoyancy is to consider a regular solid—a cylinder, for convenience—immersed in a fluid, and try to calculate the forces that operate on it (Figure 4-10). Fluid pressure on the sides of the cylinder increases with depth, but it is always symmetrical—for any pressure at one point there is an equal and opposite pressure on the other side. Clearly then there is no net horizontal force on the sides. The pressures on the top and bottom of the cylinder, however, are not equal; because the bottom is at a greater depth it is subjected to a greater pressure. There must be a net upward pressure; this is buoyancy. Suppose the area of the cylinder is A, its height x, and that its upper surface is a distance h below the surface of the fluid; let d be the density of the fluid, and D the density of the solid. It is easy to calculate the pressures on the top and bottom of the cylinder; the former is dgh, the latter $dg(h + x)$. The corresponding forces are obtained by multiplying the pressures by the area A, and the resultant of the two forces is the net buoyant force:

$$Adgh - Adg\,(h + x) = -Adgx$$

(The sign is negative because the net force is upward.)

This force is opposed by a downward force, the weight of the cylinder; since its volume is Ax and its density Dg, the weight is $ADgx$. Obviously, then, if the density of the submerged body is greater than that of the surrounding fluid, it will sink; if less, it will rise. If the two densities are equal, the object will float at whatever depth it is submerged without any tendency to sink or rise. Fins, masks, and other equipment used by skin divers are often constructed so as to have such neutral buoyancy; if a diver drops such an item it will neither sink out of sight nor ascend to the surface but will remain within easy reach. Since density is the crucial factor in determining buoyancy, it is important to bear in mind that the density of a submerged or floating object is its total weight divided by its total volume—including parts of its volume that may be empty or filled only with air. Although steel is denser than water and will ordinarily sink, a steel battleship that encloses within its hull thousands of cubic feet of air will float; the *total* density of the ship plus the enclosed air is less than that of the water it floats in.

The derivation of a formula given above for the buoyant force has one drawback: strictly speaking it is only applicable to regular bodies with vertical sides. There is a way of confirming the fact that the buoyant force on an immersed body equals the weight of the fluid it displaces, no matter what its shape. Suppose we consider a volume of fluid exactly identical in size and shape to the body to be immersed. Since that volume of liquid is at rest, it must be in equilibrium—the vector sum of all the forces exerted on it by the surrounding fluid must exactly equal its own weight, Vdg. Now if we were to substitute the solid body for this volume of fluid, the forces upon it would be exactly the same; it would experience an upward force equal to the weight of its own volume of fluid. This law is known as **Archimedes' principle**, in honor of the Greek mathematician of the third century BC who discovered it; legend has it that the basic idea occurred to him when he noticed how the water in his bathtub rose as he entered it.

Archimedes' principle enables us to determine not only whether an object will float but also how it will float. It is an easily observable fact that no object floats wholly on the surface of a fluid; part of the object is always immersed. According to the argument given above, an object should float in equilibrium when it is displacing an amount of water exactly equal to its own weight. Given the density of the object and that of the fluid—or even given only the ratio of the two densities—it is easy to calculate how much of the floating object will be submerged. The weight of the object, which can be expressed as its density times its volume, must equal the weight of the displaced fluid, which is the density of the fluid times the submerged volume:

$$DgV = dgV_{sub}$$

Thus $D/d = V_{sub}/V;$ the ratio of the densities gives the fraction of the body that will be submerged. An iceberg, for example, has a density about .9 that of seawater; therefore .9 of a floating iceberg is underwater. The same principle applies to ships. A ship loaded with heavy cargo will have to displace a great deal of water to equal its weight, and will float low in the water, like an iceberg; empty, its weight is much less, and it displaces correspondingly less water, so it will ride high in the water. A ship riding too high is not stable in rough seas; that is why freighters and tankers must carry ballast when they are not loaded with cargo.

Bernoulli's Principle

The pressure of a fluid is affected by its motion as well as its weight, as pointed out in the discussion of streamline flow in liquids. In a perfume atomizer, for example, fluid rises in a tube because the pressure at the top of the tube is reduced. This reduction in pressure is effected simply by blowing air rapidly past the top of the tube. The moving air exerts less pressure than air at rest. This principle was discovered by the Swiss mathematician Daniel Bernoulli in 1738, and is known as **Bernoulli's principle** in his honor. It is an application of this phenomenon that enables aircraft to fly. The wings of planes are designed in such a way that the air passing over the wing moves at a higher velocity than the air moving under the wing. This creates a lower pressure above the wing than below it; the pressure differential provides the upward force, known as lift, that keeps the plane airborne. Helicopters appear to have no wings, but their rotors are actually moving wings; they supply lift in precisely the same way.

The Gas Laws

Because gases, like liquids, are fluids, they share many of the properties of liquids—although sometimes, because of their much lower densities, we are hardly aware of this. For example, we are familiar with the phenomenon of buoyancy principally through our experience with bodies in water, but air also exerts buoyant force on bodies its surrounds. Since its density is only a small fraction of that of water, however, its buoyant force on a body is proportionately less—for a man of average weight it amounts to about 4 ounces. This can be roughly calculated in the following way: we know that a human body floats but

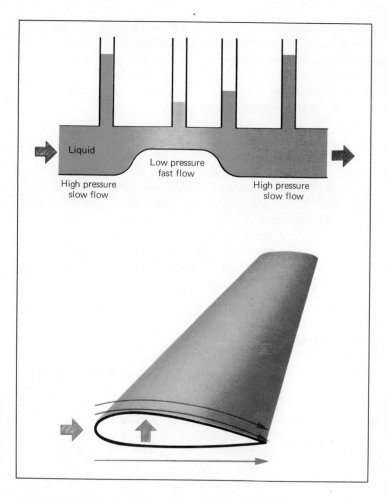

Figure 4-11
When a liquid or gas is forced through a constricted space, it flows faster and a drop in pressure is noted. In aircraft, the wing is designed in such a way that air flowing over it travels faster than air passing beneath it. The normal air pressure on top of the wing is thus reduced, relative to the air pressure against the bottom of the wing. A net upward force, or lift, is the result.

does not float very high out of the water; its density is only slightly less than that of water (mostly because of the air in the lungs). The buoyant force exerted on a body must be only slightly greater than the weight of the body—we might estimate that a man of 175 lb experiences a buoyant force in water of perhaps 200 lb. Since the density of air is about $\frac{1}{800}$ that of water, the buoyant force exerted by the air on such a man would be approximately $\frac{1}{4}$ of a pound. We are not likely to be conscious of such a small buoyant force, and most familiar objects, being denser than air, will of course "sink" in air. Still, most of us have probably seen balloons filled with helium, with a density only $\frac{1}{7}$ that of air, rise skyward, propelled by the force of buoyancy.

The most interesting and scientifically useful properties of gases, however, are those they share neither with liquids nor with solids. Gases are unique in that there are simple formulas relating their temperatures, pressures, and volumes; relationships of comparable simplicity do not exist for liquids and solids.

Gases are thus the easiest state of matter for scientists to understand; the laws governing the behavior of gases that we are about to discuss were discovered relatively early in the history of modern science—during the seventeenth and eighteenth centuries—and proved very important in the evolution of our modern understanding of the structure of all matter, as we shall see in the following chapter.

Boyle's Law

The first of the important gas laws was formulated by the English scientist, Robert Boyle (1627–1691). Building on the work of Torricelli and von Guericke, Boyle investigated the phenomenon of gas pressure in a series of classic experiments with the simple apparatus depicted in Figure 4-12. He first trapped some air in the short closed end of the J-tube by pouring mercury in through the open end. With the mercury at equal heights in both arms of the tube, the pressure of the trapped air was obviously the same as the atmospheric pressure on the mercury in the open arm. When Boyle added more mercury, the level rose unequally in the two arms. The weight of the mercury added to the open end forced mercury up into the closed end, decreasing the volume of the trapped air and raising its pressure. The amount of pressure exerted on the air by the mercury—and thus, by Newton's third law, the pressure of the air on the mercury—could easily be calculated from the difference in height of the two columns of mercury. Boyle noted that there was a simple relationship between the pressure of the trapped air and its volume. The two were inversely proportional: that is, when one increased, the other decreased in exactly the same proportion. When the level of the mercury in the open tube was 30 in. higher than that in the closed one, for example, indicating that the gas was under a pressure of two atmospheres, its volume was just half of what it had been at atmospheric pressure. This relationship can be expressed mathematically: at a constant temperature,

$$V = k/P \quad \text{or} \quad PV = k$$

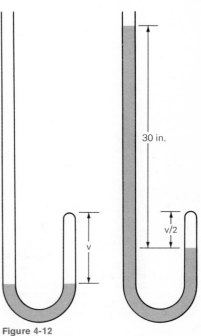

Figure 4-12
Boyle's J-tube

where P is the pressure, V the volume, and k a constant for a given temperature. This is **Boyle's law.**

The same law can be formulated in another useful way. Suppose that you have a quantity of gas initially at pressure P_1 occupying a volume V_1. If either the volume or the pressure is changed, the other will also change; we can call the new values of the pressure and volume P_2 and V_2. If the temperature remains constant, these four variables will always be related according to the equation

$$P_1V_1 = P_2V_2 \quad \text{or} \quad P_1/P_2 = V_2/V_1$$

Charles' Law

The volume of an enclosed gas varies not only with its pressure but with its temperature as well. This relationship, known as **Charles' law** after Jacques Charles (1746–1823), who, with Joseph Gay-Lussac (1778–1850) discovered it,

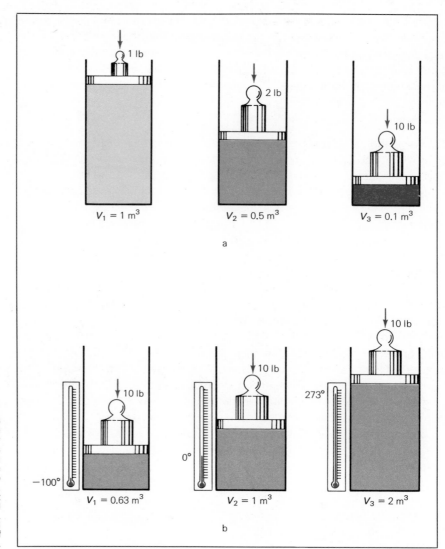

Figure 4-13
a. Boyle discovered that, in gases, the pressure and volume are inversely proportional. At a constant temperature, as the pressure on a confined gas increases (in this case from 1 lb to 10 lb) its volume decreases proportionately (from 1 m³ to 0.1 m³).
b. Charles' law states that at constant pressure, as the temperature of a confined gas increases, its volume increases as well. The volume will increase by ¹/₂₇₃ of its initial volume for each degree centigrade of temperature increase.

is also a simple one. If a fixed quantity of gas initially at 0°C is maintained at a constant pressure and its temperature is changed, its volume will increase by ¹/₂₇₃ of its initial volume for each degree centigrade it is warmed, or decrease by the same amount for each degree it is cooled. If the volume of the gas is held constant it is the pressure that will change instead, and in precisely the same way: for each degree the temperature rises, the pressure will rise ¹/₂₇₃ of its value at 0°C, while for each degree the temperature falls, the pressure will fall by ¹/₂₇₃ of its value at 0°C.

Anyone who has ever watched the smoke from a chimney or campfire knows that hot air rises. This is simply a buoyancy effect—hot air is less dense than cold air, so it will "float" upwards through a body of air at lower temperature. But why does air become less dense when its temperature rises? Charles' law provides the answer. If a body of air is unconfined, its pressure cannot rise regardless of temperature—it must remain the same as that of the surrounding atmosphere. We know from Charles' law that at constant pressure, the volume of a gas will increase as its temperature increases. As the gas expands its density, which is the mass per unit volume, must decrease. This phenomenon makes possible the simple hot-air balloons popular in the late eighteenth and nineteenth centuries, and still used by amateurs today. It is also a major factor in our weather; air warmed by contact with the earth during the daytime rises, whereas cooler air—the air over a nearby ocean, for example—rushes in to take its place. (At night, when the sea is often warmer than the land, the pattern is reversed, and the wind blows from the shore out over the ocean.) This process, called convection, is important in transmitting heat, and will be discussed in Chapter 5.

Charles' law is a very useful formula, but it raises an interesting problem: what happens at a temperature of −273°C? Charles' law implies that if the volume of a gas were kept constant, its pressure would decrease to zero. This is a strange prediction, but not so strange as the result the formula predicts if we were to keep the pressure constant: the volume of the gas would decrease to zero! Would this actually happen? Would the gas disappear? In practice, all gases liquefy before attaining so low a temperature. Furthermore, there are theoretical reasons for believing that this figure of −273°C. is an absolute lower limit for temperature—no lower temperature is possible. Even this figure has never been attained in the laboratory, though scientists have gotten to within a small fraction of a degree of it.

Absolute Temperature

The concept of an **absolute zero** temperature is a very useful one, and it is especially convenient when applied to Charles' law. Suppose we construct a temperature scale with degrees the same size as those of the Celsius scale, but with its zero set at −273°C, absolute zero, so that only positive temperatures are possible on this scale. One result of figuring temperatures on such a scale would be to simplify Charles' law even further, so that it would take on a form similar to that of Boyle's law: both the volume and pressure would be directly proportional to the absolute temperature. At constant pressure, the formula would be simply:

$$V = kT, \quad \text{or} \quad \frac{V}{T} = k$$

At constant volume, we would have

$$P = kT, \quad \text{or} \quad \frac{P}{T} = k$$

Alternately, these equations could be written in the form

$$\frac{V_1}{V_2} = \frac{T_1}{T_2} \quad \text{or} \quad \frac{V_1}{T_1} = \frac{V_2}{T_2}$$

for the relationship between volume and temperature, and

$$\frac{P_1}{P_2} = \frac{T_1}{T_2} \quad \text{or} \quad \frac{P_1}{T_1} = \frac{P_2}{T_2}$$

for the relation between pressure and temperature. In fact scientists often use such an **absolute temperature scale**; it is known as the Kelvin scale, after the British physicist Lord Kelvin (1824–1907), who initiated its use, and temperatures on this scale are expressed as degrees Kelvin (°K). The Kelvin or absolute temperature is obtained by simply adding 273 to the centigrade temperature; on the Kelvin scale, then, the freezing point of water is 273°K, and its boiling point 373°K.

The Ideal Gas Law

It is possible to combine Boyle's law and Charles' law into a single mathematical expression relating all three variables: pressure, volume, and temperature If we again let P, V, and T represent one set of values for these variables, and P', V', and T' another set, for the same quantity of gas, then

$$\frac{PV}{T} = \frac{P'V'}{T'}$$

To illustrate the working of this law, let us take as an example a quantity of gas initially at 27°C. It occupies a volume of 400 cm³ and exerts a pressure of 700 mm Hg. If this gas is heated to 57°C and reduced in volume to 350 cm³, what will be its new pressure? As a first step we must convert the centigrade temperatures to absolute temperatures:

27°C + 273 = 300°K

57°C + 273 = 330°K

Applying the combined gas law,

$$\frac{400 \text{ cm}^3 \times 700 \text{ mm Hg}}{300°K} = \frac{350 \text{ cm}^3 \times P'}{330°K}$$

Solving this equation gives a pressure of 880 mm Hg. An alternative way of expressing the same relationship is to say that, for a given quantity of gas:

$$PV/T = \text{constant}$$

This formula is known as the **ideal gas law**; though it is not a perfect description of the behavior of real gases under all conditions, it provides an excellent approximation in most circumstances. The strength of the law, however, and its value for the scientist investigating the nature of matter, is its generality—although some gases obey it more closely than others, none deviates from its predictions very substantially. The fact that all gases are compressible, and that all behave in the same way, conforming to a simple, single law, gives powerful support to the theory that matter is composed of tiny particles. It also hints that in a gas these particles are very far apart (this accounts for the low density of gases) and so do not interact with each other in the complex ways they do when more tightly packed in liquids and solids. The details of this theory, and its consequences, are explored in the following chapter.

Glossary

Adhesion is the attraction of the surface of one substance to that of another with which it is in contact.

The **atmosphere** is the standard unit of air pressure, measuring 14.7 lb/in^2 or 760 mm of mercury.

Archimedes' principle states that the buoyant force exerted on a body immersed in fluid is equal to the weight of the fluid displaced.

Bernoulli's principle states that a fluid's velocity is inversely proportional to the pressure it exerts.

Boyle's law states that at constant temperature, the volume of an enclosed gas varies inversely with pressure, such that $P = \dfrac{k}{V}$.

Charles' law states that at constant pressure the volume of an enclosed gas is directly proportional to the temperature, such that $P = kT$.

Cohesion is the attraction between the particles of which a body of matter is composed.

Hooke's law states that the deformation of an object is directly proportional to the applied force, such that $F = kS$, where k is a constant determined by the particular material.

The **ideal gas law** combines Boyle's law with Charles' law, and states that pressure varies directly with temperature and inversely with volume.

The **Kelvin scale** utilizes the same unit degrees as the centigrade scale, but sets 0°K at absolute zero or −273°C.

Miscibility is the capacity of a liquid to uniformly mix with another liquid.

Pressure is a measurement of the force per unit area, $P = \dfrac{F}{A}$.

Liquids exhibit **surface tension** due to cohesive forces at or near the surface.

Viscosity is the tendency of a fluid to resist flow over another surface and is analogous to friction in solids.

Exercises

1 Briefly define the three states of matter and their general properties. What is a crystalline solid?

2 Among the elements, why is gold used most often to create complex forms of jewelry, aside from its rarity and beauty?

3 Why does water rise in a glass capillary tube?

4 Show how Galileo's observations of the behavior of water in pipes led to the first mercury barometer.

5 A 160-lb gymnast is balancing on the ball of one foot. It the contact area of his sole is 4 in.², what is the pressure on the area of the high beam beneath?

6 A cylinder containing water has a circumferential area of 10 ft². If a downward force of 2000 lb is applied to the piston compressing the water, what will be the resulting pressure on the water?

7 A hydraulic lift contains a large piston and a small piston with circumferential areas of 40 in.² and 2 in.², respectively. Both pistons are connected to a common body of liquid such that a downward force applied to the small piston raises the large piston supporting a 2000-lb car. What force must be applied to the smaller piston to lift the car?

8 A square raft measuring 4 m on each side floats but is partially submerged 10 cm below the water surface. What is the mass of the raft? How much does the raft weigh? (Density of water = 1000 kg/m³.)

9 A rectangular barge weighing 100,000 kg (110 tons) with a bottom measuring 20 m × 10 m, is moored by a dock. How many meters below the water surface will the barge extend?

10 A bathysphere is designed to withstand pressure at great depths. Calculate the approximate pressure in lb/in.² exerted on a bathysphere at 3000 ft.

11 Convert the following temperature values from the Celsius scale to the Kelvin scale: (a) 30° (b) −50° (c) 85° (d) −160°

12 Air in a sealed cylinder has a pressure of 1 atm at 27°C. What is the pressure in the cylinder at 127°C?

13 If a thermostat is raised, changing the temperature of a room measuring 10 × 6 × 4 m from 17°C to 27°C, what volume of air escapes from the room if pressure remains constant?

14 A cylinder contains 1000 cm² of oxygen at a pressure of 760 mm mercury and at constant temperature. If a piston is lowered raising the pressure to 950 mm what will be the new volume?

15 A mass of nitrogen occupies 20 ft³ at 5°C and 760 mm pressure. What will its volume be at 30°C and 80 mm pressure?

16 As a rubber weather balloon containing 100 cm³ helium ascends, its pressure drops from 730 mm to 670 mm and temperature from 32°C to −3°C. What will be the balloon's new volume?

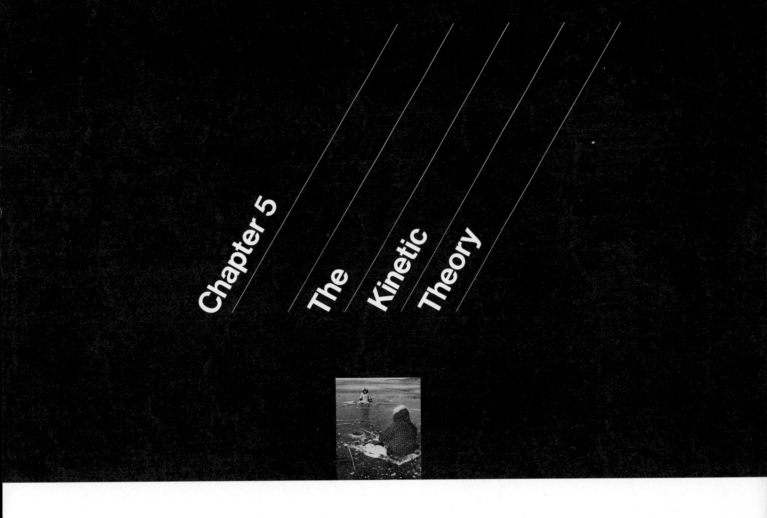

Chapter 5 The Kinetic Theory

Certain observed aspects of gas behavior suggested the molecular structure of gases long before any such structure was proved by experiment. For instance, Brownian motion (see Figure 5-2) suggests continual random movements and collisions on a microscopic level, causing the visible random movement of small dust particles. And the fact that a gas can leak through even the smallest opening indicates that it must be composed of very small particles. Gases are known to be highly compressible, so it seems plausible that the space between the particles is normally very large compared to particle size. If a gas is allowed to expand freely, it will uniformly fill any container no matter how large; open a bottle of ammonia at one corner of a room, and its odor will soon be detectable at the opposite corner.

These phenomena suggest that a gas is composed of individual particles that exert little or no force on one another, except possibly when they collide. If gas particles do collide, the collisions would have to be elastic because there is no observed loss of the total kinetic energy of the particles during collision, nothing like friction or bending or compressing of the particles to soak up energy. If this were not so, the particles of a gas, such as air in a room, would eventually lose their kinetic energy and fall, forming a layer along the bottom of the room.

Kinetic Theory of Gases

By assuming certain microscopic properties of gases, we may plausibly account for such readily observable macroscopic phenomena. The model known as the **kinetic theory of gases** is a grand hypothesis that assumes these properties in a gas:

1 Gases consist of exceedingly small particles that are in constant linear motion, equal numbers moving at random in all directions.
2 The particles are relatively far apart, and they usually exert no force on each other; they interact only when they collide.
3 The particles collide elastically, with no loss of kinetic energy.

Ideal Gases

Gas pressure is a result of particle collisions with container walls.

In 1738, about 75 years after Boyle discovered his gas law ($PV = K$), the Swiss mathematician Daniel Bernoulli (1700–1782) proposed the first kinetic theory of gases. He explained that gas pressure is due to the immense number of times gas particles hit the walls of their container each second. If the volume of the container is decreased, so that the same number of particles have less space in which to move, then the number of collisions per second with the walls will

Figure 5-1
The hot-air balloon was invented by Joseph and Etienne Montgolfier, who constructed a spherical bag about 30 feet in diameter from fabric lined with paper. It was launched before an astonished crowd in Annonay, France, on June 5, 1783. The bag was held over a smoky wood fire until it inflated; it rose high in the air and drifted more than a mile and a half before coming gently back to earth as the air inside it cooled. Etienne later gave a full report of the event to the distinguished Academy of Sciences in Paris.

increase, and pressure also increases. For instance, if volume is halved, then the number of collisions per unit time is doubled, and pressure is also doubled. Conversely, if volume is doubled, then collisions are halved and so too pressure. In 1857, a century later, Rudolf Clausius (1822–1888) established the kinetic theory of gases on a mathematical basis that holds today. He derived Boyle's law (from the assumptions listed above and the simple kinetic energy of the gas particles) as $\frac{1}{2}m\bar{v}^2$, when m is the particle mass and $\bar{v}$ is the average speed (the bar over the v indicates an average value).

An example will help explain the kinetic theory of gases. Consider a gas-filled cube whose sides are of length q. Since all of the particles are moving randomly, one may, for the sake of simplicity, assume that they all have the same average speed, $\bar{v}$. Suppose we consider the motion of one special particle whose direction is perpendicular to one of the walls. Its collisions with the wall (and the opposite parallel wall as well) are assumed to be elastic.

Under Newton's laws of motion, when the particle strikes the wall, it rebounds with exactly the opposite velocity; this is just like bouncing a perfect rubber ball on a perfectly hard floor. Its change in momentum is:

$$\Delta mv = mv - (-mv) = 2mv$$

The wall itself undergoes a change in momentum equal to $-2mv$. The time it takes for this molecule to move across the box and back to the same wall again (a distance of $2q$) is given by:

$$t = 2q/v$$

The force exerted by the particle against the wall is, by Newton's second law, $F = ma$. It can also be expressed in terms of change at momentum, Δmv, as:

$$F = \Delta mv/t$$
$$= \frac{2mv}{2q/v}$$
$$= \frac{2mv^2}{2q}$$

This is the average force exerted by each particle against the wall.

The total force on the wall is equal to the force per particle multiplied by the total number of particles striking that wall. We assume, because of random motion, that on the average an equal number of particles travel parallel to each of three directions—the length, depth, and height of the cube. Of the total number of particles, N, then, we say that $\frac{1}{3}N$ strike any wall perpendicularly. Thus, the total force on this wall is:

$$F = \frac{N}{3} \times \frac{2mv^2}{2q}$$

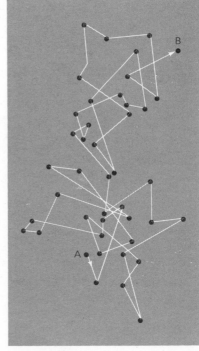

Figure 5-2
The erratic path followed by the particle as it moves from A to B is characteristic of all microscopic particles that are suspended in a liquid or a gas.

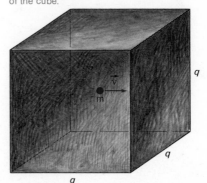

Figure 5-3
This idealized cube illustrates the case in which a particle of mass m has a velocity $\bar{v}$ that is directed perpendicularly to one face of the cube.

Pressure is force per unit area, $P = F/A$, and so $F = P \times A$.

$$A = q \times q = q^2,$$

so we rewrite our previous equation:

$$Pq^2 = \frac{N}{3} \times \frac{2mv^2}{2q}$$

Multiplying by q to get volume (q^3) rather than area (q^2) gives:

$$Pq^3 = \frac{N}{3} \times \frac{2mv^2}{2}$$

Hence,

$$PV = \frac{N}{3} \times \frac{2mv^2}{2} = \frac{2N}{3} \times (\tfrac{1}{2}mv^2) \text{ or } \tfrac{2}{3} N \times \text{(kinetic energy per particle)}$$

This is the fundamental equation for the kinetic theory of gases. It has the same significance as Boyle's law for pressure and volume, in which the expression on the right is $\tfrac{2}{3}$ of the total kinetic energy of the particles, each with an average velocity v. Through the kinetic theory of gases, we see that Boyle's relationship for pressure and volume (P and V) depends on the average kinetic energy of the randomly moving particles.

Real Gases The kinetic theory of ideal gases was initially based on the assumptions that particle size is so small that it can be ignored and that particles exert no forces on each other. This idealization was a useful fiction that permitted scientists to explain several important gas laws, but as time went by, a number of problems

Figure 5-4
The number of particles striking the container walls per unit time is less for container A than for container B, since B has the smaller volume. Consequently, the pressure which is a result of these collisions is greater in container B.

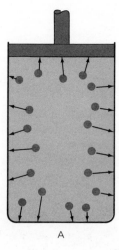

A

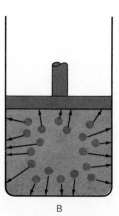

B

arose that weakened the usefulness of this assumption. For example, it was observed in laboratory tests that Boyle's law does not hold at very high pressures. Could it be because gas particles at high pressure are normally much closer together and, therefore, begin to influence each other by some electric, magnetic, or gravitational force? Another problem lay in the observation that although one gas diffuses through another, it does so only very slowly. If, as Boyle assumed, particle size can really be ignored, then gas particles should not get in one another's way and diffusion should take place rapidly, as in really empty space. Slow diffusion suggested that particle size might play some role, even if a small one, in the behavior of real gases.

In 1873, the Dutch physicist J. D. van der Waals (1837–1923) applied certain realistic modifications to the ideal gas law. He proposed that very weak forces of attraction may exist between gas particles, forces which increase with pressure. Since the mass of a gas molecule is so small, the forces cannot be gravitational; van der Waals suggested they were electrical or chemical in origin. These forces modify the force of collision between particles and walls, and therefore modify pressure. They seem to hold back a given particle as it strikes the wall, reducing the pressure effect.

Particle size and forces of interaction between particles affect gas behavior.

To arrive at his equation for real gases, van der Waals corrected for two factors: attraction between particles and particle size. He showed that the force between molecules was inversely proportional to the square of the volume of the gas; thus he denoted pressure as $P + a/V^2$, (where a is a constant of proportionality specific to each gas). To find the effective volume of a gas, van der Waals took the ideal volume, V, of the container and then subtracted from it the space, b, actually occupied by the gas particles themselves. Therefore the space really available for any particle to move into is the unoccupied space $V - b$. So the van der Waals equation reads:

$$(P + a/V^2)(V - b) = K$$

THE EFFECT OF TEMPERATURE   The ideal gas law, which combines both Charles' and Boyle's laws, is $PV = kT$. According to the equation for the kinetic gas theory this becomes:

$$PV = kT = \tfrac{2}{3} N \times \tfrac{1}{2}mv^2$$

Thus, in kinetic theory, the temperature of a gas is proportional to the average kinetic energy of the molecules of the gas, or $T \propto \tfrac{1}{2}mv^2$. Anything that increases the temperature increases the kinetic energy; conversely, anything that increases the kinetic energy increases the temperature. Temperature can be considered a macroscopic measure of the microscopic property—the average kinetic energy of the molecules of the gas. Thus an absolute zero of temperature would be that temperature at which all random molecular motion ceases—that is, the molecules no longer possess any kinetic energy. Moreover, according to this definition of temperature, all gases at the same temperature must have the same average molecular kinetic energy.

The square of the average velocity is proportional to the temperature of gas molecules.

Because molecules of different gases have different masses, their speeds may differ even though their kinetic energies are the same. But for any particular gas, with its particular fixed molecular mass, the square of the average velocity is directly proportional to the absolute temperature:

$$(\bar{v}^2) \propto T$$

In its common meaning, temperature is a measure of the degree of hotness or coldness of an object. But to ask for the temperature of an individual molecule is meaningless; it is only the speed (or its kinetic energy) that is significant. There is no such thing as a hot or cold molecule, only a fast or slow one.

Early in the nineteenth century, the Italian chemist Amadeo Avogadro (1776–1856) proposed that equal volumes of gases under the same pressure and temperature have equal numbers of molecules, regardless of the gases being compared. (For a discussion of Avogadro's hypothesis, see Chapter 13.) It remained for the kinetic theory to explain this hypothesis.

Suppose we have equal volumes of two gases—nitrogen and oxygen. If the pressures of both volumes are equal, then:

$$P_N V_N = P_0 V_0$$

According to the kinetic law, we can substitute an average energy expression for PV so that

$$\tfrac{2}{3} N_N \times (\tfrac{1}{2}mv^2)_N = \tfrac{2}{3} N_0 (\tfrac{1}{2}mv^2)_0$$

But if $T_N = T_0$, then the average kinetic energies are equal, leaving us with:

$$\tfrac{2}{3} N_N = \tfrac{2}{3} N_0$$

Therefore each gas contains the same number of molecules.

VELOCITY DISTRIBUTION Because the particles in a gas undergo billions of collisions each second, their velocities are constantly changing direction and magnitude. Under such conditions, the determination of an *average* velocity for such an enormous number of particles requires statistical methods of thinking, using the idea of probability. In 1859, the English physicist James Clerk Maxwell (1831–1879) provided just such a method. He treated the situation in which individual particles in a given volume of gas can have velocities that range from very low to very high, but most particles have velocities that are close to the average for the total volume. Such a condition is known mathematically as a normal distribution. If a graph is constructed that plots velocities against the number of particles moving at each velocity, the curve will show velocities distributed from lowest to highest, with most velocities close to an average, which will be at the peak of the curve. Such a curve is usually a slightly distorted bell shape.

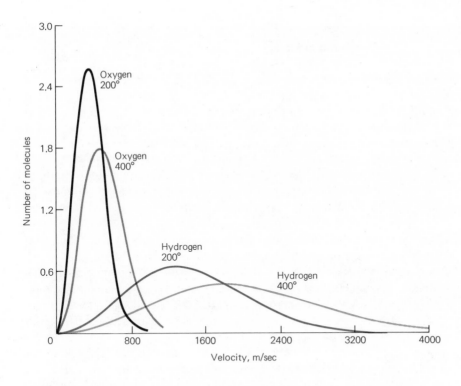

Figure 5-5
This Maxwell distribution plots the number of particles having a given velocity against velocity for oxygen and hydrogen, each at two different temperatures.

Maxwell showed that this average velocity is dependent on temperature. As the temperature increases, the peak of the curve, which represents the average velocity, shifts to the right, toward a higher velocity. So a distribution graph yields different distribution curves for different temperatures.

With Maxwell's velocity distribution, certain properties of liquids and gases can be explained. For instance, evaporation from the surface of a liquid occurs because certain molecules in a statistical distribution will have velocities far above the average (far to the right on the distribution curve), which permits them to break whatever forces hold them to other molecules at the surface.

Kinetic Theory of Heat

The disappearance of large-scale kinetic energy is usually accompanied by the appearance of heat, a rise in temperature. For example, when a car is brought to a sudden stop, the brakes heat up; a nail struck repeatedly by a hammer heats up; rub your hands vigorously together and the skin heats up quickly.

It appears, then, that heat is a form of energy. Since the addition of heat to an object causes a rise in temperature (which our theory takes to be a measure of average molecular kinetic energy), then it seems reasonable to assume that heat, like temperature, is related to molecular motion. And this is plausible, since temperature is a measure of heat intensity.

Figuring the Odds

Once, when the world's greatest expert on gambling, John Scarne, was in need of money, he asked permission to play at blackjack tables in Las Vegas—promising to leave after winning $1000. Permission granted, Scarne left the casino 15 minutes later with the money. Although the number of possible hands in blackjack is astronomical and the odds are tilted in favor of the casino, Scarne's incredible command of the probabilities involved at each point of the game permitted him to win. For this reason, he was banned from playing at all the major casinos in the world.

In one sense, probability can be defined as a ratio: how many times can a given event occur out of a total of possible alternatives? For instance, there are two possible alternatives to be realized when tossing a coin—heads or tails. One such event excludes the other, but each event has a 50 percent chance of occurrence. If 1 represents absolute certainty, or total probability, then the chance of tossing a head is $\frac{1}{2}$—that is, one chance out of two possibilities. The same is true for tails.

On the other hand, the probability of getting either a head or a tail on a single toss is 1 unless the coin lands on its edge. This obvious result is arrived at by adding the probabilities for each possibility: $\frac{1}{2} + \frac{1}{2} = 1$.

The probability of tossing heads twice in a row is another matter. It is determined by multiplying the probabilities for each event: $\frac{1}{2} \times \frac{1}{2} = \frac{1}{4}$. The probabilty of tossing 50 heads is arrived at by multiplying the probability for the single event, $\frac{1}{2}$, by itself 50 times. The probability for its occurrence is 1/1,000,000,000,000,000—for all practical intents and purposes an impossible event. This does not mean, however, that the result could not happen in the very first series of 50 tosses—only that the likelihood of it happening is very very small.

Much more complicated situations than simple coin tossing are commonplace. For instance, choosing a king out of a 52-card deck has a probability of 4/52. After one has been picked, the chance of getting another is 3/51; the chance of picking four consecutive kings is $4/52 \times 3/51 \times 2/50 \times 1/49$, or about 1 chance in half a million tries.

Applying the laws of probability depends on knowing all the possible ways in which an event can occur. More important, however, is that probabilities really concern the outcome of events in the long run. That is, although the odds are even that a head or a tail will appear when a coin is tossed, 7 heads and 3 tails in 10 tosses is not unusual. But in a million tosses, this deviation from the given odds all but disappears. This factor, known as the law of large numbers, governs most applications of probability.

For instance, John Scarne records seeing a woman throw 39 straight winning numbers in a crap game at a casino in Puerto Rico; the odds against this happening are 956,211,843,725 to 1. But Scarne points out that millions of players have rolled dice billions of times in the last 50-odd years in America, so that throwing 39 winning numbers is by no means miraculous—such an event just bears out the laws of probability.

A common method of determining probability is through recorded numerical data derived from observation of recurring events, or statistics. For instance, actuarial tables can predict the number of car accidents in a given city over a year's time, but these figures are averages based on a large number of samples. As an individual driver, you are what may be called a random variable. That is, the insurance company generally has no way of knowing under what conditions you are driving—such as weather, time of day, type of traffic. But, you can be "averaged in" with millions of other drivers.

By the same token, statistics can be applied to physical events of scientific interest that can only be measured on a large scale. The end result of molecules rushing around inside a container is well known; pressure is exerted on the walls of the container. Since there are about 10^{19} molecules in 1 cc of a gas—each undergoing billions of collisions every second—the motion of an individual molecule is completely random, and there is no practical way of predicting its velocity. But the statistical approach can provide a good theoretical picture of the behavior of a gas, by viewing it as the statistical average of countless moving molecules, as in Maxwell's velocity distribution.

In 1798, Count Rumford (1753–1814), who was boring cannon for the government of Bavaria, observed the continuous production of heat during the process. He concluded that the vibrating, twisting motion of the boring tools was somehow transformed into heat in the form of internal motions of the particles of the cannon and tools. The energy of motion became heat.

Whenever a gas is heated, its temperature rises—an increase in average kinetic energy and therefore an increase in the distribution of random motions of the gas molecules. This kind of motion along a path is called **translational** motion. Gas molecules (and those of solids and liquids as well) can move in other ways too. Diatomic gas molecules, for example—those made of two atoms per molecule, such as H_2 or O_2—can rotate about a common center of mass, with **rotational** motion. Furthermore, the atoms of such a molecule can vibrate toward each other and then away, around some average position, just as two objects connected by a spring might; this is called **vibrational** motion. Thus, heat energy can be absorbed into translational, rotational, and vibrational motions, and perhaps other forms as well. However, only the translational kinetic energy affects the temperature of a gas in our simple model gas theory.

SPECIFIC HEAT CAPACITY Bodies of equal mass may have different numbers of molecules, for some molecules are heavier than others. A gram of iron contains a small number of massive molecules, whereas a gram of water contains a much larger number of lighter molecules. If the same amount of heat energy is given to a gram of water and a gram of iron, there will be more energy on the average for each molecule of iron, because they are fewer in number. The iron molecules will acquire a higher average kinetic energy, and so the iron must reach a higher temperature than the water.

The amount of heat, H, required to raise the temperature of 1 gram of a substance by 1° centigrade is called the **specific heat capacity**, (c), of the substance. In fact, this property is used to define the calorie, a unit of heat energy. One **calorie** is the amount of energy that will raise 1 g of water by 1°C; so the specific heat of water, by definition is 1 cal/g°C. In order to calculate the heat needed to produce a given temperature change in a body, we use this equation:

$$H = mc(T_2 - T_1)$$

To raise the temperature of 30 g of iron (iron has a specific heat capacity of 0.11 cal/g°C) from 15°C to 35°C requires:

$$H = mc(T_2 - T_1)$$
$$= 30 \times 0.11 \times 20 = 66 \text{ cal}$$

To raise the temperature of the same quantity of water the same number of degrees requires 600 cal.

Since the temperature of any substance measures the average kinetic energy of its particles, we might suppose that all substances have the same heat capac-

Heat is a manifestation of the translation, rotational and vibrational motion of molecules.

Figure 5-6
Diatomic molecules such as oxygen or hydrogen exhibit three possible kinds of motion: translational (A), vibrational (B), and rotational (C). Temperature is affected only by the kinetic energy of translational motion.

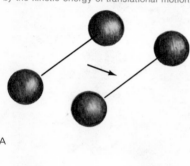

A

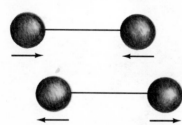

B

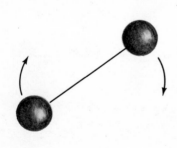

C

ity, when we consider equal numbers of particles. This is true for gases composed of single atoms. But when molecules consist of several atoms (water, for example), heat energy can be absorbed in vibrational and rotational motion as well as translational motion. In such cases, when heat energy is added, only a portion of the total heat added will increase the average kinetic energy (and hence the temperature); the rest will produce other types of motion that do not affect temperature. Therefore the heat capacity of a complex substance is always higher than that of a simple substance. More heat is needed to raise its temperature by a given amount than is needed for an equal mass of a simpler substance.

Substances composed of complex molecules have higher heat capacities than substances of simpler molecules.

Kinetic Theory of Liquids

When two gas particles approach each other at low relative velocity, the van der Waals forces between them (which increase as they come closer) may be strong enough to hold them together for a time. The association is not strong, and a fast third particle colliding with the pair may easily knock them apart again. While they are associated, though, their combined van der Waals force may capture yet other particles. When enough particles are loosely bound together in this way, a droplet of liquid results.

In this model, the difference between a gas and a liquid is simply the difference in the average distance between particles. In a gas the distance is large, and van der Waals forces are thus almost negligible. In a liquid the average dis-

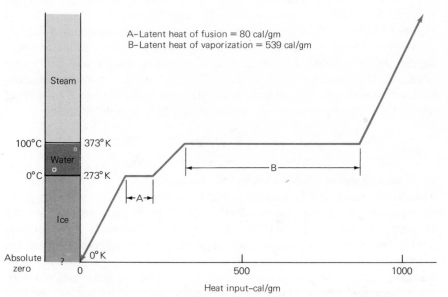

A–Latent heat of fusion = 80 cal/gm
B–Latent heat of vaporization = 539 cal/gm

Figure 5-7
The three changes in the state of water are shown on this graph. Note that no temperature change takes place at A—where 80 cal/g are absorbed by ice in changing to water, or at B—where 539 cal/g are absorbed by water in changing to steam.

Van der Waals forces are greater
between liquid molecules than
between gas molecules because the
molecules of a liquid are more tightly
bound than those of a gas.

tance is relatively small, and thus these forces are much greater. The forces are not strong enough, however, to hold two particles together permanently. A particle that escapes from one set of neighbors in a liquid is almost immediately captured by another set; thus particles of liquid slide over each other, so that the liquid has no definite shape but does have a fixed volume. Since the particles of a substance are closely packed when it is in the liquid state but well separated when it is a gas, the density of the substance (its mass per unit volume) is greater in the liquid state than when it is a gas.

A particle approaching the surface of a liquid is in an unbalanced situation. At the surface there are attractive forces pulling downward, but none pulling it up. The result is that it experiences a net van der Waals force that draws it back toward the interior of the liquid. There are also lateral forces pulling particles sideways at the surface. This combination stretches the surface like a drum head, producing surface tension. Fast-moving particles can break through this surface barrier, escaping from the liquid; slow-moving particles cannot.

The particles of a liquid, then, are loosely bound together and are in constant motion. As in a gas, the average kinetic energy of the particles is proportional to the temperature. Raising the temperature of a liquid increases the average kinetic energy of the particles; remembering the spread of slow and fast molecules in a velocity distribution, we see that a high temperature increases the rate at which fast particles evaporate across the surface of the liquid.

Kinetic Theory of Solids

The fact that it is so difficult to compress a solid or to change its shape suggests that in the solid state, particles of matter are closely and rigidly bound together. Studies of solids using x-ray diffraction (for a discussion of this technique, see Chapter 14) have confirmed this hypothesis. Most solids are held together by strong electrical forces in rigid geometric shapes, with a relatively great density of particles per unit volume. This microscopic structure is responsible for many of their macroscopic properties: the ability to retain a permanent shape; the tendency to break into geometric crystals; a slight elasticity.

Changes of State
Melting

To change the state of solid matter,
heat energy must be added to
overcome the attractive forces
between particles; during the change
of state, however, there is no change
in temperature.

When heat energy is added to a solid, the average energy of the individual particles is increased, and the temperature of the solid rises. As the particles move more vigorously about their fixed average positions in the solid structure, the spacing between particles is increased and the solid expands. If we continue adding heat energy and raising the temperature of the solid, some particles acquire enough energy to overcome the attractive forces and they move about; the solid starts to melt. At this point, adding more heat energy does not raise the temperature of the solid any further; it simply causes more melting. The heat energy required to convert a solid to a liquid, without changing its temperature, is called the **latent heat of fusion.** Exactly the same amount of heat energy must be extracted from a liquid to convert it back to a solid. For water, the latent heat of fusion is 80 calories per gram.

Evaporation and Boiling

If we partially fill a closed container with a liquid, there will be a continual exchange of particles between liquid and gaseous states. Liquid particles that have high enough kinetic energies evaporate, crossing the liquid surface to join the gas particles in the space above. Gas particles strike the surface of the liquid and are captured, becoming part of the liquid. The system will come to **equilibrium,** a state in which the separate quantities of liquid and gas remain constant, when the rate at which particles escape from the liquid is the same as the rate at which gas particles are captured by the liquid. In the equilibrium state, the temperatures of the liquid and the gas must be the same.

As we add heat energy to the system, we raise the average kinetic energy of the particles of both gas and liquid. Raising the average energy of the liquid particles increases the rate at which they escape, and the system then reaches a new equilibrium, with more gas and less liquid than before.

As the equilibrium shifts with the addition of heat, the pressure of the vapor increases. The temperature at which the vapor pressure of evaporation equals the atmospheric pressure is the **boiling point** of the liquid. At this temperature, the molecules have enough average energy to form bubbles of vapor inside the liquid. These rise to the surface where the vapor escapes, along with the vapor that forms directly on the surface.

The boiling point of a liquid obviously depends upon external pressure. At sea level, at a pressure of one atmosphere, water boils at 100°C. At low atmospheric pressure—such as occurs at the top of a mountain—the boiling point is reduced. Adding more heat energy to boiling water does not raise its temperature, but only increases the energy of the particles. Thus, it converts more water to steam. If we want to cook something in water at a higher temperature than 100°C, we must use a pressure cooker, which isolates the heated water from the atmosphere of the open air. By increasing the pressure, the boiling point of water is raised, and therefore the cooking temperature is increased. It has the same effect as taking the pot to a place well below sea level, where the atmospheric pressure is much greater.

The heat energy required to change a liquid to a gas without increasing its temperature is called the **latent heat of vaporization.** For water, at atmospheric pressure, the latent heat of vaporization is 540 calories per gram. This same energy must be extracted, as the latent heat of condensation, to convert the vapor back to a liquid. (By way of contrast, the latent heat of vaporization for ethyl alcohol is 204 calories per gram and the latent heat of vaporization for sulphuric acid is only 122 calories per gram.)

Figure 5-8
A liquid placed in an evacuated chamber begins to evaporate. As the molecules of vapor collect above the liquid, the vapor pressure increases. Once equilibrium is reached, the vapor pressure can be measured by the height (AB) of the mercury in the U-tube.

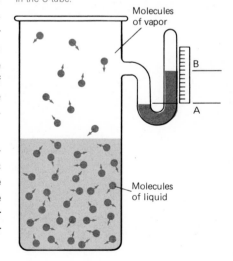

Molecules of vapor

B

A

Molecules of liquid

Thermodynamics

Thermodynamics is concerned with the relationships in energy transformations.

Thermodynamics is a branch of physics concerned with the relationship between heat and work, including all energy transformations and relationships. From a thermodynamic viewpoint, **work** is a transfer of energy between a system (such as a confined gas) and its surroundings. **Heat** is considered the energy *content* of a substance, a system, or its surroundings.

When heat is added to a gaseous system, it becomes energy of translational, rotational, or vibrational molecular motions. But in thermodynamics we are not concerned with molecular considerations; thermodynamically, the energy of a system is simply referred to as its internal energy. Since work is a method of energy transfer, thermodynamics treats *changes* in the internal energy of a system.

First Law of Thermodynamics

Consider a gas in a cylinder fitted with a piston. If a force is exerted on the piston, it will compress the gas and decrease its volume. Energy is transferred to the gas and its internal energy increases. To reverse the situation, suppose that the gas is heated and expands, raising the piston. Now work is done by the gas, and its internal energy decreases.

Energy cannot be created nor destroyed, but can be transformed.

Since this process must be consistent with the law of conservation of energy (discussed in Chapter 4), an accounting must be made of all the energy transferred in the process. Let ΔU be the change in internal energy of the system, ΔH the net transfer of heat (heat added less heat subtracted), and ΔW the net work done on or by the system. Then:

$$\Delta U = \Delta W + \Delta H$$

This is a thermodynamic statement of the law of the conservation of energy, called the **first law of thermodynamics**. It says that no energy is ever lost or destroyed in the change from one form to another.

Second Law of Thermodynamics

However, a law of conservation is not enough to explain energy changes. An egg can be scrambled but not unscrambled, although conservation of mass and energy would be satisfied in unscrambling. When an ice cube dropped in a glass of water begins to melt, the water cools. The process will never go the other way, with the cube cooling down, while the water warms up; it is one-way and irreversible.

No heat engine is 100 percent efficient. Energy is always lost to the surroundings.

Through experiments with steam engines in the eighteenth century, it was suspected that the conversion of work to heat and heat to work were not entirely reversible or efficient. The French engineer Sadi Carnot (1796–1832) explored the efficiency of heat engines in converting heat of burning fuel to mechanical energy by boiling water and using the expansion of steam to do the work. He sought to determine whether such a conversion could ever be 100 percent efficient. (In thermodynamics, efficiency is measured by work output compared to heat input.) By analyzing idealized engines, Carnot proved that the efficiency must always be less than 100 percent; the total heat input *always* exceeds the useful energy output. In modern steam engines, the maximum possible efficiency is about 35 percent.

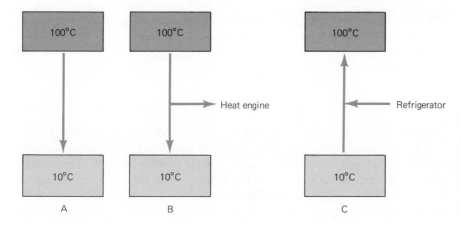

Figure 5-9
Heat can be considered as energy in transfer. It will flow of its own accord from a hot reservoir into a cold one (A). Part of this energy can be converted into work by a heat engine, while the rest is discharged to the cold reservoir (B). To transfer heat from the cold reservoir to the hot reservoir, work must be added to the system (C); a refrigerator performs this work.

A steam engine is connected to a heat reservoir, or system whose temperature remains constant when heat is added to or taken from it. A steam boiler, if constantly supplied with fuel, can give up heat to a steam engine without any appreciable change in temperature; it is, therefore, a hot reservoir. The air surrounding the steam engine can absorb heat without change in temperature; it is, therefore, a cold reservoir. A large lake can exchange heat with its surroundings—absorbing heat in the morning and giving it up in the evening—without appreciable change in temperature over a day's time, and thus it too is a heat reservoir, which tends to moderate any drastic swings of temperature in the surrounding land mass.

Clausius stated the second law of thermodynamics in this way:

No process is possible whose sole result is the removal of heat from one body at low temperature and its delivery to another body at a higher temperature.

Or to put it in more familiar terms: of its own accord heat cannot travel from a cold body to a hot body. A clearer statement of the implications of the second law came from Lord Kelvin (1824–1907), who said:

No process is possible whose sole result is the removal of heat from a single reservoir and the performance of an equivalent amount of work. The energy available for work is continually decreasing, the difference being changed into heat.

According to this thermodynamic law, heat cannot be fully converted into work. A heat engine that absorbs heat from a reservoir at one temperature will perform mechanical work, but it must reject some portion of the heat to a reservoir at a lower temperature or in frictional loss. An automobile engine is a good example. Heat at a high temperature provided by burning gasoline is partly transformed to useful work; the remaining heat is rejected through the exhaust to the surrounding air.

Figure 5-10
A steam locomotive is a self-contained unit that carries its own water supply for generating steam, which drives the piston that moves the wheels, and its own fuel (wood, coal, or oil) to heat the water. Most steam locomotives had a thermal efficiency of less than 6% and were replaced by diesel engines, which are about 4 times as efficient.

Work must be done in a system in order to transfer heat from a cold reservoir to a hot one.

The second law also implies that work must be done on a system in order to transfer heat from a cold reservoir to a hot one; this is precisely the function of a refrigerator. The reverse process—transfer of heat from a hot reservoir to a cold one—takes no work.

Entropy and Disorder

Consider two equal volumes of different gases at the same temperature and pressure in two compartments of a container separated by a partition. These gases make up an ordered system; each volume contains only one kind of molecule. When the partition is removed the gases mix, disordering the original arrangement. This new disorder will remain, because the gases will never spontaneously separate. The gases will stay in the less structured state, which is a more stable one due to the lower amount of potential energy.

Now suppose that we consider instead two equal volumes of water, one cold and the other hot. Because heat energy flows from a hot reservoir to a cold one, a heat engine can be operated between them to extract work from this transfer of heat. Yet the flow of energy cannot continue indefinitely, for sooner or later the hot reservoir will cool (as the heat energy leaves it) and the cool reservoir will heat up (as the heat energy flows into it). Eventually the two volumes of water will be at the same temperature; then no more work can be done. There has been no change in the internal energy of the system, but there has been an increase in disorder as compared to the original hot and cold arrangement of the system. It is the disorder that prevents the performance of work. We can say that the opportunity to perform work has been irretrievably lost.

All natural processes tend toward a state of disorder in which the energy available for work is irretrievably lost.

The degree to which the energy of an isolated system becomes unavailable for useful work can be measured by the amount of disorder, and the measure is called **entropy**. Clausius, who introduced the concept of entropy, stated the second law of thermodynamics by saying that the entropy of an isolated system spontaneously tends to increase.

This means that all natural processes suffer an irreversible loss in the amount of energy available for work and thus an increase in entropy. If two reservoirs in a system are at the same temperature, there can be no net transfer of heat between them. Because the temperature of the system is uniform, whatever transfer of energy occurs can only take place through random (disordered) molecular motion. Hence, despite the uniformity of being at the same temperature, such a system is very disordered. Its entropy is at a maximum, since none of the energy is available for work.

Lord Kelvin predicted that all objects in the universe would eventually reach the same temperature by the mixing process of heat transfer, radiation, and particle emission. In reaching such a state of uniformity, the energy in the universe would become completely disordered—scattered and unavailable for work. Thereafter, no natural processes dependent on temperature differences could occur, and the universe would enter a state of cold fixity, or "heat death."

The **boiling point** of a liquid is the temperature at which the vapor pressure of the liquid equals atmospheric pressure.

Brownian motion is the random motion of microscopic particles or dust suspended in a gas or a liquid.

The **calorie** (cal) is the amount of heat energy that will raise 1 g of water by 1°C. The **kilocalorie** (kcal) is the amount of heat necessary to raise 1 kg of water by 1°C.

Entropy is a measure of the disorder within a system.

Equilibrium between a liquid and its vapor is a state in which the number of liquid molecules escaping from the liquid surface is, on the average, equal to the number of vapor molecules being captured by the liquid.

Heat, according to the kinetic-molecular theory, is energy of random translational molecular motion. Thermodynamically, heat is the energy content of a substance.

A **heat reservoir** is a system whose temperature remains constant when heat is added to, or extracted from, it.

Latent heat of fusion is the heat necessary to produce a change between the solid and liquid state without causing any change in temperature.

Latent heat of vaporization is the heat necessary to produce a change between the liquid and gaseous state without causing any change in temperature.

The **melting point** of a substance is that temperature at which the substance changes from a solid to a liquid.

Specific heat capacity is the amount of heat necessary to raise the temperature of 1 g of a substance by 1°C. It is given as cal/g°C.

In the kinetic theory, **temperature** is a measure of the average kinetic energy of the molecules of a substance.

Thermodynamics is that branch of physics concerned with the transformation of energy. The **first law of thermodynamics** states that no energy is ever lost or destroyed in the change from one form to another. The **second law of thermodynamics** states that no process is possible whose sole result is the removal of heat from one body and its delivery to another body at a higher temperature.

The **vapor pressure** of a liquid is the pressure of a vapor above the liquid. It is also the pressure of the vapor bubbles formed in the liquid as it nears its boiling point.

The velocities of particles in a gas are constantly changing direction and magnitude. However, the **velocity distribution** of these particles in a given volume is a normal distribution; that is, the velocities will vary from very low to very high, but most particles will have velocities that are close to the average for the entire volume.

Exercises

1 A quantity of gaseous nitrogen is transferred from a 1-liter to a 2-liter container. According to the kinetic theory of gases, it is possible for the gas to settle in one corner of the container. Why has this never been observed?

2 How do we know that molecules in a gas are relatively far apart from each other?

3 A container with a volume of 12 in.^{3} is filled with helium to a pressure of 10 atm. What would be the effect of compressing the container to a final volume of 6 in.^{3} provided no gas escapes and the temperature remains the same?

4 What are van der Waals forces? What restrictions do they impose on the ideal gas law?

5 If, at 45°C, a 10 liter volume of gas is at a pressure of 3 atm., what would be the new pressure if the temperature is raised to 75°C and the volume is increased to 30 liters?

6 If a volume, V, of O_2 gas is compressed to $V/2$, how is the average velocity of the molecules affected, provided the temperature is constant?

7 Tank A contains 10 liters of He at 300°C. Tank B contains 10 liters of O_2 at 200°C. Which tank is more likely to contain particles with a higher average velocity?

8 By using the relationship of average kinetic energy to temperature, determine the ratio of average velocities of hydrogen and oxygen molecules in gases at the same temperature. Oxygen molecules are 16 times more massive than hydrogen molecules.

9 Why do people perspire in hot weather? How does the prespiration process cool the skin?

10 Why do we use isopropyl alcohol rather than water to rub down persons with high fever?

11 How does a liquid turn into a solid and vice versa? What process is common to both conversions? What determines whether water is found in a solid form or a liquid form at 0°C?

12 The specific heat of aluminum is .215 cal per degree while the specific heat of iron is .106. If you had to choose between the two for a pot handle, which would you choose and why?

13 How many calories are needed to raise the temperature of a 50 g aluminum bar from 10°C to 30°C? (Specific heat capacity of aluminum = 0.215 cal/g°C).

14 Which contains more heat—a 10-gallon bowl of water at 50°C or a 2-in. section of a thin iron wire at 1000°C?

15 How much heat is necessary to melt 20g of ice at 0°C into water at 0°C?

16 How much heat is required to convert the resulting 20g of water at 0°C into steam at 100°C?

17 If 1000g of water at 80°C is poured into an aquarium containing 2000g of water at 50°C, what will be the resulting temperature of the water inside the aquarium?

18 If the kinetic energy of the molecules is responsible for balancing the intermolecular forces, what generalization can be made about the boiling point temperature of a substance and the strength of its intermolecular forces?

19 List in tabular form the differences between the three states of matter in terms of intermolecular distances, van der Waals forces, and average kinetic energy.

20 Explain the difference between a refrigerator and a heat engine.

21 You are designing a powerplant which must meet the approval of ecologists who are seriously concerned about thermal pollution. To satisfy them, you must use the most efficient heat engine to power the generators. Your choice of engines is between a steam turbine, which uses steam originally at 1200°K and cools it to 373°K as it works, and a diesel engine which has a high temperature reservior of 1250°K and an exhaust temperature of 420°K. Assuming both engines are acceptable in other respects, which do you choose? Why? Give the efficiency for each engine.

22 Explain why perpetual motion machines cannot exist.

The earliest observations of electricity concern amber, which is solidified sap or resin from evergreen trees. If a piece of amber is rubbed on fur, it can attract to itself small bits of paper, straw, and other lightweight materials. About 500 BC, the Greek philosopher Thales recorded these facts about amber, which in Greek was called *elektron.*

Many other materials—glass, hard rubber, solid sulfur—were later discovered to possess the same power of electrical attraction after rubbing, apparently due to friction. Despite these discoveries, the amberlike property of these materials remained a mere curiosity for another two thousand years. But by the eighteenth century in Europe, explanations were possible. Theories based on the supposed existence of fluids were utilized. Such imaginary fluids had been invoked to explain many phenomena. For example, heat was thought to be a fluid,

Figure 6-1
William Gilbert, shown here demonstrating electrostatic attraction to Queen Elizabeth I, studied electricity and magnetism during the sixteenth century. He was the first person to use the word *electricity*.

named *caloric,* that flowed from warmer to cooler bodies; chemists believed that when an object burned, a fluid called *phlogiston* was released, diminishing the mass of the object. A fluid theory of electricity seemed a logical next step.

Electric Fluid Theory

According to the fluid theory, electricity was a weightless fluid that could pass from one body to another, particularly while the bodies were being rubbed together. When the two bodies were separated after the electric fluid flowed between them, the bodies would attract each other. But two pieces of amber that were both rubbed with fur were found to repel each other. Would one fluid be enough to explain both attractive and repulsive forces?

The French scientist and head of the royal French gardens Charles DuFay (1689-1739) made some careful observations concerning this question and concluded that there must be *two* electric fluids. In his experiment, DuFay rubbed a glass rod with silk and then brought the electrified rod near a tiny bit of cork. As expected, the cork was drawn to the rod. But after the cork actually touched the glass rod, it flew away. The attractive force had suddenly changed to a repulsive force. DuFay then rubbed a piece of amber (hard rubber or plastic would have done as well) with cat's fur. Again, the cork was initially attracted to it and then, after making contact with the amber, repelled by it. Two

Two electrical fields were presumed, at first, to be responsible for the observable electrical phenomena—mainly repulsion and attraction.

pieces of cork that had both been touched by the same electrified glass rod always repelled each other. Similarly, pieces of cork that had both been touched by the amber rod repelled each other. But if one piece of cork was touched by the amber rod and another piece was touched by the glass rod, then these two pieces attracted each other.

From these observations, DuFay concluded that there are two electric fluids. Each fluid apparently repelled itself but was attracted by the opposite fluid.

Franklin's Work

Just a few years after DuFay proposed his two-fluid theory, the American statesman and scientist Benjamin Franklin (1706-1790) gave a full explanation of all of DuFay's observations by assuming that only *one* fluid was present in all bodies. Every object had a natural capacity for this electric fluid, according to its size and the kind of matter it was made of. When the fluid was present in a normal amount, a body was electrically neutral and produced no electric phenomena. When two bodies were rubbed together, some of the fluid could flow from one to the other. The total amount of electric fluid present in the bodies did not change, but one body now had an excess and the other an equal and opposite deficit. Franklin described the body that lost fluid and had a deficit as "negatively charged." The body that gained fluid and had a surplus he described as "positively charged." We still use Franklin's designation for the sign of electric charge. However, the words *positive* and *negative* are arbitrary; they do not mean surplus and deficit.

Franklin assumed the existence of one fluid; a deficit or excess of the fluid resulted in the observed phenomena.

Electric Charge

A decisive step in understanding the nature of electric charge was taken experimentally about 150 years after Franklin proposed his theory. We know now that there are two kinds of charge: the negative charge—the main carrier of which is the subatomic particle, the electron—and the positive charge—the main carrier of which is another subatomic particle, the proton. Objects that are electrically neutral have equal numbers of negative and positive charges. In solids, the negatively charged electrons move much more readily and account for most of the phenomena observed in early electrical research.

Priestley postulated that electrical forces of repulsion and attraction obey an inverse square law analogous to that of gravitation.

Many of the aspects of electrical phenomena were discovered experimentally and described mathematically, without assuming anything about the nature of charge other than that it could be positive or negative. For instance, Franklin observed that although a piece of cork was strongly attracted to the outside of a charged metal can, it experienced no such force when lowered, at the end of a silk thread, into the can. Unable to explain this result, Franklin asked Joseph Priestley (1733-1804) to repeat the experiment. Priestley confirmed Franklin's results; he then recalled that Newton had proved that a gravitational force obeying the inverse square law would give the same null result inside a hollow planet; all the gravitational forces would cancel. Priestley argued by analogy to gravitation that the force of attraction or repulsion between electric charges also obeys an inverse square law. No assumption about the nature of the charge was necessary.

Coulomb s Law

The first direct measurements of the force between two charges were made in 1785 by the French physicist Charles Coulomb (1736-1806). To measure the force, Coulomb used a sensitive spring scale with a torsion balance. The arm of the balance was suspended by a fiber so delicate that even a very small force on the arm could twist it. The greater the twisting force, the greater the angle of twist; by measuring that angle on a graded or calibrated scale, Coulomb was able to measure the force that twisted the fiber. (Cavendish performed similar experiments with masses some thirteen years later.)

Coulomb's experimentally discovered inverse square law for electrical forces is analogous to Newton's gravitational law. Electrical forces, however, are both attractive and repulsive.

In his experiment, Coulomb attached a charged sphere to one end of the balance arm and an equally heavy counterweight to the other end. A second sphere, with the same charge, was brought near the sphere on the balance, which was repelled, causing the arm to rotate and the fiber to twist. The closer the second sphere was brought, the more the arm rotated, increasing the angle of twist.

Coulomb observed two things:

1   When he increased the amount of charge on either sphere, the force also increased in direct proportion to the product of the charges. If one of the charges is halved, the force is halved; if both charges are halved, the force is reduced to a quarter of its original value.
2   When he increased the distance between the spheres, the force decreased as its square; in other words, the force was inversely proportional to the square of the distance between the charges.

We state this mathematically in **Coulomb's law:**

$$F = k\frac{(q_1 q_2)}{r^2}$$

The letters q_1 and q_2 represent the charges, and k is a constant of proportionality whose numerical value depends on the units chosen to measure charge, distance, and force.

Coulomb's law is closely analogous to Newton's law of gravitation,

$$F = G\frac{(m_1 m_2)}{r^2}$$

with charges q_1 and q_2 replacing masses m_1 and m_2. There are two differences. First, gravitational forces are always attractive, whereas the electric force is attractive when the charges are unlike and repulsive when the charges are alike. Second, within the atom, electrical forces are enormously more powerful than

Figure 6-2
Two equally charged spheres in Coulomb's torsion balance repel each other. The repelling forces cause the balance arm to rotate; this in turn twists (produces a torsion in) the delicate fiber from which the balance is suspended. From the angle of twist, the force can be determined.

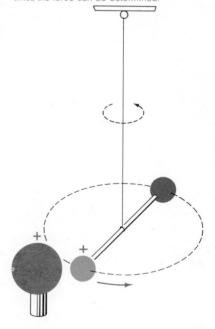

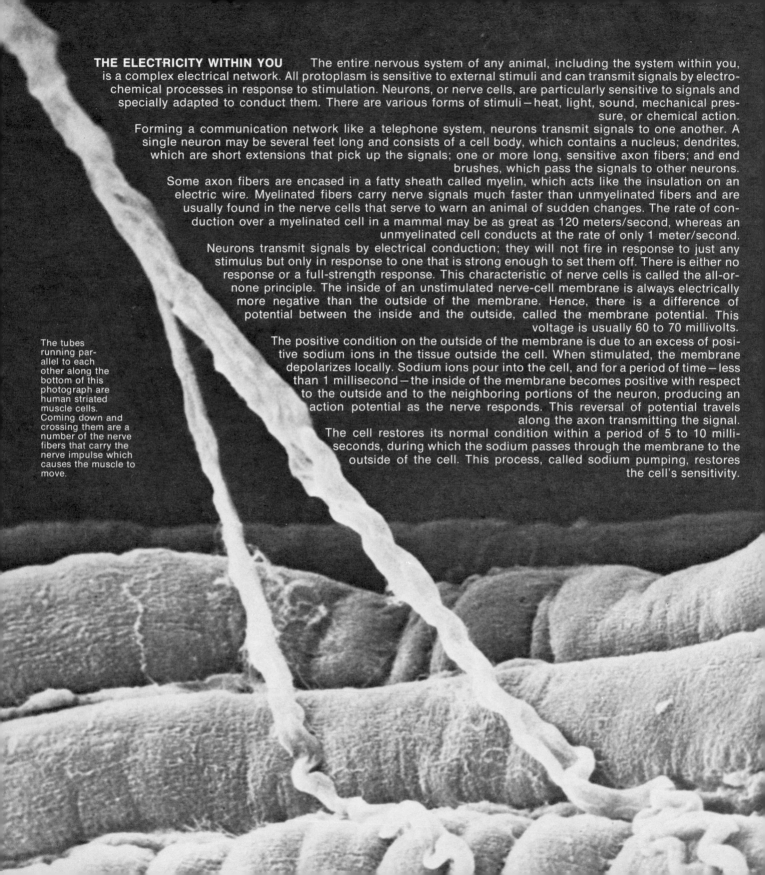

THE ELECTRICITY WITHIN YOU The entire nervous system of any animal, including the system within you, is a complex electrical network. All protoplasm is sensitive to external stimuli and can transmit signals by electro-chemical processes in response to stimulation. Neurons, or nerve cells, are particularly sensitive to signals and specially adapted to conduct them. There are various forms of stimuli—heat, light, sound, mechanical pressure, or chemical action.

Forming a communication network like a telephone system, neurons transmit signals to one another. A single neuron may be several feet long and consists of a cell body, which contains a nucleus; dendrites, which are short extensions that pick up the signals; one or more long, sensitive axon fibers; and end brushes, which pass the signals to other neurons.

Some axon fibers are encased in a fatty sheath called myelin, which acts like the insulation on an electric wire. Myelinated fibers carry nerve signals much faster than unmyelinated fibers and are usually found in the nerve cells that serve to warn an animal of sudden changes. The rate of conduction over a myelinated cell in a mammal may be as great as 120 meters/second, whereas an unmyelinated cell conducts at the rate of only 1 meter/second.

Neurons transmit signals by electrical conduction; they will not fire in response to just any stimulus but only in response to one that is strong enough to set them off. There is either no response or a full-strength response. This characteristic of nerve cells is called the all-or-none principle. The inside of an unstimulated nerve-cell membrane is always electrically more negative than the outside of the membrane. Hence, there is a difference of potential between the inside and the outside, called the membrane potential. This voltage is usually 60 to 70 millivolts.

The positive condition on the outside of the membrane is due to an excess of positive sodium ions in the tissue outside the cell. When stimulated, the membrane depolarizes locally. Sodium ions pour into the cell, and for a period of time—less than 1 millisecond—the inside of the membrane becomes positive with respect to the outside and to the neighboring portions of the neuron, producing an action potential as the nerve responds. This reversal of potential travels along the axon transmitting the signal.

The cell restores its normal condition within a period of 5 to 10 milliseconds, during which the sodium passes through the membrane to the outside of the cell. This process, called sodium pumping, restores the cell's sensitivity.

The tubes running parallel to each other along the bottom of this photograph are human striated muscle cells. Coming down and crossing them are a number of the nerve fibers that carry the nerve impulse which causes the muscle to move.

gravitational forces. Thus, in the hydrogen atom (which consists of a positive proton and its negative electron orbiting like a planet around it) the force of electric attraction between the proton and electron is about 10^{39} times stronger than the force of gravitational attraction between them.

The most commonly used unit of electric charge is named the **coulomb**, and it is a rather large unit. For example, the charge on one electron is only 1.6×10^{-19} coulomb. So it takes about 6.25×10^{18} electrons to make one coulomb of negative charge. On the other hand, there are about 6×10^{23} electrons in a gram of hydrogen. If all these electrons could be collected (and if they were not neutralized by neighboring positive charges on the hydrogen atomic nuclei), they would add up to about 96,000 coulombs of negative charge.

If, in Coulomb's equation, force is given in newtons, charge in coulombs, and distance in meters, then the constant of proportionality, k, can be determined experimentally. Its value is:

$$k = 9 \times 10^9 \ \frac{\text{newton-meter}^2}{\text{coulomb}^2}$$

Using this value, we can calculate the force of repulsion between two electrons a billionth of a centimeter apart (a substantial distance on the atomic scale):

$$F = \frac{9 \times 10^9 \ \text{newton-m}^2/\text{C}^2 \times 1.6 \times 10^{-19} \ \text{C}^2}{(10^{-11}\text{m})^2} = 1.4 \times 10^{-6} \ \text{newtons}$$

Conductors and Insulators

Materials, such as copper wire, through which charge flows freely are called **conductors**. Those through which charge does not flow readily, such as rubber, are nonconductors, or **insulators**. Metals are good conductors because they contain a great number of electrons that are not tightly bound by electrical forces to atoms in the metal. These free electrons move readily throughout the material, almost like a gas or liquid flowing through a pipe or channel. Such an assembly of electrons is called an electron gas (see Chapter 14). Nonmetals have very few such free electrons. Their charges are tightly bound to the atoms. For this reason, nonmetals are usually good insulators.

Insulators or conductors are usually determined by the lack or abundance of free electrons they contain.

A perfect conductor would be one in which there would be no resistance to the flow of charge. A perfect insulator would be one through which no charge could flow at all. This state of perfection does not normally exist for either category. Rather, there are only relatively good conductors and relatively good insulators—the degree of each being a measurable quantity. However, under conditions of extremely low temperature (close to the absolute zero), with very little heat and energy in a substance, some metals lose practically all resistance to flow of charge and become superconductors.

DISTRIBUTION OF CHARGE Excess charge placed on an insulator tends to remain localized. In a conductor, such an excess charge will be distributed

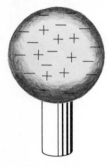

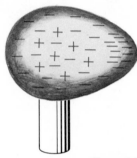

over the surface, because the charges are free to move. The way in which the charge distributes itself depends on the geometry of the conductor. For instance, on a hollow spherical conductor the charge will be distributed evenly. On an egg-shaped oval conductor there will be a greater concentration of charge on the narrow end. If the narrow end is sharp enough to cause high concentration, charge can leak off into the gas particles of the atmosphere. The more pointed a charged conductor is, the greater will be the concentration of charge at the point, causing charge to leak off more rapidly, and overcoming the insulating properties of air.

The knowledge of this property of pointed conductors led Franklin to the invention of the lightning rod, a pointed metallic rod placed on the highest parts of buildings. Because the earth is a good conductor and acts as a huge reservoir for absorbing or supplying charge, the rods are connected directly to it. During thunderstorms, clouds become electrically charged. If a cloud is positively charged, it attracts a large negative charge at the earth's surface. If the charge becomes large enough, the attractive forces between cloud and earth are strong enough to overcome the insulating characteristics of the air in between, and there is a huge rush of charge from the earth to the cloud, resulting in a flash of lightning. The cloud may be negative; then, lightning flows down to earth. The cause of lightning is quite similar to the cause of the spark you feel when you touch a brass doorknob after rubbing your feet on a rug.

Figure 6-3
For a charged sphere, the distribution of electrical forces of repulsion is uniform, but for an egg-shaped conductor, the charge accumulates closer to the pointed end.

Figure 6-4
600 flashes of lightning occur every second somewhere on earth. At its peak the power of a lightning stroke can reach a billion kW, which heats the path of the bolt to nearly 60,000°F, a temperature six times hotter than the surface of the sun. The building (left) being struck by lightning is a model used in lightning-rod research.

A rubber comb rubbed with fur will attract bits of paper. The comb has been negatively charged, so it would appear that the paper must be positively charged to be attracted to the comb. Yet the paper has not been rubbed, and it has no net charge; it is electrically neutral. Based on the equation for Coulomb's law, if one of two charges is zero, the product $q_1 q_2$ is zero, and the force of attraction or repulsion between the two charges is also zero. How, then, can the attraction of the comb to the paper be explained?

Suppose that a negatively charged rod is placed in the vicinity of a metallic sphere mounted on a nonconducting base. The negative charges in the rod will repel the negative charges in the sphere and also attract the positive charges. The sphere will essentially be divided into two regions—positive charges on the side closest to the rod and negative charges on the opposite hemisphere. The positive charges are closer to the negative rod than the negative charges; by virtue of the inverse square relationship, the force of attraction will be greater than the force of repulsion because the opposite charges are closer to one another. If the sphere were free to move, it would move toward the rod.

No overall net charge exists on the sphere. But by attraction and repulsion the charged rod has induced a separation of charge on the sphere, giving each hemisphere its region of charge. This process is known as **electrostatic induction**. If the rod is removed, the charges on the sphere will attract each other and make the sphere neutral again.

Electrostatic Induction

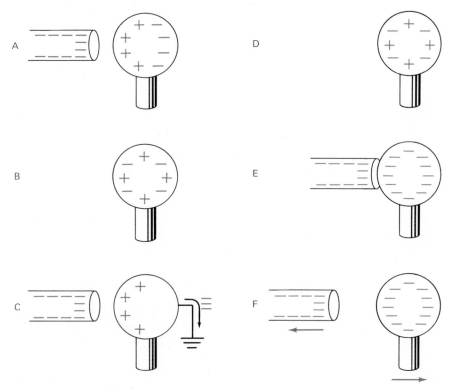

Figure 6-5
Left: A charged rod brought close to, but not in contact with, an uncharged conductor causes a separation of charge based on repulsion and attraction for the charge on the rod (A). If the rod is removed, the charge redistributes over the conductor according to its shape (B). If the conductor is temporarily grounded in the presence of the rod, the conductor assumes an *opposite* charge from that of the rod (C).
Right: A *similar* charge to that of the rod is produced on the uncharged object (D) by bringing the rod into direct contact with it (E). In this case (F), the rod and the object repel each other.

If, however, the sphere were connected to the earth (grounded) by a conductor, then, because of repulsion, negative charge would flow off the negatively charged hemisphere into the ground. If the rod is removed, after breaking the ground connection, the charges will once more redistribute. But now the sphere has too little negative charge, having lost its induced negatives to the ground. It has been given a net positive charge.

It should be noted that the induced charge is opposite to that of the inducing charge. The rod was negative, and the sphere became positive. A more direct method of charging is possible; the rod could have been brought into direct contact with the sphere, and they would share the charge. In this case, both the sphere and the rod would have a negative charge.

Force Fields   It is easy to see and understand what happens when a mechanical force is exerted on some object. It may take a lot of effort to lift a ton of bricks, but there is nothing mysterious about the process of using a winch, pulleys, and a rope attached to the container of bricks; the way in which the force is exerted on the bricks is clear to any observer. It is much harder to conceptualize the gravitational force that is also acting on the bricks. We know that it is there. Thanks to the work of Newton and the experiments of Cavendish, we can even measure the force accurately. But how does it cause an effect? The earth is not attached to the bricks by a rope; it does not send out waves like a radio transmitter; it does not even send out rays of any sort. Yet it exerts a concrete observable measurable force.

The concept of a force field accounts for forces that act at a distance.   To explain the mysterious forces—gravitational attraction, electrical attraction, magnetic attraction—that seem to be able to act on an object over a distance without any apparent physical connection or emission, scientists use the concept of the force field. A **force field** is simply a region of space in which an object will experience a measurable force. Thus we speak of the gravitational field of the earth, meaning the spatial area in which objects experience the gravitational pull of the earth; or the magnetic field of a bar magnet, meaning the surrounding space in which iron filings will experience the pull of the magnet.

Gravitational Field   The most familiar vector force field is the gravitational field, or region of space in which a mass is acted on by a gravitational force. The equation for the gravitational force is:

$$F = G \times \frac{Mm}{r^2}$$

where M is the mass of the earth, m the mass of the test body, and r the distance from the test body to the center of the earth. Suppose, however, that we want an expression for the strength of the field at a given point that is dependent only on the mass of the source (the earth) and not on the mass of the test body. This would be the field force per unit mass. If we divide both sides of the above equation by m, we get:

$$\frac{F}{m} = \frac{GM}{r^2}$$

The right side of this equation, independent of m, can be called g, the gravitational field intensity:

$$g = \frac{F}{M}$$

Note that this has the same form as Newton's second law, so g is also the acceleration due to gravity on any object m when F is the gravitational force on M.

By similar reasoning we can define an **electric field** as existing in a region of space, at any point of which a charge will experience a force. The direction of the field is the direction of the force on a positive test charge.

By Coulomb's law the force between two charges is:

$$F = k\frac{q_1 q_2}{r^2}$$

To determine the intensity of the electric field q_1 at the point occupied by q_2, we want an expression not containing q_2. If we divide both sides of the equation by q_2 we get:

$$\frac{F}{q_2} = \frac{kq_1}{r^2}$$

The expression on the right, which depends only on q_1 and r, is the electric field intensity (E).

$$E = \frac{F}{q}$$

The field intensity is measured in newtons/coulomb. A body carrying the charge of a million electrons, for example, at a point in an electric field where the field strength is 1000 newtons/coulomb, would experience a force of 1000 newtons/coulomb $\times$ ($10^6 \times 1.6 \times 10^{-19}$) coulombs $= 1.6 \times 10^{-10}$ newtons.

It was the great English experimental physicist Michael Faraday (1791-1867) who first applied the field concept to the forces produced by electric charges. Observers had previously noticed that when iron filings were sprinkled on a sheet of paper placed over a bar magnet, they oriented themselves along

Electric Field

Charges experience forces in an electric field. The ratio of force to charge is the field intensity.

Lines of Force

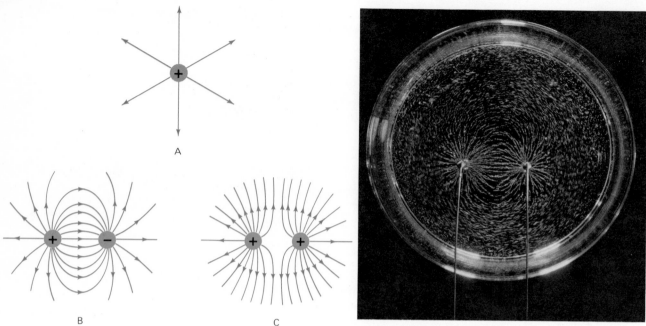

Figure 6-6
Diagrams of the field around: (A) a single positive charge; (B) two equal but opposite charges; (C) two equal and like charges. Note that the lines extend away from positive charges and toward negative charges. In the photograph, rods carrying equal and opposite charges are introduced in an insulating liquid on which grass seeds are floating. The pattern of the seeds makes the electric field visible.

curved lines running from one pole to the other; these lines were supposed to be lines of magnetic force. Faraday drew a line-of-force analogy between magnetism and electricity, and suggested that electric charges are also surrounded by lines of force.

Around a single isolated charge, the lines of force extend radially outward. Between a positive and negative charge, the lines of force curve between the charges in a manner similar to the lines between the poles of a bar magnet. Lines of force between like charges—two positive or two negative—appear to repel each other. No lines of force begin or end in the space around a charge, but only at the charge itself. Each line in the field is continuous, beginning at a positive charge and ending at a negative charge. Lines of force do not intersect. In Faraday's model, the number of lines of force crossing a unit area is proportional to the field intensity in that area; the density of lines closest to the charge producing the field indicates that the field is strongest at those points, which fits Coulomb's law.

Potential Difference

The gravitational field at the surface of the earth is directed down toward the center of the earth. To raise a test mass above the earth from a lower to a higher point requires that a force be exerted to overcome the force of gravity. If the force is exerted over a distance, work has been done—and energy has been expended—against the gravitational field. Because the gravitational field is conservative, all of the energy expended is stored as potential energy in the mass. This means that upon falling, the mass has the potential to do work equal to the work needed to raise it. The amount of work it can do is measured by the change in potential energy undergone by the mass as it was raised. This difference in potential depends on the mass, the gravitational field intensity, and the height that the mass was raised; the formula is *mgh.*

An analogous situation exists when work is done in moving a test charge from one point to another against the force of an electric field—for example, when a positive test charge is moved closer to the positive charge of the source of a field, against the repulsive force of that source. Just as a stone stores potential energy when it is placed on the edge of a roof, or a spring stores energy when its coils are compressed, the newly positioned electric test charge also stores energy by its change in position. Like a stone ready to drop, the charge is ready to move away along the lines of force.

Electrical potential is gained by a charge when work is performed in moving it against the force of an electric field.

The change in potential energy of the test charge due to its new position is dependent on the magnitude of the charge *q*, the intensity of the field, and the endpoints of motion. To get an expression that is independent of the charge *q*—potential per unit charge—we can divide the change in potential energy by *q*. The result obtained is the **electrical potential difference,** or ratio of the change in electrical potential energy of a charge *q* to the charge itself.

The symbol for potential difference *V* is given by the equation:

$$V = \frac{\Delta(PE)}{q}$$

where $\Delta(PE)$ is the change in potential energy. The unit of potential difference is called a **volt,** measured as a unit of energy per unit of charge.

1 volt = 1 joule/coulomb

Figure 6-7
Michael Faraday at work in his laboratory.

A volt represents one joule of work done on one coulomb of charge in moving between two points in an electric field. The potential difference is independent of the path taken between the endpoints but does depend on the position in the field of the endpoints themselves. If an electron, for example, is moved through a potential difference of 100 volts, the amount of work done is:

$100 \text{ V} \times 1.6 \times 10^{-19}$ coulombs $= 6 \times 10^{-17}$ joules

Change in electrical potential energy depends on the charge *q*, field intensity *E,* and the distance *d* moved through the field:

$$\Delta(PE) = qEd$$

but $\quad V = \dfrac{\Delta(PE)}{q}$

Therefore, substituting for $\Delta(PE)$ from the first equation:

$$V = \frac{qEd}{q} = Ed$$

or

$$E = \frac{V}{d}$$

Thus, the electric field intensity is expressed as volts per meter.

Currents and Circuits

The observable effects of electrical phenomena discussed so far are due to the static accumulation of electrons rather than to their movement. In order to move, electrons must be under the influence of an electric field—region where a potential difference can exist between points, the electrons moving from points of higher to lower potential. Since the direction of the field is the direction of the force on a positive test charge in the field, electrons (being negatively charged) will experience a force in the opposite direction.

If a potential difference exists between two points in such a field, then electrons can do work and expend energy in moving from one point to the other. If the points represent the ends of a conductor, such as a copper wire, then electrons in the wire will experience a force and move from one end to the other. In passing through these points of potential difference, the electrons lose potential just

Electrons move through a difference in potential in an electric field.

as a ball does when it is dropped from a roof to the ground (from a point of high potential energy to a point of low potential energy). In order for the electrons to do more work, something must first perform work against the electric field, which will raise the potential energy of the electrons once again. The electrons are then capable of repeating the cycle, and performing work, such as running an electric motor in a fan or heating the coils of a toaster.

The regeneration of electrical potential must take place at some point within an electric circuit. An **electric circuit** is a complete closed path (like a loop) over which electrons can travel. The pathway itself is a conductor such as a copper wire. The device that does work against an electric field to raise the potential of electrons can be a small flashlight battery or a giant generator at a power station; small or large, it is the source of electrical energy in the circuit.

Series Circuits

Figure 6-8 shows the simplest of circuits, in which the voltage source is a battery. The flow of charge, or **current**, is indicated by the arrow. The direction of current moves (by convention) from plus to minus in keeping with Franklin's electrical terminology. But actual physical electron flow is from minus to plus. This circuit is called a **series circuit**. The same current passing through resistor R_1 must pass through resistor R_2, because they are connected in series

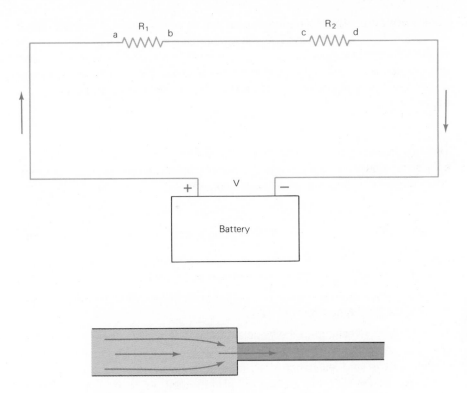

Figure 6-8
A simple series circuit showing the flow of
charge from positive to negative battery
terminals, the total charge passing through
both resistors, R_1 and R_2. Like water flowing
through pipes of varying diameters,
electricity is affected by resistances of
various sizes.

together. To visualize this, picture a water pipe forming a closed loop (circuit)
beginning and ending at a water pump. Suppose that two tubes are in
sequence connected into this water circuit. All the water that is seen passing
through the first tube must pass through the second. This occurs simply because the water can find no other path through which to return to the pump. The
same holds true for electrons in the series circuit.

Series circuits offer only one path for
electron flow.

SERIES RESISTANCE Resistance is a measure of the opposition to the flow of
electrons offered by a wire or electrical device because of collisions between
electrons and atoms in the conductor. Figure 6-8 shows two water pipes, each
offering different resistances to the flow of water because of their different
diameters. They have been hooked into a series circuit so that all the water that
flows through one must flow through the other. To measure their total resistance
to the flow of water, simply add their individual resistances.

Similarly, the total resistance to the flow of current in the series circuit of Figure
6-8 is:

$$R_t = R_1 + R_2$$

This relationship holds true for any number of resistances in a series circuit.

SERIES VOLTAGE The difference in potential or voltage across a series circuit is measured in a similar way. The voltage across the terminals of the battery in Figure 6-8 is *V*. Since points *a* and *d* are directly connected to the positive and negative terminals, the voltage between these points is also *V*. But the potential differences between points *a* and *b* across R_1 and between *c* and *d* across R_2 are given by:

$$V = V_1 + V_2$$

This relationship indicates that there is a difference in potential between the endpoints of each resistance, and the sum of these differences is equal to the total difference in potential across the whole circuit.

Ohm s Law

Current *i* measures the rate of flow of charge past a given point in a circuit:

$$i = \frac{q}{t} = \frac{\text{coulombs}}{\text{second}}$$

A flow of one coulomb per second is called one **ampere**, after the French physicist André-Marie Ampère (1775-1836).

In 1826, while investigating the resistances of different materials, the German physicist Georg Simon Ohm (1787-1854) discovered the relationship between current, voltage, and resistance in a circuit. For a given voltage the current is inversely proportional to the resistance in the circuit. As resistance increases, current decreases.

Figure 6-9
This basic parallel circuit shows the division of current between the two branches containing the resistors R_1 and R_2. Because the endpoints (a and b) are the same for both branches, the potential difference across each branch—and across every branch in a parallel circuit—is the same.

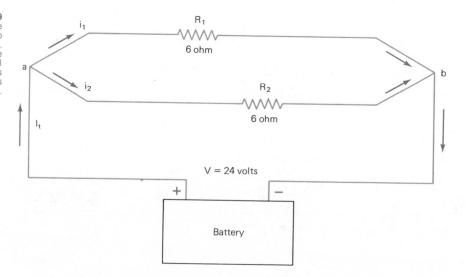

Expressed in the form of an equation, the relationship becomes:

Ohm's law relates current, potential
difference, and resistance.

$$i = \frac{V}{R} \quad \text{or} \quad R = \frac{V}{i}$$

From this last expression, the unit of resistance, appropriately called the **ohm**, can be defined as the resistance that will pass a current of one ampere when a potential difference of one volt is applied. The most familiar form of Ohm's law is:

$$V = i \times R$$

Thus if we wish to calculate the resistance of an air conditioner that draws 7.5 amps of current from a 120 volt line, we find that

$$R = V/i = 120/7.5 = 16 \text{ ohms}$$

Parallel Circuits

Figure 6-9 shows a circuit arrangement somewhat different from the series circuit of Figure 6-8; the resistances R_1 and R_2 are located in branches that are parallel to each other. For this reason such an arrangement is called a **parallel circuit**. Because both resistances are connected between the same points a and b (themselves connected directly to the terminals of the battery), the voltage across each is equal to the voltage across the entire circuit, or:

$$V = V_1 = V_2$$

This relationship holds for any number of resistances connected in parallel.

PARALLEL RESISTANCE From Figure 6-9 it can be seen that there is less resistance to current when there are two current paths rather than one. This is similar to the case in which a water pipe branches into two pipes—providing less resistance to flow than a single pipe. The total resistance in a parallel circuit is, therefore, less than the value of either resistance. Total current is:

Parallel circuits offer several paths for
the flow of charge.

$$i_t = i_1 + i_2$$

and

$$i = \frac{V}{R}$$

so

$$\frac{V}{R_t} = \frac{V}{R_1} + \frac{V}{R_2}$$

We can divide both sides of this equation by V, obtaining:

$$\frac{1}{R_t} = \frac{1}{R_1} + \frac{1}{R_2}$$

The reciprocal of the total resistance in a parallel circuit is equal to the sum of the reciprocals of each resistance. This relationship holds for any number of resistances in parallel.

As an example, suppose we calculate the resistance and current for the circuit in Figure 6-9 if $V = 24$ volts and both R_1 and R_2 are 6 ohms each:

$$\frac{1}{R_t} = \frac{1}{R_1} + \frac{1}{R_2}$$

$$\frac{1}{R_t} = \frac{1}{6} + \frac{1}{6} = \frac{2}{6}$$

$$R_t = \frac{6}{2} = 3 \text{ ohms}$$

which is just half the value of each resistance:

$$i_t = \frac{V}{R_t} = \frac{24}{3} = 8 \text{ amperes}$$

but $\quad i_1 = \frac{V}{R_1} = \frac{24}{6} = 4 \text{ amperes}$

and $\quad i_2 = \frac{V}{R_2} = \frac{24}{6} = 4 \text{ amperes}$

Thus, since:

$$i_t = i_1 + i_2$$

then

$$i_t = 4 + 4 = 8 \text{ amperes}$$

This situation is equivalent to a series circuit with $V = 24$ volts, a single resistance $R = 3$ ohms, and a current $i = 8$ amperes.

House Circuits

Parallel circuits have distinct advantages over series circuits for house wiring. In a parallel circuit, each branch has the full voltage; in a series circuit, voltage across each device drops as more devices are added to the circuit. Each branch of a parallel circuit is independent of the others, but if a series circuit shuts off anywhere, the circuit is broken. Just like a chain, a series circuit is only as strong as its weakest link.

A typical house circuit is depicted in Figure 6-10. Three devices—a light bulb, an iron, and a coffee pot—occupy three separate branches of a single circuit. This can be accomplished by plugging each into a different receptacle on that circuit or branching them off the same receptacle. In any case, each device receives a full 120 volts and a part of the total current drawn by the circuit.

Electrical energy is one of the most widely used forms of energy today. One of the reasons electrical energy is so practical is that it can be transported cleanly, cheaply, and conveniently over long distances, through wires. Another is that it is easily converted into other forms of energy at the places where it is actually used. Motors transform electrical energy into mechanical work; toasters and broilers transform it into heat; incandescent bulbs and fluorescent tubes transform it into light.

Electrical Power

As we saw in Chapter 4, power is the rate at which energy is used. Since, as we have seen, moving one coulomb of charge across a potential difference of one volt requires one joule of work, we can express electrical energy (or, equivalently, the work it can do) as the product of charge and voltage:

$E = q \times V$ (joules = coulombs × volts)

To find the rate at which work is done or energy consumed, we must divide this equation by the time:

$$P = \frac{\text{joules}}{\text{sec}} = \frac{\text{coulombs} \times \text{volts}}{\text{sec}}$$

In this equation, coul./sec is simply the current, i, so the equation becomes:

$P = iV$

The unit of electrical power, one joule/sec, is called the **watt**, W. One watt of power is consumed when a current of one ampere flows across a potential difference of one volt. Suppose we wish to know how much current flows through a 150 watt light bulb. Dividing the power, 150 watts, by the voltage (120 volts in most cities), we find that the current must be 1.25 amps.

In practice, electric companies calculate power in kilowatts (kW) — units of 1000 watts. The electrical energy consumed by customers, such as large factories or

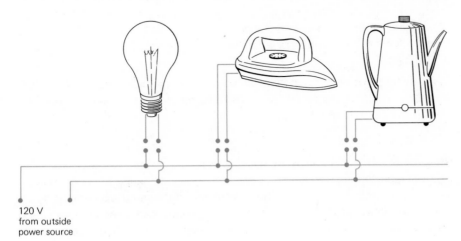

Figure 6-10
A typical house circuit, in which each branch of the parallel circuit has the full voltage.

120 V
from outside
power source

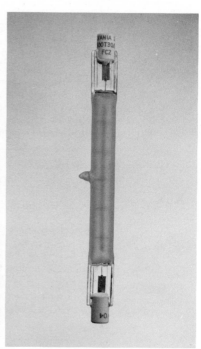

Figure 6-11
An ordinary light bulb, called an incandescent lamp, gives light when its filament glows as a current is passed through it. Shown here are three other popular light bulbs: The fluorescent lamp (left) is a glass cylinder coated with phosphor, a compound that emits light when activated by energy. The tube is filled with mercury vapor and argon, an inert gas. The high pressure sodium light (middle) produces light when an electric current passes through sodium vapor; it provides an intense golden yellow light and is often used in street lighting. The 500-watt tungsten-halogen lamp (right) is an incandescent light in which a chemical reaction between the tungsten of the filament and the halogen in the glass rod prevents the tungsten that "boils off" the filament from settling on the lamp's inner walls; instead, the tungsten is returned to the filament so that the lamp has a long life and is consistently bright so long as it burns. The tungsten-halogen lamp is used in movie lights to produce a high intensity beam.

ordinary homeowners, is measured in kilowatt-hours (kWh). Since a kilowatt is 1000 joules/sec, and an hour is 3600 sec, 1 kWh = 3.6×10^6 joules/sec.

We know that electricity can be used to produce heat. Most people are also familiar with the fact that electrical devices tend to produce heat—often in inconveniently large amounts—even when that is not their primary purpose. An ordinary electric light bulb is a relatively inefficient device; much of the power it consumes is converted into heat and a relatively small percentage into light. (Fluorescent tubes, which stay much cooler, are correspondingly more efficient.) Similarly, a high-fidelity amplifier or an electric motor generate considerable heat.

We can find out how much heat is created by an electric current flowing through a given resistance by using another form of the last equation. Since by Ohm's law $V = iR$, we can write:

$$P = iV = i^2R$$

Thus the power consumption increases with the square of the current for a circuit with constant resistance. The dependence of the heating effect of an electric current upon the square of the current was discovered experimentally by Joule. He made small wire coils of different metals and sizes and measured their temperature with a thermometer he designed to be readable to 0.1°F.

A fuse is a device that contains a metal strip that melts if the heat generated by the current exceeds a predetermined value. The fuse in the diagram is rated at 15 amperes. It appears in the circuit *before* the branches, usually at a fuse panel where power lines enter a house. A circuit breaker serves the same function as the fuse, but it is actually a switch that shuts off if the current exceeds its specified value.

Fuses and circuit breakers are safety devices that protect circuits and the house from shorts and overloads. A circuit is said to be overloaded when the number of devices connected is such that the circuit draws more current than the wiring—and so the breaker or fuse—is rated to carry. Without such protective devices as fuses and breakers, the wiring would heat up, possibly causing fires.

A short circuit occurs when an accidental connection is made between points *a* and *b,* bypassing the resistance *R.* If there is no resistance in a circuit, then by Ohm's law the current becomes infinite. In actuality, there is always some resistance in the wiring itself; nevertheless, the surge of current during a short circuit is dangerously strong and blows fuses or trips breakers almost instantly.

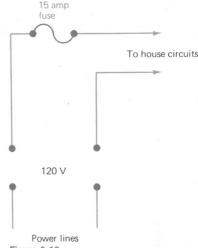

Figure 6-12
A fuse introduced at the beginning of, and in series with, each house circuit interrupts the current in the event of an overload or a short circuit.

Glossary

The **ampere** is the flow of one coulomb of charge past a given point in a circuit per second.

Conductors are substances—particularly metals—through which charge flows freely.

A **coulomb** is a unit of electrical charge. There are 6.25×10^{18} electrons in one coulomb; the electron, therefore, has a charge of 1.6×10^{-19} coulombs.

Current is the flow of charge per unit time. It is measured in amperes.

An **electric circuit** is a closed path (or loop) over which charge can flow.

Electric field intensity is force per unit charge in an electric field. It is measured in newtons/coulomb.

Electrostatic induction is the process whereby the positive and negative charges in a neutral conductor are separated (polarized) by the electrical force of a charged body in the vicinity.

A **force field** is a region of space in which an object experiences a force. The direction of the field is the direction of the force on the object at a point in the field.

Insulators are materials—usually nonmetals—through which charge does not flow readily.

The **ohm** is the unit of electrical resistance that will permit the passage of a one ampere of current when a potential difference of one volt is applied.

Potential difference is the change in potential energy per unit charge experienced by a charge that moves between two points in an electric field. It is measured in volts.

Resistance is the opposition to the flow of electrons offered by the atoms of a conductor or electrical device. It is measured in ohms.

The volt is the unit of potential difference equal to one joule/coulomb.

A watt is the unit of power equal to one joule/sec, or equivalent to a current of one ampere flowing across a potential difference of one volt.

Exercises

1 Briefly explain the role of electrons and protons in our present model of electricity. How does this theory combine the observations concerning electricity of both Franklin and Dufay?

2 During an electrical storm, why is it wise to remain in a car?

3 Insulators such as amber or glass may be charged by friction (rubbing a glass rod with silk, for example) but conductors such as metal will not become charged by friction so long as they are held in the bare hand. Why?

4 Calculate the distance separating two pith balls suspended by threads if they carry charges of $+4 \times 10^{-12}$coul. and $+2.5 \times 10^{-13}$coul. respectively, and experience a repulsive force of 2.25×10^{-11}N.

5 When an object of undetermined charge is placed 3 cm away from an iron ball bearing 6×10^{-10}coul. of charge, the ball experiences an attractive force of -1.8×10^{-5}N. Compute the undetermined charge.

6 How many electrons make up the charge of 6×10^{-10}coul. carried by the iron ball in the preceding problem?

7 If 3.15×10^{19} electrons pass through a copper wire in one second, how many coulombs/sec and amperes of charge pass through the wire?

8 What is the net charge inside a hollow iron sphere which carries 6.3×10^{-4}coul. of charge on its surface?

9 Draw two diagrams, one showing the lines of force between the field of two positive charges and the second showing the field of a positive and a negative charge.

10 Contrast the gravitational force with the electrical force between two electrons one meter apart. What is the ratio of the forces? $G = 6.6 \times 10^{11}$ Nm²/Kg²

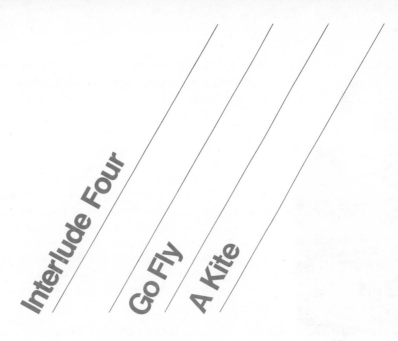

Interlude Four
Go Fly A Kite

A disillusioning result of the recent concentration on colonial American history sparked by the Bicentennial has been the discovery that many cherished beliefs about America's heroes are myths. It is now certain that George Washington did not chop down a cherry tree; Patrick Henry's immortal words "Give me liberty or give me death" are thought to be the invention of a journalist writing about the speech some twenty years later; Paul Revere never finished his famous ride because he was captured by the British somewhere along the way, and a companion completed the journey to carry the warning that the redcoats were coming. So what about Benjamin Franklin? Did he really fly that kite and draw down the lightning?

Fortunately, documents exist to confirm that this story is not a myth. A copy of Franklin's instructions for the experiment has been found:

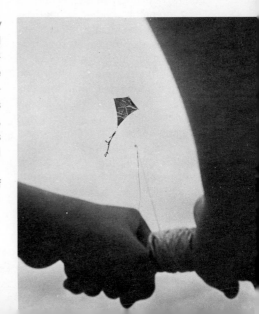

Make a small cross of two light strips of cedar, the arms so long as to reach to the four corners of a large thin silk handkerchief when extended; tie the corners of the handkerchief to the extremities of the cross so you have the body of the kite; which, being properly accommodated with a tail, loop and string, will rise in the air like those made of paper; but this being of silk is fitter to bear the wet and wind of a thunder-gust without tearing. To the top of the upright stick of the cross is to be fixed a very sharp-pointed wire, rising a foot or more above the wood. To the end of the twine next the hand is to be tied a silk ribbon, and where the silk and twine join, a key may be fastened. This kite may be raised when a thunder-gust appears to be coming on, and the person who holds the string must stand within a door or window, or under some cover, so that the silk ribbon may not be wet; and care must be taken that the twine does not touch the frame of the door or window. As soon as any of the thunder-clouds come over the kite, the pointed wire will draw the electric fire from them, and the kite, with all the twine, will be electrified.

In this letter to a friend in Boston, written on Christmas Day in 1750, Franklin recounted another accident he suffered in his experiments with electricity.

I have lately made an experiment in electricity that I desire never to repeat. Two nights ago, being about to kill a turkey by the shock from two large glass jars, containing as much electrical fire as forty common phials, I inadvertently took the whole through my own arms and body, by receiving the fire from the united top wires with one hand while the other held a chain connected with the outsides of both jars. The company present (whose talking to me and to one another, I suppose occasioned my inattention to what I was about) say that the flash was very great, and the crack as loud as a pistol; yet, my senses being

Benjamin Franklin is one historical figure who is not diminished by closer investigation. On the contrary, the more we find out about his life, the more we marvel. Franklin was a man who both represented his time and transcended it. This is particularly true when we view his scientific career.

The aspect of Franklin the scientist that has been most written about is his flair for experimentation. The kite experiment is the best known example, but there were many others. On one occasion he stood on an insulated box atop the tallest church tower in Philadelphia, to which he had attached an iron lightning rod; in a lapse from his usual practicality, he demonstrated the presence of electricity in the rod by taking the charge through his body. (After Franklin's results were made known, several Europeans were killed trying to duplicate this experiment.) Later he attached a lightning rod to his house and connected it to a bell; when electricity was present in the rod, the bell rang, somewhat to the annoyance of his neighbors, who thought Franklin's bell was worse than the clamor of a thunderstorm.

Contemporary accounts by Franklin's friends tell us that he also had a genius for devising tools. He would visit local craftsmen, such as the silversmith or the cabinet maker, and draw a complete sketch of the apparatus he wanted built. He invented a device to store electricity for relatively long periods and gave it the name by which we still call it, a battery. He rigged up devices to generate electricity, and devices to measure the amount of electricity he had generated; in short, he virtually invented an entire laboratory in which to perform his electrical experiments.

What is often overlooked about Franklin is that he was more than just a curious man who tinkered with scientific apparatus. There were, after all, other men in the eighteenth century who had the leisure time to indulge in what amounted to a social fad for scientific experimentation—it was a common custom in both Europe and America to end a dinner party with the host's display of some mysterious natural phenomenon, such as the lurid glow imparted to objects when they are rubbed with a chunk of phosphorus. Yet none of these amateurs, many of whom were both well educated and intellectually gifted, has gone down in history as a great scientist; Franklin has.

Franklin was different because he combined with his experimentation a deep interest in the theories that would explain his experimental results. He was a member of the prestigious Royal Society of London; in this group, whose members included many of the best scientific minds of the time (Newton was president of the Royal Society for many years) Franklin found a forum for discussing what he called his philosophical ideas (we would call them scientific theories). In Franklin's time, it was believed that a single individual could completely master a subject—that he could know literally everything about it. In his pursuit of complete knowledge about electricity, Franklin maintained a heavy correspondence with English and French scientists; they exchanged experimental results and, even more important, they exchanged their hypotheses about why the results had been what they were. Franklin's contribution to the investigation of electrical phenomena has had lasting value because he integrated his experimental evidence within the framework of a theory that explained the results.

Franklin's experiments with the kite brought him notoriety; his theoretical contributions to science brought him lasting fame. In his native America, he was celebrated for his discoveries, but many Europeans found it hard to believe that colonists and savages in the forests of America could produce such advances in knowledge. In France, a group of scientists and philosophers held a satirical meeting, mocking Franklin's rusticity and branding his theories nonsense. Eventually, Franklin's ideas became so widely accepted that he was given an honorary degree by Oxford University, and even the scoffers were forced to recognize the value of Dr. Franklin's achievements. By the time of his death at the age of 84 in 1790, he was esteemed on both sides of the Atlantic for his many practical and theoretical contributions to the scientific life of his time as well as for his considerable political gifts. Turgot, the French economist, immortalized him in an epigram: "He snatched the lightning from the skies and the scepter from tyrants."

instantly gone, I neither saw the one nor heard the other; nor did I feel the stroke on my hand, though afterwards I found it raised a round swelling where the fire entered, as big as half a pistol bullet; by which you may judge of the quickness of the electrical fire, which by this instance seems to be greater than that of sound, light, or animal sensation.

What I can remember of the matter is that I was about to try whether the bottles or jars were fully charged by the strength and length of the steam issuing to my hand, as I commonly used to do and which I might safely enough have done if I had not held the chain in the other hand. I then felt what I know not how well to describe; a universal blow throughout my whole body from head to foot, which seemed within as well as without; after which the first thing I took notice of was a violent quick shaking of my body, which gradually remitting, my sense as gradually returned, and then I thought the bottles must be discharged, but could not conceive how, till at last I perceived the chain in my hand and recollected what I had been about to do. That part of my hand and fingers which held the chain was left white, as though the blood had been driven out, and remained so eight or ten minutes after, feeling like dead flesh; and I had a numbness in my arms and the back of my neck which continued till the next morning but wore off. Nothing remains now of this shock but a soreness in my breast-bone, which feels as if I had been bruised. I did not fall, but suppose I should have been knocked down if I had received the stroke in my head.

The whole was over in less than a minute.

The Universe of Color

Throughout the ages, light has been associated with high and holy things: beauty, goodness, God, life itself. The fear of darkness is acquired very early in life, and darkness, to most people, suggests evil. This is easy to understand when we think how much we depend on light and our sense of vision for knowledge of the world around us. One kind of visual information that is especially important to us is color. A world without color would be a dull place—and a dangerous one, for color provides us with vital warning signals from our environment. But what is color? Like all our sensations, it is a creation of our nervous system. Color represents the way our eye and brain interpret an aspect of the physical universe: electromagnetic radiation.

Electromagnetic waves consist of regularly varying electric and magnetic fields. These are perpendicular to each other and to the direction of the wave, so that electromagnetic waves can be classed as transverse waves. Unlike mechanical waves, however, they require no medium for propagation and travel fastest in a vacuum. We know electromagnetic waves of different wavelength by different names. Radio waves (including AM, FM, television, and short-wave) have wavelengths ranging from about 10,000 meters to 0.1 meter. Still shorter are the waves in the radar and microwave bands, with wavelengths between 0.1 m and 0.001 m. Infrared waves, which are emitted by most hot bodies and thus often called heat radiation, occupy the region from 0.001 m to 4×10^{-7} m. Ultraviolet radiation, emitted by sun lamps and black light tubes, has wavelengths from

7×10^{-7} m to about 10^{-9} m. Shorter yet are the x-ray wavelengths (10^{-9} to 10^{-12} m) and those of gamma rays, ranging down to 10^{-17} m and beyond.

Of this enormous electromagnetic spectrum, the wavelengths that our eye can detect and that we call visible light occupy only a tiny region: 4×10^{-7} to 7×10^{-7} m. All the colors that we see represent wavelengths within this narrow range or mixtures of such wavelengths. White light, which is often thought of as being the purest light, is in reality the richest mixture, containing all the wavelengths of the visible range. When white light is separated by diffraction into its components, the result is a spectrum (Plate 1): red, orange, yellow, green, blue, and violet, each representing a different wavelength. And since the colors shade almost imperceptibly into each other, there are many gradations between adjacent colors; the average person can distinguish over 100. There are also hundreds of other colors not present in the spectrum. It has been found that all of them, however, can be made by appropriate mixtures of the three primary colors—red, green and blue—with the addition of varying amounts of black or white. This fact, along with other experimental evidence, tends to support the widely held theory that the eye has three kinds of receptors, one for each of the primary colors.

People who have experience with art may find it strange that green is mentioned as a primary instead of yellow. They are thinking of another kind of color mixing, one with which most of us

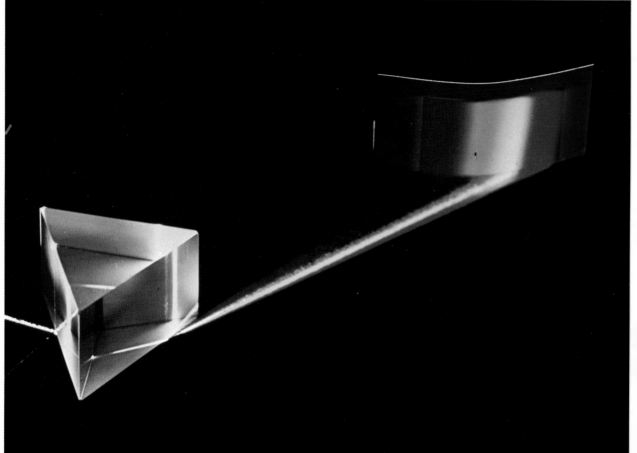

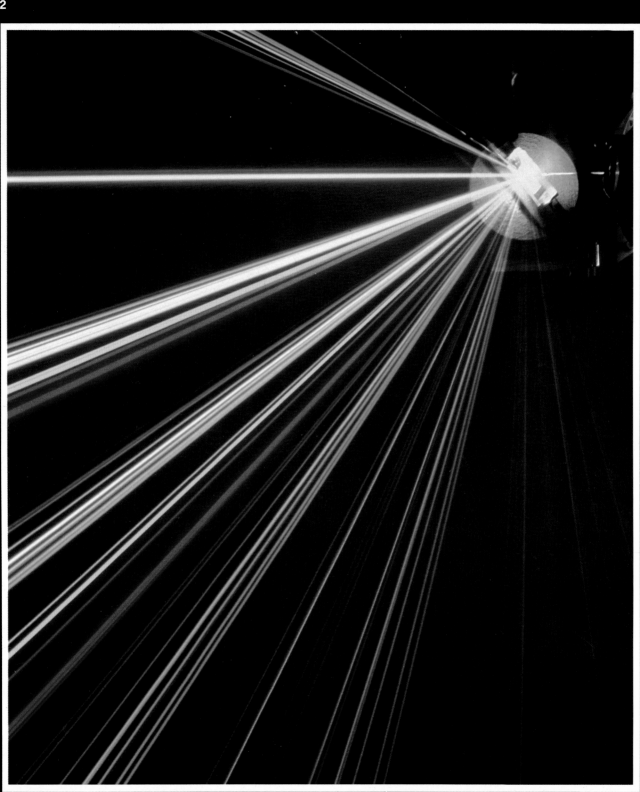

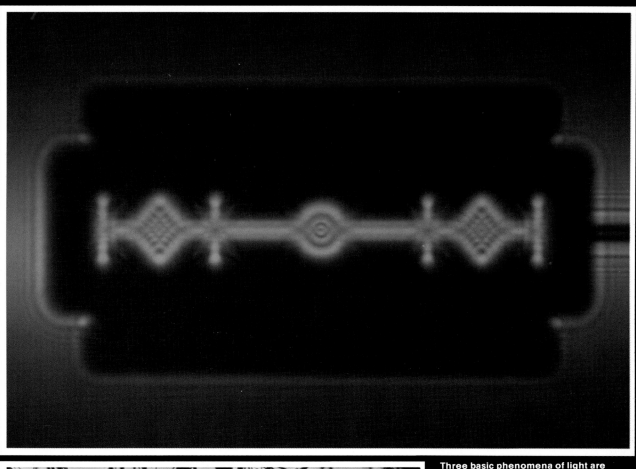

Three basic phenomena of light are illustrated here. The photograph at left shows willow branches reflecting their image in a still pool. In reflection, light from an object bounces off a shiny surface at the same angle it hit—the angle of incidence.

Refraction occurs when light partly passes through surfaces, changes speed and direction. In nature, this causes mirages and rainbows. The picture opposite shows light from a laser that has been refracted by a prism.

In the picture above, light is aimed at the slot of a razor blade. As it spills through, it spreads outward and bends around the edges. This diffraction pattern of light and dark produced is called an interference pattern. Diffraction is due to the wave nature of light.

In "Sunday Afternoon on the Grande Jatte," Georges Seurat used a technique called pointillism to achieve shadow detail and a three-dimensional effect. Painting with tiny dots of color, he relied on the human eye to integrate the points into an overall image. The top enlargement shows the dots Seurat used, while the bottom detail shows the far smaller dots of the lithographic process by which this painting has been reproduced in printing this book.

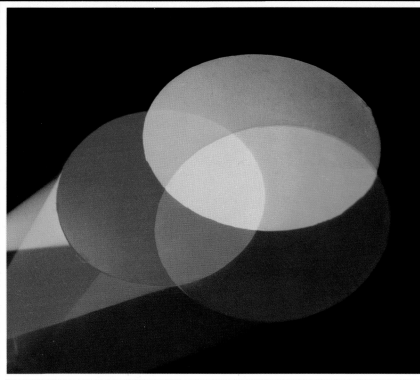

The two basic processes for the mixing of colors are called additive and subtractive. In the picture at left, white light is projected through three filters—red, green, and blue. Where the three components of the spectrum overlap, white results. Where two colors overlap, a third, new color is produced. In subtractive color, (shown below), light is absorbed by filters of the three primary colors. Here, the overlapped center is black, but the overlap of two colors still results in a third color. The combining of few basics to form many tones makes color printing possible. For example, the painting opposite, seemingly printed with many colors, was reproduced with only three inks—red, yellow, and blue, plus black for modeling and shadow detail.

A polarizing filter screens out confusing horizontal reflections that cause glare and allows only vertical light waves to pass through. Compare the two snapshots at right. The top one was made using a polarizing filter and shows more detail and clarity than the one below.

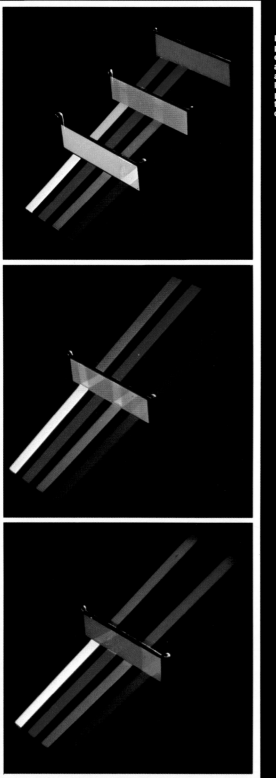

In the demonstration at left, photographic filters — glass plates coated or impregnated with pigments that absorb colors selectively — show their ability to both block and pass light. The light entering from the left side of each picture is seen as a band of white plus three bands representing the components of the whole light spectrum.

Piercing a steel razor blade like a hot awl in an ice cube, a beam of light from a ruby laser manifests its extraordinary power. This laser is of a type known as an Optically Pumped Solid Laser, and its beam is emitted only in intermittent pulses, or bursts, the result of an electronic flash tube exciting a ruby rod to give off a beam of coherent, closely focused light. The beam in this case is further concentrated by a lens, seen at the right in the photograph. A characteristic of such lasers is that they

The two lasers shown operating here produce a continuous beam of light generated by the excitation of gases in a sealed chamber. Mirrors positioned at the ends of the discharge tube reinforce and direct the beams of light. The light produced by gas lasers is similar to the light generated in a neon tube; like neon, the color of the light is determined by the vaporized gas inside the sealed tube.

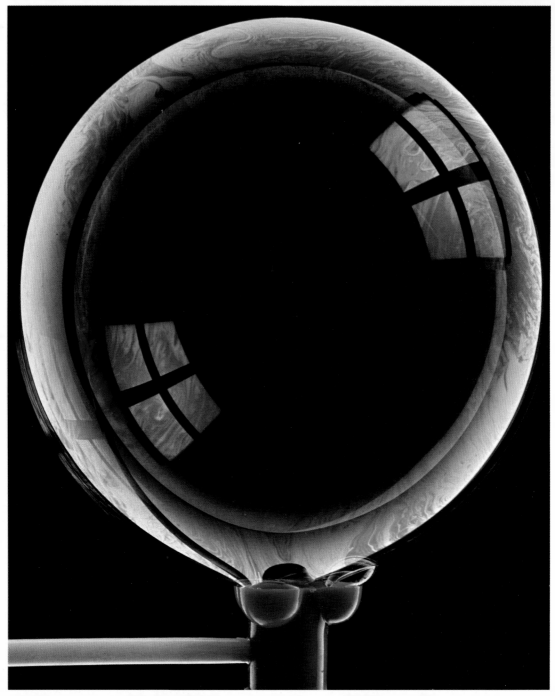

Crystals have many shapes and colors. Clockwise from upper left are organic crystals of tartaric acid, herbreide, salol, and bisphenol-A.

Certain chemical reactions have been shown to proceed in a highly symmetrical way. This sequence of three photos, taken specifically for demonstration purposes, shows one such beautifully patterned reaction.

The color of an organism is important for mating and protection. At top, the striking colors of an owl butterfly attract members of its species. Below, the coloration of a moth helps it blend in with the tree bark to avoid predators.

Colors in the depths of the sea appear lurid due to the absence of direct sunlight. Shown at left is a butterfly fish.

The greenish outline on clam at left is actually due to a growth of algae; the red clam on the right displays its own innate color.

Bioluminescence may be hidden or obvious. On this page, top photo shows railroad worm in a glass of water. Below, the same worm glows as it crawls across a photographic film, exposing it. On facing page is a chalice sponge.

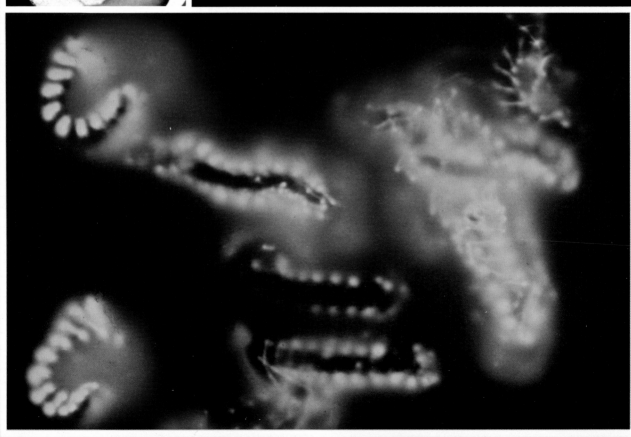

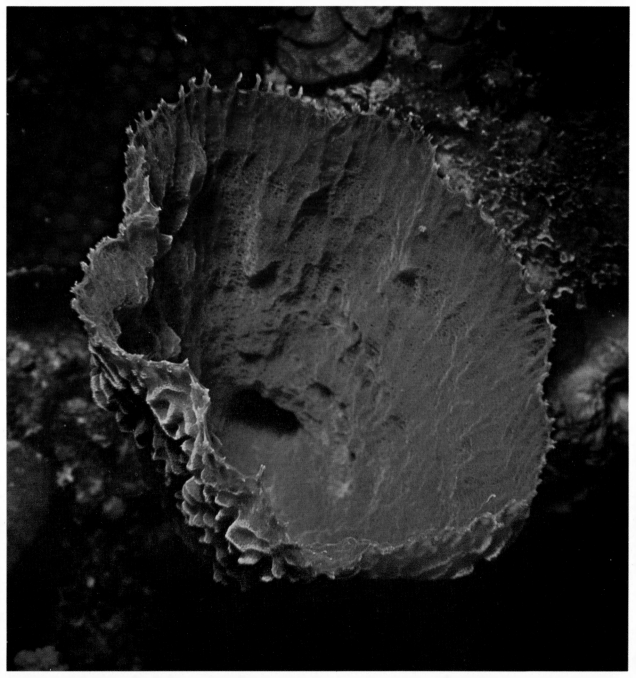

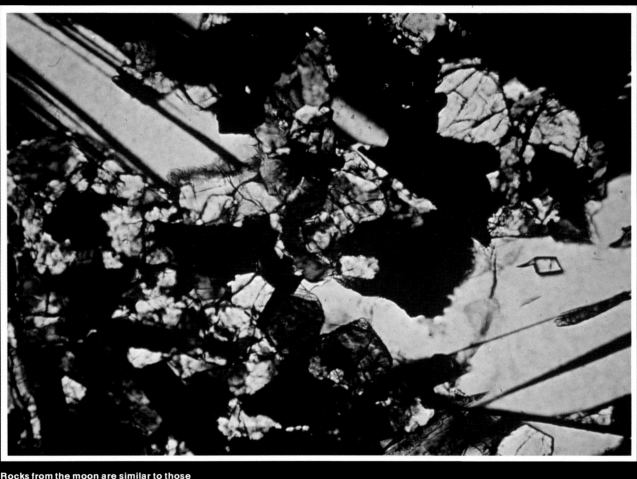

Rocks from the moon are similar to those found on earth. This photomicrograph, taken under polarized light, shows samples retrieved from the moon by astronauts. Samples include feldspar and pyroxene.

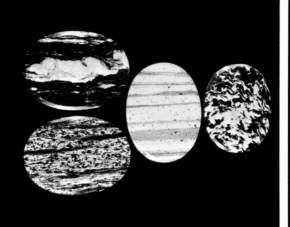

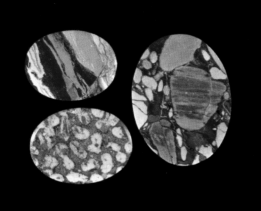

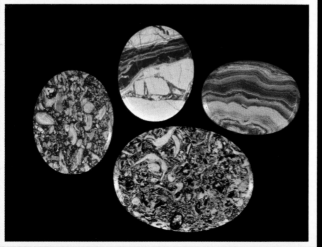

These examples of the three principal kinds of rock display different colors, textures, and compositions. Nearly colorless igneous rocks are at the top, metamorphic conglomerate limestones are in the center, and sedimentary limestone rocks are in the bottom photo

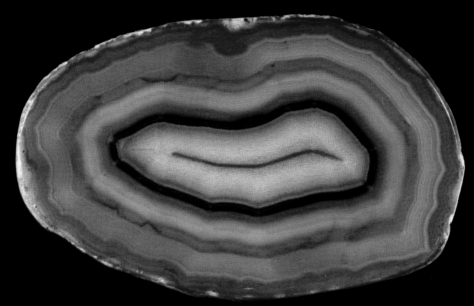

Precious and semiprecious gems are beautiful in both rough and polished states. On facing page are uncut rose quartz (upper left), tourmaline (upper right), and agate (bottom). At the top of this page is an uncut emerald; cut and polished diamonds and emeralds are below.

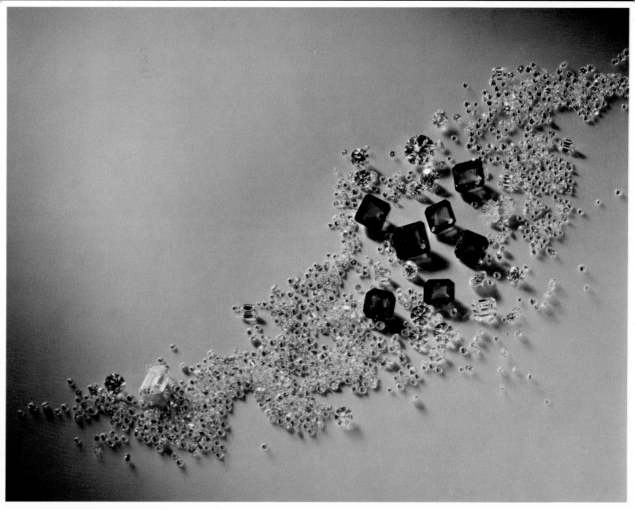

Iridescent rocks display all the colors of the rainbow. Top photo shows a sample of limonite. Below are micrographs of moon samples.

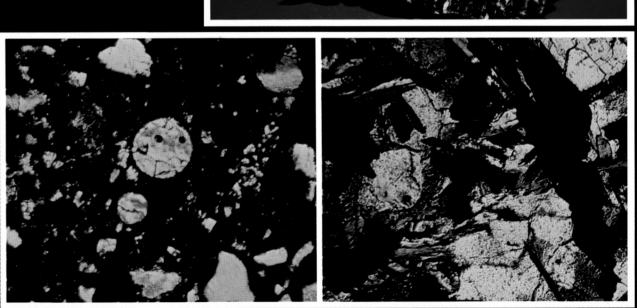

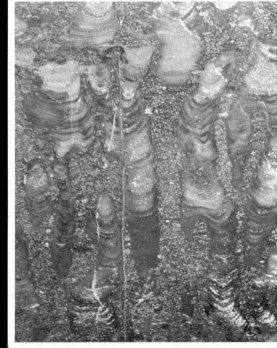

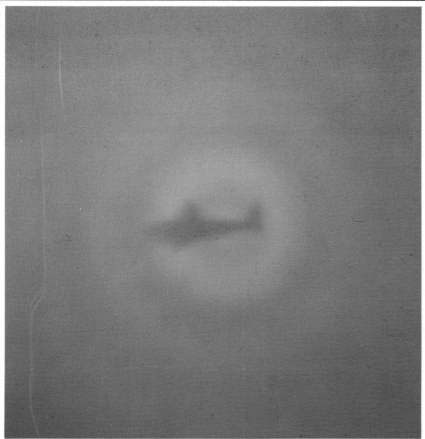

This halo of color around the shadow of an airplane cast on a cloud is called the Glory. The effect may be caused by the bending (diffraction) of light, then mirrored by the water droplets in the cloud.

Rainbow over Bridal Veil Creek in Yellowstone National Park. Sunlight, reflected in spherical water drops, is refracted and dispersed in a curving configuration of colored rings corresponding to the spectrum of visible light.

Nature can be awesomely beautiful in her most destructive moments. At right, a hurricane rages in the Atlantic Ocean. Below, dark, billowing clouds presage the coming of a storm.

The quiet before the storm. Gathering storm clouds cast an eerie glow over Chimney Rock in Colorado.

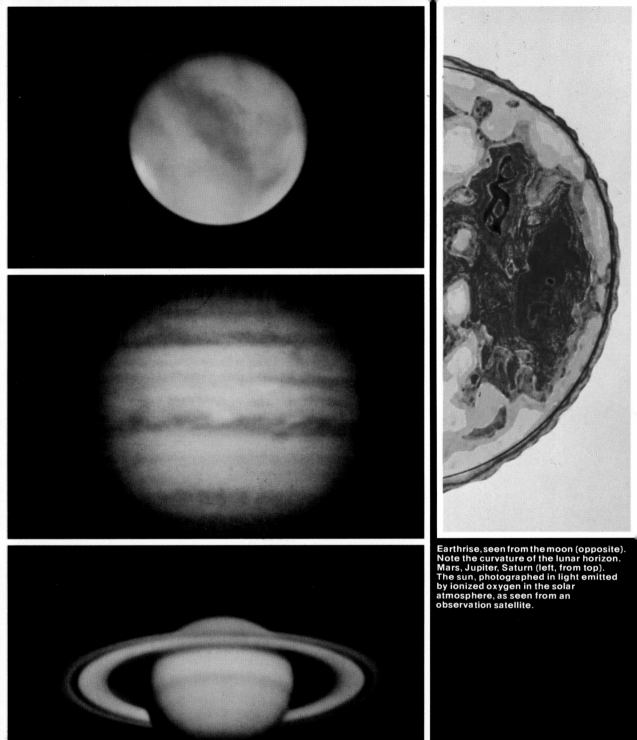

Earthrise, seen from the moon (opposite).
Note the curvature of the lunar horizon.
Mars, Jupiter, Saturn (left, from top).
The sun, photographed in light emitted
by ionized oxygen in the solar
atmosphere, as seen from an
observation satellite.

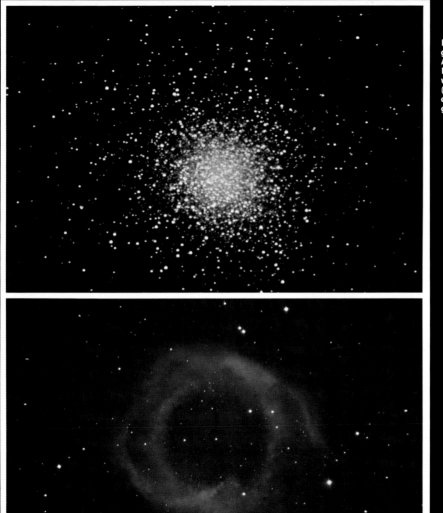

Emission nebulas, masses of glowing gas: The Veil Nebula in Cygnus (opposite). A planetary nebula in Aquarius (bottom). M-13, a globular star cluster in Hercules with several hundred thousand stars (top). A neighboring galaxy similar to our own (over): the great Andromeda spiral galaxy.

are more familiar: the mixing of pigments (paints or inks) on paper or canvas. This is called subtractive mixing, in contrast to the mixing of light of various colors, which is termed additive (Plate 5). The term is used because pigments, like filters (Plate 7), absorb certain wavelengths of the light that falls on them, and so these wavelengths are missing from the reflected rays. A molecule of blue paint, for instance, subtracts red and green wavelengths from white light and reflects only the blue. In practice, the pigments used in color printing, such as the color plates in this portfolio, are cyan (a blue-green color which absorbs only red), magenta (blue plus red, which absorbs only green), and yellow (red plus green, which absorbs only blue). From combinations of these, plus varying amounts of black, all colors can be made. Cyan and yellow, for example, will produce green when mixed together. The cyan will absorb all the red, while the yellow absorbs the blue, leaving only the green rays of the spectrum. In color reproduction, the creation of the different colors is accomplished largely by the eye of the viewer. The printed image does not consist of areas of continuous tone, but a fine pattern of tiny cyan, magenta, yellow, and black dots, which the eye fuses to produce an image (Plate 4). The image on a color TV screen is also a mosaic of colored dots, integrated by the eye.

In nature, colors are produced in many ways besides simple reflection from colored surfaces. Tiny water droplets in the air can serve as miniature prisms, producing a rainbow by reflecting and refracting the sun's rays (Plate 25). The smooth facets of many crystals also constitute natural prisms; the elaborate cutting of diamonds is designed merely to enhance this effect. Diffraction and interference, too, can produce color effects of great beauty. The delicately hued lustre of mother-of-pearl is actually produced by hundreds of microscopic ridges which break up white light into a large number of individual wave fronts. Depending upon the angle of reflection and the spacing of the ridges, different wavelengths are cancelled or reinforced by interference, producing a shimmering array of colors. Much the same thing happens when we look at a soap bubble (Plate 10) or a thin oil slick. Light is reflected from two closely spaced surfaces—the top and bottom layers of the oil film, or the inside and outside layers of the soap film. The angle at which we view the film and its precise thickness at that point determine which colors are eliminated by destructive interference and which ones heightened by constructive interference. All three processes—reflection, refraction, and diffraction—probably play a part in the iridescence of the rocks in Plate 22.

Another source of color in the natural world is fluorescence. This phenomenon underlies the glow of the fossils in Plate 23, the distant nebulas in Plates 30 and 31, the aurora in Plate 24—and all the neon signs so common on our streets. Excited atoms, boosted to high energy levels by fast-moving electrons or energetic radiation (ultraviolet or gamma rays), radiate away some of their energy in the form of visible light. Many chemical reactions also release energy, which may be dissipated in the form of visible radiation (Plate 11). Such chemical luminescence is not uncommon in the biological world (Plates 16 and 17); the precise chemical reactions involved, however, are not yet fully understood in many cases.

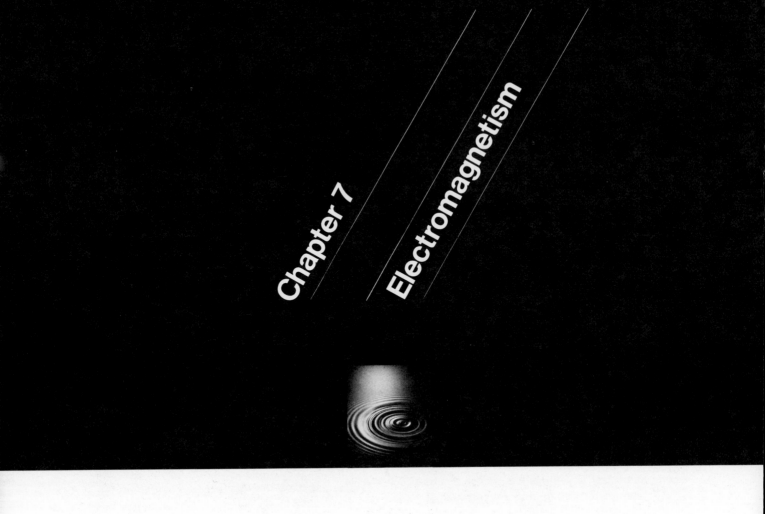

Chapter 7 / Electromagnetism

The ancient Greek philosopher Thales, the first person to leave recorded observations concerning electricity, was also the first to study magnets. His observations concerned **lodestones**, pieces of iron ore with the mysterious power of attracting any object made of iron. Because lodestones were first discovered in the region of Magnesia in Asia Minor, they were also called **magnets**.

Thales observed that the attractive power of the lodestone for iron resembled the attractive power of a rubbed amber rod for small bits of straw and other lightweight matter. But there were important differences. The amber rod attracted a great variety of substances; a lodestone attracted only iron or minerals containing iron. Moreover, the electric power of amber disappeared a few minutes after it was rubbed with fur, but the magnetic power of a lodestone was permanent and could be destroyed only by heating to a high temperature.

Magnetic attraction was observed by Thales to be similar to electrostatic attraction; magnets, however, attract only iron and magnetic forces are permanent.

Figure 7-1
Magnetite is the modern name for the
naturally occurring, permanently
magnetized iron ore that Thales called a
lodestone. Here the magnetic field around
the stones can be seen by the orientation of
the iron filings that surround them.

Perhaps the most extraordinary of Thales' discoveries was the fact that one magnet could be used to make another. An iron needle that was repeatedly rubbed along a lodestone in only one direction acquired magnetic power. This power was equivalent to that of the original lodestone and just as permanent. Because of these strange properties, magnets were once considered magical. In fact, Thales regarded the magnet's ability to attract a piece of iron as evidence for the belief that the magnet had a soul.

The Earth as a Magnet

Long before magnetism was really explained, people found that it had an extremely useful application. This early example of scientific technology is the magnetic compass. If a lodestone or magnetized needle is suspended so that it is free to rotate, it always lines up in the same direction. It was found that this direction is invariably close to the north-south line at the point of observation. One end of the needle seems to seek the earth's North Pole, and the other end seeks the earth's South Pole. This principle of the magnetic compass was first discovered in medieval China and put to practical use on sailing ships and desert caravans long before it was brought to the West. Not until the 1100s was the compass common on European ships.

A landmark in the history of science occurred in 1600, when the English scientist William Gilbert (1540–1603) published *On the Magnet.* Gilbert, who was a

Gilbert constructed a model of the earth's magnetic field and used it to explain geomagnetism.

Figure 7-2
The modern marine magnetic compass is contained in a nonmagnetic glass-topped bowl. The U. S. Navy standard No. 1 compass, shown here, consists of a compass card, to the bottom of which are attached two or more magnets aligned with the north-south axis on the card. The card rests on a vertical pin called the pivot, which enables it to rotate freely. The direction of the ship is marked on the inside of the bowl; it is 355°.

physician at the court of Queen Elizabeth I, pursued scientific studies in his spare time. Toward the end of his life, he compiled all that had been learned about magnetism up to that time, added new experimental evidence from his own discoveries, and presented the whole in his treatise, a classic of scientific observation.

Gilbert refers to the polarity of a magnet. If a magnet is dipped into a dish of iron chips, the chips stick to it when it is lifted up. The densest clusters of chips are at the opposite ends of the magnet, called the **magnetic poles**. It remained to explain why a magnetized needle lined up along the north-south direction. Gilbert guessed that the earth itself might be a giant magnet. To test his idea, he constructed a model of the earth by shaping a lodestone into a sphere. When a small magnetized needle was moved about the surface of the sphere, the needle behaved just like a compass needle in the earth's magnetic field. This magnetism did not, of course, explain the earth's *gravitational* attraction, but it did explain the north-seeking compass phenomenon.

Forces Between Magnets

Figure 7-3
The magnetic poles of the earth act as if the earth has an internal bar magnet, which is the source of the geomagnetic field. The geographic poles of the earth mark the points at which the earth's axis of rotation passes through its surface. The magnetic poles, those two points along which a compass needle aligns, move in geologic time; they are now displaced so that the south-seeking magnetic pole is near the geographic North Pole.

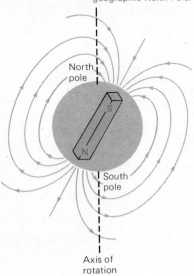

Just as the electric force between a charged body and a neutral body is always attractive, a magnet always exerts an attractive force on an unmagnetized piece of iron. But it is known that the electric force can sometimes be repulsive rather than attractive; might magnetic forces also be either attractive or repulsive?

The analogy between static electricity and permanent magnetism is close, but not exact. The basic law for permanent magnets is almost the same as that for electrical charges: *Like magnetic poles repel; unlike magnetic poles attract.* Instead of using "positive" and "negative" to designate the two kinds of magnetic pole, we use the names "north" and "south." A more accurate designation would be "north-seeking" pole and "south-seeking" pole. Since unlike poles attract, the magnetic pole of the earth near the earth's geographic North Pole is a magnetic south pole. To add to the confusion, the geographic and magnetic poles of the earth do not exactly coincide. (The geographic poles are the two points where the earth's axis of rotation passes through the earth's surface; the magnetic poles are the two points on the earth where a compass needle aligns itself vertically.)

Unlike electric charges, magnetic poles have always come in pairs—one north and one south pole. No attempt to produce a single, isolated magnetic pole has ever succeeded. If you break an electrically polarized object into two pieces, one piece will have a net positive charge and the other piece will have a net negative charge. But if you break a magnet in two, at the point where the break occurs a new pair of poles appear—one north and one south pole. Each piece of the original magnet is now a separate magnet with its own north and south pole. The dividing process can be repeated again and again, and the resulting smaller fragments will always be tiny magnets with two poles.

This suggests that the ultimate magnets may be the atoms themselves. In this view, it is thought that there are countless north and south poles of atomic magnets throughout a piece of magnetized material. Adjacent atomic north and south poles cancel one another; only near the ends are the poles not cancelled, thus creating a macroscopic magnet. The standard theory of magnetism was called to question, however, in the summer of 1975 when the track of a single magnetic pole was found among cosmic-ray photographs taken in the high atmosphere.

Atoms themselves may be the ultimate source of magnetism.

Like electric lines of force to and from a charge, magnetic lines of force leave the magnetic pole at the north end of the magnet and return at the south end. The magnetic field is strongest near the poles of a magnet where the lines of force are bunched together. The field pattern of a bar magnet, similar to that of the electric field with two equal and opposite charges, is called a **dipole field**. Magnetic lines of force form closed loops, with no ends and no beginnings. The loops pass out of the magnet at the north pole, swing around in space back to the south pole, then continue in the same direction through the magnet itself from the south pole back to the north pole again.

The strength of the electric field at any point is measured by the force on a test charge placed at that point. Similarly, the strength of a magnetic field is measured by the force on a tiny test magnet at some point on the field. If the field is uniform in the space occupied by the magnet, there will be equal and opposite magnetic forces acting on the two poles of the magnet. The net force

Magnetic Lines of Force

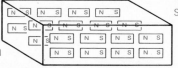

Figure 7-4
In an unmagnetized piece of iron, the individual atoms (domains) are randomly oriented. But in a magnetized piece, the domains all line up with similar orientation.

Figure 7-5
Magnetic lines of force form closed loops that begin at the north magnetic pole and end at the south magnetic pole. In the photograph here, a horseshoe magnet creates a characteristic magnetic-field pattern with iron filings that have been placed on a thin sheet above the magnet's two ends.

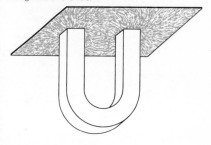

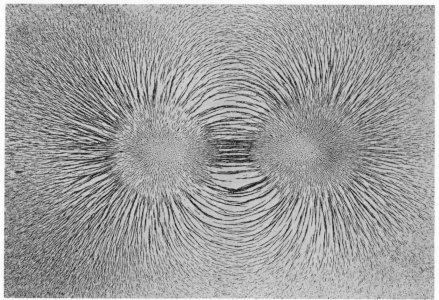

on the magnet is zero, but the two forces do produce a twisting action, or torque. A **torque** is a pair of equal and opposite forces acting on different points of the same object. The torque exerted by the magnetic field rotates the magnet so that it lines up with the field. Hence a test magnet moving in response to the torque exerted upon it can indicate the direction and strength of the magnetic field.

Magnetic forces are inversely proportional to the square of the distances between them.

Coulomb, who set down the formula for force of attraction or repulsion between electrostatic charges, did the same for forces of attraction and repulsion between magnetic poles. The formula is:

$$F \propto \frac{P_1 P_2}{d_2}$$

F is the force; P_1 and P_2 are the pole strengths, and d is the distance between them.

The Magnetic Field of an Electric Current

Since the time of Thales, scientists have felt that there might be some fundamental connection between electricity and magnetism. The real discovery of this connection was made by accident in 1820 by the Danish physicist Hans Christian Oersted (1777–1851). During a classroom lecture demonstration, he had been using a wire carrying an electric current and a suspended magnetic needle. At the end of the lecture, Oersted attempted to place the needle parallel to the wire. In the words of a student present at the time, "Oersted was quite struck with perplexity to see the needle making a great oscillation." When brought near the wire, the needle came to rest perpendicular to the current-carrying wire. When the direction of the current in the wire was reversed, the needle rotated 180° to the opposite direction from its original position, but still perpendicular to the wire. It was not attracted to the current-carrying wire, nor repulsed from it.

Figure 7-6
The compass needle (placed below the wire) is oriented at right angles to the direction of the current in the wire. As Oersted found, if the current is reversed, or if the compass is placed above the wire, the needle reverses direction.

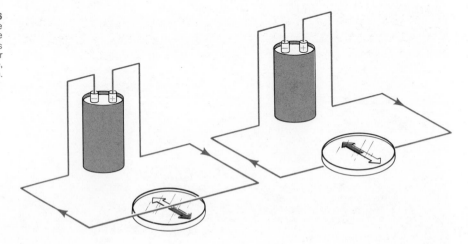

This was the first time a force of nature had been discovered that was **transverse,** or directed at right angles to the line joining the objects acted upon. The current is ringed by loops of magnetic lines of force. Both Coulomb's law of electrostatics and Newton's law of gravitation described a force that is directed along the line joining the two objects. Faraday realized that such a transverse magnetic force could cause a magnet to rotate continuously around a current-carrying wire following the loops of magnetic force lines, if the right electrical connections were made. He actually constructed such a device in 1822. It was the first electric motor, a machine to convert electrical energy into mechanical motion by the electromagnetic effect.

MAGNETIC FORCE BETWEEN TWO CURRENTS It was André Marie Ampère who in 1825 produced a mathematical theory of the magnetic effects produced by electric currents. Ampère realized that Oersted's observation of the deflection of a magnet near an electric current meant that the current itself produces a magnetic field in space. The magnetic lines of force surrounding a long wire carrying an electric current are concentric circles centered on the wire. The circles are closest together nearest the wire, where the magnetic field is strongest. Like the lines of force for a permanent magnet, the lines of force for a current-carrying wire are closed loops; unlike the former, they are entirely outside the wire.

A current-carrying wire is surrounded by a magnetic field.

The direction of the magnetic field lines produced by an electric current is given by Ampère's right-hand rule. Encircle the wire with the fingers of the right hand so that the thumb points in the direction of the electric current; the fingers then point in the direction of the magnetic field. This is the circular direction along which the north pole of a magnet will be deflected. Remember that the direction of the current is specified by the direction of positive charge flow (and is opposite to the direction in which the negatively-charged electrons actually move).

Ampère thought that if an electric current produced a magnetic force on a magnet, it seemed likely that two currents would exert a force on each other since each has a magnetic-field property. He eventually observed such an effect and determined the law for this new force. The force per unit length of each of two parallel current-carrying wires is proportional to the product of the currents and inversely proportional to the distance between the wires in meters:

$$F/L = K(i_1 \times i_2)/d$$

The force is given in newtons, the currents i_1 and i_2 in amperes, and the distance, d, in meters; the constant K equals 2×10^{-7}; L is the wire length.

Ampère's law for forces between current-carrying conductors states that the force for a unit of length of each of two parallel current-carrying wires is proportional to the product of the currents and inversely proportional to the distance between the wires.

This law of magnetic force between two currents is much simpler than the law of the force between two magnets. In fact, it is much closer in form to Coulomb's law. The principal difference is that Ampère's law gives a force that varies in-

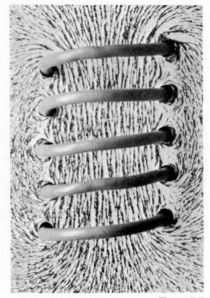

Figure 7-7
The magnetic field of a current in a series of wire coils.

versely with distance, whereas Coulomb's law gives a force that varies inversely with the *square* of the distance. But Coulomb's law describes the force between two point charges at rest. Using Coulomb's law, we can calculate the force per unit length acting between two very long lines of charge. This electrostatic situation is analogous to the case of the two currents. The electrostatic force per unit length between the lines of charge does vary inversely with the distance, as does the magnetic force between currents.

Ampère concluded his work by finding the force acting between any other arrangements of currents. The most important case he considered was a current flowing in a closed circular loop. He showed that such a loop is magnetically equivalent to a disk-shaped magnet. One face of the loop acts as a north magnetic pole and the other face as a south magnetic pole. A coil of wire wound in a helix is essentially the same as a series arrangement of single loops placed close together along a common axis. Ampère called such a coil a **solenoid**, from the Greek word for hollow pipe. The magnetic field produced by a solenoid when current passes through the coil is identical to the field produced by a bar magnet. The modern electromagnet is basically a solenoid; to make the magnetic field much stronger, a soft iron core is usually placed inside the hollow coil. The iron itself becomes magnetized and then greatly increases the field produced by the current alone.

Permanent Magnetism

Ampère's success in building magnets from current loops inspired him to formulate his theory of permanent magnetism as due to atomic-size current loops. He proposed that every permanent magnet contains circulating electric currents inside. But rather than flowing through wires from sources in external batteries, the currents within permanent magnets are endlessly flowing within the atoms themselves. The currently accepted explanation of permanent magnetism depends on the fact that the electron itself is a tiny magnet.

The magnetism of the electron occurs because the electron is spinning on its own axis as it revolves around the nucleus. The electron resembles the earth, which spins on its north-south geographic axis as it revolves around the sun. The axis of the spinning electron, which (unlike the earth) is also the direction of the electron's magnetism, must point in one of two directions—up or down. Which direction is up and which is down must be determined by reference to something outside the electron, and that is an external magnetic field.

Permanent magnetism is the result of the alignment of unpaired electrons in the atoms of the material.

In most materials, the electrons are paired off—one up and one down. Their magnetism cancels out and they produce no net magnetic field. But certain elements (for example, iron, nickel, and cobalt) have *unpaired* electrons. These unpaired electrons are inner electrons, not the outer electrons that take part in chemical bonding. So the electrons can remain unpaired even in the solid bulk form of the material, producing a permanent magnetic effect.

The work of Oersted and Ampère established that an electric current can produce magnetism. Almost immediately, the search was begun for the opposite effect—using magnetism to produce an electric current. The first successful attempt to produce an electric current by using a magnet was made by the American scientist Joseph Henry (1797–1878). Henry found that when the current in an electromagnet was turned on, a momentary electric current was produced in a nearby coil. When the electromagnet was turned off, a momentary current was also produced in the coil, but in the opposite direction. There was no wire or any other electrical connection between the electromagnet and the coil. In Henry's own words: "Thus, as it were, electricity [is] converted into magnetism and this magnetism is converted back into electricity." A few weeks later in England, Michael Faraday independently discovered the principle of electromagnetic induction. Since Faraday published his results before Henry, the principle of electromagnetic induction is called Faraday's law.

Faraday first observed Henry's effect, a momentary induced current in one coil when the current produced by a battery in a nearby coil was started or stopped. But in a second experiment, Faraday found that the induced current could also be produced by relative motion of the two coils. In Faraday's second experiment, one coil carried a steady current produced by a battery. A second coil, not connected to a battery, was placed near a suspended magnetic needle; deflection of the needle would indicate a magnetic force due to current flowing in the second coil. When the second coil was moved toward the first coil, an induced current was produced in the second coil. When the second coil was moved away from the first coil, the direction of the induced current reversed. Moving the first coil away from the second coil had the same effect as moving the second coil towards the first coil (induced current in the same direction). It did not matter which of the two coils was moving and which was still.

Faradays' third experiment on electromagnetic induction was the simplest. He produced a momentary induced current in a coil merely by thrusting a perma-

Electromagnetic Induction

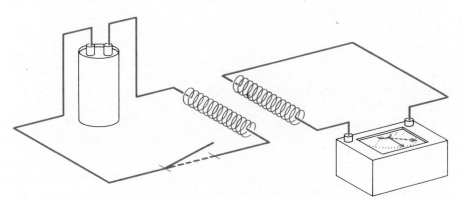

Figure 7-8
The surge of current in the first coil produces a changing magnetic field whose line of force pass through the second coil. This changing magnetic field produces a momentary current in the second coil. This is the basis of electromagnetic induction.

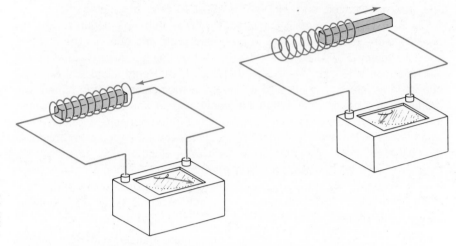

nent magnet within the coil. When the magnet was quickly withdrawn, a momentary current was induced in the opposite direction. Again, only the relative motion of the coil and magnet was important.

Faraday concluded that a changing magnetic field induces an electric field, which in turn creates an electric current in a conductor which may be located in that field. It makes no difference what causes the change in the magnetic field—moving a magnet or changing the current flowing in a loop or solenoid.

The currents produced by electromagnetic induction are only byproducts of the induced electric field. Faraday's basic discovery was that *a changing magnetic field produces an electric field.* The size of the induced electric field is proportional to the time rate of change of the magnetic field. Moreover, an induced electric field produces a force that is not conservative; in other words, a charge that moves in a closed loop through a region where the magnetic field is changing may either lose or gain energy. The induced electric field around a loop times the circumference of the loop is called the **electromotive force,** or **EMF.** EMF is measured in volts.

A changing magnetic field produces an electrical field.

Magnetic Flux

Let us imagine a circular loop placed between the poles of an electromagnet. The **magnetic flux,** or field through the loop, is related to the total number of magnetic lines of force passing through the loop. If the current in the electromagnet changes, both the magnetic field and the magnetic flux through the loop also change. **Faraday's law of induction** states that *the rate of change of the magnetic flux through the loop is equal and opposite to the EMF around the loop.*

The EMF in a loop is equal and opposite to the rate of change of magnetic flux.

Because the induced EMF is opposed to the change in magnetic flux, the current produced in the loop by the EMF tends to resist the change in flux. The induced current itself produces a magnetic field, and this magnetic field compensates at least in part for the change in the external magnetic field. This phenomenon was first formulated by the Russian scientist Heinrich Lenz (1801–1865): "A current brought into action by an induced electromotive force always produces effects which oppose the inducing action." Lenz' law is similar to Newton's third law of motion: "For every action, there is an equal and opposite reaction." And Lenz' law shows that electromagnetic induction obeys the law of conservation of energy.

Electromagnetic induction is the principle behind both the electric motor and the electric generator. A motor produces mechanical work from electrical energy. A generator does the opposite: it produces electric energy from mechanical work. An electric generator is an electric motor run backwards.

The Electric Motor and the Electric Generator

How does an electric motor work? Assume there is a conductor placed in a magnetic field. If a battery is connected so that an electric current flows through the conductor, Ampère's law says that a magnetic force will act on the conductor. This force can move the conductor, therefore doing mechanical work. However, because the conductor is now moving in a magnetic field, there is an induced EMF which, according to Faraday's and Lenz' laws, *opposes* the current flowing in the conductor. The source of the current (the battery) must do work to make the current flow against the induced EMF. This work is in addition to the work against any electrical resistance the conductor may have. Even if the conductor had no resistance, it would experience the same induced EMF and require work to be done by the battery. In the ideal case, the electrical energy expended by the battery is exactly equal to the mechanical work done by the moving conductor.

The force on a current-carrying conductor in a magnetic field is the basis for the electric motor.

To convert the motor into a generator, replace the battery with a resistor. Unlike a battery, a resistor consumes electrical energy rather than creating it. Now apply a mechanical force to the conductor and make it move through the magnetic field. Again an induced EMF will be produced. But there is no current initially flowing in the conductor for the EMF to oppose. Instead, the EMF *creates* a current, which flows in the direction of the EMF, instead of against it as in the case of the motor. The current flows through the resistor and consumes electrical energy according to the law that power equals i^2R.

Figure 7-10
Placed within a magnetic field, a current-carrying conductor experiences a force whose direction is perpendicular to the direction of both the current and the magnetic field.

What is the source of this energy? It is the mechanical work done on the moving conductor by some external force. The current in the generator flows in a direction opposite to that in the motor. Instead of assisting the motor, the magnetic force opposes the motion in the case of the generator. Work must be done against this magnetic force by the external force. In the ideal case, the force done by the external force is exactly equal to the electric energy created by the generator and consumed in the resistor.

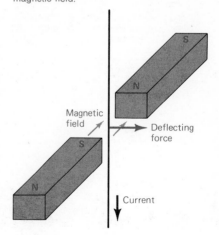

Magnetic field

Deflecting force

Current

160

Figure 7-11
In this modern generator, a set of giant coils provides the magnetic field and a current is induced in the coils of the turning armature.

Motors and generators are built with coils of wire that are free to rotate between the poles of a magnet. Sliding electrical contacts, or brushes, connect the rotating coils to the external power supply in the case of a motor or to the load (the device consuming electrical energy) in the case of a generator. In a generator, a source of mechanical energy, such as a steam or water turbine, makes the coil rotate.

In an electric generator there is a current induced in a conductor that moves in a magnetic field.

A ROTATING LOOP OF WIRE The EMF induced in a rotating coil is easily illustrated by considering a single loop of wire rotating about a line through one of its diameters. Let the loop begin in a position where the magnetic field is perpendicular to the plane of the loop. In this position, the magnetic flux through the loop is at its maximum positive value. When the loop rotates through 90°, the flux is zero, because the plane of the loop is now parallel to the magnetic field and no lines of force are passing through the loop. After rotation through 180°, the flux is at its maximum negative value. The flux is negative because the lines of force are now passing through the loop in the opposite direction. After rotation through 270°, the flux is again zero. After a full rotation of 360°, the flux has returned to its original value.

During the rotation from 0° to 180°, the flux was constantly decreasing and the induced EMF was positive. During the second half of the rotation (from 180° back to 0°), the flux was constantly increasing and the induced EMF was negative. The induced EMF is essentially the voltage produced by the generator.

Now let us consider the electric generator operating backwards as an electric motor. Alternating electric current at 60 cycles per second is now being supplied to the rotating loop. The loop is thus a small electromagnet. Which side of the loop is the north pole and which the south pole depends on the direction of the current in the loop. When the current changes direction, the north and south poles of the loop electromagnet change places.

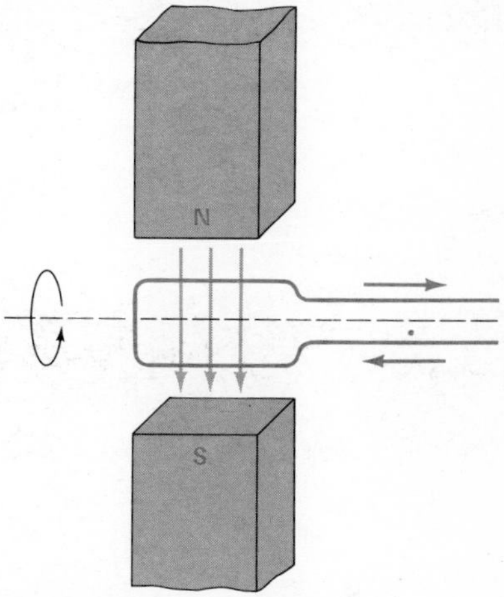

Assume the coil is rotating at 60 cycles per second and that the current changes direction when the loop is directly facing the magnetic field. Every time the loop electromagnet gets lined up with the magnetic field (as required by Oersted's law), its north and south poles switch places. So now the electromagnet is pointing in exactly the wrong direction; it is antiparallel to the field. By the time the loop rotates through 180° (half a cycle) to get parallel to the field, the current and magnetic poles of the loop reverse again. Hence the loop continues to rotate in the magnetic field as long as the alternating current is supplied.

Two points must be made concerning the discussion above. First, no reference has been made to Faraday's law of electromagnetic induction. This is because the emphasis has been on the work done by the motor. Electromagnetic induction is responsible for the electrical energy that must be consumed by the motor in order to make the current flow through the loop against the induced EMF. Second, the motor works only if the loop rotates at exactly the frequency of the current in the loop. Otherwise, the rotation of the loop and the changes of its magnetic polarity will get out of phase (out of step), and the magnetic force will begin to oppose the rotation of the coil instead of aiding it.

Figure 7-12
An electromotive force, and thus a current, is induced in the loop as it rotates in the field of a magnet. The current changes direction in the loop as its orientation in the field changes.

The induction motor works on a different principle. Instead of a constant magnetic field, there is an alternating field produced by a set of stationary electromagnets (the **stators**). A piece of soft iron (the **rotor**) is mounted so that it is free to turn in the magnetic field of the stator. The rotor becomes magnetized by the field of the stators. There are no electrical connections between the rotor and stator. Instead the field of the stators magnetize the rotor by induction. This magnetic induction is the same as the magnetic attraction of iron filings by a permanent magnet. It is merely a polarization or lining up of the magnetic domains in the rotor. As the magnetic field of the stators changes direction, the magnetism induced in the rotor changes so that the force on the rotor is always attractive, keeping the rotor moving.

The Induction Motor

Another important application of electromagnetic induction is the **transformer**. The transformer does not convert electrical power into another form of energy; instead, it transforms one form of electrical power into another form of electrical power.

The Transformer

A transformer increases or decreases voltage but does not affect power.

The transformer changes the properties of a source of AC electric power. Electric power is voltage times current ($P = Vi$). A small current across a small voltage can have the same power as a large current flowing across a small voltage. Which of these power sources is most convenient depends on the particular use being made of the power. For example, a television picture tube requires high voltage to produce the light signal on the picture tube and low current; a loudspeaker requires low voltage and high current.

A transformer can change a source with low voltage and high current into a source with high voltage and low current, with no net increase in energy or power. A device used to increase voltage and decrease current is called a step-up transformer. A step-down transformer does the opposite; it decreases voltage and increases current.

In the ideal transformer, there is no loss of power. The current is divided by exactly the same factor by which the voltage is multiplied. In a real transformer, there is always some energy loss, so that the power out is somewhat less than the power in.

Figure 7-13
The changing field in the first coil induces a changing electromotive force in the second. The ratio of the voltages in the two coils is proportional to the ratio of turnings in the coils. Transformers, such as the one pictured below, use this principle to step power up to a high voltage for efficient energy transportation.

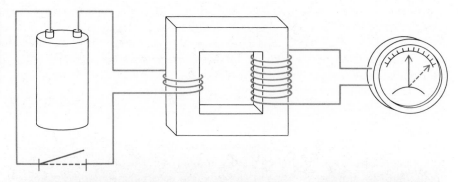

The idea behind the transformer was revealed in the very first electromagnetic induction effect discovered by Joseph Henry. He observed that when a current is stopped or started in one coil, a momentary current flows in a nearby coil. Thus an alternating current flowing in one coil will induce a second alternating current in a nearby coil. The original and induced currents will have the same frequency. Two such coils wound on a common iron core constitute a transformer. The coil to which power is supplied is called the primary. The coil from which power is drawn off is called the secondary. If the primary and secondary coils have an equal number of loops (turns of wire) and are in all other respects the same, the voltage induced in the secondary will be equal to the voltage supplied to the primary. If there are N_2 turns in the secondary and N_1 turns in the primary, then the ratio of the voltages in the secondary (V_2) and the primary (V_1) will be equal to the ratio of N_2 to N_1:

$$V_2/V_1 = N_2/N_1$$

For example, if the primary has 200 turns and the secondary has 5000 turns, the voltage ratio of the secondary to the primary is 5000/200 = 25. If the primary voltage is 100 volts, the secondary voltage will be 2500 volts. The current in the secondary will be reduced by the factor of 25. If 5 amperes are flowing in the primary, then 5/25 = 0.2 ampere will flow in the secondary. In both the primary and secondary, the power is 500 watts.

The most important application of the transformer is in the transmission of electrical power. Power lines may extend hundreds of miles from the source of power (such as a hydroelectric plant) to the consumer (a city or industrial complex). According to Ohm's law, power is wasted in the form of heat produced in the power lines. Since the wasted power is given by $P = i^2R$, the lower the current, the smaller the waste of electrical energy. Therefore, the power from

Figure 7-14
The generator hall in the hydroelectric power plant of the Bratsk Dam in Siberia. This plant is equipped with 18 generators and produced 24 billion KW hours in 1972.

a generator is stepped up to a high voltage and low current form. The voltages may be as high as tens of thousands of volts. The current is correspondingly low, minimizing the energy lost in transmission. When the power arrives at the consumer, it is stepped down to the more convenient American voltage for a typical house circuit of about 100 volts and 20 amperes.

The Galvanometer

Oersted established that forces exist between a conductor carrying an electric current and a permanent magnet. If the strength of the magnet is known beforehand, the size of the current can be determined by the force that acts on the magnet. For any given magnet, the size of the force is proportional to the size of the electric current.

This fact is the basis for an instrument used to measure electric currents. The instrument is called a **galvanometer**. A coil of wire is mounted on delicate springs between the poles of a horseshoe-shaped permanent magnet. When current flows through the coil, the coil becomes an electromagnet and tries to line up with the magnetic field of the horseshoe magnet. The springs resist this motion of the coil. Therefore, the coil rotates until the electric force of the springs and the magnetic force are balanced. The angle by which the coil had rotated indicates the size of the current; the larger the angle, the larger the current flowing in the coil.

Because the galvanometer is so sensitive, it can be used only for very small currents. To measure large currents, an ammeter is used. An **ammeter** is essentially a galvanometer with a resistor connected so that most of the current passes through the resistor. A known small fraction (usually 1/1000 of the current) actually passes through the galvanometer coil. So if the galvanometer reads 5 microamperes (5×10^{-6} ampere), the total current is a thousand times larger—5 milliampere (5×10^{-3} ampere).

The galvanometer is also the heart of another important electrical instrument—the voltmeter. The **voltmeter** is used to measure the voltage (potential difference) between two points in a circuit. In the voltmeter, a small current passes through a resistor and then through the galvanometer coil. According to Ohm's law, the voltage is equal to the current times the resistance. Since the resistance is known and the current is measured by the galvanometer, the voltage can be computed. In practice, the scale of the voltmeter is calibrated to read directly in volts.

Motion of Charges in a Magnetic Field

The force acting on a conductor carrying an electric current in a magnetic field has been described at some length. But you will recall that a current is nothing more than charges in motion. Therefore, even a single free charge moving through a space where a magnetic field exists will also experience a force.

The force on a moving charge is proportional to the product of three factors: the magnitude of the charge, the intensity of the magnetic field, and the velocity of the charge in the direction perpendicular to the magnetic field. A charge moving parallel to a magnetic field experiences no force at all. The force is largest when the charge is moving at right angles to the magnetic field. The direction of the force is at right angles to both the magnetic field and the velocity of the charge. If the charge is moving at right angles to the field, then the field, the velocity, and the force are in three mutually perpendicular directions.

Because the magnetic force is perpendicular to the velocity, a charge in a magnetic field will move in a circle with uniform speed. The magnetic force is then always directed towards the center of the circle. The frequency of revolution depends on the charge, the mass of the particle, and the magnetic field. This kind of circular motion occurs in the **cyclotron**, a device for accelerating subatomic particles to very high energy for use in atom-smashing experiments. Huge electromagnets are used to produce a uniform magnetic field in a circular region as large as several meters in diameter. Charged particles such as protons begin with low velocity near the center of the region. AC electric fields accelerate the particles. The magnetic field confines the motions of the particles so that they move in a spiral path outwards to the edge of the circular region. When the particles emerge from the cyclotron, they have very high energies.

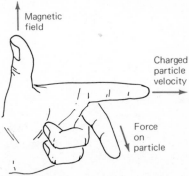

Figure 7-15
When the velocity of a charged particle is perpendicular to the direction of the magnetic field, the force on the particle is perpendicular to both the velocity and the field. The relationship of the three can be illustrated in the three-finger rule, illustrated here: the thumb and first finger are held at right angles to each other to represent the force of the magnetic field and the velocity of the particle; the middle finger, held at right-angles to the first finger, then represents the force on the particle.

Glossary

An **ammeter** measures current in a circuit.

A **cyclotron** is a device that accelerates elementary particles to high energies.

A **dipole field** is the magnetic field associated with a bar magnet. The lines of force form closed loops.

Electromotive force (EMF) is the induced electric field in a conductor multiplied by the length of the conductor.

Electromagnetic induction occurs when a current is induced in a conductor that is under the influence of a changing magnetic field.

A **galvanometer** is a coil of wire free to move in the field of a permanent magnet. The degree to which it moves measures the size of the current in the coil.

A **magnet** is a substance (usually iron) that attracts pieces of iron.

Magnetic flux is the total number of lines of force passing through a conductor.

Magnetic poles are the ends of a magnet where there is the strongest attraction for iron objects. One pole is north-seeking, the other south-seeking.

A **solenoid** is a coil of wire that has a magnetic field identical to that of a bar magnet when current passes through it.

A **torque** is produced by equal and opposite forces acting on different points of the same object.

A **transformer** increases or decreases the voltage of an AC current by utilizing the principles of electromagnetic induction.

A **voltmeter** measures the potential difference between two points in a circuit.

Exercises

1 If a coil is suspended in a magnetic field, what will happen if a current is passed through the coil? What happens if the current is reversed?

2 If a charge is moving in a magnetic field, in what direction will it experience the greatest force? In what direction will it experience the least?

3 Explain what happens when a conductor in the shape of a loop is rotated in a magnetic field.

4 If an iron bar magnet is broken into pieces, what happens to the magnetic poles? Why?

5 Do magnetic lines of force ever cross? Explain.

6 Describe Oersted's discovery of electromagnetism.

7 Explain the principle of electromagnetic induction. How is it applied to the electric motor? How is it applied to the electric generator?

8 A current in a horizontal conductor is flowing from left to right. How would you determine the magnetic lines of force around the conductor?

9 A transformer has a primary with 2400 turns; the secondary has 800 turns. If the voltage in the primary is 120 volts, what is it in the secondary?

10 If the power in the primary is 1200 watts, what is the current in the primary? What is the current in the secondary?

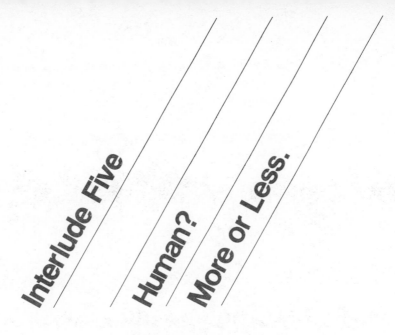

Interlude Five

Human? More or Less.

Cyborg. The word smacks less of science than of science fiction. A combination of the words *cybernetics* and *organism*, it was coined in 1960 by Drs. Manfred Clynes and Nathan Kline to describe the union of human and machine. The term *cybernetics* comes from the Greek word *kubernetes*, meaning "steersman," which also gives us the word *governor*. The term was invented by the American mathematician Norbert Wiener, who in 1948 published *Cybernetics*, a pioneering book dealing with self-regulating or *feedback* mechanisms.

A common example of a feedback mechanism is the thermostat described in Chapter 3. Say you set your thermostat for 70°F, when the room temperature is at 65°F. The thermostat receives this temperature as an instruction to turn on the furnace, which it does. When the temperature reaches 70°F, the thermostat will take this as an instruction to turn off the furnace. Thus the thermostat influences the furnace, which influences the temperature, which influences the thermostat. This circle of influence is called a feedback loop, and it's at the heart of cybernetics.

Norbert Wiener defines cybernetics as the study and comparison of communication and control in humans and machines. A cyborg—the union of a person and a machine—answers the need for a machine as versatile and imaginative as a human, or a human as durable and efficient as a machine. People are relatively weak animals. The mightiest weight-lifter is no match for a small crane; Ivory Crockett, the world's fastest human, couldn't outrace even a common family automobile. Electronic brains now have memory capacities greater than human brains, but the human brain is unsurpassed by the best electronic brains yet made—in its ability to recognize patterns, to adjust to unforeseen circumstances, and to come up with original solutions to problems.

The relationship between humans and machines began when a caveman picked up a stick or a rock and found he could do things with it that he could not do alone. Human work is mostly work with tools. In fact, there are some who say that people did not invent tools, rather tools invented people in their distinct character among the animals. In the long rise to civilization the human species has relied increasingly on tools and machines. Today some people depend wholly on machines to keep them alive—machines such as the pacemaker, surgically implanted to regulate the heartbeat, and the artificial kidney. Medicine is not alone, however, in the effort to unite humans and machines.

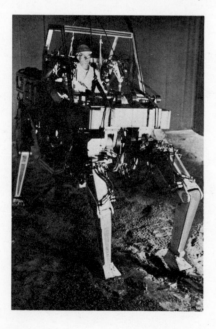

The "walking truck" was developed by the General Electric Company.

One of the first attempts to fuse human and machine is the "walking truck" developed by the General Electric Company to travel over terrain where vehicles with wheels or tractor treads cannot be used, an 11-ft-tall, four-legged monster of aluminum beams and hydraulic tubing that can "walk" along at a rate of up to 5 mph. The operator puts feet into ski boots attached to hydraulic devices that control the rear legs, and with hands grasps controls for the front legs. When the operator's right leg lifts up, the monster lifts its right hind leg. When the operator moves the left arm, the monster moves its left front leg. With practice a person can lumber over a boulder-strewn field almost with ease and pick up 500-lb loads with one of the front legs almost effortlessly. The machine is powered by a 75-hp engine that pumps liquid through the hydraulic tubing, and it has electronic sensors that convert the controller's motions into parallel mechanical motions.

A breakthrough that may enable cyborgs to become real partners with humans was the discovery of a way to provide feedback—in this case, the ability of the operator to feel resistance encountered by the mechanical arm. This ability has been refined to the point that the controller of the "walking truck" can stamp one of its front legs to produce 1500 lb of pressure, or rest it lightly on an egg without cracking the shell.

Machines that mimic the actions of human masters have many possible uses. The first was a simple arm and grip that handled radioactive materials while the controller was safe in another room behind a shield. The descendent of that first remote arm is now in use in industry and enables a single operator to move loads of up to 20,000 lb within an inch of where it is wanted while exerting less than 12 lb of human force.

Discoveries made by medical science are being incorporated into newer, more sophisticated versions of these human amplifiers. The human nervous system functions as a low-power electrical system whose individual units are cells known as neurons. The electrical potential inside the cell wall of each neuron is different from its outside potential, and when these potentials equalize, the neuron is said to "fire." When the nervous system sends a message to the hand to move, the message is carried via a chain of firing neurons along this electroneural circuit. These minute electrical charges can be detected, magnified, and put to use to control electro-mechanical devices.

The energy of the brain, too, is being tapped in the attempt to control sophisticated machines. Scientists have detected brain waves associated with expectancy and intention. When you decide to lift your arm, for instance, your brain emits a recognizable burst of energy. These waves are present even when less mechanical decisions are made—such as a decision to grow your hair long. They enable people to operate machines simply by "willing". In such a process, sensors detect the brain wave created by the decision, an electronic signal is sent to the control panel of a machine, and the machine responds. People have learned to control their brain waves by controlling their mental attitudes and decisions well enough to send a code that causes a computer to operate a typewriter, typing out the message intended by the sender, and the nervous system has been employed to operate artificial limbs for amputees.

Research work has been done on an "exoskeleton"—a frame of aluminum tubing to surround a person. Linked by sensors to the human body, the outer skeleton amplifies movements, enabling a person to lift 750-lb weights with one hand, or to perform other superhuman tasks. Such a device could be useful for walking on planets with gravitational fields stronger than the earth's, where human muscle power could not even enable the body to lift itself.

The next logical step in this line is to build remote-controlled cyborgs—machines that resemble people and mimic the actions of their "masters" from some remote point. Fires could be fought by machines that walk into the heart of the blaze, heat-resistant glass protecting a miniature television camera to provide a picture of what is happening for the control person, far from the scene. Mine rescues could be performed by cyborgs capable of lifting beams and boring through rock with drills attached to their bodies, unaffected by smoke and poison gas. Brain surgery could be performed by a team of cyborgs, operated by the world's best surgeons from their own hospitals thousands of miles away from the patient and from the other surgeons. Their incisions could be made without the trembling that is present in even the surest human hand. Each movement of the surgeons could be electronically modified to provide the best combination of human judgment and nonhuman precision.

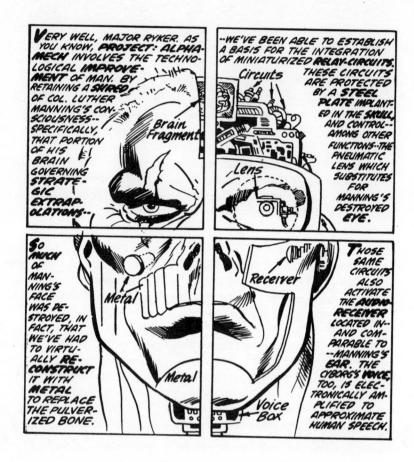

Like all of the creations of science, cyborgs have potential for dangerous as well as beneficial uses. Here Major Ryker, the evil mastermind of *Marvel* comics, demonstrates his cyborg creation—a human being who has been "improved" into a super-slave.

Sending such cyborgs on space flights could free human beings from confinement and boredom within the spacecraft. There would be no need to worry about making the living quarters comfortable and airtight or to carry a supply of food or oxygen. There is, however, a serious problem with such a plan. Radio signals travel at the speed of light, which is "only" 186,000 miles per second. At such speeds a signal from Mars would take 3 min to reach Earth. Thus, if a cyborg on Mars started to fall into a hole, at least 6 min would pass before its master could influence the cyborg's response. By then it might be too late.

The solution to such problems may be to copy the solution devised by the human nervous system. When you touch a hot stove, you don't have to wait for a conscious decision to move—if you did, you would suffer much more than you do. Your hand jumps back automatically, as a reflex action. It may be possible to provide space-exploring cyborgs with small computers programmed to respond automatically to certain situations—and thus to simulate the brain in one of its functions. Thus, if the cyborg's center of gravity began to fall out of line with its base, its automatic pilot would take over to set things right.

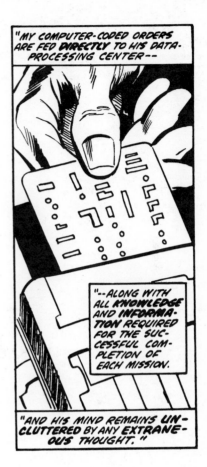

As we learn more about the human brain, it may become unnecessary to use control devices on our bodies to direct the cyborgs. Just as it is possible now to direct a computer to type messages by controlling your brain's alpha waves, it may become possible to control much more complex cyborg operations by the brain alone. The communication between the human brain and the electronic brain may eventually be reversed, enabling humans not only to give orders directly to an electronic brain, but to receive information directly from it. Speculation has it that with direct access to a computer via personal brain waves, a physicist could do in seconds what would have taken years without one, and even now takes hours to accomplish using standard programming.

Human memory could be multiplied a billionfold with such a system, and one could have instant recall. Technology now exists that makes it possible to store all the information produced in the last 10,000 years in a 6-ft cube. With human brains in direct communication with such an electronic brain, advances would accelerate in every intellectual field that depends on accumulated information.

Once it is possible to communicate directly with a computer, it will be possible to have it make records of human sensations. Thus tapes could be made of every sight, sound, smell, and feeling encountered by men and women as they climbed Mount Everest, ran the Olympic marathon, or played a leading role in a movie. This could lead to the greatest entertainment revolution yet. People could simply plug their minds into the entertainment circuit.

As more and more functions are performed in this way, evolution may take a new turn. Over the eons the human brain might become increasingly important, the body less, until human beings evolve into large collections of gray matter in control of a universe full of cyborgs. Although this may not sound appealing, remember that all our sensations are actually perceived in our brains. With feedback from cyborg to human brain, we would feel all the sensations now available (except perhaps pain) in addition to sensations now inconceivable. Our cyborgs could be built with "eyes" capable of seeing infrared and ultraviolet wavelengths now invisible to the human eye. They could "hear" sounds above and below the normal range of hearing. They could have a perfect sense of time and direction and also of radioactivity and magnetic fields—all communicated directly to the brain of the master. Of course our brains would then envolve new areas of sensitivity in response to these new perceptions.

Humans and machines have already been closely united. There are people with artificial ears implanted in place of damaged ears, augmented by small power packs and transmitters outside the body, which are connected through the nervous system into the brain. Scientists are experimenting with several types of electronic substitutes for the human eye for blind people. One involves a camera connected directly to the optic nerve in the eye; another uses a camera that translates its pictures into a pattern of pressures on the blind person's back. An atomic-powered heart is under development, a tiny closed-system steam engine heated by a minature atomic reactor, the steam of which is condensed and recycled through the system again and again. Such a system implanted in a person's chest would have enough power to operate not only an artificial heart, but artificial lungs and other organs as well. Some artificial organs are powered by long-lasting batteries, and some are controlled from outside the body. Solar energy is another possibility, with solar cells located outside the body that convert the sun's energy directly into forms usable by the mechanical devices augmenting the body. Fuel cells, which created electricity from biochemical reactions could also be used.

All these developments make it possible to conceive of an ultimate cyborg. Artificial ears more sensitive than a human could hope for, a heart strong enough to last centuries and withstand loads that would kill a "mere human," lungs capable of storing a month's supply of oxygen, a human brain in direct communication with an electronic brain—these could be the attributes of the person of the twenty-fifth century.

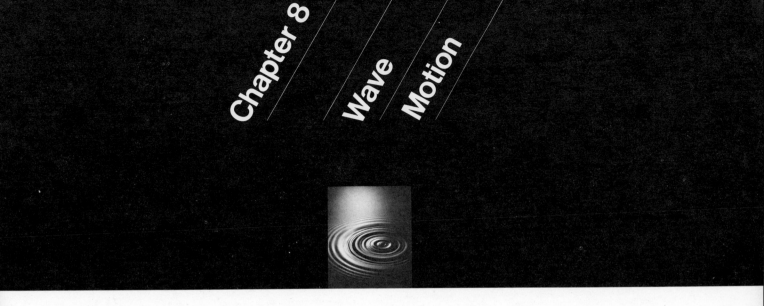

Chapter 8 Wave Motion

Anyone who has ever dropped a stone into a pond and watched the circular ripples spread out over the water's surface is familiar with wave motion. In addition to such common examples, there are many other forms of wave motion that are of great importance to our daily lives—though their wave nature is not always apparent to a casual observer. We hear because our ears are sensitive to sound waves that travel through the air. We see because our eyes respond to a range of electromagnetic waves that we call light. Other kinds of electromagnetic radiation, such as infrared, radio waves, and x-rays, have become increasingly useful to us through the developments of modern technology. There are other waves in nature that we have so far not learned to control, such as earthquakes and tidal waves.

Wave phenomena include sound, light, infrared radiation, radio waves, and x-rays.

The Nature of Waves

What is a wave? Let us imagine a pool of still water with chips of wood floating on its surface. When a pebble is dropped into the pool, small circular waves expand outward from the point where it enters the water. As the waves pass over the surface, the floating chips will bob up and down. They will not, however, move in the direction of the expanding waves to any appreciable extent. Evidently, then, the waves do not carry any matter with them—neither the chips nor the water itself. In fact, wave motion involves no net transfer of matter from one place to another. This can be seen most clearly if we consider another example of wave motion—a wave traveling down a rope that is fixed at one end (Figure 8-1). The wave is generated by holding the rope nearly taut and quickly jerking the free end to the side for an instant. A single wave pulse will travel down the length of the rope—a moving shape that is called the waveform. But if you mark a single point on the rope, you will see that it moves only from side to side, not at all in the direction of the wave.

Though matter does not move with the wave, we certainly perceive *something* moving along the rope as the wave travels. What is it that we see? It is a process—something that *happens to* the rope. More specifically, the wave is a traveling *disturbance;* a portion of the rope that in its normal undisturbed state lies in a straight line has been pulled out of its normal position. This disturbance—a displacement of a segment of rope from its usual position—travels along the rope and constitutes the wave pulse. When the first segment or particle of the rope is pulled to the side, it exerts a force on the second particle, pulling it too out of line. The movement of the second particle acts on the third in the same way, and so the disturbance is propagated along the rope. As the first particle returns to its normal equilibrium position, its movement sets up a similar chain reaction, causing each particle in turn to be pulled back into line.

A mechanical wave is a disturbance that travels through a material medium.

Kinds of Waves

The waves we have been describing so far are mechanical waves, all of which involve the propagation of a disturbance through a material medium—water in the first example, the rope in the second. In order to displace the particles of the medium, work must be done against the attractive forces that link them together. External energy must be supplied to generate a wave, and this energy travels with the wave. As the medium is disturbed it acquires elastic potential energy. For example, the stone hits the surface and pushes water up all around it. As the individual particles of the medium are set into motion they acquire kinetic energy. There is a constant interchange of these two forms of energy while the wave travels. This transmission of energy is a general property of wave motion. There are waves that do not need a medium for propagation, and indeed travel most rapidly in a vacuum. These are electromagnetic waves, which include visible light, infrared, radiant heat, ultraviolet, radio waves, x-rays, and gamma rays. These waves too, however, are carriers of energy, as we shall see in Chapter 9.

All waves transmit energy.

Not all disturbances in a medium produce mechanical waves, however. In order for wave motion to take place, the medium must be reasonably **elastic**. This means that any disturbance will give rise to internal forces that tend to restore the material to its undisturbed state. Any material that restores itself to its original form after a deforming stress is removed is elastic. If you pull on a piece of taffy and on a rubber band, and then let each one go, only the rubber band will return to its original form; the taffy will remain distorted. Therefore, the rubber is elastic, whereas the taffy is not. Jello, once it has set, is elastic; soup is not. Solid metals are also elastic; even though they cannot be stretched as much as rubber, they still restore themselves to their original form. A disturbance that exceeds the elastic limit of a material, however, cannot give rise to wave motion, since it will permanently deform the medium and destroy its elastic properties. At some point the rubber band breaks and the steel ball cracks.

As a disturbance travels through a medium the particles of the medium **oscillate**, or **vibrate**, about their equilibrium (undisturbed) positions, much as a pendulum swings to either side of its rest position. Waves can be classified according to the direction of this oscillation with respect to the direction in which the wave itself propagates. The movement of the individual rope particles in Figure 8-1 is always perpendicular to the direction of the wave's travel. This is characteristic of **transverse waves**. A different kind of wave motion is illustrated in Figure 8-2. When a segment of the spring is compressed so that its coils in

To transmit a wave, a medium must be elastic.

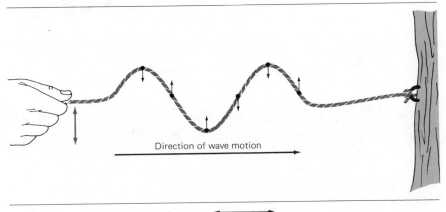

Direction of wave motion

Figure 8-1
Transverse wave motion along a rope that is fixed at one end.

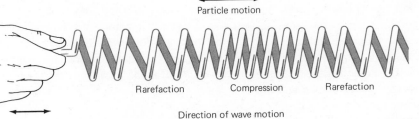

Particle motion

Rarefaction   Compression   Rarefaction

Direction of wave motion

Figure 8-2
Longitudinal wave motion along the length of a spring.

that segment are squeezed together, and then released, the compression will travel down the length of the spring. On either side of the compression are regions of rarefaction, where the coils of the spring are stretched apart, and these too propagate along the spring. As this disturbance, consisting of alternating compressions and rarefactions, passes along the spring, an individual coil will oscillate back and forth, parallel to the direction in which waves are moving. This is an example of **longitudinal**, or **compressional**, wave motion. Sound waves are the most commonly encountered examples of compressional waves. Air molecules squeeze together and bounce apart in regular rhythm.

Sound waves are an example of longitudinal waves.

Longitudinal waves can travel through all three states of matter—solids, liquids, and gases. Transverse waves, by contrast, can generally propagate only through solids. Transverse waves depend on the existence of powerful intermolecular forces which tend to pull displaced molecules back toward their equilibrium positions. Such forces are not strong in liquids and gases, which consequently lack the ability to restore themselves when disrupted by a transverse disturbance. Thus, liquids and gases are not sufficiently elastic to carry transverse waves. In the light of this fact, the discovery that earthquake waves cannot pass through the center of the earth provides evidence to support the idea that they are transverse waves, and that our planet has a liquid core (see Chapter 18). There is a significant exception to this general principle of liquids, however; because of the strong elastic forces of surface tension, surface waves in liquids are largely transverse.

Describing Waves

It is easiest to describe waves traveling in one dimension, such as the rope wave of Figure 8-1. All such waves can be represented graphically in the manner shown in Figure 8-3. The horizontal axis of the graph represents the distance along the direction of wave propagation. Points on this axis correspond to points in the medium through which the disturbance will pass. The vertical axis of the graph measures the magnitude of the wave disturbance, called **amplitude**. For a transverse wave moving along a rope, the vertical axis measures the displacement of points on the rope from their normal equilibrium positions. In this case the graphic representation of the wave resembles a snapshot of the wave itself; it looks exactly like the wave as it actually appears traveling down the rope.

For other types of waves the vertical axis can represent less visible disturbances, such as air pressure in a sound wave or electric and magnetic field strength for an electromagnetic wave. Such a graph describes the amount of disturbance at every point in the medium along a single direction at a given instant of time. Because of its applicability to all kinds of wave motion, we can use it to discuss the general properties of waves.

PERIODIC WAVES All waves whose graph shows a series of regularly spaced disturbances are called periodic. This category includes all disturbances pro-

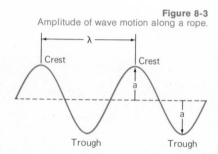

Figure 8-3
Amplitude of wave motion along a rope.

duced by regularly vibrating sources, such as the vibrating air column in a blown flute or the plucked string of a guitar. In fact, most of the sounds you commonly hear are periodic—the chief exceptions are harsh noises produced by explosions or the collision of two objects.

The graphic shape of any single disturbance in a periodic wave is called the **waveform,** or simply the wave. The distance between two corresponding points of successive waves is the **wavelength** of the wave (denoted by the Greek letter λ). For example, we could take the distance between two successive troughs in Figure 8-3 to determine the wavelength. The time which any particle of the medium takes to move through a complete waveform before starting to repeat its motion—say from crest to equilibrium position to trough to equilibrium position and back to crest—is called the **period** of the wave (denoted by T). Thus, if a periodic wave travels down the rope in Figure 8-1, a point on the rope will go through a complete vibration, a cycle of oscillation, every T seconds. Another way of saying the same thing is that the point will go through $1/T$ cycles each second. You can see from Figure 8-3 that the number of cycles a point in the medium goes through each second is equal to the number of waves passing through that point each second. This number is known as the **frequency** of the wave (denoted by f). It is evident from these definitions that the frequency of a wave is the reciprocal of its period:

$$f = 1/T$$

The frequency of a wave is the number of cycles it completes in a given unit of time.

Figure 8-4
"The Great Wave" by Hokusai (1760–1849), one of the masterpieces of Japanese printmaking.

Wave frequencies are commonly measured in cycles per second, or Hertz (Hz).

One cycle per second is called the hertz (Hz), after Heinrich Hertz (1857–1894), the German physicist whose experiments demonstrated that there are electromagnetic waves and they behave exactly like light waves. One hundred cycles per second is thus a frequency of 100 Hz. Frequency corresponds very closely to the sensation of color for visible light and of pitch in music. Sounds that we call "high," such as the upper register of the soprano voice or the piccolo, have high frequencies; sounds we call "low," such as those produced by a baritone voice, a tuba, or double bass, have low frequencies. The frequency sensitivity of the normal human ear covers roughly the range from 20 to 20,000 Hz. Dogs and cats, however, can hear tones up to about 50,000 Hz, and certain species of bats can detect sounds as high as 120,000 Hz.

Another important characteristic of a wave is its **amplitude**—the height of its crests and troughs. We can define it in terms of Figure 8-3 as the maximum value of the vertical displacement from equilibrium. Since the amplitude measures the amount of disturbance in the medium, amplitude measures the energy carried by the wave. If you increase the amplitude of vibration of a guitar string by plucking it more forcefully, you will create a more energetic wave, and hence a louder sound, without changing the pitch.

If it were possible to watch the waveforms advancing on the graph through successive moments in time, we would see that after the duration of one period, T, each waveform would have moved up to occupy the exact position that had been held originally by the waveform in front of it. That is, each waveform advances the distance of one wavelength during one period. The waveforms, then, travel with the same velocity, the **wave velocity**, v_w. (We designate it this way to distinguish it from the instantaneous velocity, v, of the individual particles of the medium as they oscillate back and forth about their equilibrium positions.) We can see that:

$$v_w = \frac{\text{distance traveled}}{\text{time}} = \frac{\lambda}{T}$$

Combining this with the equation for wave frequency, we can express the wave velocity in terms of the wavelength and the frequency:

$$v_w = f\lambda$$

The wave velocity generally depends on the elastic or other properties of the medium through which it is traveling. Sound, for example, travels at about 1100 ft/sec through air (slightly faster in warm air than cold air); through water its velocity of propagation is some 5000 ft/sec. Other examples of wave speeds in various substances are given in the accompanying table. In any medium, however, the wave velocity will always equal the product of the wavelength and frequency. This means that, for a given wave propagating through a particular medium, each frequency is associated with one wavelength.

Table 8-1
The Speed of Sound Through Different Materials

Material	Speed
Rubber	100 ft/sec
Air	1100
Lead	4000
Sea Water	4800
Iron	15000
Granite	20000

This relationship enables us to calculate the wavelength of a wave if we know its velocity and frequency. As an example, let us calculate the wavelength of a musical tone of 256 Hz (middle C on the piano):

$$\lambda = \frac{v_w}{f} = \frac{1100 \text{ ft/sec}}{256 \text{ Hz}} = 4.3 \text{ ft}$$

Conversely, given the velocity and wavelength, we can easily compute frequency. Suppose the wavelength of an FM radio wave is 3 meters; what is its frequency?

$$f = \frac{v_w}{\lambda} = \frac{3 \times 10^8 \text{ m/sec}}{3 \text{m}} = 10^8 \text{ Hz}$$

WAVEFRONTS What we have so far said about one-dimensional waves is largely true of waves in two or three dimensions as well. Figure 8-5 shows two-dimensional waves expanding from a point source, such as a pebble dropped into a pond. This figure could also be thought of as a two-dimensional cross-section of a spherical wave. It is easy to see that all points that are the same distance from the origin of the disturbance at any given moment will be "in step"—that is, at the same point in their cycle of oscillation. If, at time t_1, a point 3 meters from the origin is at its maximum displacement, all other points 3 meters from the origin will likewise be at their maximum displacements. A circle 3 meters in radius will then represent the crest of the wave. If the wavelength is 1 meter, all points $2\frac{1}{2}$ and $3\frac{1}{2}$ meters from the origin will be part of the trough of the wave at that same moment.

As time passes and the wave expands, the position of the crests and troughs will change. After an interval of time equal to half the wave's period ($\frac{1}{2}T$) the points that were formerly on the crest will be in the trough, and vice versa; but all points at the same distance from the origin will still be in step at the same point in their cycle together. Such points are said to be **in phase**. From this example we can also observe that all points along a wave separated by an integral number of wavelengths (that is, by λ, 2λ, 3λ, . . .) will also be in phase. Points separated by an odd number of half wavelengths ($\lambda/2$, $3\lambda/2$, $5\lambda/2$, . . .) are completely **out of phase**—that is, at opposite points in their oscillatory cycles. An imaginary line drawn so as to pass only through points that are in phase is called a **wavefront**.

In the previous example, the wavefronts are concentric circles. If we imagine the wave expanding in three dimensions from a point source, like sound from a gong or light from a candle flame, the waveforms will be nearly concentric spheres. By the time they have traveled very far from their source, short segments of a circular wavefront (such as that produced by a point disturbance in water) become almost straight lines, just as the earth's curved surface appears flat locally because the radius of its curvature is so great. Similarly, sectors of spherical waves that have traveled over great distances—such as light waves

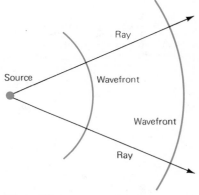

Figure 8-5
Two-dimensional waves expanding from a point source.

reaching earth from the sun—become nearly flat planes; such waves are called **plane waves**. In all these instances we can see that the wave front is everywhere perpendicular to a straight line drawn outward from the source of the wave in the direction of its travel. Such lines are called **rays** and are sometimes a more convenient means for representing the path of a wave than drawing the actual wavefront itself.

If a wave is the same in all directions, as in these examples, then describing its propagation in any one direction is sufficient to describe the entire wave. A diagram like Figure 8-3 can thus represent not only a one-dimensional wave, but also any one dimension of a wave propagating identically in two or three dimensions. In such a case each point on the waveform represents a point on the two- or three-dimensional wavefront. Nevertheless, the wave in three dimensions may still have important physical properties not shared by a one-dimensional wave. For example, let us consider the **intensity** of different waves, defined as the amount of energy transmitted through a unit area perpendicular to the wave in a unit of time. A plane wave will continue to have the same intensity as it propagates. A spherical wave with the same one-dimensional waveform, however, will decline in intensity, since the area of an expanding spherical wavefront increases as the square of its radius. When the radius of the wavefront doubles, for example, the energy it carries will be spread out over four times the initial area. Thus, the intensity of a spherical wave diminishes inversely as the square of its radius. This decreasing intensity takes the form of a decrease in the amplitude of the wave as it expands—a result that can easily be verified by measuring the brightness of illumination as we approach or recede from a small light source, such as a candle or light bulb. Frequency and wave length remain unaffected.

One-dimensional waves continue to have the same intensity as they propagate, but the intensity of three-dimensional waves declines as they propagate.

Wave Behavior

It is useful to compare the behavior of waves with that of individual particles, as we have studied it in previous chapters. One similarity has already been touched on. We know that when a single particle is given an initial velocity and is not acted on by any outside force, it will continue to move in a straight line with constant velocity. A plane wave behaves in much the same way; provided there are no energy losses or transformations, such a plane wave traveling in a uniform medium will also move in a straight line with constant velocity.

Reflection and Refraction

Another similarity concerns reflection. When a ball bounces off a smooth surface, the initial path and the final path always lie in the same plane. Moreover, the **angle of incidence** is equal to the **angle of reflection**, where both are measured with respect to a line perpendicular to the surface at the point of impact. The same principles hold when a wave strikes an opaque boundary (one that the wave cannot penetrate). In such a case it is most convenient to follow the paths of rays rather than of wavefronts; the rays behave exactly as if they represented the paths of particles bouncing off a hard surface. A flashlight beam directed at a mirror gives a good illustration of this phenomenon.

When a wave arrives at a boundary between two media in which its speed of propagation differs, however, something happens that has no exact analogy in the behavior of particles. The wave splits up into two parts, each of which is a wave of the original frequency but with a reduced amplitude and a different speed. One of these parts is transmitted into the new medium, and the other is reflected backward into the original medium. Figure 8-6 shows that the angle of the reflected wave is still equal to the angle of incidence. The transmitted wave, however, forms an angle that is *not* equal to the angle of incidence. As the wave enters the second medium, its direction of propagation is shifted somewhat, and the wave continues in a different direction. Such a change in wave direction is called **refraction**.

The figure shows how the magnitude of this effect depends on the relative velocities of the wave's propagation in the two media. Because the wave moves faster in medium 1 than in medium 2, the portion of the wavefront that is still in the original medium gains ground on and starts to catch up with the adjacent portion that is now traveling more slowly through the new medium. The phenomenon resembles a column of soldiers making a turn; those on the inside mark time or march slowly, while those on the outside move at their normal speed until the column has wheeled around to face in a new direction.

When the entire wavefront has entered the new medium, the wave continues in a straight line once again, but in a new direction. Notice that while the frequency remains the same in the two different media, the wavelength is shorter in medium 2 than in medium 1. This is consistent with the wavelength–frequency equation, which tells us that for a given frequency the wavelength of a wave is proportional to its velocity of propagation. Similarly, when a wave moves from medium 2 to medium 1, the emerging part of the wavefront will outstrip the adjacent part of the wavefront still in the slower medium. The wave will bend in the opposite direction, and its wavelength will increase.

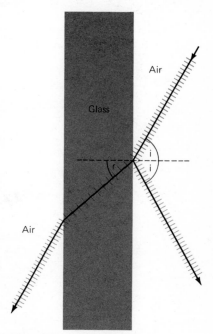

Figure 8-6
Refraction of a wave passing through two different media.

Perhaps the most profound difference between particles and waves becomes evident when we compare what happens when two particles collide with the corresponding events when two waves collide. Contact forces between particles, we know, will cause them to bounce off each other upon collision, their paths deflected in different directions. Particles also affect each other by repulsive or attractive forces at a distance, as, for example, when they are electrically charged (see Chapter 6). Nothing of the sort happens with waves. There are no forces between the traveling periodic disturbances we call waves, so waves pass right through one another, emerging completely unaffected by the encounter. The nature of the relation between two waves is given by the princi-

Superposition and Interference

A

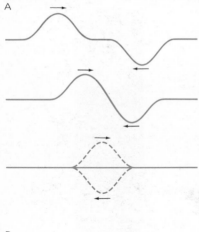

B

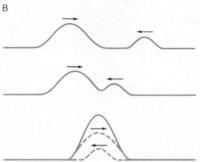

Figure 8-7
(A) Destructive interference caused by
superposition of two waves which are out of
phase. (B) Constructive interference caused
by superposition of two waves which are
in phase.

Interference changes the amplitude of
two periodic waves of the
same frequency.

ple of **superposition**: when two or more waves pass a given point at the same time, the instantaneous displacement from the normal undisturbed condition at that point is the sum of the several displacements due to the separate waves.

For one-dimensional waves, the resultant displacement is simply the arithmetical sum of the individual displacements. Figure 8-7 shows the case in which two identical, symmetrical pulses traveling in opposite directions meet. At the instant when the two pulses exactly coincide, the displacements cancel each other, and there is no disturbance at all in that region of the medium.

The superposition of two periodic waves is called **interference**. If two one-dimensional waves of the same frequency are superposed so that their crests coincide, then the two waves will be in phase. The effect of the superposition will be to reinforce the crests and troughs of both waves. The result will be a single wave with the same frequency and wavelength as the two component waves, but with a larger amplitude than either. This is an example of **constructive interference**. On the other hand, if the two waves are superposed out of phase by half a wavelength so that the crests of one coincide with the troughs of the other, they will cancel each other out entirely, and the result will be no wave at all. This phenomenon is known as **destructive interference**.

Interference becomes more interesting when it occurs in two or three dimensions. Consider the two sets of circular waves in Figure 8-8, generated by separate point sources vibrating at the same frequency. As the waves spread out from the point of origin, they will begin to interfere with each other. In the figure, crests of wavefronts are represented by red circles, and troughs of wavefronts by grey circles. Wherever the crests and troughs intersect, complete destructive interference takes place. Notice that these intersections lie on a regular array of curved lines, called **nodal lines**, running between the two wave sources. At any point along a nodal line, the superposed wavefronts are out of phase with each other by exactly half a wavelength. The nodal lines represent the locus of points satisfying the equation:

$$x_1 - x_2 = (n - \tfrac{1}{2})\, \lambda \qquad n = 1, 2, 3, \ldots$$

where x_1 and x_2 are the distances to the two sources, and n is an integer different for each nodal line. This is the equation for a hyperbola. At all points along these hyperbolas, the medium is completely undisturbed because the two waves are exactly out of phase. This phenomenon is in sharp contrast to what we would expect using particles, for there is obviously no way that we could add or superimpose two streams of particles and get no particles at all! Destructive interference is easy to observe at home or in the lab; simply spread two fingers apart and dip them rhythmically into a pan of shallow water.

Diffraction Another striking property of waves is their ability to travel around corners and obstacles in their paths. This property is called **diffraction**. It can be seen in

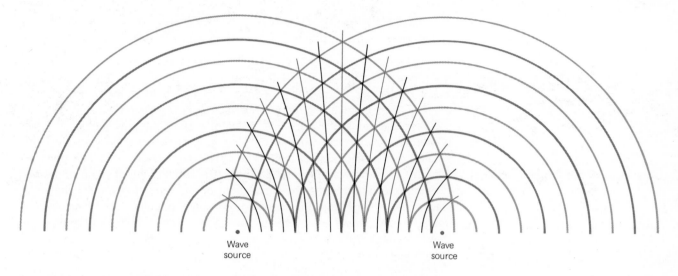

Wave
source

Wave
source

this simple example: two people separated by the corner of a building cannot see each other, but they can hear each other's voice. Evidently the sound waves bend around the corner of the building. The idea of reflection cannot be used to account for this bending because there are no objects in the figure which would suitably deflect the waves. Neither is this a case of refrac-

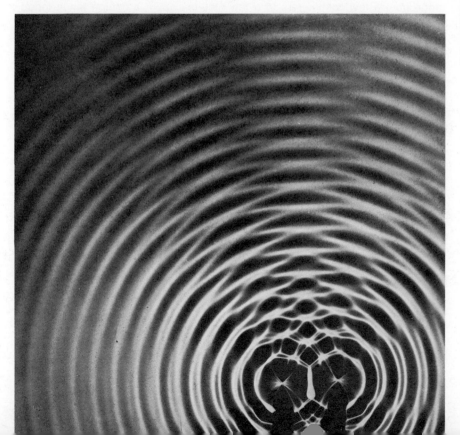

Figure 8-8
Interference between two sets of circular waves, generated by separate point sources vibrating at the same frequency. The pink circular arcs represent crests, the grey arcs troughs. Where crests and troughs coincide, destructive interference causes the cancellation of the wave, and the medium is undisturbed. The black lines passing through these points are the nodal lines. The red antinodal lines run through points where constructive interference takes place—crests reinforce crests, and troughs reinforce troughs. Both nodal and antinodal lines are hyperbolas. The photograph below shows the same interference pattern generated in a tank of water.

tion, since the speed of sound does not change over the path traveled by the sound waves. Clearly, diffraction must be a separate phenomenon altogether.

This provokes further questions: if the two people can hear each other, why can't they see each other? Since light also travels as waves, why do sound waves bend around the corner of the building while light waves do not? Why do the two waves behave differently? The nature of this difference can be seen in the three diagrams of Figure 8-9. In each, a straight-line wavefront of water waves strikes a barrier with an opening in it. Notice that as the wavefronts emerge on the other side of the barrier openings, the portions of the wavefront near the edges of the openings are bent. The width of the opening in the first drawing is comparable to the wavelength of the impinging waves. The water waves that emerge from this opening are nearly circular—*as if there were a source of circular waves located at the center of the opening.* In the next two drawings, the barrier widths are larger and the emerging wavefronts are correspondingly less curved.

It can be shown that in general the extent that waves are diffracted when they pass through an opening depends on the ratio of their wavelength to the width of the opening. If the opening is very large with respect to the wavelength of the incident wave, the wave is diffracted only slightly. The reason that the diffraction of light around buildings is not noticeable is that its wavelengths are far shorter than those of sound waves. The wavelengths of visible light are on the order of 5000 Å, or 5×10^{-7} m, whereas the wavelengths of audible sounds range from a few centimeters to a few meters—about a million times longer.

Figure 8-9
Behavior of straight-line wavefronts striking barrier with an opening in it.

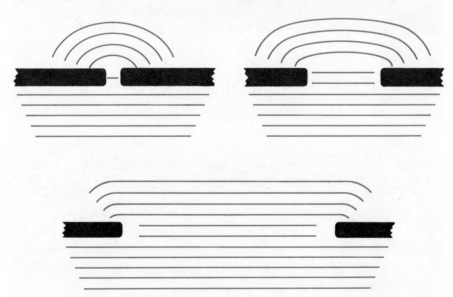

Thus everyday objects with sizes measurable in centimeters or meters can cause considerable diffraction of sound, but they are far too large to have very much effect on light waves.

Diffraction of light can indeed take place, but only when light passes through holes that are extremely small. Optical diffraction was first observed by the Italian scientist Francesco Grimaldi (1618–1663), who noticed that when light shines through a tiny pinhole, the image formed on a screen is very slightly larger than would be expected if the light traveled in a perfectly straight line. As light passes through the hole, diffraction causes the waves to spread out a little, thereby increasing the size of the image.

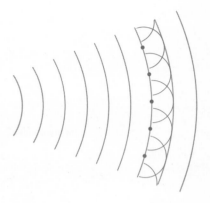

Figure 8-10
Generation of circular wavelets to form a new wavefront.

HUYGENS' PRINCIPLE That an obstacle can cause waves to bend around it is a rather puzzling phenomenon, with no counterpart in particle behavior. To explain diffraction we must utilize the imaginative insight of the Dutch physicist Christian Huygens (1629–1695), known as **Huygens' principle**. This principle has importance well beyond the phenomenon of diffraction, since it plays a key role in the analysis of wave propagation in general.

To approach Huygens' principle, we can return to the stone dropped into still water. When the stone strikes the water it creates a series of circular wave disturbances—ripples—that travel outward from the initial point of impact. Very quickly the stone falls to the bottom of the pool, but the waves continue to move. Since the stone resting on the bottom of the pool cannot have anything to do with the ongoing activity of the wave disturbance, it is only natural to assume the disturbances along the wavefront at any moment in time are the cause of new disturbances at the next instant. Disturbed particles of water as they fall back toward their normal positions are responsible for disturbing their neighbors. In this way the wave disturbance progresses outward in the pool.

Figure 8-11
Wavelets produced by a point on a wavefront impinging on a barrier opening

Huygens' principle generalizes this idea to all forms of wave propagation by asserting that *each point* on a wavefront acts as a new source of waves. This can be visualized for a straight-line wave as shown in Figure 8-10. Each point on the wavefront at *A* acts as a generator of circular wavelets. The new wavelets interact by superposition, interfering constructively and destructively with each other to form the new wavefront located at *B*, also a straight line. The points along the wavefront at *B* will in turn generate additional wavelets which will combine to form a wavefront at *C*, and so on.

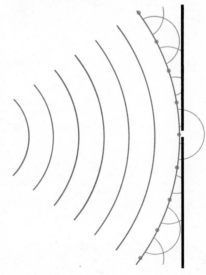

We can use Huygens' principle to explain the diffraction of a wave by a narrow opening in a barrier (Figure 8-11). Figure 8-11 shows the secondary wavelets generated by a point along a wavefront at the opening. Notice that the interference of these wavelets causes the new wavefront to be straight over much of its length. However, near the two edges of the barrier opening, the interference pattern is different due to the absence of any wavelets from the regions occupied by the barrier. It is this that gives rise to the diffraction effects. If the opening is small compared to the wavelength of a wave, the sources of the secondary wavelets inside the opening are very close together. The wavelets that they generate very nearly coincide, and the net effect is not very different from that produced by a single source of circular waves located in the center of the opening.

Standing Waves

Another effect of interference is the creation of standing waves, which have particular importance for musical instruments. Let us consider again the simple arrangement of a rope fastened to a wall, depicted in Figure 8-12. When a single wave pulse traveling along the rope reaches the wall, it is reflected backwards, but with its displacement reversed—that is, it returns half a wavelength out of phase with the original wave. What started out as a crest becomes a trough in the reflected wave, and a trough is reflected back from the fixed end as a crest. Now suppose the free end of the rope is moved from side to side uninterruptedly, so that a continuous series of periodic waves is produced. In a short time, after several of the initial wave pulses have been reflected, there will be two trains of waves traveling in opposite directions along the rope. These two wave trains interfere with each other to produce the resulting movement of the rope.

Under these conditions, waves of certain frequencies produce a pattern of movement in the rope that does *not* have the appearance of traveling waves. It is true that at any instant the rope has the static *form* of a wave disturbance, with a wavelength equal to that of the original traveling waves, but this waveform does not move along the rope. For this reason the fixed wave pattern is called a **standing wave**. There is, however, a good deal of vibrational movement taking place, for the rope is divided into one or more segments, each of which oscillates up and down in a loop—forming a crest shape at one moment and a trough the next—with the frequency of the original disturbance. Separating these segments are fixed points along the rope, called **nodes**, which do not undergo any movement or displacement from normal position. Thus, the nodal points have zero amplitude at all times; they are one-dimensional equivalents of the nodal lines formed by the interference of two-dimensional waves. Points halfway between the nodes, where the maximum rope displacement occurs, are called **antinodes**, or **loops**.

Figure 8-12
(Top drawing) A standing wave. (Bottom four drawings) Various standing waves which may be excited in a string stretched between two fixed points. N shows the location of the nodes.

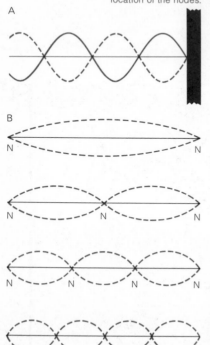

Figure 8-12A shows a standing wave and the locations of the various nodes. From the figure it is clear that the distance between consecutive nodes is equal to half the wavelength of the original traveling wave. Figure 8-12B shows various standing waves that can be excited in a string stretched between two fixed points. Since the two endpoints of the string do not move, they must be nodal points in each case. The first example shown is one in which the only nodes are the endpoints. If the distance between the two fixed points is L, then we must have $\frac{1}{2}\lambda = L$ because this is the lowest frequency standing wave that can be excited in the string. In the second example of Figure 8-12B there is an additional nodal point at the center of the string, and therefore there is one entire wavelength fitted into the distance L, that is, $\lambda = L$. In the third example we have $\frac{3}{2}\lambda = L$.

Since the two endpoints must always be nodes, we can see that, in general, there must always be an integral number of half-wavelengths fitted into the distance L. So we can write the general condition required for standing waves as:

$$n(\lambda/2) = L \qquad \text{or} \qquad \lambda = 2L/n$$

where n is an integer.

For any given stretched string, the wave velocity is fixed, so we can use the above equation to get an expression for the select frequencies that give standing waves:

$$f = \frac{n v_{\text{w}}}{2L}$$

The lowest frequency at which a standing wave can be set up for a particular vibratory element is called its **fundamental frequency**. For a string, whose end points must be nodes, this corresponds to the frequency at which $n = 1$. For an open column of air, such as is found in an organ pipe, where only the closed end must be a node and the open end can be a loop, it corresponds to the frequency at which $n = \frac{1}{2}$. Frequencies which are integral multiples of the fundamental frequency are called **harmonics**, or **overtones**.

From the above equations we see that a stretched string can vibrate either at its fundamental frequency or at any of its overtones. In general, when a stretched string is set to vibration (for example, when a piano string is struck), it does not vibrate at a single, pure frequency. Instead, the vibrations consist of a mixture or superposition of the fundamental frequency and several harmonics, the exact number and intensity of which are unique to each vibrating system and constitute its particular timbre, or tone color. For this reason a given note played on the piano sounds quite unlike the same note played by a violin, a clarinet, or a trombone. Figure 8-13 shows the superposition of a fundamental frequency and some harmonics. The resulting wave pattern has the same overall period as the fundamental, but its detailed waveform can be quite complex.

Although we have discussed standing waves in terms of only one dimension, it is not difficult to see that standing waves may result from any kind of wave motion confined to a close space. For example, the sounds produced by a vibrating drumhead are due to standing waves in two dimensions with the rim serving as nodal line. The creation of sustained notes, even brief ones, may require an instrument to vibrate hundreds of times per second in the same way. This is achieved by setting up standing wave patterns in the instrument's vibra-

A
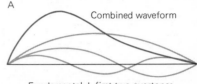
Combined waveform

Fundamental + first two overtones

B
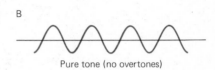
Pure tone (no overtones)

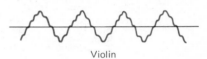
Violin

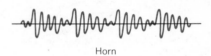
Horn

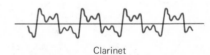
Clarinet

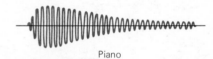
Piano

Figure 8-13
(A) A fundamental frequency and its first two overtones. (B) Waveforms of different musical instruments.

tory element—a column of air in a trumpet, a stretched skin in a drum, or a taut string in a guitar.

Sound

Now that we have surveyed some of the general properties of waves, we will discuss sound in more detail. Sound waves are compressional wave disturbances that can be detected by our sense of hearing. These compressional waves are created by vibrating objects, such as the tines of a struck tuning fork. The back and forth movement of the fork alternately compresses the air molecules together immediately in front of it and then presses in the other direction, leaving relatively empty space. This results in a series of pressure disturbances, alternate compressions and rarefactions, which propagate through the air at a characteristic wave velocity. When this pressure wave is described graphically (Figure 8-14), the vertical axis represents the pressure of the air. Where the amplitude of the pressure disturbance is represented as positive, it means that the pressure is slightly higher than in the surrounding undisturbed air. A negative amplitude means that the pressure is below normal.

The speed at which sound waves travel through the air depends on both the pressure and the density of the air. At sea level the speed is about 1100 ft/sec, or 750 miles per hour. This speed is rapid by ordinary human standards, but it is far slower than the speed of light (186,000 mi/sec). Listening to thunder during an electrical storm, often you hear a clap of thunder several seconds after you see the flash of lightning that caused it. Even sitting in the outfield bleachers at a baseball park, only a few hundred feet from home plate, you will see that ball being hit before you hear the crack of the bat. Light is so fast that in everyday matters we can consider it virtually instantaneous; therefore, what we see always represents a more up-to-date version of reality than what we hear by means of the relatively slow sound waves. The speed of sound decreases as the air pressure drops. Consequently, at high altitudes the speed of sound is less than at sea level—some 13% less at 40,000 ft, for example.

Physicists study the mechanical, thermal, and other physical properties of sound waves; psychologists and physiologists study the perception of sound by living organisms and the mechanisms of ear and brain that produce these sensations upon interaction with sound waves. We have already mentioned that our perception of pitch is related to frequency. It has been experimentally determined that a sequence of sounds whose frequencies increase by successive equal ratios—for example, 67 Hz, 100 Hz, 150 Hz, 225 Hz, and so on, each of which has a frequency of 1.5 times the previous tone—will be heard by the normal listener as a series of equally spaced notes. It is also true, at least for listeners brought up in our Western musical culture, that sounds whose frequencies can be expressed as the ratios of small whole numbers—1:2, 2:3, 3:4, 3:5, 4:5, and so on—are perceived as pleasant, "harmonious," or, to use

Figure 8-14
The Doppler effect. Wavefronts ahead of the moving source are crowded together, resulting in a shorter wavelength and a higher frequency. Behind the source, the wavelength is greater and the frequency lower.

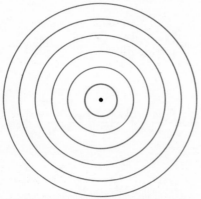

Source at rest

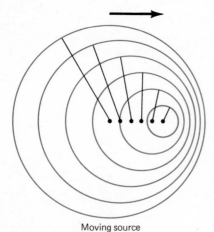

Moving source

the musician's term, **consonant**. Of these, the most important is the ratio of 1:2 (for example, notes with frequencies of 220 Hz and 440 Hz), called the **octave**. Tones related in this way are sensed to have a similar tonal quality and are thus designated with the same letter name—220 and 440 Hz are both A. Our system of scales is based on this interval, with each octave divided up into 12 evenly spaced notes.

The intensity of a sound wave, as we have mentioned, determines the perceived loudness of the sound. Because the human ear is sensitive to such a wide range of intensity levels, it has been found convenient to express this characteristic on a logarithmic scale, employing units called the bel (named after the American scientist and inventor Alexander Graham Bell, 1847–1922) and the decibel (dB), equal to 0.1 bel. A difference of 1 bel (10 dB) between the loudness of two sounds means that the first is 10^1, or 10, times louder than the second. A difference of 2 bels (20 dB) represents a ratio of 10^2, or 100 times the loudness, and so on.

Zero bels corresponds to the threshold of human hearing—an energy of 10^{-16} watts/cm². Table 8-2 shows various typical noises with their relative intensity levels expressed in decibels above the threshold of hearing. The loudest sound that can be tolerated without pain is about 120 dB above the threshold of hearing, which is about 10^{-4} watts/c_2. That means that it has an intensity that is 10^{12}, or one trillion, times greater than the intensity of the softest sound that can be heard.

Table 8-2
Sound Intensities

Sound	Intensity Relative	dB Level
Threshold of hearing	1	0
Breathing at rest	10^1	10
Leaves rustling	10^2	20
Two people talking	10^6	60
Busy street traffic	10^7	70
Jackhammer	10^9	90
Propeller plane takeoff	10^{12}	120
Machine-gun fire	10^{13}	130
Rocket lift-off	10^{17}	170

The Doppler Effect

When a train with a continuously sounding whistle or horn passes, a marked change occurs in the pitch of the sound as it goes by. The pitch of the horn seems to be shifted much lower when the source is moving away than when it is approaching. This effect is common to all types of wave motion, and is called the **Doppler effect**, or **Doppler shift**, after the Austrian physicist, Christian Doppler (1803–1853), who studied it. The Doppler effect is the change in the observed frequency of waves whenever the source of the waves and the observer are in motion relative to each other. Since the frequency of a sound wave determines its pitch, the Doppler shift for sound is perceived as a change in pitch. For light waves, the frequency shift is perceived as a change in color, though because of the great speed of light the effect is usually too small to be detected except in astronomy, where it plays an important role in the determination of stellar velocities (Chapter 23).

The change in the observed frequency of a wave by an observer who is moving in relation to the wave source is called the Doppler effect.

Figure 8-14 illustrates the nature of the Doppler effect for circular waves, such as those on the surface of a body of water. Notice that as the waves spread out from a moving source, the wavefronts are crowded together in the direction of movement and spread out in the opposite direction. Thus the movement of the source brings about a decrease in the wavelengths of the waves ahead of it and an increase in the wavelength of the waves behind it. Alternately, we can

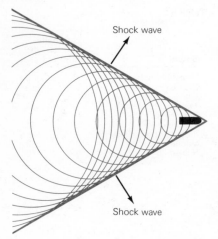

Shock wave

Shock wave

Figure 8-15
Shock waves produced by a high-speed
rifle bullet.

consider the situation from the point of view of an observer moving toward a source of waves. Because of his own motion, he will encounter more than the f waves per second that would reach him were he standing still. Conversely, if he were moving away from the source, the waves would have to catch up with him, and fewer than f waves per second would succeed in reaching him. The amount of the difference is proportional to the relative speed of the observer and the wave source; the greater the velocity, the larger the shift in frequency and wavelength.

SHOCK WAVES What happens when the velocity of a moving source is so great that it exceeds the wave velocity? No waves will spread out in front of the source, so the conditions for the Doppler effect will not exist. A quite different phenomenon results, which is illustrated in Figure 8-15. The circular wavefronts add together by the principle of superposition to form the two tightly compressed diagonal wavefronts shown in the figure. These two wavefronts are called **shock waves**. The most familiar example is the bow wave produced in water by a speedboat. Shock waves in air are produced when objects travel at speeds faster than the velocity of sound (supersonic). High-speed rifle bullets produce shock waves that are heard as loud "cracks" when they strike the ear drum. The energy of a shock wave increases with the size and speed of the supersonic object. The shock waves produced by supersonic aircraft, the sonic booms, have enough energy to shatter windows and cause other damage.

Glossary   The **amplitude** is the magnitude of the wave disturbance and is equal to the maximum displacement from the equilibrium position (half the distance from the top of the crest to the bottom of the trough).

An **Angstrom** is a unit of wavelength equal to 10^{-10} meters. It is abbreviated Å.

The ability of waves to bend around the edge of an obstacle is known as **diffraction**.

The **Doppler effect** is the change in the frequency of a sound wave when either the listener or the source of the sound moves—for example, the change in pitch of a train whistle as the train approaches.

The **frequency** of a wave is the number of cycles it completes in a unit of time and is commonly measured in cycles per second, or Hertz (Hz).

The lowest frequency at which a standing wave can be set up in a particular vibratory element is called its **fundamental frequency**.

Huygen's principle states that every point on a wavefront may be considered as a new source of circular waves.

Regions of a wave that are at the same point in their cycle of oscillation at the same time—for example, all at the crest together—are **in phase**.

Wave **intensity** refers to the amount of energy transmitted through a unit area perpendicular to the wave in a unit of time.

Wave **interference** occurs when two or more waves of the same type overlap, resulting in increased or decreased intensity.

Longitudinal waves, such as sound, consist of alternating regions of compression and rarefaction in the medium through which they pass.

The **period** of a wave is the time necessary for it to complete one cycle.

Wave **refraction** is the bending of a wave due to a change in its velocity as it passes through different media.

A **transverse wave** is a disturbance which causes the particles of the medium to vibrate in a direction perpendicular to that of the wave's movement.

A **wavefront** is an imaginary line drawn so as to pass only through points that are in phase.

The **wavelength** of a transverse wave is the distance between two successive crests or troughs. The wavelength of a longitudinal wave is the distance between two successive compressions or rarefactions.

Exercises

1 What is the difference between a transverse and a longitudinal wave? Give examples of each.
2 Draw a transverse wave and label each of the following: amplitude, wavelength, crest, and trough.
3 Why is the period of a wave equal to the inverse of its frequency?
4 Porpoises use high frequency sounds for echolocation (sonar). If such a wave bounces off a coral reef and returns to a stationary porpoise 0.42 sec after being emitted, how many feet is the porpoise from the reef?
5 What is the frequency of a sound wave with a wavelength of 2.5 ft moving through a lead wall?
6 How long is the wavelength of a note produced by a guitar string with a fundamental frequency of 270 Hz, if the speed of sound in air equals 343 m/sec?
7 If the range of frequencies detectable by the human ear lies between 20 Hz and 20,000 Hz, what are the upper and lower limits for the wavelengths of audible sound?
8 A seismograph records the first wave of an earthquake fifteen minutes after it strikes a small town. If $\lambda = 3.7$ m, and $f = 52$ Hz, what is a) the velocity of the wave and b) the distance between the earthquake site and the seismographic station?
9 What is the shift in frequency that occurs in a sound wave with $\lambda = 2.5$ ft as it moves from iron to air?
10 Two children are playing 20 ft and 80 ft respectively from their mother when she calls to them. What is the ratio of the sound intensities that reach the two children?
11 What is meant by sonic boom and how is it produced?

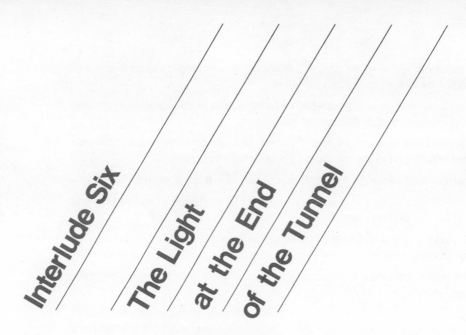

Interlude Six

The Light
at the End
of the Tunnel

The term *laser* is an acronym for *l*ight *a*mplification by *s*timulated *e*mission of *r*adiation. A laser is a device that produces a very intense, highly directed, monochromatic, coherent beam of light.

Most atoms and molecules will emit light if they are energized by an outside source, usually heat, electricity, or other radiation. Ordinary light—from a candle, a light bulb, the sun, or any other source—comes from such energetically excited atoms, which emit photons, the short bursts (quanta) of light energy, in a random fashion without any further outside influence. The photons are out of phase with each other, may be of different wavelengths and travel in various directions. Such random light waves are known as the *spontaneous* emission of radiation.

But suppose that just as an atom reaches its excited state, it is bombarded by an additional photon of that atom's specified frequency for one of its forms of spontaneous emission. Since the atom is already energetically excited, it cannot absorb the new photon. But Einstein showed that the new bombarding photon will stimulate the excited atom to emit a second photon identical to it, one that is of the same wavelength precisely and is exactly in phase (step) with it. Moreover, the second photon emission occurs more quickly than any subsequent spontaneous emission would. There are then two photons for the original one. Any input signal placed on the bombarding photon will be amplified exactly, with no distortion.

Thus, the incoming light wave will be amplified by the new *stimulated* emission, the two joining together exactly in phase to produce a stronger light. Intensifying this process sufficiently, it should be possible to produce a beam of light that is monochromatic (all of the same wavelength), coherent (in phase), powerful, and highly directional. The predicated effect was first demonstrated by the American Charles Townes nearly 40 years after Einstein provided the theory. In 1953, Townes constructed the first maser, which uses microwave frequencies. The laser, which amplifies visible light by stimulating emission, followed in 1960.

A ruby laser may be used as an example of how any laser produces its beam. The coils of a photoflash lamp are wrapped around a cylinder of ruby crystal, which is composed of aluminum oxide with a small amount of chromium oxide impurity that gives the crystal its normal ruby red color. Both ends of the ruby crystal, which acts like a light tunnel, are polished and mirrored, one com-

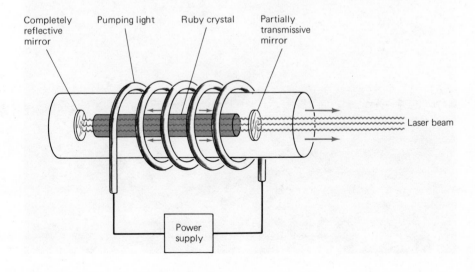

pletely silvered to allow total reflection and the other incompletely silvered to allow partial reflection and partial transmission of the light striking it.

Green light, which excites the chromium atoms, enters the ruby from the flash lamp. The chromium atoms then emit photons by spontaneous emission in two stages, the second of which is a photon of the ruby red light. The crystal begins to glow at its characteristic color (6934 Angstrom units wavelength). The red light photons stimulate emission from other nearby excited atoms, each amplifying a bombarding photon by two, which then stimulate emission from other atoms in a growing surge of amplification. Most of the red light escapes out through the long crystal sides, but some reflects back from the mirrored ends into the crystal again. The crystal is designed so that its length is an integral number of half-wavelengths of the ruby laser light, and so the reflected waves are reinforcing in a typical resonance wave pattern. The light grows with ever increasing intensity, until it finally escapes out of the half-silvered end of the

The laser reflector disk placed on the moon by Apollo astronauts in 1971. The rows on top of the disk are the mirrors used to reflect the laser beams. Near the disk are several astronaut footprints.

crystal; this escaping light is fully coherent. The whole process takes less than 10^{-6} sec, and it repeats at once. Since they are coherent, the photons do not interfere, nor do they scatter; the light beam stays in a tight, narrow band that is highly directional.

The great coherence and high directionality of laser light can be used to measure distances with astonishing precision. Recently a laser has been used, for example, to measure the distance between the earth and the moon. The experimental design was simple. A reflective disk was placed on the moon's surface during the 1971 Apollo landing. Later a beam from a laser on earth was aimed at the disk. A portion of the laser beam eventually struck the moon disk and bounced back to earth, where it was received by photosensitive equipment installed in a telescope. The distance between the disk and the telescope was determined by measuring the time it took for the light to come back to earth (2.6 sec) and then multiplying this time by 3×10^8 m/sec (the speed of light).

Geologists intend to use a variation of this procedure to measure continental drift, the horizontal movement of continents in relation to one another. Laser pulses from two sources on the earth's equator will be transmitted to the moon over a period of years. By comparing the change in the time it takes for the beams to return, the amount of drift between continents can be accurately determined.

Laser light can be focused on a very small spot with extraordinary power. In fact, the light emitted by the earliest lasers had a greater power density (the rate at which energy is delivered to a unit of area) than that emitted by an equivalent surface area of the sun! It is estimated that the total power output of all wavelengths of light from the sun is equal to 7000 watts per square centimeter, whereas the power density of a light beam from an argon-gas laser is equivalent to 1 million watts per square centimeter. Of course, a laser cannot deliver such powerful light for unlimited periods or in the amount equivalent to that given off by the sun's huge surface. But the highly intense and narrowly focused laser light has found many applications in medicine and industry.

One early and successful use of the laser involved eye surgery on detached retinas. The retina is the light-sensitive area at the back of the eye. Thousands of people every year become blind when their retinas completely detach from the optic nerve. Lasers are now used to correct this condition. A laser beam is focused through the eye's lens onto the back of the eye, where it burns the edges of the retina and actually welds it back into place. Since the wavelength of the laser light is less than a millionth of a centimeter and the diameter of a laser beam can be narrowed to a distance equivalent to that of a single human cell, the possibilities of laser surgery without damage to healthy tissue are enormous. Some other possibilities for medical applications of the laser include suturing and cauterizing wounds, genetic manipulation, laser x-rays, and painless dental drilling.

This microdrill uses a beam of laser light to make small, uniform holes in thin metals and ceramic. It is capable of drilling 60 uniform 0.002-inch-diameter holes per second through a sheet of nickel and steel alloy 0.003 inches thick.

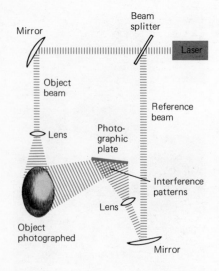

Beam splitter

Mirror

Laser

Object beam

Reference beam

Lens

Photo-graphic plate

Interference patterns

Lens

Object photographed

Mirror

In 1972 a famous Fifth Avenue jeweler displayed $100,000 worth of diamonds in a hologram projected from one of the store's display windows. The jewels appeared to be suspended in midair over the sidewalk in front of the store.

Another exciting use for lasers is in the entertainment field through a new medium called holography. Holography is three-dimensional, lensless photography which employs lasers. The technique was first conceived by Dennis Gabor, an English engineer, in 1948.

The procedure for making a hologram is relatively simple. The object to be photographed is placed in a large box with a photographic plate several inches in front of it. A beam splitter is then employed to divide a laser beam into two distinct parts. One beam is spread by a lens so that it covers the photographic plate fully and uniformly; this is the reference beam. The other beam, the object beam, is focused by mirrors to illuminate the object to be photographed. The light from the object beam is scattered from the various surfaces of the object and then falls onto the photographic plate. The film records the interference patterns of the combined reference and object beams rather than the picture image of the object being photographed. When this film plate is later illuminated by a coherent laser light beam, it produces a fully lifelike, three-dimensional reproduction of the object's optical surfaces in space.

Viewing a hologram image through the photographic plate is almost the same as looking through a window. For example, imagine a hologram of a pencil. As you look through the plate and shift your eyes or your head, your view of the pencil and its apparent position in space will alter just as it would if the pencil were actually there since the light coming to your eyes behaves just as light waves from the actual pencil. Holograms seen through a plate are called transmission holograms. Recently, image-plate holograms have been developed; these require less than fully coherent light for illumination and production of the image.

Holography will find spectacular uses in film and television. It may be twenty years before we will be able to view three-dimensional color TV programs and films, but most industrial engineers believe that the technological refinements required can be achieved within that time. Holography will undoubtedly increase the entertainment value of these media. For example, ballet enthusiasts frequently complain about the loss of the sense of three-dimensional space in viewing dance films; holography can restore that sense. And the depth realism of sports events on film and TV will be extraordinary.

Some of the potential uses of lasers are not so benign. Because of its intense power, the laser beam may be used for devastating weapons, not very different from the ray-guns of science fiction. So far these weapons are publicly said to be only a frightening theory, but other military uses for the laser are already known. Antiballistic systems now use lasers to track down and strike given targets, both fixed and moving, with great precision. The uses to which laser techniques will eventually be put depend on the human agencies involved. New scientific developments are harnessed to suit the social and political needs of those who control them in the same way that individuals use their own personal energy and creativity to meet their needs. It is in this broader arena that decisions about applications of the laser will be made.

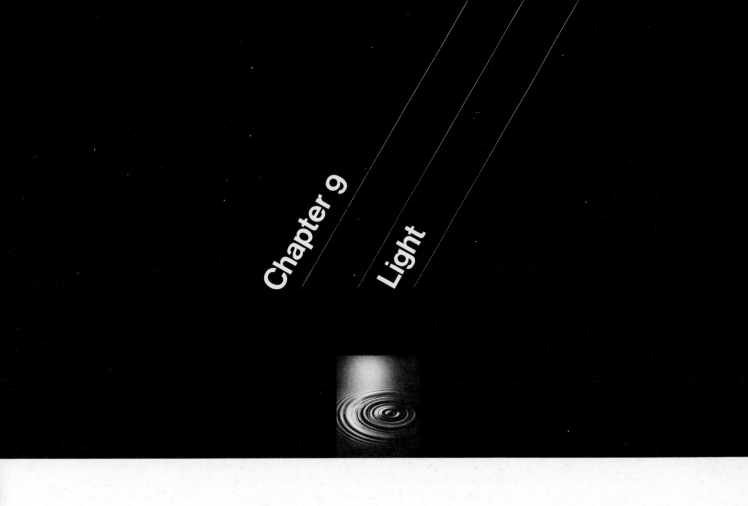

Chapter 9 Light

It was known for centuries that light travels in straight lines, and that it can be reflected. It was also known that objects appear bent when observed through water. To understand these phenomena, it is helpful to think of light as being composed of rays. The branch of physics dealing with light as rays, or straight lines, is called **geometrical optics**. The properties of light that can be explained by geometrical optics are reflection and refraction.

Geometrical Optics

The study of light rays is called geometrical optics.

Hero of Alexandria (a Greek who lived in the first century **AD**), described how light is reflected from surfaces. The angle at which a ray is reflected from a surface is equal to the angle at which it strikes the surface. The angles are measured from a line, perpendicular to the surface, called a **normal**.

Reflection

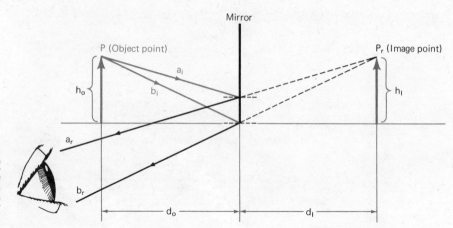

Figure 9-1
A ray diagram illustrates the fact that the virtual image of an object seen in a mirror is formed behind the mirror. The height of the object, h_o, equals the height of the image, h_i. The distance from mirror to object, d_o, equals the distance from mirror to image, d_i.

Reflection may be either regular or diffuse.

Picture a handball being thrown at a wall, and construct a normal at the point where it strikes the wall. The angle that the incoming ball makes with the normal is called the **angle of incidence**. The angle it makes with the normal after reflection is the **angle of reflection**. The law of reflection simply states that *the angle of incidence is equal to the angle of reflection.* Handball players automatically make use of this principle to anticipate where the returning ball will bounce—but watch out for the spin that a clever opponent will put on it! This law holds true for a light ray as well as a ball. The law also states that the normal, incident, and reflected rays lie in the same plane. There are two types of reflection—regular and diffuse.

REGULAR REFLECTION Regular reflection occurs when light rays bounce off a smooth regular surface, such as the highly polished surface of a mirror. The normals are parallel to each other and perpendicular to the plane of the reflecting surface; therefore, all angles of incidence and reflection are equal, and the reflected rays remain parallel.

Regular reflection produces an image; a ray diagram (see Figure 9-1) shows how. Think of an arrow standing vertically in front of a flat (plane) mirror. Incident rays a_i and b_i are drawn from the head of the arrow at point P. By the law of reflection, it is easy to construct the reflected rays a_r and b_r that reach the eye. Point P_r—the head of the arrow—is visually determined by extending the lines behind the mirror until they appear to intersect. It is possible to construct similar rays for every point on the arrow, thus producing its entire image.

Any image seen in a plane mirror—the image of the arrow, for example—is virtual, or imaginary, not real. It only *appears* to be formed by rays of light that come from behind the mirror. If a screen is placed behind the mirror at the actual points of intersection of the backward-extended rays, no image would appear on the screen, and nobody would expect to see light there.

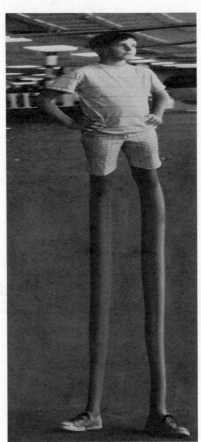

Figure 9-2
A lighted candle behind the mirror (far left) is moved so that it coincides with the image of the bottom half of an unlighted candle—demonstrating that the virtual image lies behind the mirror. The round mirror (near left) is made of small pieces of glass glued to a convex core; it reflects the right eye of a person looking in the mirror. If the pieces of glass were even smaller, the mirror would become a paraboloid surface, like the mirror of a telescope. In a fun-house mirror (below), convex and concave surfaces are combined.

This virtual image is upright; the image appears as far behind the mirror as the object is in front of it; the image is as tall as the object. Mirror images are reversed from left to right around a vertical line of symmetry running through the object—such as an imaginary line that divides your face in half.

CURVED MIRRORS Curved mirrors, as well as flat ones, can form images, so long as the curve is regular. **Concave** mirrors are curved inward in a spherical or parabolic shape—like the inside of a bowl. Such a mirror collects (converges) light, causing *parallel* incident rays to be reflected inward to meet at a point in front of the lens called the **focus**. Concave lenses are used in telescopes to gather light. The image formed is real, since light actually passes through the points where the image is seen, and can be focused on a screen. A light source placed at the focus of a concave mirror will be reflected outward as a parallel beam, as is the case with car headlights.

Convex mirrors are curved outward, like the outside of a bowl. A convex mirror spreads (diverges) light so that parallel incident rays appear to be coming from a point behind the mirror called the **virtual focus**—virtual, because no light is actually behind the mirror. The images thus formed are also virtual. Such lenses are used on trucks as rearview mirrors, because light from a large area can be gathered to be seen in the small area of the mirror.

Fun-house mirrors, which combine concave and convex surfaces, produce distorted images by alternately converging and diverging the incident light from different parts of the reflected body to make a crazy and fantastic image.

DIFFUSE REFLECTION **Diffuse reflection** occurs when light rays bounce off an irregular surface, such as rough wood, paper, or unpolished metal. Here, the normals are not parallel but varying; each is perpendicular to some little rough bit of the surface. Thus, for a set of parallel incident rays, although each ray obeys the law of reflection, the group of reflected angles are not equal. Clearly, the spreading of reflected rays from an irregular surface would not permit the formation of an image. Two rays extending from a point behind the surface would diverge and would not reach the eye together.

Refraction

The law of refraction describes a relationship between incident and refracted rays.

Place a penny at the bottom and to the far side of a shallow opaque bowl. Sight across the top of the bowl so that you can see into it but not to the penny at the bottom. Now, slowly pour water into the bowl, and watch the penny become visible. The puzzling phenomenon you have just observed is the same one that is responsible for the apparent angular sharp bending of a stick placed in a tank of water. It is called **refraction**, or the bending of light as it passes from a transparent medium of one optical density to a transparent medium of a different optical density.

Refraction was described mathematically in 1621 by the Dutch mathematician, Willebrod Snell (1591–1626). Consider a ray of light incident on a glass plate whose opposite sides are parallel. Snell explained that as the ray enters the glass, it bends toward the normal drawn at the point of entry. Upon leaving the glass at the opposite face, it bends away from the normal. Rays entering the surface of the glass parallel to the normal experience no bending at all.

Snell noted that as the light rays passed from a less optically dense medium to a more dense medium, they bent toward the normal. Conversely, if they passed from a more dense medium to a less dense one, the rays bent away from the normal. Snell also learned that the angle of bending experienced by a light ray depends on the optical properties of the material that it passes through. Thus, light rays passing from the water in the bowl containing the penny are bent away from the normal to the water surface as they enter the air. The apparent direction of the ray makes the coin seem to float higher in the bowl.

It should be noted that the bending discussed here takes place only at the boundary where the two media are in contact. As the ray leaves the first medium (air) and enters the second (water) its direction is changed, but within each uniform medium the ray travels in a straight line.

THE LAW OF REFRACTION Based on his observations, Snell formulated a mathematical expression governing the refraction of light between two media. Snell's law uses the mathematical expression *sine* (abbreviated *sin*). *Sine* indicates the ratio formed by the side opposite an angle in a right triangle divided by the side opposite the right angle (which is called the hypotenuse). The sine of an angle is the same no matter what the size of the triangle is, since the two sides will change together as the triangle gets larger or smaller. The **law of refraction** states that *the ratio of the sine of the angle of incidence to the sine of the angle of refraction is constant.* (See Figure 9-5.) If we take air as one of the media in question—through which the incident ray travels—this constant value

Figure 9-3
Wind ruffles the surface of flood water, causing a diffuse reflection of the poplar trees. The calmer the surface of the water, the more regular the reflection is apt to be.

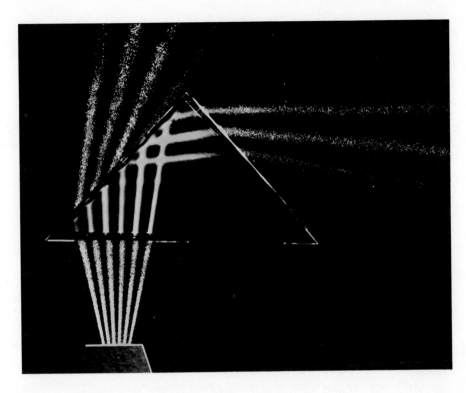

Figure 9-4
In this photo, pencils of light are refracted and internally reflected through a glass prism; the angle of refraction is that characteristic of glass.

is the **index of refraction**, n, of the other medium. (The index of refraction of air is considered equal to 1.) Snell's law is then written as:

$$\frac{\sin i}{\sin r} = n$$

where i is the angle of incidence in air. Since this angle of incidence is greater than the angle of refraction, sine i is greater than sine r, and, thus, n is greater than 1 for all transparent media.

TOTAL INTERNAL REFLECTION Figure 9-6 shows three light rays bouncing back from a fish swimming under water. The entering rays are both refracted and reflected; as the rays leave the dense water for the less dense air, they bend away from the normal. Ray *1* is refracted normally, producing an image that appears higher in the water than the object. Ray *2* is incident at an angle i_c, called the *critical angle*. At this angle, the angle of refraction, r_2, is equal to 90°. The critical angle depends on the index of refraction of the substance. Ray *2* skims the surface of the water—no light can be seen from above. Ray *3* is incident at an angle i_3, greater than the critical angle. At this value, the light ray is not refracted; instead it is reflected entirely under water, and the image it forms can only be seen under water. This phenomenon is called **total internal reflection**.

Figure 9-5
This schematic diagram shows the relation of incident and refracted rays.

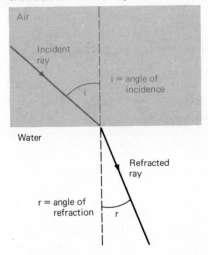

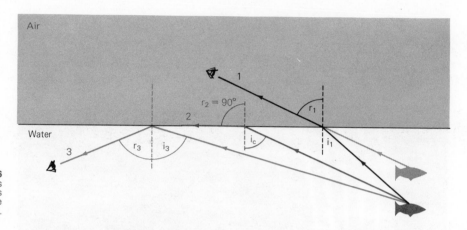

Figure 9-6
In this diagram, the black line illustrates
normal refraction; the dark red line shows
the critical angle, and the light red line
shows total internal reflection.

Figure 9-7
The top drawing shows refraction from a
convex lens; the bottom drawing shows
refraction from a concave lens.

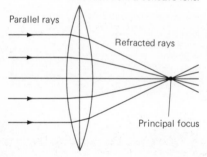

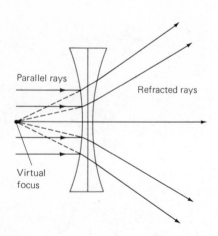

FIBER OPTICS A whole technology, called **fiber optics,** is based on the principle of total internal reflection. A tube of transparent material—usually of plastic—is constructed with such a diameter that, for its index of refraction, all light entering through one end will always be incident on the inner surfaces at greater than critical angles. This condition prevails regardless of how the tube (often called a light pipe) is bent. Therefore, the light is totally internally refracted and can only exit at the other end. Fine strands or fibers of such materials can be arranged in parallel bundles whose ends are perpendicular to the fibers. Light from an object entering at one end will produce an undistorted image at the other end. One fiber carries light; other fibers carry the image of what is lit up. Such a fiber bundle is called **coherent** because the fiber arrangement maintains the exact light pattern that was incident on the object end as it is transmitted through the light pipe. These light pipes are used in medicine as a diagnostic tool, allowing doctors to see into otherwise inaccessible areas.

LENSES One important application of the property of refraction by transparent materials is the bending of light by lenses. A **lens** is a plate of transparent material—often glass—whose opposite sides are curved surfaces, either convex or concave.

A convex lens is thicker in the middle than at the ends. It refracts parallel rays so that they converge at a point called the principal focus, and is often called a converging lens. A magnifying or burning glass is an example of a convex lens. With such a lens, real images of an object—that can be focused on a screen—are formed. The size and orientation of the image depends on the distance of the object from the lens.

A concave lens is thicker at the ends than in the middle. It refracts parallel rays so that they *appear* to be coming from a point called the virtual focus—a point

Figure 9-8
Lenses are still employed in even the most complex modern microscopes and telescopes. They are not the only focusing devices used, however. The largest telescopes today use parabolic mirrors to gather and focus light from distant objects. Such mirrors can be made larger and more distortion-free than lenses. The 200-inch mirror of the Mount Palomar telescope (near left) enables astronomers to photograph fine detail of extremely faint galaxies. Electron microscopes (far left) use beams of high-speed electrons instead of light. The electrons behave like light waves, and can be made to form an image (Chapter 11). They must be focused by magnets instead of lenses, however.

on the same side of the lens as the object. These rays diverge upon refraction by the lens; thus a concave lens is a diverging lens. All images formed by concave lenses are virtual and cannot be focused on a screen. Both convex and concave lenses are used in eyeglasses to correct defects in vision—convex for farsightedness, and concave for nearsightedness.

Physical Optics

Early observers, such as Snell, were able to report on the way light behaved when it struck an object, but little progress was made in trying to explain the nature of light. Its physical properties were baffling: seemingly instantaneous speed; capable of producing heat yet without observable bodily impact; not solid or liquid or gas; able to create images and cast shadows. The ray model served to describe phenomena of geometrical optics, such as reflection and refraction, but it did not help to explain these puzzles of physical optics. What *was* the nature of light?

Newton and Huygens proposed conflicting theories of light.

Wave Theory

According to the Dutch scientist Christiaan Huygens (1629–1693), light was a wave phenomenon. He suggested that light waves travel out spherically from, and are concentric with, the light source—considered as an idealized point source such as a pinhole. These spherical surfaces are called wave fronts and represent all points at which all the vibrations are in phase, with all crests and troughs coinciding. Within a single plane, these fronts become concentric circles, and any radius drawn through them becomes a ray and indicates the direction of the wave. Furthermore, Huygens postulated, in what came to be called Huygens' principle, that *every point* on a wave front may be regarded as a new source of waves or wavelets. According to Huygens, when light passes from a medium of lesser optical density—such as air—to a medium of greater optical density—such as glass—the waves slow down and bend toward the normal. Thus, light has varying speed. As the light leaves the denser medium and enters air again, it speeds up, bending once more—away from the normal.

Corpuscle Theory

Newton disagreed with Huygens. He proposed that light is composed of luminous particles, or corpuscles, which are attracted more strongly by the matter of a denser medium—in a manner similar to gravitational attraction—than by the matter of a less dense medium. The particles in the denser medium speed up and bend toward the normal. Conversely, as they enter air again, the attraction decreases, the particles slow down and bend away from the normal.

Both Newton's and Huygens' theories on the nature of light adequately explain reflection and refraction; their explanations for refraction, however, are diametrically opposed. Both theories could predict what would happen when conditions for reflection or refraction prevail.

One way of testing the two theories was to see how they accounted for the observable behavior of shadows. Huygens' theory would suggest that shadows would be fuzzy and diffracted, since the light waves would just curve around the object in the path, like water waves around a rock. On the other hand, Newton's theory demanded sharp distinct shadows, because of the movement of particles in a straight line.

Figure 9-9
This laboratory demonstration of a wave front (moving up from the bottom of the picture) is a visual analogy of the way light can behave as a wave phenomenon, bending around the surface of the objects it strikes.

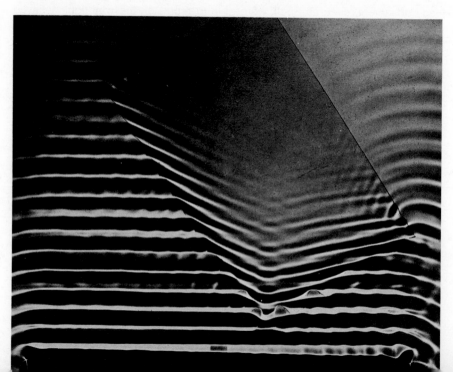

Upon ordinary observation, the shadow of an object, such as a straight edge or ruler, appears to have a sharp outline. It was in part because of this observation, and in part because of Newton's awesome reputation, that the corpuscular theory held sway for over 100 years.

By the early nineteenth century, there was a new bit of evidence that seemed to discredit Newton's particle theory of light and support Huygens' wave theory. Physicists had discovered that overlapping of light rays produced **interference**, the same phenomenon earlier seen in mechanical waves and sound waves (see Figure 8-8).

One of the best known demonstrations of light interference was that performed in 1803 by an Englishman, Thomas Young (1773–1829). A light source such as a candle or a light bulb emits light waves from an infinite number of points—the waves overlap so that all combinations of phases are possible. The net effect of this overlapping is to cancel all the interference phenomena so that only light of uniform intensity is observed. Such light that is partially or totally out of phase is called incoherent. In order to observe interference, light that is emitted from one point with all waves in phase—coherent light—is necessary. Young predicted that interference could be observed by a pattern of light and dark bands on a screen placed in the path of the interfering waves.

Young passed light through a pinhole, creating, in effect, a point source of coherent light all of whose waves were in phase. This coherent light was then passed through two more pinholes in a second screen—each of these acting as a separate source of light—according to Huygens' principle. The waves from these two sources overlapped in the region behind the screen. When allowed to fall on a third screen, these waves produced a series of light and dark bands, indicating areas of constructive and destructive interference.

The final bit of evidence for the wave model of light was so overwhelming that even the strongest advocates of the corpuscle theory yielded. In 1815, Augustin-Jean Fresnel (1788–1827) performed decisive experiments that demonstrated the diffraction of light, and explained why it had previously been so difficult to observe this phenomenon. When Fresnel cast light on a straight edge, extremely close observation showed that a very narrow band of light could be seen *inside* the edge of the shadow; this proved that light waves did indeed spread into a shadow. Fresnel suggested (and demonstrated) that most of this light interferes destructively, leaving a dark region beyond the edge of the obstacle. This accounts for the apparent rectilinear propagation of light.

Interference

Young's experiment on the interference of light supported Newton's theory.

Diffraction

Polarization

Light waves can be polarized because they are transverse.

That light was a wave phenomenon was now well established, yet the type of wave form it took—that is, whether it was transverse or longitudinal—was not known. Huygens himself thought that the waves were longitudinal. But the discovery of a new aspect of light behavior eventually led to the establishment of the transverse nature of the waves, and in fact confirmed the wave nature of light itself.

Iceland spar (calcite) crystal was found to produce a double image of what was viewed through it. For instance, a dot on a page was seen as two dots when viewed through the crystal. Rays passing through the crystal were apparently split into two rays—each refracted by a different amount—producing what is known as **double refraction**. Furthermore, as the crystal was rotated, one dot remained stationary while the other rotated around it. The ray producing the stationary dot was called the ordinary ray, and the other one the extraordinary ray.

Huygens explained this phenomenon by saying that there must be two different wave forms, one to produce each type of ray. Newton, however, disagreed; he explained the phenomenon in terms of the properties of light corpuscles. Corpuscles, according to Newton, were probably not spherical but had different sides, just like the poles of a magnet. They would respond to the properties of the crystal—that is, take different paths—depending on their orientation.

Figure 9-10
This highly magnified photo of a paper clip shows alternating bands of light and shadow inside and outside the clip, demonstrating the phenomenon of diffraction.

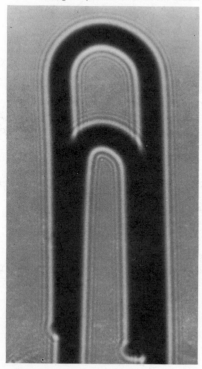

Nothing further was made of this phenomenon until 1808, when Etienne Malus (1775–1812) noted that light reflected from the glass of a window did not produce the characteristic double image when viewed through calcite. He theorized that the reflected light contained only one of the poles of light that Newton spoke of, and named the effect **polarization**.

Not too long afterward, Fresnel and D. F. Arago (1786–1853), a French astronomer and physicist who worked with the interference of polarized light, established that only transverse waves could account for polarization. If light is transverse, the vibrations occur at right angles to the direction of propagation. But the vibrations can occur in any direction in a plane parallel to the vibrations; that is, the vibrations can be vertical (up and down), horizontal (side to side), or along any diagonal—regardless of its angle—and still remain perpendicular to the direction of motion. Such light having random direction of vibrations is called **unpolarized**.

If light is limited to vibration in one direction it is said to be **plane polarized**. Longitudinal waves vibrate in the same direction in which the waves travel. Thus, passing a longitudinal wave through a polarizer has no effect on it; there is no plane of polarization for it. Particles travel in straight lines. They, too, have no plane of polarization and would not be affected by a polarizer. This last fact gives further credence to a wave theory of light.

To picture the mechanism of polarization, imagine that a rope attached to a wall at one end is being disturbed at the other end so as to produce transverse

Figure 9-11
These two photos of the New York City skyline were taken at the same time with identical lighting. The photo on the right was taken through a polarized lens and illustrates how polarization cuts down the amount of light that passes through the lens.

waves vibrating in a vertical plane. And suppose that the rope passes through a vertical slot cut in a plywood board. Since the vibrations move in the same direction as the slot, they will not be hampered in any way. Now as the slot is rotated, the vibrations will be inhibited so that only the component of the wave that lies in the direction of the slot will get through. When the slot is rotated through 90°, so that it is horizontal, no part of the vertical transverse wave can get through. The rope will appear stationary beyond the board.

Polarizing materials—such as crystals of tourmaline—have atomic and molecular arrangements that act like slots by allowing transverse vibrations in only one plane to pass through the material. All other vibrations are absorbed. If this polarized ray then passes into another polarizing material, it will pass through only if its direction and the polarizing direction of the material coincide. If this second material is rotated, changing its alignment, less and less light will pass through. When the polarizing angle of the material is perpendicular to the direction of the plane polarized ray, all the light will be absorbed (Figure 9-11). The doubly-refracting calcite, in fact, absorbs all light but the horizontal and vertical waves, refracting these at slightly different angles, thus producing a double image.

Lenses work by refracting light. In passing through the lens, light rays are bent so that they converge or diverge. A convex lens causes rays to converge; the convergent rays form an inverted image of the original source at their focus. Such an image will be real—that is, it can be projected onto a screen and seen. (When we bring our eye closer to a convex lens than the point of focus, we seen an enlarged, right-side-up image. Such an image, also encountered when we use a magnifying glass, is called a virtual image; it cannot be projected onto a screen.)

The camera is perhaps the simplest adaptation of the lens. A convex lens is used to form a real image on chemically treated film which reacts to light in a permanent and visible way. The human eye is optically not much different from a camera. The

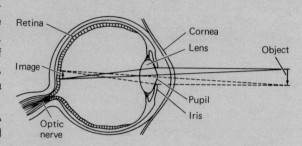

image formed by the lens is projected onto a thin layer of light-sensitive receptor cells, called the retina, at the back of the eyeball. The stimulated retinal cells send chemical signals by way of nervous impulses to the brain, which receives the signals as a meaningful image.

There are two kinds of receptor cells in the retina: cone cells, concentrated in the center of the eye, provide us with detailed color vision; rod cells, distributed around the periphery of the retina, provide a black-and-white image that is less clear, but they are far more sensitive to dim light so the rod cells provide our night vision. (Dim-light observers, such as astronomers, learn not to look directly at faint objects, but to see them out of the corner of their eyes instead.) There are three different kinds of cone cells, each sensitive to a different portion of the color spectrum— orange, green, and blue. Our visual world is seen in color—an ability not shared by many species of animals.

Both camera and eye must have some device for adjusting their focus, so that a clear image is formed on the film or retina, and for controlling the amount of light that they allow to enter. In the human eye the size of the pupil, the opening that admits light, is regulated by the surrounding muscular ring called the iris. The iris usually responds automatically by reflex to the amount of light falling on the eye, opening wide in dim light and closing down in bright light. A cat's eye is very clear in this reflex. In a camera, a mechanical diaphragm performs the same function. The human eye and camera differ, however, in their focusing mechanism. The eye has special muscles that change the thickness of the lens when objects at various distances are viewed. The light rays from nearby objects diverge sharply; they require a thicker lens to bring them to focus, whereas a thinner lens is needed to correctly focus the more nearly parallel rays from far off objects. In a camera (and in the eyes of certain other organisms, such as fish) the lens itself moves closer to or further away from the film.

Many objects, such as stars and planets, are too far away to be seen clearly; others are too small to be visible—bacteria, for instance. To transcend the limitations of the naked eye and open more of the universe to our sight, optical instruments such as the telescope and the microscope have been developed during the past 350 years. These are quite alike in principle. A convex objective lens is used to form a real image of the object being viewed, and this image is then magnified by another convex lens or series of lenses—the eyepiece—which produces an enlarged virtual image just as an ordinary magnifying glass does. The telescope objective, unlike that of the microscope, must be large, so as to gather the maximum amount of light. In this way it makes visible objects that are too dim to be seen with the unaided eye. Most large modern astronomical telescopes use a parabolic mirror instead of a lens as the objective. Such reflecting telescopes are less expensive to make in the large sizes required by present-day astronomers—the Mount Palomar telescope, for example, has a mirror 200 inches in diameter.

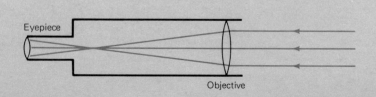

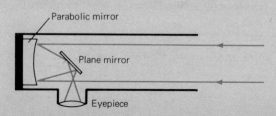

Certain crystals split light into two plane polarized beams, like Iceland spar; but unlike Iceland spar, they absorb one beam and transmit the other. Edwin H. Land (1909–   ) discovered that tiny crystals of this type could be embedded in plastic sheets. When the sheets were stretched, the polarizing planes of all the crystals would be aligned—making the sheets into polarizers. He named these sheets *Polaroid*. If such Polaroid sheets are used in glasses or on automobile headlights, they cut glare by eliminating all light not polarized in a given plane.

Many early experiments in geometrical optics indicated that light traveled at different speeds under differing conditions. But what were those speeds? In 1675, the Danish astronomer Olaf Roemer (1600–1700) tried to answer this question by measuring the velocity of light from observations of one of the moons of Jupiter. He was observing the eclipsing of the moon as it passed out of view behind Jupiter. He noted that when the earth and Jupiter were on the same side of the sun, the eclipse occurred earlier than when the earth and Jupiter were on opposite sides of the sun. Roemer concluded that it is the distance equal to the diameter of the earth's orbit that accounted for the time difference. Based on available data about the diameter of the earth's orbit, Roemer obtained a value of about 2.1×10^8 m/sec (about 130,000 mi/sec). Although his value was low, Roemer did succeed in establishing that the speed of light had a finite value.

In 1849, the French scientist H. L. Fizeau (1819–1896) used another method of determining the velocity of light. His apparatus used a toothed wheel that could be rotated to allow light to pass through the gaps between the teeth. The light was reflected between two mirrors separated by the wheel, so the returning beam could be viewed only after it passed through one of the gaps. If the rate of rotation of the wheel was adjusted so that the light struck a tooth, no light would be seen. From the rate of rotation and the distance between the two mirrors, Fizeau was able to determine the velocity of light.

A year later, in 1850, another French scientist, Leon Foucault (1819–1868) tried to measure the speed of light in different media. Using apparatus similar to that of Fizeau, Foucault added a tube of water placed in the path of the light whose speed he was measuring; he found that the speed of light was indeed *slowed* as it passed through the water—providing experimental proof for Huygens' theory of refraction and, hence, the wave theory, almost 175 years after Huygens proposed it.

Measurements of the velocity of light—one of the most fundamental constants in nature—have continued ever since. In 1923, the American physicist Albert A. Michelson refined Foucault's methods with electrical apparatus. Using a distance between two mountains of 22 miles, measured to an accuracy of an

The Velocity of Light

The first significant measurement of light was made by astronomical observations.

The speed of light is one of the fundamental constants in nature.

less than an inch, he determined the velocity of light to be 2.99796×10^8 m/sec. For all practical purposes, we take the velocity of light to be 3×10^8 m/sec, or about 186,000 mi/sec.

Refractive Index

Once Foucault established that light travels more slowly in a medium of greater optical density, the refractive index of a given material could be defined as *the ratio of the velocity of light in a vacuum to its speed in the material.* Because the velocity of light is greatest in a vacuum, the index of refraction in any substance is always greater than 1. For all practical purposes, the index of refraction of air is also 1 because the velocity of light in air differs so little from than in vacuum. For example, the speed of light traveling through a diamond is 77,000 mi/sec. Its index of refraction is:

$$n = 186,000/77,000 = 2.42$$

This high index of refraction indicates that light is bent considerably upon entering a diamond from the air—more than for any other substance (which accounts for the brilliance of diamonds).

Wavelength and Frequency

From the interference properties of light, Young and others after him were able to measure its wavelength quite accurately. We now know that the wavelength of visible light ranges from 4×10^{-7} m to 7×10^{-7} m. Frequency f can be determined by using the equation for wave motion, $v = f\lambda$, which can be rewritten:

$$f = c/\lambda$$

where c is the velocity of light in a vacuum—equal to 3×10^8 m/sec. The frequency then ranges from:

$$f = \frac{3 \times 10^8}{4 \times 10^{-7}} = 7.5 \times 10^{14} \text{ cycles/sec (cps) for violet light}$$

to:

$$f = 4.0 \times 10^{14} \text{ cps for red light}$$

One cps is usually referred to as one Hertz (Hz).

The Electromagnetic Nature of Light

By 1850 no one could doubt that light was a transverse wave. But it was not clear just what sort of wave it could be. Mechanical waves, as we have seen, are disturbances in a physical medium such as air or water. But light could travel through a vaccum. The true nature of light was first proposed by the Scot James Clerk Maxwell (1831–1879), a theoretical physicist who was working on a seemingly very different problem. Maxwell was attempting to unify what was known about electricity and magnetism within a single mathematical framework. It was known that a changing magnetic field always produces an electric field in space. Maxwell proposed an additional, new principle for which there was no experimental evidence at the time: *a changing electric field always produces a magnetic field in space.* He was able to express these relationships between electric and magnetic fields in a series of equations, which are now regarded as the greatest achievement in theoretical physics of his day.

Maxwell drew an important conclusion from his new principle. Suppose that we have a rapidly changing electric field—such as might be produced by a sudden spark of electricity jumping between two charged metal terminals. The changing electric field will produce a changing magnetic field. But this changing magnetic field will in turn produce another electric field, which will generate an additional magnetic field, and so on. The result will be a fluctuating *electromagnetic field* in which the changes of the electric and magnetic components are synchronized so that they vary in unison. This field is not static but will expand outward in all directions in the form of an *electromagnetic wave.* The direction of the electric and magnetic forces will be perpendicular to each other and to the direction of the advancing wave, so an electromagnetic wave is transverse. Maxwell was able to calculate the velocity at which such a wave must travel. It turned out to be very close to the experimentally measured speed of light.

Maxwell realized that this could hardly be a coincidence. He suggested that light in fact consisted of electromagnetic waves. He further predicted the existence of other such waves, with frequencies much higher and lower than those of visible light. In 1887 the German physicist Heinrich Hertz (1857–1894) verified these predictions. He used a spark apparatus like that described above to generate electromagnetic waves (of the type now called radio waves) and showed that they had properties similar to those of visible light—they could be reflected, refracted, and diffracted. Maxwell was dead by this time, but the discovery of electromagnetic waves fully confirmed his theory and opened the way for the development of radio, radar, television, and x-ray technology.

Color

The range of wavelengths and frequencies of visible light corresponds to the range of colors that comprise light. Until the time that Newton experimented with color, white sunlight was thought to be "pure light," and color was thought to be the effect of impurities picked up by the light upon reflection from a colored surface or transmission through colored glass. Yet a rainbow or the color observed at the edges of ordinary glass could not be explained by this premise. By Newton's time, scientists began to suspect that these phenomena were a result of refraction. In 1666, Newton confirmed the connection between refraction and the production of color. He allowed a narrow shaft of light to enter an otherwise darkened room through a small circular hole in a window shade. He passed this beam through a prism, where it was refracted and then

Newton demonstrated that white light is composed of colors.

made to fall on a white screen. Instead of casting a circular image of the white light on the screen, the prism produced a wide band of colors that faded into each other. The colors were arranged as in a rainbow—red, orange, yellow, green, blue, and violet—forming a spectrum.

To show that white light was not pure but composed of these colors, Newton then passed the color spectrum through another prism placed upside down in relation to the first, thus reversing the refraction. When the light was allowed to fall on a screen once more, it produced a circle of white light. To demonstrate that each color was not simply an impurity introduced by the prism, Newton passed one color from the spectrum through a hole in a screen and refracted it through another prism. When allowed to fall upon a second screen, the light was spread out but no new colors appeared; the prism did not add anything to the light.

Newton concluded that white light was not pure but was a mixture of all the colors of the rainbow. It appears white only because the brain interprets it in this way.

Newton found that red light is refracted least, while violet light is refracted most. The separation of light into its component wavelengths is called **dispersion**. The angle between the red and blue rays is a measure of the dispersion, a property of the optical refracting material being used.

Different colors of the light spectrum move at different speeds in different media. For this reason, the index of refraction of a transparent medium is arrived at by using the angle of refraction of yellow light, which lies roughly in the middle of the light spectrum. In practice, the yellow light of a sodium flame is used for this purpose.

Red Sunsets and Blue Skies

As sunlight enters the earth's atmosphere, it strikes molecules of the gases in the atmosphere. Because the size of these molecules is on the order of wavelengths of light, the law of reflection does not apply to rays incident on these molecules. That is, the angle of reflection no longer equals the angle of incidence. Thus, the rays are *scattered* randomly in all directions. But different wavelengths of light are scattered differently. The scattering is greatest for the shortest wavelengths—blue and violet. Because these colors are scattered across the sky most strongly, the sky appears blue (the eye is not as sensitive to violet lights).

The remaining sunlight that reaches the eye lacks the blue and violet rays scattered by the air and appears yellowish during hours when the sun is well above the horizon. As the sun sets, its rays must travel through an ever thickening layer of air as it approaches the horizon. Higher wavelengths left in the sunlight become increasingly scattered. The light least affected by scattering—owing to its greater wavelength—is red. Thus, the setting sun appears red because more of its red light reaches our eyes directly than any other color. (See the Color Portfolio.)

The **angle of incidence** is the angle between an incoming light ray and the normal; the **angle of reflection** is the angle between the normal and the ray that bounces off the reflective surface.

Critical angle is an angle of incidence in a medium of high optical density that produces an angle of refraction of 90° when a ray passes to a less dense medium.

Concave mirrors and lenses are those that curve inward in a spherical or parabolic shape, like the inside of a bowl. **Convex** mirrors and lenses curve outward, like the outside of a bowl.

Diffuse reflection occurs when parallel light rays are no longer parallel after reflection.

The **focus** is the point in front or behind a mirror or lens where reflected or refracted rays meet, or appear to meet, forming a real or a virtual image.

Geometrical optics is the study of the behavior of light made by tracing the paths of rays of light.

The **index of refraction** is the ratio of the sine of the angle of incidence to the sine of the angle of refraction.

A **normal** is an imaginary line that is perpendicular to a reflective surface.

Physical optics is the study of the properties and nature of light.

Polarization is the limiting of light waves to vibration in one direction.

Regular reflection occurs when parallel light rays remain parallel after reflection.

Total internal reflection occurs when the critical angle is exceeded by an incident ray.

1 Contrast the strengths and weaknesses of the two major theories of light proposed by Newton and Huygens.
2 Why does a simple mirror reverse handwriting reflected on its surface, from left to right? Why does the mirror not reverse top and bottom?
3 Using straight beams of light and a simple figure, diagram the formation of a real image in front of a concave reflecting surface.
4 How might refraction of light through "swamp gas" (methane) account for UFO-like apparitions?
5 Why does the velocity of light decrease when it enters a high-density medium?
6 Briefly outline Newton's theory of color. How did Newton test his hypothesis?
7 Propose a theory to account for each of the following phenomenon in the sky: red sunsets, rainbows, haloes around the moon.
8 Why are polaroid lenses used to make sunglasses, and how do they reduce the passage of light?

Classical mechanics, the physics of Galileo and Newton, is the science of matter in motion. But what is motion? To measure motion we must measure two fundamental quantities: position in space and duration of time. Classical physics was based on certain assumptions about space and time and how they could be measured. These assumptions seemed so basic and so sensible that they were taken for granted for more than 250 years.

But towards the end of the nineteenth century scientists began to find experimental results and conceptual puzzles that classical physics did not explain. Many wished to ignore these discrepancies rather than tamper with the foundations of a magnificent system that had been so successful for so long. But Albert Einstein (1879–1955) chose instead to reexamine the old assumptions underlying classical physics; the result of this reexamination was the theory of relativity, which has reshaped modern physics and changed the way we think about space and time.

Suppose you climb to the top of the mast of a boat moving at constant speed in a straight line and drop a hammer towards the deck: what path will it pursue, and where will it land? The answer seems fairly obvious to us, a matter of common sense. We know that the hammer shares the forward motion of the boat and will continue to share it even after it leaves your hand. Thus it will fall parallel to the mast and land at the foot of the mast. Its path, *as seen by an observer on the boat,* will be a straight line. An observer floating in the water near the boat, however, will see the forward motion of boat and hammer relative to the water as the hammer falls, and to him the hammer's path will appear to be curved, a parabolic path downward rather than a straight one.

Yet Aristotle, known for his common sense, answered this question differently. He thought that heavy objects always fell straight down towards the earth because that was their "natural" place. Therefore the hammer (in Aristotle's view) would simply fall straight downwards, without sharing in the forward motion of the boat. Since the boat would have moved forward somewhat by the time the hammer reached the deck, it would not fall at the foot of the mast but several feet further back. But this is not what really happens.

This example suggests that common sense is not universal and unchanging; rather, it varies for different people and different times. Common sense is a combination of ordinary life experience with last year's—or last century's—physics, as it has trickled down into the general knowledge. Common sense is an essentially conservative expression of ideas and experiences we are used to. This is its limitation for the scientist, who must always be prepared to deal with the unexpected.

Figure 10-1
Aristotle believed that objects fall straight down no matter what their state of motion is before they start to fall. In Newtonian mechanics, however, a falling body will continue to possess any horizontal velocity it had before it started its fall. The resulting trajectory—a combination of the constant horizontal component of velocity and the increasing vertical component caused by the acceleration of gravity—is parabolic to a stationary observer. To an observer moving with the same original velocity as the body itself, however, the body is initially "at rest," and its path as it falls is a straight vertical line. The measurement of motion is relative—that is, it depends upon the relative motion of the observer and that which he is observing.

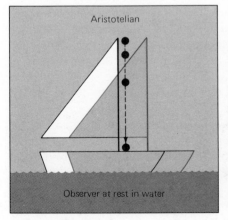

Aristotelian

Observer at rest in water

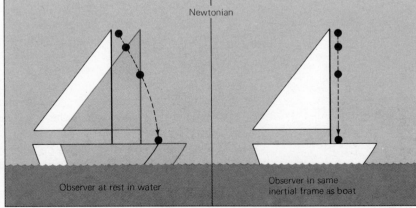

Newtonian

Observer at rest in water

Observer in same inertial frame as boat

Motion and Reference Frames

The example of the falling hammer illustrates a fundamental principle of classical mechanics: the measurement of motion depends on the relative motion of the observer with respect to that which he is observing. To the observer on top of the mast, the hammer falls in a straight line; to the observer at rest in the sea, the same path is a parabola. And since a parabolic path between two points is longer than a straight one, but the time of falling is the same for both observers, the observers will also disagree about the speed of the falling hammer. This is because the stationary observer sees a horizontal component of the hammer's velocity; since the shipboard observer is moving along with the hammer, the hammer has no horizontal motion at all relative to him.

Inertial Frames

Inertial reference frames are those that move uniformly in a straight line, without acceleration.

A description of movement only makes sense if we specify the frame of reference against which it is being measured. The boat moving at constant velocity is one reference frame, while the water at rest defines another. Both are **inertial reference frames**—frames moving with uniform straight-line motion. Newton's second law applies in all inertial frames, and classical mechanics then provides a simple set of equations relating the motion of a body with respect to different frames. To take an example, suppose a railroad car is moving at 70 ft/sec. An archer in the car shoots an arrow which leaves the bow at a velocity of 80 ft/sec. To find the velocity of the arrow with respect to the track, we need only add the velocity of the arrow with respect to the car and the velocity of the car with respect to the track: 70 ft/sec + 80 ft/sec = 150 ft/sec. (Similarly, if the arrow were fired backwards, its velocity with respect to the track would be 10 ft/sec in the opposite direction.) This example involves motion in only one dimension, but a similar principle applies in three-dimensional space. We use such *transformations* unconsciously every time we take a basketball lay-up, throw a newspaper from a moving bicycle, or jump off a moving ski lift.

Figure 10-2
According to the classical principle of addition of velocities, the velocity of the arrow with respect to the track (V_3) is simply the vector sum of the velocity of the arrow with respect to the archer (V_1) and that of the archer with respect to the track (V_2).

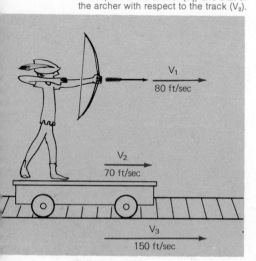

V_1
80 ft/sec

V_2
70 ft/sec

V_3
150 ft/sec

Which description of the hammer's fall or the arrow's motion is the truer one? Does the hammer really travel in a straight line or a parabola? Is the arrow really traveling at 80 ft/sec or 150 ft/sec? The answer is that all of these statements are equally true to the facts. The differences depend solely on the observer's state of motion. It is tempting to assume that a description of motion from a stationary frame of reference is the most accurate. But the concept of "stationary" raises problems in classical physics, for how can we determine what is "stationary"? In the example of the hammer, one frame of reference is stationary with respect to the boat, another with respect to the water. Even the observer who is at rest in the water may not be stationary with respect to the earth's surface—perhaps he is being borne along by a strong current. For convenience in everyday life we generally take the earth's surface as a fixed frame of reference. But we know that the earth is in reality no more at rest than the boat. It would seem, then, that there is no such thing as absolute rest or absolute motion anywhere.

Newton believed that empty space constituted an absolute frame of reference, relative to which a body might be in a state of absolute rest or absolute motion. But in Newton's time there seemed to be no way of observing or detecting this "absolute space"; its existence was merely a useful assumption, although Newton himself offered a serious argument to support his belief. He pointed out that the surface of water in a rotating bucket is curved, climbing up the walls and deepening at the center, unlike the flat surface when the bucket is at rest. The curvature is greater when the bucket rotates faster. No state of motion of an observer will change the observed curvature, even if the observer sits on the edge of the bucket, rotates with it, and thereby is at rest relative to the bucket.

Absolute Motion

The observed curvature seemed to Newton to be clear evidence that the bucket was rotating relative to absolute or fixed space rather than relative to any observer's state of motion. Therefore he argued there must be an absolute space. And yet scientists wanted a more direct way to demonstrate that absolute space exists—and with it a way to measure absolute motion. Later, when the wave theory of light became established in the early nineteenth century, scientists thought that light might provide a solution to the dilemma of absolute motion. Instead, investigations into the nature of light made the problem far worse, for it was the study of light that revealed an unsuspected but basic flaw in the beautiful structure that Newton had built.

It is not hard to see why physicists hoped that light would solve the problem of absolute motion. Light had been shown to be a wave. A wave is a traveling disturbance, and if there is a moving disturbance, something must be there to be disturbed. Every other wave known to science required some medium for propagation. Sound, for example, is a vibration that travels through air and through most other material substances, whether solid or liquid; but it cannot propagate through a vacuum. Light was unique in that it apparently could travel through a vacuum, but to imagine the mechanics of such a thing would be like trying to imagine ocean waves without water. So physicists postulated something easier to imagine—some invisible, intangible medium filling "empty" space. This odd substance, called the **ether**, would be the medium through which light is propagated. At the same time the ether, fixed in space, could provide the absolute frame of reference relative to which all motion could be measured.

Existence of the Ether

The ether, convenient as it was for theory, presented rather unpleasant practical problems for experimenters. For one thing, it had to have properties that were quite strange. It had to be rigid, fixed, incompressible, unaffected by gravitation. It had to have the elastic behavior of solids, since light could be shown to be transverse and transverse waves cannot propagate through a fluid. A second difficulty was the elusiveness of the ether. The ether was everywhere, but unlike the air, it could not be felt. Most physicists, even the most theoretically minded, felt more than a little uncomfortable in guessing at the existence of something

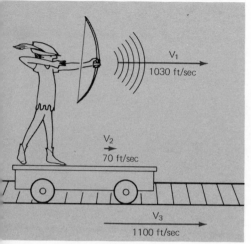

Figure 10-3
The velocity of a wave through its medium is independent of the motion of its source. Thus the velocity of the bow's sound with respect to the air is a constant 1100 ft/sec (V_3). Because the archer is in motion at a velocity of 70 ft/sec with respect to the air (V_2), he would find that the velocity of sound in the forward direction is 1030 ft/sec (V_1).

Figure 10-4
Michelson's experiment. If the earth were moving through the ether, the two beams would travel with different velocities, depending on the orientation of the instrument with respect to the earth's direction of motion. No such effect was observed.

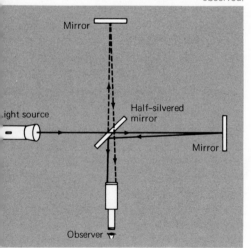

that is completely undetectable. How could its existence be either confirmed or refuted?

In the 1880s the American physicist A. A. Michelson (1852–1931) devised an experiment to demonstrate the existence of the ether by measuring the earth's speed through this substance. Michelson assumed that light should travel with constant velocity through the ether. But the earth itself must be traveling rapidly through the fixed ether as a result of its various motions — its rotation on its axis, its revolution about the sun, the movement of the entire solar system through space, and so on. Hence, the velocity of light should appear to vary to observers on earth, depending on whether it is measured in the direction of the earth's motion, at right angles to it, or against it.

To see this, consider again the example of the archer in the railroad car. This time, let us try to determine the velocity, not of the arrow, but of the sound of the bowstring as the arrow is shot. Sound waves, we know, travel through the air at sea level at about 1100 ft/sec. This is the velocity of sound with respect to the air, and it is independent of the motion of its source. One of the fundamental properties of a wave is that its velocity depends solely on the medium that transmits it. The velocity of the source that emits the wave is irrelevant.

Because of this property of waves, the apparent velocity of a wave such as the twang of the bowstring will be different for receivers who move in various ways in the transmitting medium. This effect is useful to meteorologists who can take the difference in measured speeds of sound from different directions to measure the speed of the wind. The differences in observed wave velocity should also make it possible to measure differences in the motion of the receiver relative to the medium. We know this is possible for sound waves transmitted through air; Michelson thought he might achieve the same result measuring light waves transmitted through the ether.

Michelson built an apparatus that split a beam of light into two rays and sent them traveling equal distances along paths lying at right angles to each other. Through this experiment, he hoped to measure the difference in their velocities and thus determine how fast the earth was moving through the ether. But the result of this ingenious experiment, which was carried out with great thoroughness and precision, shocked the scientific world. No matter in what direction the apparatus was oriented, no variation in the velocity of light could be detected. Moreover, the orientation of the earth in space made no difference, as was proven by experimenting at different times of day (rotational orientations) and at different seasons of the year (orbital orientations).

For two decades the fate of the ether remained in doubt. Some perplexed physicists tried to find an error in the Michelson experiment; but science seldom advances by trying to deny disturbing new observations. The experimental results had to be accepted; it was rather the preconceived theoretical notions that had to be sacrificed.

The man who advanced physics past the puzzle of the Michelson experiment was Einstein, who in 1905 (at the age of 26) published the first part of his theory of relativity; interestingly, he had not at the time heard of Michelson's work. Einstein realized that it was necessary to abandon the clumsy and unworkable notion of the fixed ether. But if the ether had to go, so did the concept of absolute motion, for then we have simply no preferred inertial frame of reference which is at rest in space. This led directly to Einstein's first postulate of relativity:

All the laws of physics are the same in any inertial frame of reference.

The first postulate applies not only to the laws of mechanics, but to light and other electromagnetic phenomena as well. As Michelson's results suggested, the wave behavior of light cannot be used to define a state of absolute motion or absolute rest. This was an important step; but Einstein went further. The second postulate of relativity theory states:

Light is always propagated in empty space with a fixed velocity, c, which is independent of the state of motion of the source.

With this postulate, the theory of relativity leads to many strange conclusions. For an observer moving along with a light source, the light will seem to behave like the archer's arrow: it will share the motion of its source and travel with the same velocity in all directions relative to the observer, just as common sense tells us it would if the source and observer are at rest. And, of course, they *are* at rest—relative to each other. But an observer in a different inertial frame from the source will find the light behaving like sound in the air: the velocity of light will appear independent of the motion of the light source and the same in all directions. If this is true, the classical law of the addition of velocities does not hold for light. (Nor, it turns out, does it hold for any physical objects, but the effect becomes significant only when their velocities approach that of light.)

The speed of light in a vacuum is constant for all observers regardless of their state of motion.

Let us again consider a concrete example. A spaceship, approaching earth at a velocity of .5c, sends a light signal ahead of it. The second postulate tells us that an observer on earth will find that the signal travels at a velocity of c with respect to earth. But an observer on the ship will find that the light is moving at a velocity of c with respect to the ship! Each might think the other's measurement quite incomprehensible. For to an observer on earth, the velocity of the light with respect to the ship should be .5c, while to the observer on the ship, the velocity of light with respect to earth should be 1.5c.

How can two observers, in motion relative to each other, see and measure the same thing traveling at the same speed relative to each of them? It seems absurd. But Einstein saw that there was a way to accept this strange conclusion without contradiction, and Einstein's insight opened up a whole new world of physical laws. Measuring motion means measuring time and space. Two observers in motion relative to each other can see the same thing traveling at the same speed, if (and only if) their perceptions of time and space differ.

Conclusions from the Theory

To explain the theory of relativity, Einstein devised what he called "thought experiments"—hypothetical situations in which the consequences of his ideas could be explored. We have already used several simple ones in the previous discussion. Let us employ another to demonstrate one of the most important corollaries of the theory: the relativity of simultaneity.

THE RELATIVITY OF SIMULTANEITY Imagine a train x meters long traveling at a constant velocity, v, along a straight track. At the front and rear of the train we will place two observers, whom we can call Mike (at the front) and Mack (at the rear). Now let us station two observers exactly x meters apart along the track; we'll call them Sid (forward) and Sam (rear). We can mark the point midway between Sam and Sid and call it O. Suppose that precisely in the middle of the train there is a woman with a lamp; we can call her Lucy. The train speeds down the track, and at the exact moment that she passes point O, Lucy turns on her lamp. What do the four observers see?

Figure 10-5
Sid and Sam, in the inertial frame of the tracks, will conclude that the light reaches them simultaneously but that Mack will see it before they do and Mike after they do. Mike and Mack, in the inertial frame of the train, will reach a very different conclusion. They will think that the light reaches them simultaneously but that Sid will see it before they do and Sam later. Evidently the concept of simultaneity is relative, not absolute; it depends upon the observer's frame of reference.

To the men on the track, the situation seems simple. Lucy was equidistant from them when she lit the lamp. Both are in the same inertial frame, that of the tracks, (which we are calling "stationary") and so both will see the light reach them with a velocity of 300,000 km/sec. The fact that Lucy was moving does not affect this observation at all. Since Lucy was equidistant from them when she lit the lamp, they will conclude that they must see the lamp at the same instant.

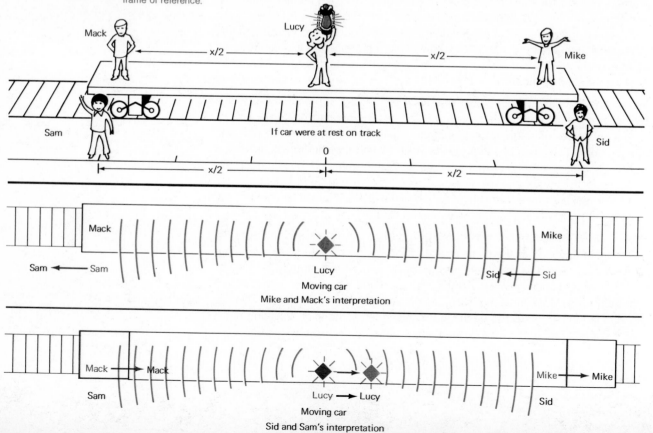

But what will they expect the men on the train to see? Evidently Mike and Mack were also equidistant from the light when it flashed. But while the light was traveling towards them with uniform velocity with respect to the track, they themselves were also moving—Mack rushing toward the light, Mike away from it. Thus Sid and Sam would expect Mack to see the light flash before they did, and Mike to see it afterward. But Mike and Mack would disagree vehemently! They will see the light moving with velocity c, not with respect to the track, but with respect to their own inertial frame, the train. As Einstein's second postulate suggests, the fact that the train is moving does not affect this result at all. Relative to their point of view, it is the tracks that are moving and the train that is at rest. Since the light flashes from a point halfway between them, they will conclude that it must reach each of them at the same time, and their measurements will bear this out. From their point of view, Sid is streaking towards Lucy and her lamp, Sam away from it. If light travels at a velocity of c with respect to the train, and the train is traveling at velocity v, the light must be traveling with a velocity of $c + v$ with respect to Sid, and a velocity of $c - v$ with respect to Sam—or so it seems by Newtonian common sense. Mike and Mack will suppose that the light will reach Sid before reaching them, but reach Sam only after they have seen it.

Which observers are right? They are both right. In classical physics, where time is taken to be unvarying, the concept of simultaneity has a simple and unambiguous meaning. Relativity shows, however, two events (distant from one another) that occur "at the same time" for one observer will not appear to take place at the same time for another observer moving in a different inertial frame.

Two events in different places that are simultaneous for an observer in one inertial frame may not be simultaneous for an observer in another frame.

TIME DILATION But if this is true, observers in different inertial frames must measure duration of time differently. It can be shown, from thought experiments like the one we have just constructed, that an observer will always find that his own clock runs more rapidly than any identically constructed clock in motion relative to him. The greater the other clock's velocity with respect to the observer, the more slowly it will run in comparison with his own. This situation too is completely symmetrical; the "moving" observer will find that his own clock appears normal but that of the other man is running slow.

It is easy to construct, in imagination, a simple clock that will demonstrate this relativistic effect. Suppose we measure time by bouncing a beam of light back and forth between two mirrors. (Fig. 10-6) Suppose that an observer is watching two such clocks, one in his own inertial frame (he calls it "at rest") and one in a different frame ("moving"). We can see from the diagram that the light ray in the moving clock will seem to the fixed observer to have a longer path to travel for each trip from mirror to mirror. But the observer will always see the two beams of light traveling at the same speed. In order for the speed to be measured the same, it will always take the light in the moving clock longer to make the trip, and the moving clock will always run more slowly than the clock in the observer's own inertial frame. It can be shown that a time interval on the clock at rest, t_0, is related to a time interval on the moving clock, *t*, by the formula:

$$t = \frac{t_0}{\sqrt{1 - \dfrac{v^2}{c^2}}}$$

This phenomenon is called **time dilation**. Time dilation is not merely a peculiarity of certain kinds of mechanical clocks; it is a property of time itself, and it will be an apparent factor in any process by which time can be measured. If you are in rapid motion with respect to an observer he will see you breathe more slowly; you will, in his eyes, age more slowly than he does. If you travel at .997*c*, you will seem to him to age only one month for each year he experiences. But to you, he will seem to live only one month for each of your years.

The same phenomenon can be measured using subatomic particles as clocks. Certain of these particles have been found to decay into other particles after an average lifespan of 2×10^{-8} sec when moving at relatively low velocities. When the same particles are accelerated to .8*c*, relativistic time dilation slows down the decay process, and the particles seem to last 3.3×10^{-8} sec by the laboratory clock, though an observer moving with the particles would still record the elapsed time as 2×10^{-8} sec on his clock.

Figure 10-6
The first clock, using a bouncing ball to mark units of time, is assumed to operate by Newtonian laws. When the clock is moving, the ball must cover a longer path. The ball, however, shares the clock's motion; its velocity, V_R, is thus greater than its vertical velocity when the clock is at rest, and it completes each cycle in the same period of time. The ray of light in the relativistic clock, by contrast, can only travel with a velocity of *c* whether the clock is moving or not. It takes more time to cover the longer distance when the clock is moving; thus the clock runs more slowly.

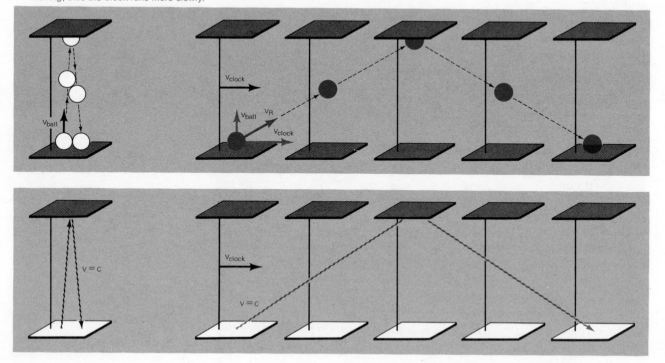

LENGTH CONTRACTION If time is measured differently by observers in different inertial frames, such observers will also find their measurement of length to disagree. Suppose, for example, we have two synchronized clocks located 600,000 km apart. (In order for them to be synchronized, of course, they must be in the same inertial frame—that is, at rest relative to each other.) A rocket ship traveling at .8c — 240,000 km/sec — flies from one to the other, taking 2.5 sec to complete the trip. Because of the time dilation effect, a clock on the ship will have ticked off during the passage only $2.5\sqrt{1 - .8^2} = 1.5$ sec.

From this measurement, the shipboard observer will conclude that he has covered a distance of only 360,000 km. (He would also, incidentally, find the synchronized clocks to be out of synchronization, since, as we saw in the example involving the train, events that are simultaneous to one observer cannot be simultaneous to an observer in a different inertial frame.) Once again, the effect is symmetrical; an observer stationed at one of the clocks watching the ship streak by would find that it appeared only ⅗ as long as it did when at rest.

Objects in motion relative to an observer always seem smaller in the dimension of their motion than they would in the observer's inertial frame.

We can see, then, that lengths and distances always seem shorter in a moving frame of reference than they would in the observer's rest frame. This shortening only takes place, though, in the dimension parallel to the direction of the motion. The ship in the above example would appear to have the same width to an ob-

Figure 10-7
The length contraction predicted by the special theory of relativity suggests that if the dachshund runs fast enough, it will come to look like a beagle to an observer who does not share its motion. To a flea on the dog's back, however, the dog's length will be unchanged, since the flea is in the same inertial frame as his host.

server on board as it does to one stationed outside. If l_0 represents the rest length of an object, its apparent length to an observer moving at velocity v with respect to it is given by the formula:

$$l = l_0 \sqrt{1 - \frac{v^2}{c^2}}$$

In common language we can say that objects shrink in the direction of motion when they move with velocities approaching c. Balls tend to become discs. This is a trifle misleading, however. We must remember that an observer moving with the object will perceive no alteration. Nor would we expect him to, for he will be making all his measurements with a ruler that seems, to an observer in another inertial frame, to have shrunk by exactly the same percentage. This explains why, despite the distortions of space and time that accompany rapid motion, all observers, regardless of their frame of reference, find the velocity of light to be the same. They measure time with clocks that run at different speeds, and distance with rulers of different lengths; but when they determine the velocity of light the two effects always cancel each other out exactly, and all observers get the same result.

RELATIVISTIC MASS INCREASE One of the most surprising consequences of the theory of relativity is the fact that mass too is affected by velocity. If Einstein's first postulate is correct, then all observers must find that laws such as the conservation of momentum, the conservation of energy, and $F = ma$ hold for any event they observe, regardless of their state of motion relative to the event. This can no longer be taken for granted, however, since observers in different frames of reference will have such different perceptions of space and time. Einstein was able to show that for the above laws to hold for all observers, regardless of their frame of reference, the mass of a body can no longer be considered an unvarying quantity, as in classical physics, but must increase with velocity. The formula for the relativistic mass, m, of a body moving with velocity, v, in terms of its rest mass, m_0, is:

Objects gain in mass as their velocity increases.

$$m = \frac{m_0}{\sqrt{1 - \frac{v^2}{c^2}}}$$

Relativity and Experience Here again we find the term $\sqrt{1 - v^2/c^2}$ familiar to us from the formulas for time dilation and length contraction. We can see from these questions that unless v is fairly close to c, v^2/c^2 is a very small fraction and $\sqrt{1 - v^2/c^2}$ is very close to 1. This means that for velocities that are not close to that of light, the relativistic changes in time, length, and mass are vanishingly small. At a velocity of $.1c$—an enormous velocity, over 18,000 mi/sec—the term $\sqrt{1 - v^2/c^2}$ is only .995. For an example using more familiar velocities, let us calculate the change in mass

The Paradox of the Twins

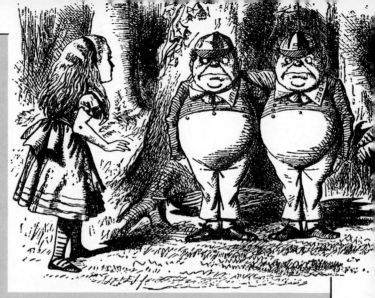

Many relativistic effects are commonly described as paradoxical in the sense that they seem strange, even illogical, by everyday standards. In the first years after publication of the special theory of relativity, however, scientists came upon certain situations which seemed paradoxical in a more literal sense—that is, they seemed to involve a contradiction within the framework of relativity itself. The most famous of these is the paradox of the twins. This paradox, which arises from an apparently simple thought experiment, is based on the time-dilation effect predicted by the special theory. We have seen that an observer in any inertial frame will find that the passage of time is slower in all other inertial frames. The greater their velocity with respect to him, the more slowly everything seems to happen in such frames. Clocks run more slowly, eggs cook more slowly, atoms vibrate more slowly, even the process of aging is less rapid.

Suppose, then, that Tweedledum leaves the earth in a rocket capable of traveling at nearly the speed of light while his twin brother, Tweedledee, remains behind to observe his flight. If Dum reaches a velocity of .995c, he will seem to Dee to age only one year for every ten earth years that pass. Returning to earth after ten years' flight, therefore, he should be nine years younger than his twin, Dee. This may seem very strange; still, it involves no logical inconsistency. Stranger things have been predicted by the theory of relativity and confirmed by experiment. Suppose, however, we look at the same situation, not from the point of view of Dee on earth, but through the eyes of Dum in space. To him, the inertial frame of the rocket can be considered at rest; he sees the planet earth, and with it his brother Dee, traveling at a velocity of .995c, and he sees time on earth passing ten times more slowly than it does for him. Dee seems to age only one year to his ten; when Dum returns to earth after a decade in flight, Dee should be nine years younger than

Dum! Now we are presented with a situation that is not merely strange but contradictory: How can *each* of the two twins be younger?

The answer, as is so often the case, lies in certain details of the situation we have overlooked. We have been assuming that the relationship between the twins is entirely symmetrical, but this is not the case. Dee has remained in an inertial frame throughout the episode, but Dum has undergone acceleration at least three times. This is of crucial importance because the special theory applies only to inertial—that is, unaccelerated—frames of reference. To attain a velocity of .995c, however, Dum's rocket had to be accelerated. In order for it to return to earth, it had to reverse its course at some point in its flight, and any change in the direction of motion also represents a form of acceleration. Finally, to land on earth the rocket had to undergo still another form of acceleration (deceleration, in this case). Using the general theory of relativity to take into account the times when the rocket was not in an inertial frame, it is possible to show that earthbound Dee does in fact age more rapidly than his more adventurous twin. Space travel, it turns out, enables us to "stretch" time, but only from the point of view of an observer left behind on earth; subjectively, the space traveler experiences no difference, and time seems to him to pass quite normally.

Table 10-1
Sample Values of m/m_0

V/c	$\dfrac{1}{\sqrt{1-\dfrac{v^2}{c^2}}}$
0.0001	1.000 000 6
0.001	1.000 05
0.01	1.001
0.1	1.005
0.2	1.021
0.3	1.048
0.4	1.091
0.5	1.155
0.6	1.250
0.7	1.400
0.8	1.667
0.9	2.294
0.95	3.203
0.98	5.025
0.99	7.089
0.995	10.013
0.998	15.819
0.999	22.366
0.9999	70.72
0.99999	223.6

and size of a golf ball as it leaves the tee after a drive. Suppose the ball is traveling at 135 mi/hr—about 60 m/sec. (Such calculations are usually easiest to perform in metric units since the velocity of light in this system is a convenient 3×10^8 m/sec.)

There is a mathematical shortcut we can use to simplify the computation further. When v is less than about .3c, the following approximations can be used with very little loss of accuracy:

$$\sqrt{1-\frac{v^2}{c^2}} \cong 1 - \tfrac{1}{2}\frac{v^2}{c^2} \quad \text{or} \quad \frac{1}{\sqrt{1-\frac{v^2}{c^2}}} \cong 1 + \tfrac{1}{2}\frac{v^2}{c^2}$$

The mass of the ball, then, will be approximately its rest mass times $1 + \tfrac{1}{2}(60/3 \times 10^7)^2 = 1 + (2 \times 10^{-14}) = 1.000\ 000\ 000\ 000\ 02$. Similarly, its diameter in the direction of its flight will be less than its normal diameter by a factor of .999 999 999 999 98. It is hardly surprising, then, that the changes in length, time, and mass predicted by the theory of relativity had not been observed before Einstein published his work and are never noticed in everyday life.

This explains, too, why our common sense is so reluctant to accept the conclusions of relativity. Common sense is the product of ordinary experience, and velocities approaching that of light are not a part of our ordinary experience. Einstein did not prove that Newton was factually wrong so much as he showed him to be incomplete. For the range of phenomena they were intended to describe, Newtonian mechanics works perfectly. His equations become inapplicable only for objects moving with much higher velocities than Newton could ever have encountered. Then Newtonian equations must be replaced by those of relativity. Einstein's ideas are more satisfying because they work at all velocities.

The Limiting Velocity

We can see from the relativistic mass equation that, as v approaches c, the mass of the moving body increases more and more rapidly. If v were to equal c, the mass would become infinite. But there is no way for a material particle to reach the velocity of light, for from the equation $F = ma$, an infinite force would be required to accelerate it to that speed. The fact that c is the absolute limit for velocities of material objects, as well as the constant speed of all electromagnetic radiation in empty space, is one of the most fundamental implications of relativity. It follows from this limitation that the law of addition of velocities that we are familiar with from classical physics cannot be applicable when velocities close to that of light are involved.

If a spaceship traveling past us at .75c were to launch a missile in a forward direction with a velocity of .50c, for example, classical mechanics would lead us to expect that the missile would have a velocity of 1.25c relative to us. But relativity asserts that this is impossible. For the addition of two velocities, v_1 and v_2, where either is close to c, we must use the formula:

$$v = \frac{v_1 + v_2}{1 + (v_1 v_2 / c^2)}$$

No material object can travel faster than light.

When the two velocities are small compared to c, this reduces to the classic formula $v = v_1 + v_2$. But when the velocities are large, this formula ensures that the sum can never be greater than c. Thus in the above example, the sum of $.75c$ and $.50c$ turns out to be $.91c$.

From the relativistic mass equation Einstein derived a much more famous equation, $E = mc^2$. Transposing the equation for the mass of a moving body, and using the approximation for low v, easily gives:

Mass and Energy

$$m - m_0 = \tfrac{1}{2} m_0 v^2 / c^2$$

Multiplying both sides by c^2 gives us:

$(m - m_0)c^2 = \tfrac{1}{2} m_0 v^2$ (The result can be proved for all values of v.)

The right-hand term is the kinetic energy of the moving body. When a body is in motion, in other words, its mass increases, and it acquires kinetic energy; the kinetic energy is seen to be a function of the gain in mass. But the mass gained through motion times c^2 equals the kinetic energy, this implies that the normal rest mass of the body multiplied by c^2 may also represent a kind of energy. This inference turns out to be correct. Even at rest, a body may be said to have an energy given by the formula $E = mc^2$. Moreover, it was later shown that matter can be transformed into energy: this is what takes place in nuclear bombs and nuclear reactors. The square of the velocity of light is an enormous number; therefore a very small amount of matter represents a very large amount of energy. From Einstein's formula we can calculate that 1 g of matter, if transformed entirely into energy, would supply:

Mass can be transformed into a quantity of energy given by the formula $E = mc^2$.

$$10^{-3} \text{ kg} \times (3 \times 10^8 \text{ m/sec})^2 = 9 \times 10^{13} \text{ joules}$$

Converted into more familiar units of energy, this is the equivalent of 25 million kilowatt hours — enough energy to power a million average household toasters for an entire day.

The special theory of relativity deals only with observers in inertial frames — that is, moving in straight lines with uniform velocities relative to each other. In 1915, ten years after the formulation of the special theory, Einstein published his general theory of relativity, which takes up the problem of accelerated motion. At the same time, the general theory introduces a new interpretation of gravitation.

The General Theory of Relativity

Gravitation and Inertia

We saw in Chapter 2 that in Newtonian mechanics there are two different definitions of mass. Inertial mass, or resistance to acceleration, is measured by the equation $F = ma$, Newton's second law of motion. Gravitational mass is the attraction of bodies for one another, as given by Newton's equation for the law of gravitation, $F = G\ m_1 m_2/r^2$. Gravitational mass and inertial mass have always been found to be exactly equal, but there seems to be no theoretical reason in Newtonian mechanics why this should be so, and classical physics was never able to explain this remarkable and convenient connection.

Einstein took a characteristically fresh and radical approach to this problem by imagining another hypothetical situation: a scientist-astronaut making observations in the laboratory of his spaceship. The lab is equipped with all the apparatus necessary for measuring distance, time, and force, but it is sealed and windowless with no way for the observer to see anything of his surroundings. Suppose the observer finds that everything in his laboratory seems completely weightless. Objects float freely in space, with no tendency to fall towards the floor, or they move steadily in a straight line if given the slightest push. The observer would naturally assume that his ship is in interstellar space, far from the gravitational field of any star or planet.

He might also make some assumptions about his state of motion. Whether he is in motion or at rest, how fast he may be moving, and in what direction, he cannot say; for as the special theory of relativity shows, these terms are all meaningless except in relation to some external frame of reference, and the observer is cut off from any such frame. But he can conclude that his lab is an inertial frame—its motion must be linear and uniform, not curved or accelerated. If the lab were accelerating, decelerating, or traveling in a curved path, inertial or centrifugal forces would cause the objects in the lab to behave quite differently.

Everyone is familiar with the feeling of being thrown back against the seat when a car accelerates suddenly or thrown to the side as it rounds a sharp turn. Such effects would be detectable in the lab if it were in nonuniform motion of any kind. Suppose, for example, that the ship was experiencing an acceleration of 32 ft/sec² in an upward direction. The floor of the lab would be rushing up to meet any object in the lab with that acceleration. To an observer in the lab, taking the walls of the lab as his frame of reference, it would seem that all the objects in the lab were falling towards the floor with an acceleration of 32 ft/sec².

Would the observer conclude that his ship was accelerating? Not necessarily, for there is an equally plausible alternative. The observer could just as logically conclude that his ship has entered a gravitational field. In fact, if the acceleration of the ship were equal to g, as in the instance we have been describing, he might believe that he had never left earth, for he would observe objects accelerating towards the floor, which means falling exactly as they do on the surface of our planet.

This situation is completely reciprocal. If the ship were actually in a gravitational field, the observer might conclude, with equal logic, that objects were falling towards the floor because the ship was actually accelerating. In short, Einstein suggested that the physical effects of gravitation and acceleration are equivalent and indistinguishable. There is no difference between the behavior of objects in an accelerated frame and their behavior in an inertial frame which is subject to a uniform gravitational field; the inertial reaction to acceleration (inertial mass) and the gravitational acceleration due to objects acting on one another at a distance (gravitational mass) are equivalent. This **principle of equivalence** is the heart of the general theory.

The effects of gravitation and acceleration cannot be distinguished.

We can now solve the old riddle of why gravitational and inertial mass are identical. There is an old humorous saying: "If it looks like a duck, walks like a duck, and quacks like a duck, it's a duck." For scientists this is not just a joke but a fundamental principle. If two things are indistinguishable in principle and in experiment, they must be considered one and the same thing. Einstein's theory shows that inertia and gravitation are simply different aspects of the same phenomenon.

The general theory of relativity has proven harder to test experimentally than the special theory. Some of its consequences, however, can be and have been confirmed by observation. Consider the following situation: a bullet is fired through the spaceship-laboratory described previously. What will the observer see? If

Testing the Theory

Figure 10-8
The principle of equivalence states that the effects of gravitation and acceleration cannot be distinguished. The scientist in the first rocket, traveling with an acceleration of *a*, will observe exactly the same results in his experiments as the scientist in the second rocket, at rest on the surface of a planet where the gravitational acceleration, *g*, is equal to *a*. Even the behavior of light will be identical.

Planet X

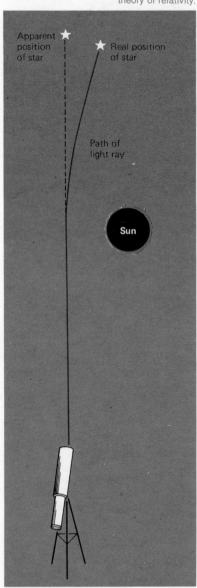

Figure 10-9
During an eclipse of the sun, it is possible
to observe that stars whose light must pass
close to the sun to reach us appear slightly
out of position. This is due to the bending of
their light rays by the sun's gravitational
field—an effect predicted by the general
theory of relativity.

Apparent
position
of star

Real position
of star

Path of
light ray

Sun

the laboratory is at rest on earth, it constitutes for the observer an inertial frame in a uniform gravitational field. Under the influence of gravity the path of the bullet will be parabolic. If the lab is far outside the earth's gravitational field but is accelerating with an acceleration equal to g, the principle of equivalence assures us that the bullet will pursue an identical parabolic path as seen by an observer in the lab—though to an external observer in an inertial frame the path will appear to be a straight line.

But suppose that a beam of light rather than a bullet crosses the lab. If the ship is accelerating, the laboratory will move during the time it takes the beam of light to cross it, and the beam of light, like the bullet, will strike the opposite wall of the lab lower than its point of origin. Because light is so much faster than a bullet, the curvature of the light ray's path will be less marked, but it too will appear parabolic to an observer in the lab.

So far these thought experiments would not be particularly surprising to a student of classical mechanics. Now, however, we can invoke the principle of equivalence and obtain a rather unexpected result. Einstein's theory asserts that acceleration and gravitation are identical in their effects on all natural phenomena, electromagnetic as well as mechanical. If light will travel in a parabolic path as a result of acceleration, it must also travel in a parabolic path through a gravitational field. Light, in other words, is subject to gravity!

Though this is a remarkable conclusion, it is consistent with the results of the special theory of relativity, which also implies that light has mass. Light is a form of energy, and the special theory shows that energy and mass can be considered equivalent. Though light has no rest mass (light is never at rest in any case), it should have associated with it a mass of E/c^2, where E is its energy.

Can we test this conclusion? Light is so low in mass and travels so fast that gravitational influence on its motion is usually too small to detect. Einstein calculated, however, that the gravitational field near the surface of the sun is sufficiently strong to bend light coming from distant stars. Normally, of course, stars cannot be seen near the sun because of its glare, but during the solar eclipse of May 29, 1919, astronomers measured the position of several stars near the darkened solar disc. In each instance the stars appeared slightly out of place compared to their normally observed positions, just as would be the case if the path of their light had been bent. The amount of deflection by the sun's gravitational field agreed very well with Einstein's prediction.

Another observation that tends to confirm the general theory is the influence of strong gravitational fields on the frequency of electromagnetic radiation. Since a gravitational field produces the same effects as acceleration, it should give rise to time dilation similar to that described by the special theory. Time dilation means that all physical processes occur more slowly. Since light is an electromagnetic vibration, this vibration too should be slower in a gravitational field. Thus an atom in a strong gravitational field—in the outer layers of a star, for instance—should emit light of lower frequency and longer wavelength than it would in the absence of such a field. Most stars, however, do not have gravitational fields powerful enough to influence greatly the wavelength of the radiation they emit, and when Einstein made this prediction no such effect had yet been observed. But in 1924 the American astronomer W. S. Adams (1876–1956) obtained the spectrum of a white dwarf—an unusually small, dense star with a correspondingly huge surface gravity—and found precisely the changes in wavelength Einstein had anticipated.

The general theory of relativity shows that light is affected by gravitational fields.

Space–Time Continuum

Einstein's theory of gravitation attempts to clear up another problematical aspect of Newton's theory of gravitation—the idea that gravitational force can act directly on distant bodies through empty space. The nature of such action at a distance had never been satisfactorily explained in classical physics. The theory of the ether had been one attempt to deal with this mystery, but special relativity had dealt the ether a death blow. Einstein discarded both the concept of action at a distance and the concept of the ether. He substituted the notion—hard to visualize but mathematically powerful—of a four-dimensional space–time, comprising the three familiar dimensions of space and a fourth dimension, time.

This four-dimensional space–time is not necessarily straight and uniform, like the space of Euclidean geometry; it can be curved. Just as the theorems of Euclidean plane geometry do not hold on the surface of a sphere, they will be inapplicable to any region of space that is curved. The concept of curved space underlies Einstein's theory of gravitation. In Einstein's view, any matter in space distorts the four-dimensional fabric of space—time around it much as a heavy weight will distort the surface of a trampoline. Calculating the trajectories of bodies in space requires that we take into account the distortion of space by the masses in it—rather like a golfer allowing for the "break" of a putt on an undulating green.

To elaborate further, imagine a game of marbles played on a lake of jello. If there are no objects resting on the jello, it will remain a flat surface, and a marble will roll along it in a straight line. Suppose, however, that there are several heavier marbles scattered over the surface. Each one will cause the jello to sag at that particular spot, creating a small localized depression. The heavier the marble, the deeper and wider will be its distortion of the surface. A marble rolled along such a surface will no longer travel in a straight line but will tend to roll downhill towards the closest marble. If it is moving fast enough, it will veer somewhat towards its neighbors, but its momentum will carry it safely by. If it is moving more slowly, it will fall into the valley made by one of the other marbles and collide with it. If the velocity and direction of its motion are just right, however, it may fall only partway down one of the valleys and roll around it, like a golf ball rimming a cup or a basketball rimming a hoop. Indeed, if it could roll without any friction at all, it might fall into orbit and follow such a circular or elliptical path forever.

This game of marbles presents a fairly good analogy for Einstein's interpretation of the dynamics of bodies in space. In our analogy, however, the marbles still fall down into the depressions because of gravity. It is important to bear in mind that in the actual four-dimensional reality, there is no up or down and no gravitational force. The gravitational attraction of bodies is simply the result of their tendency to follow the contours of space, and those contours are created by the presence of mass in space. Thus the concept of gravitational force acting at a distance disappears and is replaced entirely by the geometry of space. Masses no longer exert forces, but rather shape the space about them, and the shape of the space determines the free motion of bodies through it.

Figure 10-10
According to the general theory of relativity, the presence of mass warps the four-dimensional fabric of space–time, producing the effect we call gravitation. (The photograph of Einstein on the bicycle was taken in California in 1933.)

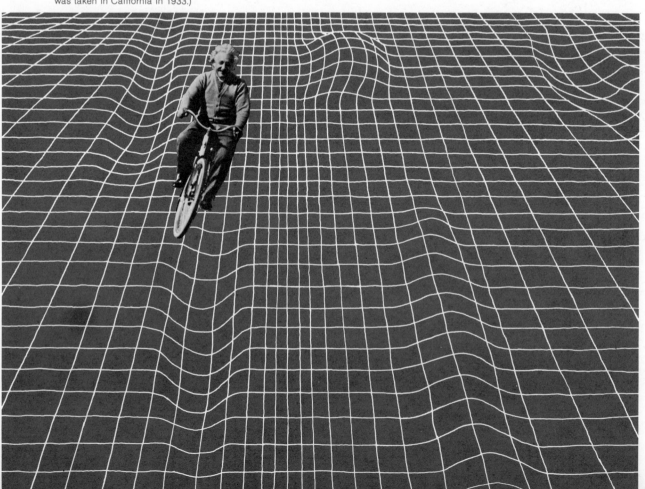

There is no such thing as **absolute motion,** according to the theory of relativity; only the relative motion of two objects can be measured.

The **ether** was assumed to be an imperceptible elastic substance filling all space that served as the medium through which light and other electromagnetic radiation propagated.

Inertial frames are frames of reference moving in uniform straight-line motion, without acceleration of any kind.

Length contraction is the relativistic effect which causes an object in motion relative to the observer to appear shorter in the direction of its motion.

The special theory of relativity asserts that the velocity of light is a **limiting velocity,** which no material object can exceed.

Relativistic **mass increase** is the gain in mass observed for objects in motion with respect to the observer.

The **Michelson experiment** was an attempt to determine the earth's motion through the ether by measuring the difference in velocity of two light beams traveling in different directions.

The **principle of equivalence** in general relativity theory states that the effects of gravitation and acceleration cannot be distinguished by any experiment.

Time dilation is the relativistic phenomenon whereby time appears to pass more slowly in inertial frames that are moving with respect to the observer.

1 What is meant by the assertion that all inertial frames are equivalent? If each of a number of persons in uniform relative motion claims that *he* is fixed and the others moving, it is possible to settle this argument by conducting an experiment?

2 Can a person in a completely sealed room carry out experiments to determine whether the room is rotating?

3 What is the theory of ether and why was it postulated? What were the objections to this theory?

4 Describe the Michelson experiment. What result was expected when the experiment was undertaken? Why was the negative result so important to our understanding of physics?

5 It is occasionally asserted that Einstein's theory of relativity proves that everything is relative. In fact, however, the special theory is based on the realization that there is something in nature that is *not* relative. What?

6 If a friend of yours switched on a flashlight, the beam would travel away from him at the speed of light. Could you chase after the beam and catch up with it if you had at your disposal a rocket ship capable of immense acceleration?

7 The neutrino is a massless particle that travels at almost the velocity of light. What would be the velocity of such a neutrino as seen by an observer who is traveling towards the neutrino source at $0.6c$?

8 A rocket ship travels past you at a uniform velocity of 0.8c. A person in the rocket measures its height to be 10 ft and its length to be 40 ft. If you made accurate measurements of the rocket's dimensions as it sped past you, what figures would you get for its length and height? If it took you ten minutes, by your watch, to complete your measurements, and the person in the rocket had an identical watch, how much time would your measurements appear to take when measured by his watch?

9 Some elementary particles have such short lifetimes that if they traveled at the velocity of light for their entire lifetime one would expect them to cover a distance of only about 10^{-10} cm—much too short to be photographed. In fact, however, these particles are often photographed making much longer tracks. How is this possible?

10 In modern accelerators, electrons often have relativistic masses that are 10,000 times their rest masses. What does this imply about their velocities? By what factor does the mass of an object increase when its velocity is 0.7c?

11 Why should the predictions of the special theory of relativity resemble those of Newtonian mechanics for velocities less than about 5 percent of the speed of light?

12 Is it possible, in considering relativistic effects, to distinguish between what is observed and what "really" happens?

13 What is the principle of equivalence?

14 What effects are predicted by the general theory of relativity and how have they been tested? Why are eclipses of the sun so important in testing the theory?

15 If a person lived on a very dense star that had an immense gravitational field, how would his reckoning of time differ from that of his twin living on a planet with a far less intense gravitational field?

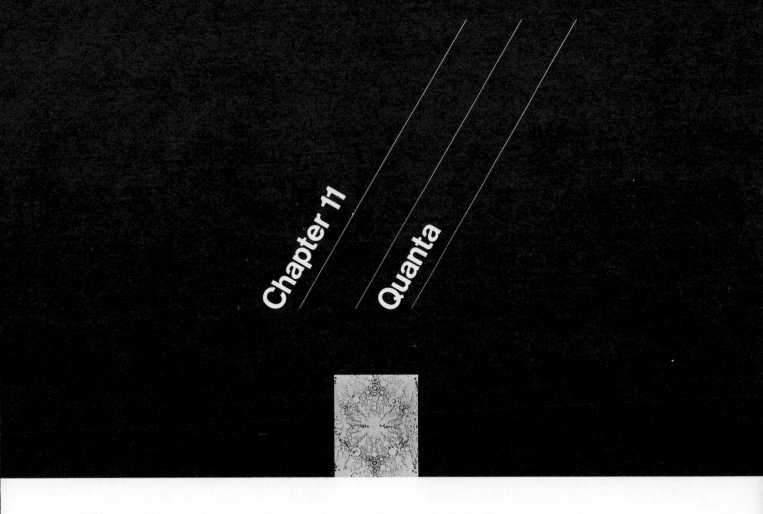

Chapter 11

Quanta

In 1905, a landmark year for modern physics, Einstein published his special theory of relativity, which radically changed our ideas of time and space. Astonishingly, in the very same year he produced another paper that was destined to have an equally great impact on the scientific thought of this century. In his second paper Einstein set forth a simple explanation for a puzzling experimental phenomenon known as the photoelectric effect. His explanation, however, did much more than account for the experimental results; it implied a fundamental change in our way of thinking about light and electromagnetic radiation. This new theory soon proved to have far-reaching implications, for it led to a breakthrough in our understanding of energy, and of the structure of the atom as well.

The Quantum Nature of Light

The Photoelectric Effect

The photoelectric effect was not considered very significant when first discovered. Certainly no one thought that it would point the way towards a great theoretical revolution. In 1887 the German physicist Heinrich Hertz (1857–1894) found that when ultraviolet light fell on a metal surface, an electric current was emitted from the metal. Later it was established that, with some metals, visible light could produce the same effect. The electron was discovered by J. J. Thomson in 1897, and it was then quickly seen that the photoelectric current was made up of electrons.

According to classical theory, a beam of light could transfer some of its energy to an electron in a metal and ultimately dislodge it from its place in a lengthy process similar to evaporation. Moreover, classical theory predicted that the kinetic energy of the electrons should increase with the intensity of the light. The brighter the light, the faster the electrons and the more violently they would be torn from the metal. The frequency of the light, on the other hand, should not matter at all; the result ought to be the same whether red, blue, or ultraviolet light is used.

The experimental results did not agree with these predictions, however. In the laboratory it was found that the photoelectrons were emitted almost instantaneously when light fell on the metal. Moreover, the intensity of the light was related only to the *number* of electrons emitted, not to their energy. A brighter light would produce more electrons than a dim one, but each individual electron would have the same kinetic energy. The energy of the electrons was related instead to the frequency of the light. Each metal was found to have a threshold frequency (sometimes in the visible range, sometimes in the ultraviolet). Light of a frequency below this threshold produced no photoelectrons at all, no matter how bright the beam of light. At the threshold frequency, electrons were released but with a minimum of kinetic energy. As the frequency increased, however, the kinetic energy of the electrons increased proportionately.

Figure 11-1
A simple apparatus for studying the photoelectric effect. Light falling on a metal plate causes the emission of electrons. If the electrode at the opposite end of the evacuated tube is kept positively charged, a current of photoelectrons will flow through the tube; measurement of this current will reveal the number of photoelectrons released. If the electrode is given a negative charge, the kinetic energy of the photoelectrons can be determined by measuring the potential difference needed to keep the electrons from reaching it.

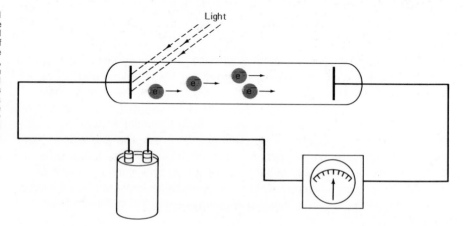

PLANCK'S CONSTANT Einstein soon realized that the key to explaining this phenomenon was already at hand. This was the **quantum theory** of light, proposed by the German physicist Max Planck (1858–1947) in 1900, but not yet widely understood or accepted. Planck had been working on a very different problem—the radiation emitted by hot solid bodies, another experimental phenomenon the laws of classical physics were unable to explain. Everyone is familiar with the way hot bodies radiate light—the coils of an electric broiler, for example, or the filament of an incandescent light bulb. The amount of light radiated at various frequencies, however, turned out to be very different from what classical physics predicted.

By trial and error Planck found that the experimental results could be accounted for if one assumed that light was given off not as a continuous flow, as everyone had previously thought, but only in small fixed quantities, like individual packets of light. Planck named these atoms of energy **quanta**. A radiating body may emit one quantum, or ten billion quanta, but never 1½ or 99.1; only whole numbers of quanta can be radiated. Furthermore, Planck found that the energy of these quanta is proportional to their frequency. A quantum of violet light has more energy than one of red light, and a quantum of still higher frequency ultraviolet light has more than either. In mathematical terms:

$$E = hf$$

The constant of proportionality, h, called **Planck's constant**, is now recognized as one of the most fundamental constants in nature. Its value is 6.63×10^{-34} joule-sec.

THE PHOTON Planck's hypothesis led to equations that described the radiation of hot bodies perfectly, and so solved the original problem, but it left scientists with a deeper problem: why was radiation quantized? Since no one knew how to interpret Planck's theory, it received relatively little attention. Einstein, however, grasped the fundamental and revolutionary significance of Planck's quanta, and saw how the concept could be extended to explain the photoelectric effect. Though Planck had proposed that radiant energy was emitted in discrete quanta, he had not suggested that these packets of light were themselves any different from the ordinary electromagnetic waves described by Maxwell's equations. He thought they were little packages or bunches of waves. Einstein went further. If light was emitted in the form of discrete packets of energy, he reasoned, it seemed plausible to assume that it traveled through space as discrete packets and interacted with matter as discrete packets. Light, in other words, could be thought of as consisting of particles, called **photons**, each carrying a quantum of energy equal to hf.

This interpretation would explain the photoelectric effect very neatly. Suppose that when light falls on a metal surface, a single electron absorbs the energy of a single photon. Let us call w the work necessary to dislodge the electron from

Light can be radiated and absorbed only in discrete packets called photons, whose energy is directly proportional to the frequency of the light.

its place in the metal. If the energy carried by the photon, *hf,* is less than *w,* it does not matter how many photons per second bombard the metal and collide with electrons; no electrons will be liberated because none of the photons has enough energy to do the job. This explains the existence of the threshold frequency. But if light with precisely the threshold frequency strikes the metal, the individual photons will have just enough energy to dislodge electrons, and if the frequency of the light is above the threshold, there will even be some energy left over. This surplus energy gives the electrons their velocity. The kinetic energy of the emerging photons, therefore, must equal the difference between the energy carried by the photon and the energy necessary to pull the electron free of the metal:

$$\tfrac{1}{2}mv^2 = hf - w$$

where *m* and *v* are the mass and velocity of the electron.

Experimental Verification Einstein's explanation of the photoelectric effect was at first greeted with suspicion, primarily because it went beyond known experimental fact to predict new phenomena. But experimental work confirmed Einstein's theory. In 1916 Robert Millikan carefully measured the kinetic energy of photoelectrons produced by light of various frequencies, and his results agreed closely with the values predicted by Einstein's equations. Millikan's measurements also led to a value for Planck's constant very similar to that which Planck himself had derived from his original work on radiation. The fact that a theory derived from the phenomenon of radiation of light by hot bodies should be so successful in explaining the very different phenomenon of emission of photoelectrons was strong evidence in its favor. No longer could the quantum theory be regarded as a misguided speculation; it had to be taken seriously.

Additional support for Einstein's idea that light could be regarded as consisting of individual photons soon followed. X-rays, for example, had been discovered in 1895 but only in 1912 was it determined that they were actually electromagnetic radiation of very high frequency. X-rays can be generated by bombarding a metal target with fast-moving electrons, a kind of photoelectric effect in reverse: instead of a shower of photons giving up their energy to knock electrons from a metal, a stream of electrons give up their kinetic energy to knock out very high energy photons. When experimenters studied this inverse photoelectric effect, they found that Einstein's quantum interpretation again was applicable. The number of electrons bombarding the metal determined the intensity of the x-rays produced — that is, the number of x-ray photons — but had no effect on their frequency. Instead frequency was related only to the energy of the electrons, with slow-moving electrons producing low frequency x-rays and faster electrons producing x-rays of higher frequency. If an electron stopped by the target atoms gives up its kinetic energy in the form of a photon and if each photon has an energy proportional to its frequency, as the quantum theory asserts, this is just the result we would expect.

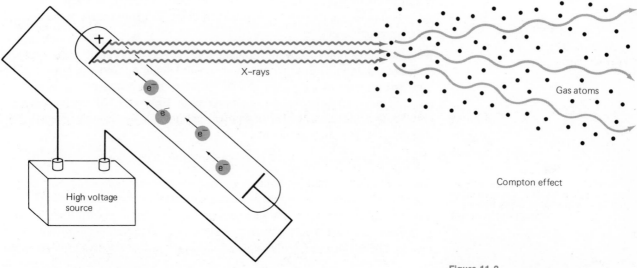

Inverse photoelectric effect

X-rays

Gas atoms

High voltage source

Compton effect

Figure 11-2
Two demonstrations of the quantum nature of electromagnetic radiation. In the inverse photoelectric effect (left), electrons are accelerated across a potential difference. When they strike the metal target they give up their kinetic energy, generating x-rays—electromagnetic radiation of very high frequency. The frequency of the x-rays emitted is determined by the kinetic energy of the electrons, not by their number. X-rays that collide with gas atoms (right) lose some of their energy to the atomic electrons. In quantum theory, a lower energy corresponds to a lower frequency; thus the scattered x-rays are found to be lower in frequency and longer in wavelength than those of the original beam. This is the Compton effect.

Another x-ray effect offered further support for the quantum theory of light. In 1923 the American physicist Arthur H. Compton (1892–1962) found that when x-rays pass through a gas, some of them are scattered in various directions. Moreover, the scattered x-rays are lower in frequency than those of the original beam. If x-rays are regarded as waves, this phenomenon, called the **Compton effect**, cannot be explained, for waves passing through a gas stay at the same frequency. But suppose, as Einstein suggested, we consider the x-ray beam to be a stream of photons, each with an energy of hf. When a photon collides with an electron, it gives it a shove. The electron, set into motion, has acquired some kinetic energy at the expense of the photon, which must lose some energy.

If the collision were between bits of ordinary matter, we would expect the photon to bounce off, like a rubber ball when it hits the pavement, with a smaller velocity than it had initially, since some of its energy has been imparted to the electron. But the photon is a quantum of light, so it cannot move with a velocity less than c. The quantum theory helps to explain how it loses energy. The energy of a photon, hf, is directly proportional to its frequency. If it loses energy, its frequency must diminish. If the bombarded electron acquires a kinetic energy of K from the bombarding photon, then the law of conservation of energy assures us that the photon must lose the same amount of energy. If the initial frequency of the photon is f and its frequency after the collision is f', the energy lost to the electron will be given by the formula:

$$K = hf - hf'$$

This formula was confirmed in Compton's experiments, for which he won a Nobel Prize in 1923. Einstein himself had been awarded a Nobel Prize two years earlier for his pioneering application of the quantum theory of radiation to the photoelectric effect.

Why was this not discovered before 1905? The answer is that, at most frequencies, an individual photon is simply too small a bundle of energy to be detected by then existing instruments. Ordinary light in the orange portion of the spectrum, for example, has a frequency of about 5×10^{14} cycles per second. To find the energy of a photon of orange light we simply multiply this frequency by Planck's constant:

$$5 \times 10^{14} \text{ sec}^{-1} \times 6.6 \times 10^{-34} \text{ joule-sec} = 3.3 \times 10^{-19} \text{ joules}$$

Photons of visible light represent such a tiny amount of energy that an ordinary light source emits trillions of them each second.

This is an extraordinarily minute amount of energy. In other words, an ordinary beam of light such as we might use in a macroscopic physical experiment consists of trillions upon trillions of photons. The amount of energy carried by a single one is far too small to have been observed with nineteenth-century techniques, but the new x-ray techniques made Compton's observations possible.

The Quantum Structure of the Atom

Just eight years after Einstein published his 1905 paper on the photoelectric effect, a young Danish physicist, Niels Bohr, (1885–1962) showed that quantum ideas were the answer to the riddle of atomic structure. Atoms were not, of course, a completely new concept; the idea had been anticipated by the ancient Greek philosophers. But solid scientific evidence about their composition had come only in the nineteenth century. In 1897 Thomson had proven that electrons were a basic constituent of all matter, and that they were nearly 2000 times lighter than even the hydrogen atom. Presumably, then, atoms consisted in part at least of electrons.

No one, however, knew anything precise about the structure of the atom. Where, for example, were the electrons when they were parts of atoms? Thomson himself favored a raisin-pudding model, in which the atom was a diffuse sphere of positive charge, within which the individual negative electrons were embedded like raisins. Such a model accounted for the fact that the atom is electrically neutral under ordinary circumstances, even though it contains negatively charged electrons.

Atoms and Radiation

Any model advanced had to explain one of the most important facts about atoms: they emit light and other forms of electromagnetic radiation. The study of atomic spectra (visible and invisible light of different wavelengths emitted by substances which have been heated or electrically excited to incandescence) had been under way since the last third of the nineteenth century. Much had

Measuring The Electron

In 1897, the great British physicist J. J. Thomson (1856–1940) carried out a series of experiments to determine whether cathode rays were electromagnetic waves or a stream of charged particles. He found that they could be deflected both in magnetic fields and in electric fields; therefore they must be charged particles and not waves. The directions of the deflections indicated that the charge was negative. Most striking, the properties of these particles were the same no matter what material the cathode was made of, and the particles were immensely lighter and smaller than the lightest atoms. Thomson's fundamental particles were later named *electrons*. His discovery of such subatomic particles was an extraordinary event, much resisted by his contemporaries. Thomson later wrote: "I had myself come to this explanation of my experiments with great reluctance, and it was only after I was convinced that the experiments left no escape from it that I published my belief in the existence of bodies smaller than atoms."

The experimental data permitted Thomson to calculate the relative sizes of charge and mass of his particles, namely the ratio q/m. The value of q/m was constant at 1.76×10^{11} coul/kg. This is some 1800 times the q/m for ionized hydrogen atoms (which we now know to be single protons). If the electron charge were roughly equal (though opposite in sign) to the hydrogen ion's charge, then Thomson's electrons were drastically less massive than the lightest atoms.

Measuring q/m was fairly simple, but it required several steps. First, Thomson observed the curved path of the cathode rays in a known uniform magnetic field, H, to be the arc of a circle with radius R. To produce such a circular acceleration, v^2/R, of particles of mass m and velocity v, a force (the familiar centripetal force) would be needed, mv^2/R. Now the force must be supplied by the magnetic field, since otherwise the cathode rays are not deflected into the curved path. That deflecting force was known to be Hqv. And so, $Hqv = mv^2/R$. From this, $q/m = v/HR$.

Now the magnetic field H can either be measured directly, or it can be calculated from the design of the electromagnete which produces it (the size and number of the coils and the current flowing through them). The particle's path has a radius R which is directly measurable from the amount of deflection of the rays at the standard visible screen at the end of the cathode-ray tube. The velocity v can be determined by balancing out the magnetic deflection with an electric field E perpendicular to H. The electric force on a charge q is Eq, and that must equal the magnetic force, so $Eq = Hqv$. Therefore $v = E/H$. Since E can be measured or calculated from the geometry of the two electrically charged plates which produce it and the voltage across them, v is easily obtained. With v, H, R all known, the ratio q/m was too.

If q were known, then the electron mass could be determined more precisely. Even more was at stake since the charge on the electron is a fundamental constant of nature, repeatedly found in the structures and properties of nuclei and atoms. After some years, during 1909 and 1916, the American physicist Robert A. Millikan (1868–1953) devised a clever and beautiful experiment to measure the electronic charge. He sprayed a mist of fine oil drops between two horizontal charged metal plates, the drops being charged by friction as they leave the nozzle. Millikan knew the density of the oil, and he could measure the size of the drops by visible estimate through a strong microscope and so he knew the mass of the drops (although this matter of calculating the mass of the drops was troublesome when precise results were wanted). Millikan's task was to measure the electric charges on the drops! What he did was to balance the downward force of gravitation on the drops with an upward electric force, so that the drops remained suspended in air. For this, $mg = qE$, and $q = mg/E$. The electric force was known from the voltage across the two plates and the distance between them.

The result was startling: the charges on the drops were always a small integer times a small value, 1.6×10^{-19} coul. It seemed unavoidable that the drops were being charged by one or more electrons, and that the single electron charge was simply that smallest value. It has since been uniformly observed that electric charge comes in multiples of the charge on the electron, q_e, 1.6×10^{-19} coul.

With this value for q_e, Thomson's ratio q/m provides an electron mass of 9.1×10^{-31} kg. The way to the physics of elementary particles was open.

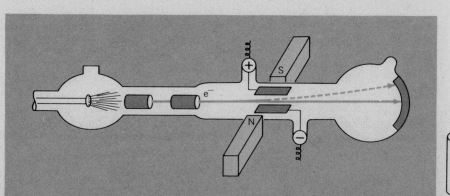

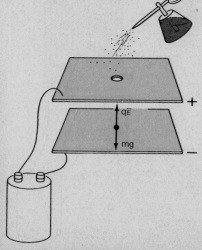

been learned, but little was well understood. It had been established that light from hot dense bodies—solids, liquids, and even dense gases—could be broken up by a prism into many different colors. These colors were not separate and discrete but merged into each other in imperceptible gradations, forming a continuous spectrum much like a rainbow in which an enormous number of different wavelengths were present.

The spectra of rarefied gases were very different. Instead of a continuous spectrum, each element (in gas form) emitted a number of sharply defined thin bright lines; each element had its own set of such wavelengths, which identified it as accurately as a fingerprint identifies a person. In between these bright lines—that is, at intermediate wavelengths—there was no radiation. A reverse phenomenon was also observed. When a continuous or rainbow spectrum was passed through a cool gas, the gas absorbed light of precisely the same characteristic wavelengths that it radiated when it was itself incandescent. The result was a spectrum broken by numerous thin, dark absorption lines—the dark-line spectrum.

In a rarefied gas, the atoms are far apart. Collisions cannot be very frequent, and so the atoms do not interfere with each other's movements most of the time. Under such conditions, the radiation frequencies emitted must be determined by properties of the individual atom itself, unaffected by interactions between atoms. If so, the line spectrum of an atom should give us some clues to its structure. But the clues proved very difficult to interpret. Each element did produce its own unique bright-line spectrum, and hence each could be detected by means of what could be called its own individual spectral fingerprint. Molecules also show a characteristic spectrum. Thus the atomic hydrogen spectrum is different from the spectrum of molecular hydrogen. But the number of spectral lines varied mysteriously from element to element. Some had only one or two in the visible range, while iron, for example, had over 6000. Nor could the spacing of the lines be easily explained, though they could be measured with the greatest precision. Those of many elements—hydrogen, for example—hinted at some regular pattern, but the pattern remained elusive for many years.

BALMER'S FORMULA In 1885, however, Johann Balmer (1825–1898), a German mathematician of no great scientific reputation who taught at a girl's school in Switzerland, had discovered at least part of the pattern. Balmer found, strictly by trial and error, a mathematical equation which fitted known frequencies and predicted many other frequencies of the lines of the hydrogen spectrum with amazing accuracy. The most useful form of Balmer's equation is:

$$\frac{1}{\lambda} = \frac{f}{c} = R\left(\frac{1}{2^2} - \frac{1}{n^2}\right) \qquad n = 3, 4, 5, \ldots$$

where n is any whole number greater than 2, and R is the Rydberg constant $1.1 \times 10^7 \ m^{-1}$, named after the Swedish spectroscopist who extended Balmer's

Red
6563A H_α

Green
4863A H_β

Blue
4340A H_γ
Violet
4101A H_δ

3646A   Series limit

Figure 11-3
The Balmer series of hydrogen spectral lines.

work. (Notice that for any electromagnetic wave the product of the frequency, f, and the wavelength, λ, must equal the speed of light:

$$f\lambda = c$$

so we can write Balmer's formula in terms of either f or λ.)

The Balmer formula worked too well to be accidental. Although Balmer had based his work entirely on the measured position of only four lines in the hydrogen spectrum, his formula proved successful in predicting the frequencies of many others which were found in the following years. All the lines whose wavelengths are given by Balmer's original formula make up the Balmer series of hydrogen. Other whole families of lines in the H spectrum were later discovered with wavelengths specified by similar formulas such as:

$$\frac{1}{\lambda} = R\left(\frac{1}{1^2} - \frac{1}{n^2}\right) \quad n = 2, 3, 4, \ldots; \qquad \frac{1}{\lambda} = R\left(\frac{1}{3^2} - \frac{1}{n^2}\right) \quad n = 4, 5, 6, \ldots; \qquad \text{etc.}$$

The lines of other series were not discovered in Balmer's time because most of them lie outside the visible portion of the spectrum, in the infrared and ultraviolet regions. With these, it is possible to generalize Balmer's original formula:

$$\frac{1}{\lambda} = R\left(\frac{1}{n_1^2} - \frac{1}{n_2^2}\right) \qquad n_2 > n_1$$

There remained two problems. One was that the spectra of other elements, though they seemed to exhibit regularities like those found by Balmer, were too complicated to be analyzed by his kind of simple mathematical sequences: Balmer's formula worked only for hydrogen. Second, no one knew why Balmer's formula worked at all. It was purely empirical, a great descriptive generalization with no theory of atomic structure behind it. What was it in the structure of the atom that gave rise to the bright spectral lines? With the pudding model of the atom, some physicists thought they were on the way to solving this problem.

Maxwell had shown that electromagnetic waves are produced when an electrical charge is accelerated in a regular, periodic way. Presumably the electrons in the pudding atom normally occupied certain stable positions in which all electrical forces were in equilibrium. If an electron was jostled out of its equilibrium position, however, it might vibrate or be set into oscillatory motion, like a little weight attached to a spring. This motion would create electromagnetic waves to form the spectral lines. The frequency of the oscillation would determine the frequency of the emitted light.

So, at least, it seemed to Thomson and his colleagues. There were others who favored a model of the atom similar to a solar system. In this model the negative electrons were thought to circle a positively charged central nucleus in planetary orbits. This theory, however, had one fatal flaw. When an accelerated charge radiates electromagnetic waves, as Maxwell's equations showed it must, it loses energy. This would not be a problem for the pudding atom since

When excited to luminosity, the atoms of each element emit their unique pattern of characteristic spectral lines.

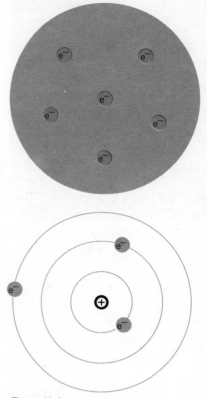

Figure 11-4
In Thomson's raisin pudding model of the atom, the electrons are scattered through a diffuse sphere of positive charge. In the nuclear model of the atom, the electrons circle a small positive nucleus in much the way that the planets circle the sun. Such an atom would not be stable by the laws of nineteenth century physics, however.

the electrons would simply oscillate with decreasing amplitude, like a ball giving up energy on each bounce. Eventually the motion would die out unless the electrons were jostled again by some outside force, but throughout the process the frequency of their vibration would remain unchanged.

The same would not be true for electrons in orbit, however, because their velocities would decrease as they lost energy and their orbits would become smaller and smaller, like the orbit of an earth satellite being gradually slowed by friction with the atmosphere. And the final result would be the same—the electrons would spiral more and more rapidly in towards the central nucleus. During this process the radiation they emitted would change in wavelength, and calculations showed that this inward spiral collapse should take less than a millionth of a second. Obviously such an atom could not be stable. But almost all atoms are stable for millions of years.

THE NUCLEAR ATOM   Theories, no matter how plausible, must always bow to experimental facts in the end. In 1911 the pioneer of nuclear physics, the New Zealander Ernest Rutherford (1871–1937), began bombarding thin layers of gold foil with alpha particles—atoms of helium from which two electrons had been stripped, leaving them positively charged. If the pudding model of the atom was correct, there would be no reason to expect that the course of the alpha particle would be affected greatly by their passage through many layers of atoms, for the mass and charge of each atom would have to be distributed over its full volume. Nowhere could they be sufficiently concentrated to deflect a heavy, fast-moving alpha particle from its course.

To Rutherford's surprise, some of the alpha particles were found to be scattered through wide angles; a few were even bounced back toward their point of origin. Years later, describing his reaction when told of this discovery, he wrote:

Figure 11-5
Ernest Rutherford in his laboratory.

Figure 11-6
In Rutherford's scattering experiment, alpha particles were directed at a thin gold foil target. Some were bounced back at very large angles, indicating that they had encountered a heavy, positively charged mass. This suggested that the atom did in fact have a compact nucleus.

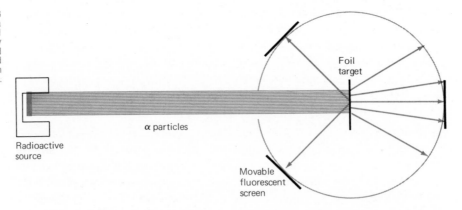

Foil target

α particles

Radioactive source

Movable fluorescent screen

It was quite the most incredible event that has ever happened to me in my life. It was almost as incredible as if you fired a 15-inch shell at a piece of tissue paper and it came back and hit you.

Rutherford realized that such a huge deflection could only mean that the alpha particle has passed very close to something very heavy carrying a positive charge. Thomson's pudding atom had nothing dense enough to produce such a result. The mass and charge of the atom, Rutherford concluded, were concentrated in a tiny nucleus. Presumably the electrons occupy planetlike orbits about this nucleus. But what prevented them from losing energy by radiation and eventually falling in toward the nucleus?

This was the problem that Niels Bohr pondered during the spring of 1912. Bohr was convinced that the quantum theory of Planck and Einstein could provide the answer. If the emission of light by the atom was quantized, which is what the Balmer formula suggested, then the structure of the atom itself had to be quantized in some way too. Bohr's progress was slowed by his unfamiliarity with Balmer's formula. When a friend showed Balmer's work to him in early 1913, everything fell into place; Bohr had the answer. Balmer's spectroscopic formula, Rutherford's nuclear atom, the quantum theory of Einstein and Planck: they could all be combined to produce a new model of atomic structure.

ELECTRON ORBITS Bohr's approach was successful because he realized the necessity for a break with classical ideas. He postulated that Maxwell's theory of radiation by accelerated charges had to be modified when dealing with subatomic particles such as the electron. Instead, Bohr assumed, the electron can exist in certain planetary orbits without radiating away its energy. Each of these stable orbits would represent a different energy state of the atom, since for each orbit the distance of the electron from the nucleus and also, therefore, its orbital velocity would be different. Thus the possible energy states of the Bohr atom are quantized. Atoms can only exist in certain discrete energy states, which correspond to the different stable orbits.

Bohr further suggested that only in switching from one stable orbit down to another at lower energy could the electron radiate electromagnetic energy. How the electron made such a jump Bohr did not attempt to say. He merely postulated that it took place instantaneously and was accompanied by the emission of a single photon of light. The energy of the photon must be exactly equal to the difference in energy states of the two orbits. Since the energy of a photon is hf, we can write:

$$hf = E_2 - E_1$$

where E_2 is the energy of the atom in the final state and E_1 its energy in the initial state. If the electron were to jump upward to a higher energy state, it

The mass and positive charge of the atom are concentrated in a tiny, dense nucleus.

Bohr's Model

Figure 11-7
Two of a series of cartoons representing
Bohr's career, drawn by his fellow physicist
George Gamow. In these, Bohr, depicted as
Mickey Mouse, is shown putting the
electron in orbit about the atomic nucleus.
(Bohr proposed his atomic model
in 1913.)

would have to absorb a photon of the right energy, hf. What were the actual allowed energy states which characterized the stable orbits? Bohr had little basis for his answering hypothesis but intuition. He speculated that it was the angular momentum of the electron in its orbit about the nucleus that was quantized. The formula for angluar momentum is mvr (where mv is the ordinary momentum, r is the radius of the orbit, m the mass of the electron, and v its velocity). Bohr assumed that the angular momentum could only have values that were integral multiples of $h/2\pi$, so that:

$$mvr = nh/2\pi$$

where n is any whole number.

In the Bohr model of the atom, the electron can occupy only certain stable orbits, each of which represents a permitted value of angular momentum.

Bohr's results confirmed that this was indeed the right hypothesis, for he was able to calculate almost exactly the experimentally known size of the hydrogen atom. Even more impressive, Bohr's model concluded by giving the Balmer formula for the hydrogen spectrum. This was a great triumph, since nothing from the Balmer formula had entered into Bohr's assumptions for his model.

ENERGY LEVELS The Bohr atom has often been compared to a ladder (instead of the solar system), with the various rungs representing different energy states. An electron can drop from one rung to any lower rung—that is, to a smaller orbit—by emitting a photon of the appropriate wavelength. It can also move to a higher rung by absorbing a photon with exactly the amount of energy corresponding to the difference between two energy levels. This accounts for the dark absorption and bright emission lines at the same wavelengths in the hydrogen spectrum.

Thus all the wavelengths present are those of photons with energies exactly equal to the difference between two energy states of the hydrogen atom. The lines of the Balmer series represent photons radiated by electrons dropping down from higher energy levels to level 2 ($n = 2$). The drop from $n = 3$ to $n = 2$ produces the red line of hydrogen radiation at wavelength 6563 A; the drop from $n = 4$ to $n = 2$ produces the green line at 4861 A, and similarly for all the other lines of the series, extending into the ultraviolet portion of the spectrum. In the ultraviolet there is another series, the Lyman series, whose lines are caused by electrons falling from higher energy states to the ground state ($n = 1$). And there are other series in the infrared.

How much energy does it take to pull an electron completely free from its nucleus? This is called the **ionization energy** of the atom, since it leaves the atom positively charged or ionized. In Bohr's theory, it corresponds to the energy necessary to boost the electron from the ground state ($n = 1$) to the energy state where $n = \infty$ (thus n is indefinitely large). The solution is easy; setting

Atoms emit radiation when an electron drops from one orbital to another of lower energy.

Figure 11-8
Two ways of representing the electron orbitals of the hydrogen atom. The figure at right depicts circular electron orbits (though today this model is not taken literally). The figure at left gives a graphic depiction of the "ladder" of energy levels.

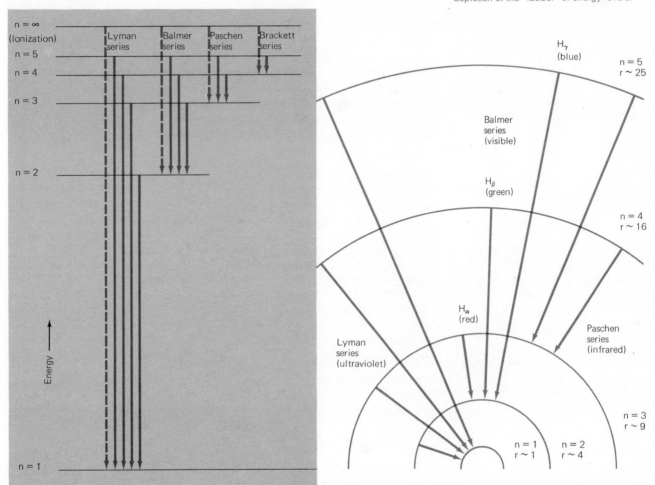

n equal to ∞ in Bohr's energy equation, we find the ionization energy of hydrogen. The value obtained agrees very closely with experimentally derived values for ionizing hydrogen, another triumph for the Bohr model.

Quantum Numbers

Bohr's original model of the atom was quite simple. The energy states of the atom corresponded to different electron orbits, which were initially thought to be concentric and circular. But even the simplest atoms other than hydrogen turned out to be more complex. The wavelengths of the lines in the spectra of larger atoms often did not match the predictions of Bohr's theory. Moreover, there were too many of them—more than Bohr's circular orbits, determined by a single quantum number for each electron, could explain. The Bohr model soon had to be modified and elaborated. The first change was proposed by the German physicist Arnold Sommerfeld (1868–1951) in 1915. He suggested that elliptical as well as circular orbits were possible, and used the mathematics of special relativity to calculate their properties.

It turned out only some elliptical orbits were possible. The more elongated the elliptical orbit, the smaller the angular momentum of an electron occupying it; the angular momentum is greatest when the orbit is circular. Sommerfeld preserved Bohr's quantization of angular momentum by adding a new quantum number, *l*. The principle quantum number, *n*, was retained to specify the size of the orbit—the radius of the circle or the major axis of the ellipse—while quantum number *l* was used to indicate the angular momentum of the electron (and thus the shape of the orbit). *l* can have any integral value from zero to *n* − 1. If the principle quantum number *n* equals 1, for example, *l* must equal 0; if n = 2, *l* can be either 0 or 1; for n = 3, *l* can be 0, 1, or 2, and so on. The larger the value of the angular momentum of quantum number *l*, the more cigar-shaped is the electron orbit.

This modification increased the number of possible atomic energy states and made it possible for chemists to begin to understand the periodic table of the elements in terms of atomic structure. The electrons of each atom could now be assigned to shells, each one representing different values of *n* and *l*. The number of electrons in each shell determines the chemical properties of the various elements, as we shall see in Chapter 13.

It soon became evident that even more complex quantization of atomic structure was necessary. The Dutch physicist Pieter Zeeman (1865–1943) observed in 1896 that individual spectral lines always split into three or more lines when the emitting atoms were subjected to a magnetic field. This could only mean that certain energy levels within the atom were changed, and increased in number, by the interaction of the magnetic field with the electron orbits.

An electron moving about an atomic nucleus is in effect a tiny electric current, and like any moving charge, it generates a magnetic field. The energy level of the atom would be slightly affected by the angle which the external magnetic

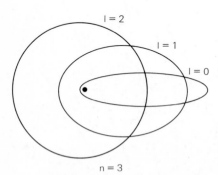

l = 2

l = 1

l = 0

n = 3

Figure 11-9
Electron orbitals need not be circular; Sommerfeld showed that elliptical orbits were also possible, provided the total angular momentum was quantized. The shape of the orbital was specified by a second quantum number, l.

field made with that of the orbital electron. With this in mind, Sommerfeld easily explained the Zeeman effect by imagining that the orientation in space of the electron orbits could also have only certain permitted values. A third quantum number, *m,* was introduced to specify the allowed orientations. This magnetic quantum number can have any value from $-l$ to $+l$. When $l = 2$, for example, m can be -2, -1, 0, $+1$, or $+2$. Thus for every value of l, there are always $(2l + 1)$ magnetically determined energy substates.

The set of quantum numbers necessary to specify the energy state of an electron was eventually completed with a fourth, the spin quantum number, s. It corresponds to the further angular momentum possessed by the electron in addition to that deriving from its orbital motion around the nucleus. If the electron were a little sphere spinning on its axis it would possess such angular momentum of spin, and scientists at first used this analogy. But such visual models can be misleading; the electron is a point charge, not a ball, so it could not spin in any such literal way. We can only say that the electron possesses some intrinsic angular momentum, *as if* it were spinning, the magnitude of

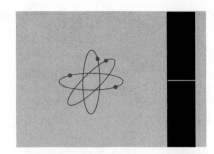

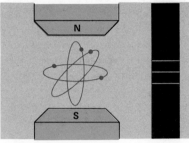

Figure 11-10
Normally, orbitals in different spatial orientations have the same energy. In a magnetic field, however, the energy will vary slightly, according to the angle of the orbital plane with respect to the field. As a result, the spectral lines of an atom in such a field will split up into several components, each with a slightly different wavelength.

Figure 11-11
Wolfgang Pauli and Niels Bohr contemplate a spinning top.

which is $\frac{1}{2} \times h/2\pi$ or $h/4\pi$. Since this angular momentum can be in the same direction as the electron's orbital angular momentum, or in the opposite direction, the spin quantum number can have a value of $+\frac{1}{2}$ or $-\frac{1}{2}$.

The exclusion principle states that no two electrons can have exactly the same quantum numbers.

One important question remained: What determines the energy states of the various individual electrons in the heavier atoms? The answer was supplied by the great Austrian physicist Wolfgang Pauli (1900–1958) in 1925. Pauli's **exclusion principle** states that, in any atom, no two electrons can exist in the same energy state. This is equivalent to saying that no two electrons can have all the same quantum numbers. Therefore the different energy levels had to fill up, one after the other, as more electrons were added to the atom to form the different chemical elements. This principle proved essential in explaining the chemical properties of the various atoms.

The Double Nature of Matter

The complete quantization of the atom met with great practical success, but it presented scientists with new theoretical problems. The Bohr theory of the atom rested on an odd mixture of classical and quantum ideas. Moreover, it was becoming evident that the visual model of central nucleus and planetary orbits could not be taken literally. Electrons, for example, did not spin like little tops. It was also difficult to understand how the electron could jump instantaneously from orbit to orbit, going from one quantum state to another with no intermediate stage. But the greatest problem was that of the quantized orbits themselves: why were only certain orbits stable?

The Electron Wave

A new way of looking at this problem soon came from an unlikely quarter. Louis de Broglie (1892–   ), a member of one of France's oldest and most distinguished aristocratic families, and a late starter in science, published his doctoral dissertation in 1924. In it he set forth a revolutionary hypothesis. Einstein had shown that light, for so long considered a wave phenomenon, could also be thought of as consisting of particles (photons). Could not particles, such as electrons, de Broglie suggested, also have wave properties? Using the basic formulas of relativity and the quantum theory of light, it can be shown that the wavelength, λ, of a photon of light is related to the momentum p by the formula $\lambda = h/p$. De Broglie suggested that a moving electron has wave properties and that the wavelength of the electron is given by the same formula. Since the momentum of a particle is the product of its mass and its velocity, the de Broglie wavelength can be written in the form:

Particles also exhibit wave properties.

$$\lambda = \frac{h}{mv}$$

De Broglie had no proof for this idea; it was based entirely upon analogy and a belief in the symmetry of nature. But de Broglie's intuition proved sound, and proof of his hypothesis was soon forthcoming. Investigators in America reported in 1927 that electrons were diffracted from the atoms in crystals—something that could only be explained if the electron behaved like a wave. Ironically, much of the important work that demonstrated the wave properties of electrons was done by G. P. Thomson, the son of J. J. Thomson, who first observed that electrons acted like particles.

De Broglie's hypothesis of the wave properties of particles turns out to be very general in its applicability. All particles behave like waves as well as like particles. If we try to calculate the wavelength of a macroscopic particle, however, we can see why the wave properties of relatively large, massive, and slow-moving bodies were never suspected before de Broglie published his theory. Consider, for example, a baseball traveling at 90 mph—about 40 m/sec. The mass of a baseball is about .150 kg. Using de Broglie's equation, we find that the wavelength of the baseball is:

$$\frac{6.6 \times 10^{-34} \text{ joule-sec}}{.150 \text{ kg} \times 40 \text{ m/sec}} = 1.1 \times 10^{-34} \text{ m}$$

This is so incredibly minute compared to the dimensions of the ball that no wave properties could ever be observed. But consider an electron moving at about $.01c$—a relatively modest velocity for an electron. Its wavelength is:

$$\frac{6.6 \times 10^{-34} \text{ joule-sec}}{9.1 \times 10^{-31} \text{ kg} \times 3 \times 10^{6} \text{ m/sec}} = 2.4 \times 10^{-10} \text{ m}$$

This is a significant wavelength on the atomic level; it is only about 2½ times the diameter of the hydrogen atom in its ground state. Accordingly, the wave properties of the electron can be detected in many atomic phenomena.

THE BOHR MODEL RECONSIDERED   De Broglie's theory proved extremely useful for understanding the stability of the Bohr orbits. If the electron can be considered a wave, an electron in orbit about a nucleus might be treated as a standing wave. This imposes certain limitations on the circumference of the orbit. If the orbit contains a whole number of wavelengths, the electron wave will reinforce itself. If not, destructive interference will take place, and the wave will annihilate itself. In mathematical terms, this condition can be expressed as:

$$n\lambda = 2\pi r$$

where λ is the wavelength of the electron. If we substitute de Broglie's expression for the electron wavelength:

$$\lambda = \frac{h}{mv}$$

Figure 11-12
De Broglie's hypothesis of the wave nature of the electron made it possible to interpret the electron orbitals as standing waves. Unless the orbital can accommodate an integral number of wavelengths, destructive interference will occur, and the wave will die out. Such an orbit would not be stable.

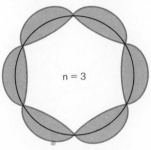

n = 3

Stable orbit

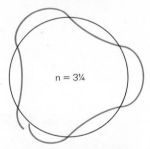

n = 3¼

Interference-orbit not possible

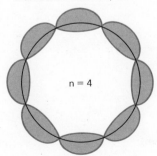

n = 4

Stable orbit

in this equation, we obtain:

$$\frac{nh}{mv} = 2\pi r$$

Transposing, we find this equation is identical to Bohr's formula for the quantization of the angular momentum of the electron:

$$mvr = n\frac{h}{2\pi}$$

Thus, from de Broglie's hypothesis it is possible to derive Bohr's seemingly arbitrary postulate in a very simple way.

This too was great support for de Broglie's idea that matter is *both* wavelike and particlelike. We may still wonder, what is waving? What is analogous in these electron waves to the water in water waves, the vibrating air molecules in sound waves, or the electric and magnetic field strengths in electromagnetic waves? Nevertheless, this breakthrough marked the beginning of the end for Bohr's model of the atom, with its visualizable electron orbits about the nucleus. De Broglie's theory called for an electron that had wave properties, and a wave cannot be localized in space the way a particle can be; it exists spread out in many places at once. It was not long before the Austrian physicist Erwin Schrödinger (1887–1961) developed a remarkable mathematical model in which an atomic electron was treated as a three-dimensional standing wave surrounding the atom's nucleus. Schrödinger's **wave equation** gives a formula for the amplitude of the wave at any point in space. The advantage of Schrödinger's approach was that it avoided the concept of electron orbits and with it the mysterious quantum jump from orbit to orbit that Bohr had never been able to explain.

Particle or Wave?

Bohr had based his model of the atom on the concept of the electron as a particle. Researchers could point to dozens of experiments that showed the electron behaving like a particle, and more were being performed every day. In the face of all this evidence, was it possible to accept this new theory that treated the electron entirely as a wave? What had become of the particle nature of the electron? Schrödinger's wave equation worked well and accounted for all the atomic spectra, but the physical reality behind it was more confusing than ever.

The issue was sharpened when in 1926, just months after Schrödinger had published his work, the German physicist Max Born (1882–1969) proposed an interpretation of the Schrödinger wave equation that seemed to restore the electron to its status as a particle. Born suggested that Schrödinger's equation described, not a physical wave, but rather the statistical probability that the electron would be found at any particular point in space. In this view, the electron can be considered a particle, but a particle whose exact location at any given time cannot be determined. All we can tell is its likelihood of being in various places. The areas where the wave amplitude is great are areas where the electron is most likely to be; the areas where the wave amplitude is low are places where the electron is less likely to be found.

THE UNCERTAINTY PRINCIPLE Born's theory soon found powerful support in the work of a brilliant young physicist, the German Werner Heisenberg (1901–1976). Heisenberg's uncertainty principle both explained and deepened Born's statistical interpretation of quantum mechanics. Like so many of the key ideas of twentieth-century physics, it rests upon the fundamental insight that phenomena in the realm of the atom are not bound by the same laws that apply to macroscopic events and processes. When we observe macroscopic events, we commonly assume that our observations do not significantly alter what we see. Measuring the velocity of the moon, for example, does not change the moon's speed or the direction of its motion by very much. Theoretically, however, there is necessarily some interaction between the observer and the thing observed. We could not see the moon at all if it were not illuminated by light or radiation of some sort. Each photon that bounces off the moon and into our telescope alters the path of the moon slightly. But obviously this effect is so tiny that it could never figure significantly in our results.

Heisenberg realized, however, that the same is not true on the atomic level. Suppose that we want to find the position and velocity of an electron, for example. An electron is a small, light, and fast-moving body. We can only find out where it is by observing the results of its interaction with some other small particle or wave. We can say that we observe an electron just as we observe the moon, by bouncing something off it—another electron, say, or a photon of light. But what happens to the electron when we do this? A photon of light is the most delicate probe we can use for this purpose. Even a photon, however, carries a

We cannot observe events on the subatomic level without altering them substantially.

Figure 11-13
This experiment illustrates the element of uncertainty introduced into all observations on the subatomic level by the act of observation itself. If we use a low-energy photon to observe an electron, the interaction between the two is small, and the electron is disturbed only slightly. Because of the long wavelength of the photon, however, it will be impossible to pinpoint the exact position of the electron. Using a high energy photon improves our knowledge of the electron's position, but changes its momentum greatly, so that we do not know where it will be in the future.

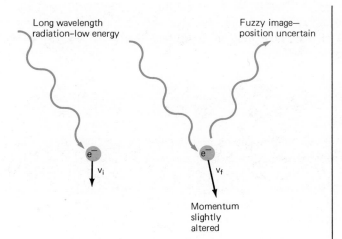

Long wavelength radiation–low energy

Fuzzy image— position uncertain

v_i

v_f

Momentum slightly altered

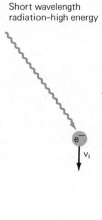

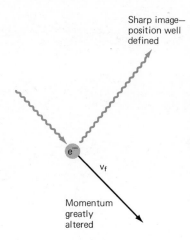

Short wavelength radiation–high energy

Sharp image— position well defined

v_i

v_f

Momentum greatly altered

significant amount of energy. Some of that energy—at least one quantum—must be exchanged with the electron if there is any interaction between them. And without some interaction we cannot make an observation. The electron, then, must receive a "bump" from the photon, which will change its velocity somewhat. We cannot know exactly where it is going after the collision, or exactly how fast. In establishing the present position of the electron, we have paid a price; we have sacrificed some of our knowledge of its future position.

We can see from this argument that the greater the energy of the photon, and therefore the shorter its wavelength, the more violently it is likely to disturb the trajectory of the electron we are trying to observe. Observing an electron with energetic photons such as x-rays or gamma rays is rather like observing a balloon by bouncing rocks off it and watching how they rebound. Each observation will tell us something about the balloon's position but will also disturb it so much we will be unable to determine its velocity and plot its subsequent position.

Why not use photons of very long wavelength, and thus low energy, so as to minimize the impact of the photon on the electron? But such photons will have wavelengths considerably larger than the wavelength of the electron we are trying to observe. This limits their value in determining the position of the electron, for the position of the photon itself cannot be known to an accuracy greater than its own wavelength. The photon, in other words, may bounce off the electron, but its image will be too large and blurry to define the position of the electron with any precision. The longer the wavelength of the photon used, the more gentle its interaction with the electron, the less it will disturb the electron's path, but the greater will be our uncertainty about the exact location of the electron. What we gain in knowledge of the electron's velocity, we lose in knowledge of its position.

This principle, worked out rigorously by Heisenberg from very basic laws governing the behavior of particles and waves at the quantum level, can be formulated as follows: the product of our uncertainty about the position (Δx) and the momentum (Δp) of a particle can never be less than Planck's constant divided by 2π. In mathematical form:

$$\Delta x \; \Delta p \geqslant \frac{h}{2\pi}$$

The uncertainty principle shows that we can never have precise knowledge both of where an electron is and how fast it is going.

THE PRINCIPLE OF COMPLEMENTARITY One of the consequences of the uncertainty principle is that we cannot devise any experiment that will determine once and for all whether the electron—or for that matter, the photon as well—is really a particle or really a wave. All we can say is that matter and electromagnetic radiation exhibit both particle and wave behavior. Which one we observe depends entirely upon the kind of experiment we do and the kind of observations we perform. If we set up a diffraction experiment, for example, we will find that the electron acts like a wave. Yet we can devise other experiments to measure its mass and charge, and find it acting just like a particle. Both descriptions, seemingly contradictory as they are, are necessary for a full account of reality. This is the principle of **complementarity**, advanced by Niels Bohr in 1928.

The uncertainty principle is one of the most far-reaching ideas in all of modern science, for it defines a limit that nature itself seems to place on our knowledge of properties of nature at the quantum level. The imperfections and uncertainties in our measurements do not stem from any technical shortcomings in our instruments. Even with ideally perfect apparatus, we will always be faced with the same choice. We can know where the electron is, or where it is going, but not both—how energetic it is, or where it is, but not both. The more accurately we know one, the less precisely we can know the other.

Does this lack of precise knowledge for the individual particle mean that the laws of classical physics are not really valid? Not necessarily. Macroscopic bodies are made up of atomic and subatomic entities obeying the statistical laws of quantum mechanics. But the more particles are involved, and the greater the time during which we observe them, the closer such probabilities come to being certainties. Remember that even the smallest everyday object consists of trillions upon trillions of subatomic particles, and that even a thousandth of a second is a long time on the subatomic scale. We can see, then, that the uncertainties found on the quantum level become vanishingly small in the macroscopic world.

The laws of classical mechanics are valid descriptions of the macroscopic world; quantum mechanics is necessary to describe the subatomic world.

Glossary

Bohr's atomic model explains that the electron of a hydrogen atom is confined to discrete and stable orbits about the nucleus.

The **Compton effect** refers to the increase in wavelength of X-rays or other strong electromagnetic radiation when it is scattered by electrons.

The **exclusion principle** states that no two electrons in an atom can have identical quantum numbers.

The amount of energy needed to remove an electron completely from its orbital is the **ionization energy**.

A **photon** is a discrete packet of light; each photon carries a quantum of energy equal to *hf*.

Planck's constant expresses the proportionality between the frequency of a quantum of radiation and its energy.

The **principle of complementarity** states that the photon exhibits both particle and wave behavior.

A **quantum** is an individual packet of electromagnetic energy.

Four **quantum numbers** are needed to specify the energy state of an atomic electron.

Heisenberg's **uncertainty principle** states that we cannot measure precisely both the position and the momentum of a particle.

Schrodinger's **wave equation** is a formula for the amplitude of the electron wave at any point in space.

Exercises

1   Describe the photoelectric effect. Suppose that a bright green light and a faint ultraviolet light fall on two pieces of metal; in which case will the electrons emitted have a higher energy?

2   How did Einstein's analysis of the photoelectric effect lead him to the concept of the photon? How did his interpretation explain the existence of a threshold frequency?

3   Describe some of the experimental evidence that helped confirm Einstein's theory of the photon.

4   What is the energy of a photon with a wavelength of 5000Å?

5   A strobe unit is rated at 100 joules per flash; the light emitted has a frequency of 6×10^{14} Hz. How many photons are there in each flash?

6   What was Balmer's formula and how was it arrived at? Why did scientists believe the formula to be much more than a lucky coincidence?

7   What experimental result led Rutherford to postulate the existence of a compact atomic nucleus? What were the shortcomings of the nuclear atomic model?

8   How did Bohr modify Rutherford's model? Why did these modifications seem somewhat arbitrary and unsatisfying even to Bohr himself?

9   How were photons absorbed and emitted according to the Bohr theory? How did this model explain the Balmer formula? What was the special feature of the permitted orbits?

10 Considering only the first six energy levels of the hydrogen atom, how many different spectral lines could be produced by transitions from one level to another?

11 What is the exclusion principle? How many quantum numbers are required for a full description of an electron energy state?

12 What was de Broglie's contribution to quantum theory? How did his ideas provide a rationale for some of Bohr's unexplained assumptions?

13 Why are the wave aspects of a macroscopic object such as a golf ball difficult to perceive? Calculate the de Broglie wavelength of a golf ball with a mass of 100 g moving at a velocity of 1 m/sec?

14 An electron microscope uses moving electrons instead of light waves to form an image. Why does this permit very high magnifications to be attained? If protons were used instead of electrons, would the maximum feasible magnification be greater or less?

15 What is the wavelength of an electron traveling at half the speed of light?

16 Is an electron a particle or a wave? Explain.

17 What is the uncertainty principle? Describe how it enters into an experiment to determine the position and momentum of a subatomic particle. Does the uncertainty principle undermine our notions of cause and effect in everyday situations? Why does the principle not make baseball impossible, since the batter must know the position and momentum of the ball in order to hit it?

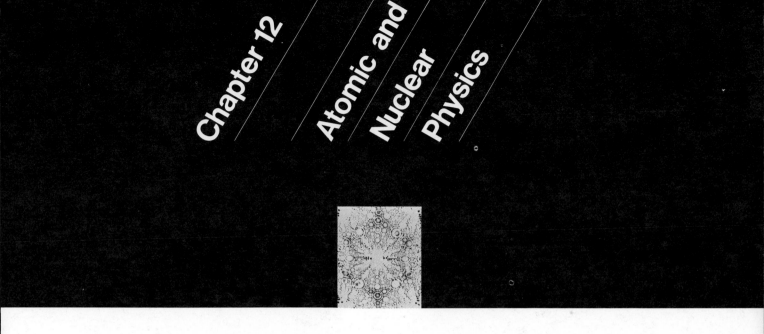

Chapter 12 / Atomic and Nuclear Physics

The French physicist Henri Becquerel (1852–1908) was interested in the fluorescence of uranium salts; they emit a visible colored glow when exposed to light. In 1896, he discovered (quite by accident) that the salts were also phosphorescent. Even when not exposed to light, they emitted some kind of radiation that could darken photographic plates. Becquerel concluded that the radiation was not the result of some outside source of energy, but a spontaneous phenomenon inherent within the salts. He called this phenomenon **radioactivity.**

Could there be other radioactive elements? In the next two years, Marie and Pierre Curie found three—thorium, and two new elements they named polonium and radium. Since the property of radioactivity in all four elements was independent of physical state or chemical combination, Pierre Curie concluded it must be a property of the atoms themselves.

Figure 12-1
Marie and Pierre Curie stand in their
laboratory. They shared the Nobel prize with
Becquerel in 1903; after Pierre's death,
Marie went on by herself to isolate radium
and polonium, and won another Nobel prize
in 1911.

Radioactive materials can emit alpha,
beta, and gamma rays.

By 1900, data had been collected about the rays emitted by radioactive materials. The rays proved to be of three major types. **Alpha (α) rays** are a kind of
radiation, easily absorbed by a paper-thin sheet of metal or a few cm of air. The
α ray is actually an emission of a stream of alpha particles. Each particle has a
positive charge of +2e (twice the charge of a single electron) and a mass of
6.62×10^{-24} g, four times the mass of the hydrogen nucleus and about 7500
times the mass of a single electron. The α particle is, in fact, a doubly ionized
helium atom—a helium nucleus with no orbital electrons.

Beta (β) rays, or the emission of β particles, is a more penetrating kind of radiation. Both the mass and the charge of a β particle turned out to be the same as
those of the electron. β rays are streams of electrons. These smaller, lighter,
negatively charged particles can actually reach velocities of .995 c.

Gamma (γ) rays, highly penetrating rays that are unaffected by electric or magnetic fields, were identified as very short wavelength x-rays, in the high
frequency range of the electromagnetic spectrum. γ rays, which have the speed
of light, can penetrate thick lead sheets.

Rutherford used α rays in his experiments to determine the structure of the nucleus. He bombarded thin sheets of gold foil with α particles, and as we saw in Chapter 11, he discovered that the foil scattered the particles in all directions. He explained the backward scatter effect by hypothesizing that the scattering was the result of an intense positive electric field deep within a single atom. For such a large field to exist within the atom, which is ordinarily electrically neutral, the entire positive charge of the atom would have to be concentrated in a very small nucleus, itself surrounded by electrons that balance it electrically. Calculations from further scattering experiments supported Rutherford's assumption of a subatomic nuclear charge. These experiments were not sufficiently accurate to determine the quantity of the charge on the nucleus, but they established the nuclear radius as being about 10^{-14} cm (as compared to 10^{-10} cm for the atomic radius); the nucleus was found to be 1/10,000 of an atom.

Until 1913, the periodic table of the chemical elements was arranged in rough order of increasing atomic weight (see Chapter 13 for a full discussion of the periodic table). Then H. G. Moseley (1887–1915), a student of Rutherford, began to analyze x-ray spectra obtained by bombarding different metals with β particles. His analysis showed that the shift to shorter wavelengths was correlated with increasing atomic number, or quantity of electrical charge on the nucleus. This x-ray data seemed independent of other atomic properties (such as chemical behavior and visible light spectra). Moseley concluded that the atomic nucleus was the source of the x-ray emissions, which were a response to the bombarding electrons. But what then did change with atomic number? The atomic number expresses quantity of the nuclear property of positive electrical charge; the higher the number, the greater the electrical charge of the nucleus. Since the ordinary atom is electrically neutral, for each unit of positive charge on the nucleus there must be one unit of negative charge—one electron—elsewhere in the atom. Thus the atomic number of an element also defines the number of electrons contained in a neutral atom.

But what carried the positive charge? From his work with hydrogen atoms that had lost their one electron, Rutherford knew that the hydrogen nucleus is a single positive charge; he conjectured that the hydrogen nucleus, which he called the **proton**, might be the fundamental unit of nuclear charge for all atoms. He confirmed this hunch experimentally in 1919 by bombarding nitrogen atoms with α particles. Since the atomic weight of nitrogen was 4.5 times that of helium, he expected that the α particles would be slowed or stopped by collisions with the heavier nitrogen nuclei. Surprisingly, he found that the experiment produced the emission of particles that traveled long distances (up to 30 cm) and thus had to be smaller and lighter than even the bombarding α par-

The Structure of the Nucleus

Nuclear Charge

The Proton

Protons in the nucleus carry the atom's positive charge.

ticles. Rutherford concluded that these particles must be hydrogen nuclei, or protons, produced through the breakdown of the nitrogen nuclei by the α bombardment. Rutherford's experiment was significant because it was both the first nuclear reaction and the first laboratory-induced change of one element into another. His report was prophetic:

> . . . we must conclude that the nitrogen atom is disintegrated under the intense force developed in a close collision with a swift alpha particle, and that the hydrogen atom which is liberated formed a constituent part of the nitrogen nucleus. The results on the whole suggest that, if alpha particles— or similar projectiles of still greater energy—were available for experiment, we might expect to break down the nuclear structure of many of the lighter atoms.

As Rutherford guessed, this was the pathway to serious studies of nuclear fission and the atomic bomb 25 years later.

All atomic weights are measured relative to a fixed value for carbon.

NOTATION SYSTEMS Atomic weights, as we have seen, were one of the earliest properties of the atom to be measured. These weights are all relative; since 1960 they have been based on a scale that compared all elements to carbon, which has the designated value of 12.0000. On this scale, the weight of hydrogen is 1.0080 and the weight of oxygen is 15.9994. For ease in calculation, these awkward atomic-weight numbers are rounded off to the nearest whole number, which is called the mass number. Thus hydrogen has a mass number of 1, oxygen of 16. In conventional usage, the atomic number is written as a subscript to the left of the symbol for the element, and the mass number is written as a superscript to the right. For example, hydrogen is denoted as $_1H^1$—atomic number 1, mass number 1. Carbon is denoted as $_6C^{12}$—atomic number 6, mass number 12.

Nuclear Reactions

Using this notation system, we can write an equation for the reaction that led Rutherford to discover the proton. The α particle is a helium nucleus, $_2He^4$; the proton is a hydrogen nucleus, $_1H^1$. The reaction is:

$$_2He^4 + {}_7N^{14} \longrightarrow {}_8O^{17} + {}_1H^1$$

In such a reaction process, several conservation principles must be taken into account:

1 Conservation of charge. The same total charge must appear on each side of the reaction.
2 Conservation of mass number. The same number of protons, for example, must appear on each side of the reaction.
3 Conservation of energy and momentum.

In the Rutherford equation, we see that according to conservation principles the product of the bombardment of the nitrogen nucleus by the α particle must have atomic number $(7 + 2) - 1 = 8$ and mass number $(14 + 4) - 1 = 17$. Oxygen has mass number 8, so the product must be $_8O^{17}$, an oxygen isotope.

MASS–ENERGY EQUIVALENCE The mass of a nuclear particle is so small that direct notation is cumbersome. Therefore a more convenient unit of measurement has been established, called the **atomic mass unit** (amu). The amu, which like the mass number takes the carbon atom as the fixed reference point, has been derived from Avogadro's number (see Chapter 13) in the following way:

6.023×10^{23} atoms of $C^{12} = 12$ g

1 atom of $C^{12} = 12/6.023 \times 10^{23}$ g

1 amu $= 1/12$ ($12/6.023 \times 10^{23}$ g)

1 amu $= 1.66 \times 10^{-24}$ g or 1.66×10^{-27} kg

On this scale the proton has a mass of 1.0073 amu; the electron, 0.000549 amu.

Based on Einstein's equation for the relationship of mass to energy ($E = mc^2$), these masses can be expressed in terms of their energy equivalent. Since $c = 3 \times 10^8$ m/sec, and $m = 1$ amu $= 1.66 \times 10^{-27}$ kg:

$$E = mc^2 = 1.66 \times 10^{-27} (3 \times 10^8)^2$$
$$= 1.492 \times 10^{-10} \text{ joules} = 1 \text{ amu}$$

Again, this is an unwieldy number to deal with; a much smaller energy unit is needed. A unit that is appropriate for nuclear masses is the **electron volt** (ev) — defined as the kinetic energy gained by an electron when it is accelerated through a potential difference of 1 volt (joule/coul). Since $V = E/q$, and $q = e = 1.6 \times 10^{-19}$ coul:

1 ev $= E = Vq = 1$ joule/coul $\times 1.6 \times 10^{-19}$ coul

1 ev $= 1.6 \times 10^{-19}$ joule

1 Mev $= 10^6$ ev $= 1.6 \times 10^{-13}$ joules

The **Mev** (one million electron volts) is the unit of energy commonly used in nuclear physics.

$$1 \text{ amu} = 1.492 \times 10^{-10} \text{ joules}/ 1.6 \times 10^{-13} \frac{\text{joules}}{\text{Mev}}$$

1   amu $= 931$ Mev

The mass of the electron is equivalent to 0.511 Mev.

The Neutron

In all elements except hydrogen, the mass number is at least twice the atomic number. Since the atomic number gives the number of protons in the nucleus, what is it that makes up the rest of the mass? Rutherford guessed that this might be one or more tightly-bound combinations of a proton and an electron that are somehow different from the usual association of proton and electron that produces an H atom. He called this electrically neutral unit a **neutron**.

Rutherford's guess was confirmed by a series of experiments in 1930 by the German physicists W. G. Bothe and H. Becker. Using a cloud chamber, they bombarded boron and beryllium with α particles. No tracks of condensation droplets appeared to come from the collision, but unexplained tracks simultaneously appeared from other starting points some distance away. The distribution of these new tracks was not influenced by electric or magnetic fields. Perhaps these photons were produced by collisions of nonionizing radiation (either neutral particles or γ rays) in a reaction like this:

$$_4Be^9 + _2He^4 \longrightarrow _6C^{13} + \gamma$$

The problem was that the effect could be observed even through a photon-absorbing lead shield. So the particles emitted must be at greater energy than any γ rays known, in an energy range of about 10 Mev.

In 1932 Irene Curie (1897–1956) and her husband, Frederic Joliot (1900–1958) discovered that this radiation could knock large numbers of protons out of paraffin or other hydrogen containing materials. The emitted protons had energies of about 5 Mev. By applying conservation rules for energy and momentum, the Joliot-Curies calculated that the proposed gamma ray would need to have an energy of 50 Mev to knock out a proton of 5 Mev, and this was too great for the 10 Mev gammas observed in the experiments with absorption in lead following Bothe and Becker. At that point, there were disagreements between different experimental results, and a puzzling difference between the simple theory of nuclear processes that produced the protons and the new experiments that also produced protons but through an intermediate process of immensely great energies. Could the conservation rules be wrong? Rather than accept such a fundamental revision of physics, the scientists searched further.

Table 12-1 Comparison of Particles

Symbol	Particle	Charge (coul)	mass (amu)	mass (kg)	$\times m_e$	Mev
$_{-1}e^0$	Electron	-1.6×10^{-19}	0.000549	9.107×10^{-31}	1	0.511
$_1H^1$	Proton	$+1.6 \times 10^{-19}$	1.007276	1.673×10^{-27}	1836	938
$_0n^1$	Neutron	0	1.008665	1.675×10^{-27}	1839	939

Chadwick discovered that irradiated elements (he used beryllium) emit neutrons.

In 1932, James Chadwick (1891–1974) using other nuclei such as lithium and carbon, confirmed the Joliot-Curie results. To resolve the difficulty, Chadwick hypothesized that the irradiated beryllium emitted an uncharged particle of a mass roughly that of the proton. This particle is Rutherford's neutron, which makes up the difference between the atomic number and the mass number in an atom.

The correct way to write Rutherford's reaction is:

$$_2He^4 + _4Be^9 \longrightarrow _6C^{13} + _0n^1$$

where $_0n^1$ is the neutron, with no charge and a mass number of 1. Because of its electrical neutrality and its small size, the neutron can easily penetrate a nucleus. It can also slip easily through dense matter like lead with no loss of energy because it has no electromagnetic interaction. A neutron can only be slowed or stopped by a direct hit with a nucleus, as in the paraffin experiments.

THE POSITRON At the same time Chadwick discovered the neutron, the American Carl D. Anderson (1905–   ) was studying cloud-chamber photographs of particles produced by cosmic rays, the extremely penetrating radiation from beyond the earth's atmosphere. The photographs showed tracks of particles with a mass and charge equal to those of the electron; the puzzling difference was that they curved in the wrong direction in a magnetic field, meaning that they must have a charge opposite that of the electron. Anderson called these positively charged electrons **positrons** ($_{+1}e^0$).

Unbeknownst to Anderson, the existence of the positron had been predicted earlier by the English theoretical physicist P. A. M. Dirac (1902–   ). The basis of Dirac's prediction was his assumption of a natural symmetry of charges and spins; for every known particle there might be a kind of mirror-image particle with the opposite charge and the opposite spin. The positron, then, is an antielectron—the first particle of antimatter to be discovered.

It was discovered that when a positron collides with an electron, both particles disappear, to be replaced by 2 photons, in a transformation of matter into energy:

$$_{-1}e^0 + {_{+1}}e^0 \longrightarrow 2\gamma$$

The opposite effect, changing energy into matter, is called pair production. It is obtained by bombarding heavy nuclei (lead) with γ rays. The ray gets close to the strong electric field of the nuclei and simply disappears. Emerging from the collision, in a materialization of energy, are pairs of oppositely charged particles:

$$\gamma + Pb \longrightarrow {_{+1}}e^0 + {_{-1}}e^0$$

Recently such pairs have been produced from the interaction of γ rays with very strong electric fields, in an amazing process that uses no matter.

ANTIPROTONS If there is an antielectron, could there be an antiproton? Because of its large mass, the antiproton would require 1840 times as much energy as the positron for its production. The first clearcut evidence of the antiproton's existence came in 1955 from scientists working with the Berkeley particle accelerator (called the Bevatron because it can generate 6 billion electron volts—6 Bev—of energy). Bombarding copper with 6-Bev protons, the Bevatron produced antiprotons in abundance.

Figure 12-3
This bubble chamber photo shows 2 antiprotons entering at the bottom; both collide with protons, causing the 4-pointed star tracks. In the upper star, the curved track shows a pion that goes on to collide with another proton, producing a positron (large counter-clockwise spiral) and an electron (smaller clockwise spiral) in an example of pair production.

As Becquerel noted, radioactivity occurs independently of physical or chemical changes in an element. It is a nuclear process of change, or spontaneous disintegration of nuclei. The disintegration sometimes produces different elements by change in nuclear components. In other cases it produces **isotopes** of the original elements, variants with the same atomic number but different mass numbers.

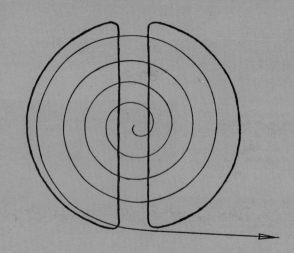

THE CYCLOTRON AS SEEN BY THE INVENTOR

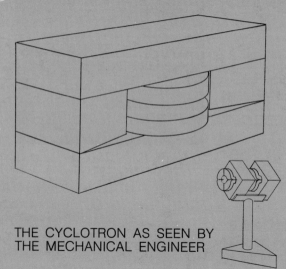

THE CYCLOTRON AS SEEN BY
THE MECHANICAL ENGINEER

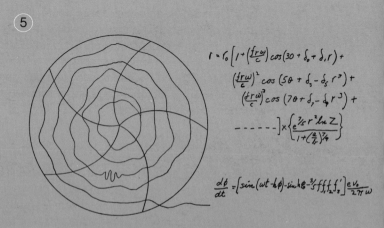

THE CYCLOTRON AS SEEN BY THE OPERATOR

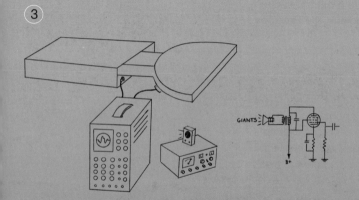

THE CYCLOTRON AS SEEN BY
THE ELECTRICAL ENGINEER

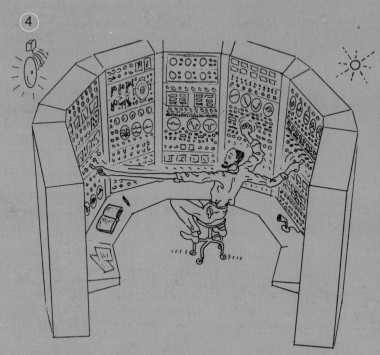

THE CYCLOTRON AS SEEN BY
THE THEORETICAL PHYSICIST

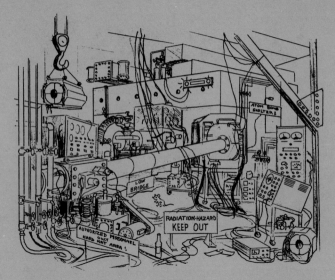

THE CYCLOTRON AS SEEN BY THE VISITOR

THE CYCLOTRON AS SEEN BY
THE LABORATORY DIRECTOR

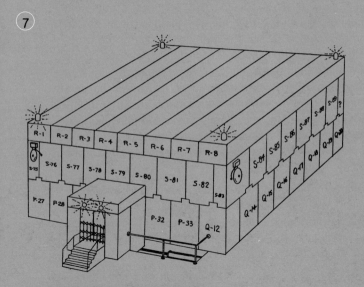

THE CYCLOTRON AS SEEN BY
THE HEALTH PHYSICIST

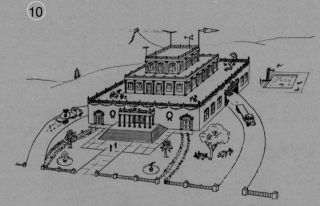

THE CYCLOTRON AS SEEN BY THE
GOVERNMENTAL FUNDING AGENCY

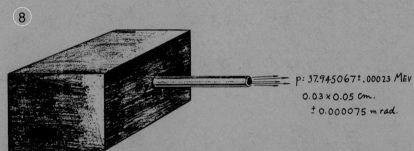

THE CYCLOTRON AS SEEN BY THE EXPERIMENTAL PHYSICIST

THE CYCLOTRON AS SEEN
BY THE STUDENT

The rate of radioactive emission, or decay, is measured by the element's **half-life**. The half-life is the time it takes for the rate of radioactivity to drop by one-half of its initial value. For any initial number of atoms, the half-life is the time it takes for half the nuclei in the sample to disintegrate. For instance, tritium (a hydrogen isotope, $_1H^3$) has a half-life of 12.26 years. Thus, for an original 10,000 tritium nuclei, 5000 will be left in 12.26 years. Of that 5000, 2500 will remain after a further 12.26 years.

Radioactive decay is statistically predicted by an element's half-life.

There is a close parallel between the way radioactive atoms decay and the statistical laws about human populations or about the outcomes of games of chance. We can predict the *proportion* of coins that will come up heads or tails in coin tossing with greater accuracy as the number of tosses is greater, but we cannot predict the outcome of a single toss with any greater accuracy. The same is true of predictions when we roll dice, even though the mathematics of figuring out the various combinations that can appear becomes more complicated. We can use illness frequency data from public health records to predict what fraction of the given, or similar, population will contract a particular disease, but such data cannot be used to determine whether a given person will get it. And so with the radioactive atoms. The half-life is a statistical property of the population of atoms of a given isotope. It says nothing about a single atom other than the rather limited information that the atom has a probability of $\frac{1}{2}$ that it will decay in the next half-life period. Moreover, just as in the coin tossing, each decay process is independent of what happened to other atoms previously; whatever the previous coin showed, whether heads or tails, the next toss is still at the same odds, and so for any run of tosses. The law of radioactive decay is a statistical law. And of course the shorter the half-life, the more likely that an atom will decay.

Practically all naturally radioactive elements have atomic numbers between 81 and 92; they include uranium, thorium, and actinium. Artificially produced elements (those with an atomic number greater than 92) also exhibit radioactivity. Both the natural and the artificial radioactive elements decay in patterned steps, or **decay series**, to more stable isotopes. Several such series are shown in Table 12.2.

Table 12-2 Uranium-radium decay series

Present Name and Symbol		Mode of Decay	Half-life
Uranium 238	$_{92}U^{238}$	α	4.51×10^9 years
Thorium 234	$_{90}Th^{234}$	β,γ	24.1 days
Protactinium 234	$_{91}Pa^{234}$	β,γ	1.18 minutes
Uranium 234	$_{92}U^{234}$	α	2.48×10^5 years
Thorium 230	$_{90}Th^{230}$	α,γ	8.0×10^4 years
Radium 226	$_{88}Ra^{226}$	α,γ	1620 years
Radon 222	$_{86}Rn^{222}$	α	3.82 days
Polonium 218	$_{84}Po^{218}$	α	3.05 minutes
Lead 214	$_{82}Pb^{214}$	β,γ	26.8 minutes
Bismuth 214	$_{83}Bi^{214}$	β,γ	19.7 minutes
Polonium 214	$_{84}Po^{214}$	α	1.64×10^{-4} seconds
Lead 210	$_{82}Pb^{210}$	β,γ	21.4 years
Bismuth 210	$_{83}Bi^{210}$	β	5.0 days
Polonium 210	$_{84}Po^{210}$	α,γ	138.4 days
Lead 206	$_{82}Pb^{206}$	stable	

Some radioactive elements as they decay emit α particles, each of which carries away 2 units of positive charge and 4 units of mass (often a significant portion of the original mass). Examples of α decay are:

$$_{92}U^{238} \longrightarrow {}_{90}Th^{234} + {}_2He^4 + \gamma$$

$$_4Be^8 \longrightarrow {}_2He^4 + {}_2He^4$$

$$_3Li^5 \longrightarrow {}_2He^4 + {}_1H^1$$

Other decaying elements emit β particles, each of which generally carries away 1 unit of negative charge and almost no mass. Thus thorium decays to protactinium:

$$_{90}Th^{234} \longrightarrow {}_{91}Pa^{234} + {}_{-1}e^0$$

Less frequently, β decay produces the emission of a positron rather than an electron. An example is:

$$_6C^{10} \longrightarrow {}_5B^{10} + {}_{+1}e^0$$

In effect, during β decay a neutron changes to a proton within the original nucleus, with a transmutation to the isotope of a different element with the same atomic weight. The atomic number of the isotope changes by $+1$, but the mass number remains the same. The parent and daughter nuclei are called **isobars**—chemically different elements with the same mass number.

Most radioactive elements emit either α or β particles, but not both. Either type of element may also emit γ rays. In the process of γ decay, the emitted photon removes neither charge nor mass from the radioactive nuclei. No nuclear change occurs; the γ ray just carries off energy.

Both α particles and γ rays are emitted within narrowly limited energy levels. β particles appear to range over a continuous energy spectrum; but in all cases of β decay, there is a definite limited quantity of energy released. According to the principles of conservation, the emitted β energy should be equivalent to the difference between the mass of the parent nucleus and the masses of the daughter nuclei. Yet experimentally β particles have rarely achieved the full value of this difference. Some mass, and the equivalent energy, is missing.

About 1931, the Swiss physicist W. Pauli (1900–1958) proposed that the apparently missing mass was carried off by a massless uncharged particle—a "baby" neutron, or **neutrino** (ν). The full kinetic energy of the β ray emission process is thus divided between the β particle and the neutrino.

By 1934, the great Italian physicist Enrico Fermi (1901–1954) had worked out a full theory of beta decay, using the neutrino hypothesis. Pauli's ghostly particle did have the physical property of spin, as did all the others, but how this fit its nearly zero mass, so much less than even the electron mass (in fact, about $1/5000\ m_e$) was a puzzle. Later, in the 1950s, the antineutrino was found to be necessary to account for positron emission, roughly similar to the normal negative electron problem of beta emission. How can two nearly massless, neutral particles be told apart? Again it is the singular property of spin, for the neutrino has a right-handed or anticlockwise spin, and the antineutrino has a left-handed clockwise spin. They are each the mirror image of the other. And their spins are not observed, but indirectly inferred by observations of the spins of the parent nucleus, the emitted electron or positron, and the remaining daughter nucleus. Fermi saved the conservation of energy in beta decay. But for 20 years sceptics wondered: did neutrinos exist, or is the neutrino just the name of the problem of the apparent nonconservation of energy and momentum?

Like all antimatter particles, the antineutrino is distinguished by the difference in the direction of its spin.

Detection of the neutrino was exceedingly difficult. This massless uncharged particle interacts weakly with nuclei and gives little chance for collision; it could pass through the entire planet earth with only one chance out of 10 billion of being stopped by some other particle.

It was guessed that neutrinos (along with β rays and protons) might be produced by the decay of a free neutron (one not associated with a nucleus). Actually, for reasons of spin direction, it produces **antineutrinos** ($\bar{\nu}$), in the reaction:

$$_0n^1 \longrightarrow {_1}H^1 + {_{-1}}e^0 + \bar{\nu}$$

Reversing this reaction, we could expect to get neutrons:

$$\bar{\nu} + {_1}H^1 \longrightarrow {_0}n^1 + {_{+1}}e^0$$

In 1956, this reaction was observed, verifying the existence of the antineutrino, in an experiment that bombarded protons with the vast neutron output of a nuclear reactor (assumed to emit antineutrinos in the process of neutron decay). In 1964, the neutrino itself was finally detected, from solar radiation, when neutrons were observed to capture neutrinos:

$$\nu + {_0}n^1 \longrightarrow {_1}H^1 + {_{-1}}e^0$$

Nuclear Energy

Alpha, beta, and gamma radiations are emitted with enormous energy. An emitted alpha particle, for instance, has an energy range of about 5 to 10 Mev—about a million times greater than the energy of chemical dissociation of a hydrogen molecule. The force necessary to eject the alpha particle with such energy could be accounted for by the enormous Coulomb repulsion that exists between nuclear protons—about 230 newtons per proton.

Two problems arise here. First, what accounts for the energy of beta and gamma radiation—since it is not a result of Coulomb repulsion? And second, in view of the enormous Coulomb repulsion between nuclear protons, what holds the nucleus together? What is the nuclear force?

At the beginning of the twentieth century, the English chemist and physicist F. W. Aston (1877–1945) performed a number of experiments on isotopes of neon that led him to propose a **whole-number rule**: the deviation of the atomic weight of an element from a whole number is due to a mixture of isotopes, each having a whole-number atomic weight. For example, his samples of neon contained 9 parts of Ne^{20} to 1 part Ne^{22}, and therefore the average atomic weight was $(9 \times 20.00) + (1 \times 22.00)/10 = 20.20$, or the experimentally observed atomic weight of neon.

In the process of his measurement of isotopes, Aston found another interesting anomaly. If he made his measurements of the various isotopes more precise, they too began to deviate from the whole number rule. For example, he found that the mass of the deuteron (a hydrogen isotope) nucleus is actually 2.0136 amu instead of 2.00. This difference between the actual measurable mass and the mass number is called the **mass defect**.

It was in the course of deriving an explanation for the mass defect that the forces that bind the particles that make up the nucleus were made evident. The solution to the problem of the mass defect seemed to lie in increasingly precise measurement of decreasingly small particles. By measuring the amu of the protons and neutrons that make up the nucleus it should be possible to determine whether these particles are each slightly heavier than 1 amu. The results confirmed this suspicion; both proton and neutron were slightly heavier than 1 amu—a very very small fraction that becomes noticeable only when they group together in larger numbers, as they do in an atomic nucleus. For example, based on the measurement of the exact amu of proton and neutron, the deuterium nucleus (deuteron) should have a mass of 2.0160 amu. But the mass of a deuteron, as we saw, was measured at 2.0136; it is actually lighter by about .0024 amu than it should be. Measurements on other nuclei confirmed this puzzling phenomenon.

Could this small amount of mass have been lost by being converted to energy when the nucleus was formed? Using Einstein's formula $E = mc^2$, it has been calculated that the energy equivalent of 1 amu is 931 Mev. The mass (.0024 amu) missing from the deuteron nucleus can also be converted to an energy equivalent: $931 \times .0024 = 2.2$ Mev. It happens that this is exactly the measured energy of a gamma ray photon, a fact we take as confirming the change from mass to energy.

We can now write the complete equation for the formation of the deuteron nucleus:

$$_1H^1 + {}_0n^1 \longrightarrow {}_1H^2 + \gamma$$

The energy equivalent of the γ-ray is called the binding energy (E_b) of the nucleus. It is calculated by using the formula ΔE or $E_b = (\Delta m)c^2$. Thus, for deuteron E_b is 2.2 Mev; for helium it is 28.4 Mev; for boron-10 it is 64.7 Mev; for

Binding Energy

Figure 12-4
The Brookhaven National Laboratory on Long Island uses a small vertical accelerator for subatomic research. The protons or electrons travel through the long coiled tubes and can accelerate to 3 Mev.

uranium-238 it is a whopping 1760 Mev. The reaction that forms the nucleus liberates the binding energy; in order to reverse the reaction and break the nucleus into its component parts, energy equivalent to E_b has to be added to the nucleus. It is as though the particles fall into a deep well or pit together and then it takes energy to get them moving fast enough to get out.

Nuclear Stability

Common sense suggests what experimental evidence confirms: the larger the nucleus, the greater its binding energy. But that does not necessarily mean that these large nuclei are more stable, because they also contain more particles that have to be held together. A better index to nuclear stability is the amount of binding energy associated with each proton-neutron unit (called the nucleon). The formula for the binding energy per nucleon is E_b/A. It varies from 1 for deuterium up to nearly 9 for iron, but for most elements with a mass number above 12, the binding energy per nucleon is about 8 Mev. The higher the binding

Figure 12-5
The peak of the binding energy curve comes at 8.7 Mev, for the element iron. Elements to the left of the peak undergo fusion; those to the right undergo fission.

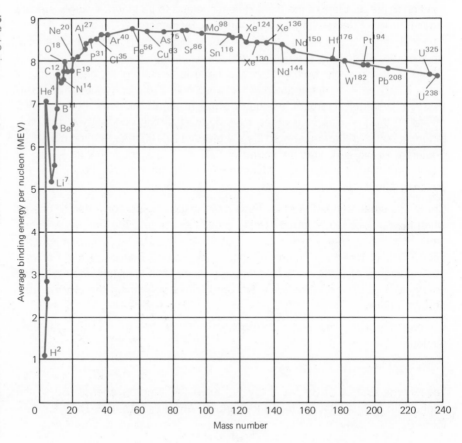

energy per nucleon, the more stable the nucleus. Thus iron with an E_b/A at 8.7 Mev has the most stable nucleus of the entire periodic table; it has given up the most mass in being formed and thus exists at a lower energy state than any other element. Its nuclear well is the deepest of all.

Another factor affecting the stability of a nucleus is its ratio of neutrons to protons. Neutrons seem to bind protons; for example the helium nucleus contains 2 protons that remain together despite their mutual repulsion, due to the presence of 2 neutrons. Although it acts over an exceedingly small range, this binding force of attraction between protons and neutrons is strong enough to overcome the Coulomb repulsion between protons.

The ratio of neutrons to protons is 1-to-1 for elements with mass numbers up to about 40. Above that level, as the total repulsive force between protons increases, the ratio becomes greater than 1, because extra neutrons are needed to hold the nucleus together. There seems to be an upper limit, beyond which no amount of extra neutrons serve to counteract the increasing Coulomb force of repulsion. The heaviest stable nucleus is an isotope of bismuth, $_{83}Bi^{209}$. Elements with higher mass number all have unstable nuclei; in other words, they are radioactive. They emit α and β rays, giving off neutrons and protons, until they eventually reach a stable ratio.

Nuclear stability is affected by the amount of binding energy per nucleon, and the ratio of neutrons to protons.

The Meson

In 1935, the Japanese physicist Hideki Yukawa (1907–   ) suggested that in addition to the gravitational and electrical forces then known to act within the nucleus, there was also a distinctive short-range nuclear force between nucleons. He proposed that the force was associated with an as-yet-undiscovered type of particle with a mass of about 200–300 m_e. Since that size would put it midway between the proton and the electron, Yukawa called it the **meson**, from the Greek word for middle.

In Yukawa's theory, the force associated with mesons is an exchange force that is the result of continuous formation and destruction of a binding linkage between nucleons via mesons. A proton becomes a neutron by emitting a positive meson, which combines with a neutron to form a proton, and the cycle repeats. Like nucleons similarly exchange neutral mesons. The exchange is so rapid and continuous that nucleons remain in prolonged close contact.

Two years after Yukawa proposed this theory, Anderson discovered, during cloud-chamber studies of cosmic rays, one kind of intermediate mass particle, called a mu (μ) meson, or muon. It has a mass of 207 m_e and it decays in 2.15×10^{-6} seconds to a positron. In 1947, a second kind of meson, the pi (π) meson, or pion, was observed. With a mass of 273 m_e and a decay time of 2.6×10^{-8} seconds, this heavier particle conformed very closely to Yukawa's prediction. The pion, which becomes a muon as it decays, is the effective generator of the nuclear binding or exchange force.

Nuclear Transmutation

In 1934, the Joliot-Curies had been working on an experiment to generate neutrons by bombarding various light metals, such as boron, aluminum, and magnesium, with α particles. They expected that the bombarded metals would emit protons and positrons; the unexpected result they observed was that positron emission continued even after the α bombardment had stopped. In other words, they had created a new isotope of the target metal that had an artificially induced radioactivity. Such artificially radioactive isotopes were unstable and had a very short half-life.

Fission

Fermi immediately began to use the Joliot-Curie technique to explore radioactive possibilities in many elements. He found that he had the greatest success in creating artificially radioactive isotopes when he bombarded elements with neutrons rather than α particles; neutrons penetrate more readily because they have no charge. Then he used the same technique to try to create entirely new elements with atomic numbers greater than the 92 of uranium, the heaviest naturally occurring element, by the bombardment of uranium. Fermi believed he had succeeded, because he began to observe emissions, and it was known that uranium was an α emitter.

Figure 12-6
Two nuclear physicists ride bicycles through the accelerating ring of a synchrotron. As the particles accelerate through the ring, their mass increases, leading to more spectacular collisions and more fruitful scientific results.

But several years later (1939) two German chemists made careful analysis of the Fermi reaction and discovered that one of the newly formed radioactive elements was an isotope of barium, with an atomic number of 56. Since the parent element (uranium-235) had an atomic number of 92, it was apparent that the bombardment must also have created another element of lower atomic number, the radioactive isotope of the element with an atomic number of (92–56) or 36 — krypton. Both barium and krypton are known to be β emitters, so the presence of these isotopes (rather than transuranium atoms) was confirmed by Fermi's observation of β emission.

The process through which a heavy nucleus disintegrates to form 2 lighter ones is called **nuclear fission**. The formula for Fermi's fission reaction is:

$$_0n^1 + {}_{92}U^{235} \longrightarrow {}_{92}U^{236} \longrightarrow {}_{56}Ba^{141} + {}_{36}Kr^{92} + 3{}_0n^1 + E.$$

Especially significant is the fact that energy is released. The amount of energy can be predicted from our knowledge of the amount of binding energy per nucleon of the elements involved. U^{235} has an E_b/A of 7.6 Mev; both Kr^{92} and Ba^{141} have E_b/A values of 8.5. So for each nucleon involved in the reaction there is an energy release of about 9 Mev; since the uranium nucleus has 235 nucleons, it means a total release of about 200 Mev per uranium atom. This is 50 million times more energy than is released by the oxygen combustion of a carbon atom. One kg of U^{235} can release 2.3×10^7 kw-hr of energy — enough to supply all the electricity a city of 100,000 people can use in two weeks.

From the fission equation for U^{235}, it is evident that while 1 neutron is needed to start the reaction, 3 are released in the process. In turn, these excess neutrons

Figure 12-7
The French Atomic Energy Commission built this experimental nuclear reactor. The end product of its power-producing fission reactions is plutonium, which can later be reused in another cycle of reactions.

can initiate new fission reactions, which release more neutrons, which start more reactions. In short, a single neutron can initiate the fission of all the available U^{235} in a block of uranium, rapidly releasing enormous amounts of energy in an explosive series of reactions. Such a self-sustaining process is called a **chain reaction**.

When a chain reaction of nuclear fission is allowed to proceed in an uncontrolled manner, the result is an atomic bomb. But the same reaction can also take place more slowly, by adding to the U^{235} a certain dispersed quantity of nonfissionable material that absorbs the extra neutrons and acts as a brake on the speed of the reaction. The first successful nuclear reaction using U^{235} to produce controlled fission was constructed in Chicago as part of the American atomic bomb project during World War II. It was operated successfully on December 2, 1942, under Fermi's direction.

Fusion

It is logical to assume that the smaller nuclei of lighter elements, such as hydrogen, nitrogen, and carbon, cannot undergo fission. Yet, like the very large nuclei, these have low amounts of binding energy per nucleon, and are therefore potentially unstable. These small nuclei can achieve a higher binding energy per nucleon by joining together to make a heavier nucleus that is more stable. The process is called **nuclear fusion**.

The binding energy curve predicts whether an element will undergo fission or fusion.

An example of a fusion reaction can be seen in the following equation:

$$_1H^1 + {_1}H^3 \longrightarrow {_2}He^4 + {_0}n^1 + 17.6 \text{ Mev}$$

This is approximately the reaction that takes place in the hydrogen bomb. But for this reaction to take place, the hydrogen and tritium atoms must be very hot, and therefore very energetic, with enough energy to collide with one another despite their repulsive electrical forces. In the H-bomb, this heat is supplied by the firing of a small A-bomb; the fission reaction provides the heat to make the fusion reaction possible.

When we compare the 17.6 Mev of energy released in the fusion reaction above with the 200 Mev released in the fission of an uranium atom, it seems as if fusion produces much less energy. But if we look at the figures for energy released per unit of mass, the comparison is very different. The mass of the uranium atom is 115 times greater than that of the hydrogen atom; so the hydrogen atom would release 17.6 Mev $\times$ 115 = 2024 Mev, compared to 200 Mev for the same mass of uranium.

The fusion of hydrogen nuclei is the main source of the sun's energy. The energy released in fusion creates the high temperature necessary to sustain the process indefinitely (until the hydrogen atoms are used up) even though the sun emits huge quantities of energy as light and heat. Scientists have not yet discovered a way to control fusion reactions so as to harness that source of energy for man's use, but many predict that this breakthrough will come before the end of the twentieth century. ·

Glossary

Antimatter is a kind of mirror image of matter—particles with an opposite charge and spin.

The **atomic mass unit (amu)** is a unit of measurement for nuclear particles, derived from Avogadro's number.

The **atomic number** of an element is the number of protons in the nucleus of its atoms; it expresses the quantity of positive charge in the nucleus.

A **chain reaction** is a self-sustaining succession of nuclear fissions in which the neutrons produced by one fission reaction induce further reactions.

An **electron volt (ev)** is a unit measuring the kinetic energy gained by an electron when it is accelerated through a potential difference of 1 volt. A unit of one million electron volts is called a **mev**; a unit of one billion is a **bev**.

Nuclear **fission** is the process through which a heavy nucleus disintegrates to form two lighter ones plus gamma rays and other particle products.

Nuclear **fusion** is the process through which small light nuclei join together to form a more stable heavy nucleus.

An element's **half-life** is a measure of the time it will take for half the atoms in any given sample to decay.

Isobars are chemically different elements with the same mass number; an example is the parent and daughter nuclei in radioactive decay.

Isotopes of an element have the same atomic number but a different atomic mass.

The **meson** is a particle, in size midway between the proton and the electron, that is associated with the exchange force that helps bind the nucleus together. The **muon** and **pion** are two types of meson.

The **neutrino** is a massless uncharged particle emitted during β decay.

A **neutron** is an uncharged particle with a mass number of 1.

The **positron** is a particle of antimatter, an electron with the opposite charge and spin.

A **proton** is a hydrogen nucleus; it has a single positive charge.

Radioactive materials are those which spontaneously emit charged particles or very high frequency electromagnetic waves or both. The three major types of rays emitted are **alpha particles** (α), which are helium nuclei with a charge of $+2e$; **beta particles** (β), which are electrons or positrons; and **gamma rays** (γ), which are very short wavelength x-rays.

Exercises

1 Does the radioactivity of an element change when it is chemically combined with another element? Explain.
2 What are the three kinds of rays emitted by radioactive substances? How do these three rays differ from each other? Which is similar to light rays?
3 What is atomic number? What is atomic weight? Does an atom with a higher atomic number always have a higher atomic weight? How did Moseley's experiments enable him to place the elements in order of their atomic number?
4 How was the proton discovered? In what way was it experimentally distinguished from α particles?
5 What does $_6C^{12}$ mean? How does it differ from $_6C^{14}$? What does $_7N^{16}$ mean? What distinguishes it from $_8O^{16}$, which has the same atomic weight?
6 What is an electron volt? What does one mean by saying that 1 amu = 931 ev? Why can one say that the mass of an electron = 0.511 Mev?
7 How do positrons and electrons differ from each other? How were positrons discovered?
8 What is meant by half-life? If the half-life of an element is 10 days and you start with 1024 grams, how much will you have of the element after a) 10 days; b) 20 days; c) 50 days; d) 60 days; e) 100 days?
9 Why may the atomic number of an element increase if a particle is emitted by the nucleus?
10 What are mesons? How do they differ from the other particles that we have studied? What was Yukawa trying to explain when he postulated them?

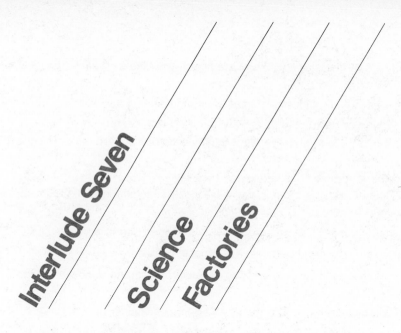

Interlude Seven

Science Factories

The twentieth century has been called the Age of Science. If the concerns of an age can be measured by the way it allocates human and material resources, then our age has surely been given the right name. In the last hundred years, scientific research has been increasing at an astonishing and ever-expanding rate. There are more scientists today than ever before in our history, and a substantial majority of all the scientists who have ever lived are still alive. New areas of scientific specialization—geochemistry, astrophysics, mathematical biophysics, molecular genetics, neurochemistry—proliferate as the need for deeper and more detailed investigation grows. The amount of information generated by scientific research and the complexity of new scientific ideas and theories require increasingly sophisticated mathematical techniques, computational hardware, and calculation language so that these contributions can be understood and used. The study of science gets bigger and more complicated all the time.

Big science is a big industry. The scale of scientific research today requires elaborate experimental arrangements and large numbers of people to support them. Laboratory instruments look like the parts of great factories: huge generators, gigantic metal parts moving on overhead rails, mile-long vacuum chambers, robot spacecraft carrying complex cameras, long banks of controlling devices and computers, panels of meters and dials. Precise engineering standards and rigorous procedures govern the development and use of these instruments. And the size of many experimental facilities is comparable to that of any heavy industry. This is especially true of elementary particle accelerators such as the Enrico Fermi National Accelerator Laboratory in Batavia, near Chicago. The ring of the particle accelerator at the Fermilab, as it is called, is 4 miles in circumference, and the peak energy for particles is at least 400 Bev.

The Enrico Fermi National Accelerator Laboratory. The outline of the main accelerator, which is 4 miles in circumference, can be seen in the background. The inset (lower left) is a view of the underground tunnel in which the main accelerator is housed. Along the left side of the tunnel wall are the magnets through which the protons travel in a vacuum tube.

The research facilities of the European Council for Nuclear Research, called CERN, were established in 1952 at Geneva, Switzerland, for studies into the structure of matter.

Dozens of scientists from several fields collaborate in the work of a single project; research papers may be signed by 5 or 15 or 55 different researchers. And their work requires the support of a background staff of engineers and other scientists who keep the laboratories running along with hundreds of electricians, machinists, plumbers, computer technicians, and other workers.

Clearly the day is over when one man, such as Benjamin Franklin, could work alone using equipment of his own design that he paid for out of his own pocket. The quantity of information needed and the scale of equipment required, in scientific study today is too great for even a genius like Madame Curie, who with her husband personally examined several tons of pitchblend ore by hand in their studies of radioactivity. This is the age of mass production by large groups researchers, technicians, designers, and managers: the age of the science factory.

Much of the research carried on today is too costly to be supported by individuals or even by university systems or private foundations. The annual budget of the Fermilab in 1974 was $40 million; it cost $250 million to build. The financial cost of big science has become a sizable fraction of the entire national budgets of small nations and a constant budgetary problem even in the greatest nations. Most countries cannot support certain major fields of scientific research alone, and so they cooperate in jointly-financed efforts, such as the European Center for Nuclear Research, called CERN, in Geneva. Big science has thus come under the influence of big government and of international conglomerates.

These funding agencies take an active part in fundamental decisions about the nature of the scientific research they support. They have a great deal to say about what will be studied, how much money will be devoted to various areas of study, and who will do the research. They influence decisions for example, about whether our search for alternate forms of energy will focus on nuclear energy or on solar energy. They help to determine whether medical research will be concentrated on curing cancer, developing means for genetic manipulation, or reversing the effects of early protein deficiency. Research facilities hire public-relations people and Congressional lobbyists to help shape the decisions of government and other funding agencies. But most of these decisions are made with very little input from the nonscientific community. The ordinary citizen has practically no voice.

Yet the research—and the decisions surrounding it—will have a profound effect on the quality of life on this planet in the decades to come. And much of the money invested, at least in this country, is public money. It is sometimes argued that the lay person is simply unable to understand the implications of scientific research and that such decisions must be left to the scientists. The scientists, however, are clearly influenced by the sources of their funding, and the values of those sources are often unclear. If the work of science is to support human values, it is essential for these decisions to be debated on a much more public scale. Otherwise, the science factories may deliver a product that serves its investors rather than the ultimate consumer.

Two views of facilities in the Brookhaven National Laboratory in Upton, New York. The view at the left is inside the tunnel of the alternating gradient synchrotron at the juncture of the linear accelerator with the main magnet enclosure. The proton beam emerges from the accelerator, behind the two men, and travels along the 4-inch pipe to the lower right, passing through a series of focusing lenses and steering magnets into the orbit of the synchrotron magnet ring, part of which can be seen on the right. The view at the right shows the control room for the cosmotron. Through the window a 2200-ton ring-shaped magnet, which is covered by a white plastic cocoon, can be seen.

Five

The Play of the Elements

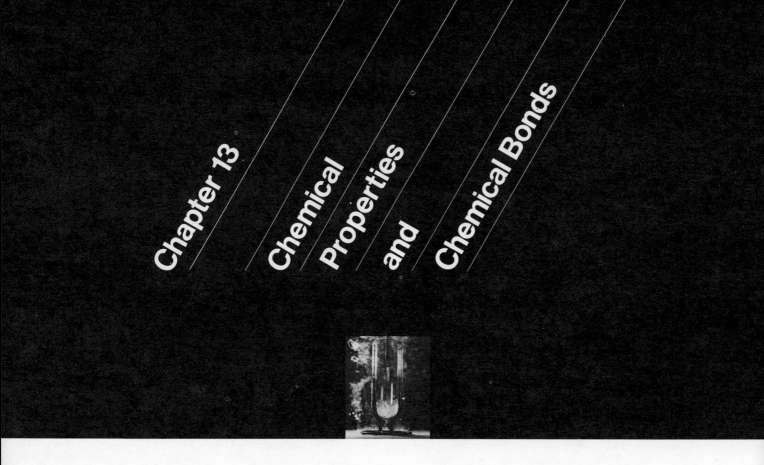

Chapter 13 Chemical Properties and Chemical Bonds

The Development of Chemistry The ancient Greeks questioned whether there was a limit to the divisibility of matter. Aristotle said that the four elements of which all matter was composed—earth, air, fire, and water—were continuous and therefore infinitely divisible. You could take the smallest imaginable piece of earth and divide it in half, and you would have a substance exactly the same in all its properties as a ton of earth; no matter how often the process was repeated, the tiny fragment would still remain essentially like the earth you started with. But Democritus, a contemporary of Aristotle, held that there was indeed a limit to the extent to which these elements, and other pure substances, could be divided. The smallest units into which matter could be divided he called **atoms** (*atomos,* in Greek, means no longer divisible). Democritus explained the differing properties of various kinds of matter as being due to the combination of atoms of the different pure substances they contain. The atoms themselves, Democritus claimed, were of different shapes and sizes for each type of matter, and these shapes and sizes were characteristic of all materials made of that matter. For example, he proposed that gold was made of one type of atom, whereas iron was made of another—a guess that later scientists confirmed. He also suggested an application of his theory, which is so far incapable of scientific test—that the perfectly shaped atom, or sphere, was the matter of which the human soul is made.

Robert Boyle (1627–1691), a seventeenth century Irishman, was the first modern scientist to try to elaborate Democritus' view. He felt that the Greek definition of an element was too general; he therefore proceeded to define an element pragmatically as any substance which could not be reduced by physical or chemical means to different or simpler substances with different properties. (It has recently been shown that elements can, under extreme physical conditions and with specialized equipment, be broken down into other substances, but these reactions involve smashing atoms.) Substances which he proved to be composed of more than one element Boyle called **compounds**.

Boyle's observations stimulated a surge of interest in chemical experiments. One process in particular that aroused interest was burning. What happened in a fire to change solid matter to ashes and smoke? In the late seventeenth century, many scholars postulated that matter, during combustion, lost an intrinsic part of its substance. The lost material was given the name **phlogiston**. It was held to be completely invisible and intangible. When a candle or any other object burned, it was supposed to give off phlogiston; when the air became saturated with this phlogiston, so that it could no longer support the burning of the candle, it was known as "phlogisticated air." The death of a mouse, kept in an ordinary jar of air for a given period of time, was attributed to the "phlogistication" of the air by the mouse's respiration.

Early Chemical Experiments

At the same time, experiments were also being carried on with metals. It was observed that when a metal was heated in air, it formed a powder and gained weight. It was subsequently discovered that the powders thus formed were the oxides of these metals, derived from the chemical combination of the metals with oxygen from the air. Early chemists, however, believed that the powders, or calxes, as they were called, were the pure metals themselves.

Heating these calxes over charcoal produced a shiny substance, which we know today to be the true metal. What the chemists of the time were actually doing was removing the oxygen from the metal oxide powders by means of the action of charcoal and heat. What they *believed* they were doing was transferring phlogiston from the charcoal, which lost weight, to the shiny metal. There was a problem with this explanation, however. The shiny metal obtained from heating the calx with charcoal weighed less than the calx; how could it both gain phlogiston from the charcoal and lose weight at the same time? It was postulated that the phlogiston that was gained by the pure metal was negative phlogiston; that is, that it was phlogiston with negative weight. Therefore, when the metal gained phlogiston, it actually lost weight. Thus the seventeenth century model of matter was "saved" by imagining this special property of negative weight.

Chemistry in the Eighteenth Century

Figure 13-1
The French chemist Antoine Lavoisier (1743–1793), shown here in his laboratory, discovered the law of conservation of mass.

Antoine Lavoisier (1743–1794), a French chemist of the eighteenth century, tested the phlogiston theory by placing metallic tin on a dish and inverting a jar over it. The rim of the jar, which was upside down, he placed in another dish containing water, sealing the reaction from the outside. He then used a magnifying glass to focus the rays of the sun on the tin. Lavoisier observed that the intense heat of the focused sunlight caused the formation of the oxide, or calx, of tin. Simultaneously, he noticed that the water in the jar rose. Lavoisier several times repeated this and other experiments in which calxes were formed from metals. He always carefully weighed both the pure metal and the air in the jar before beginning the experiment, and weighed them again after he was through; his balances, which he had commissioned to be made for him, were capable of measuring weights as small as several thousandths of an ounce. In every case, he found that the weight of the metal plus the air before the experiment equaled exactly the weight of the metal calx plus the air that remained in the jar after the experiment. By this experimental demonstration Lavoisier had disproved the theory of negative phlogiston.

From these experiments Lavoisier deduced one of the fundamental laws of nature, known as the **law of conservation of mass**. This law states that, during a chemical reaction, the gain in mass of one or more of the products of the reaction is equal to the loss in mass of one or more of the reacting substances; the total mass of all matter before the reaction is equal to the total mass after the reaction. Lavoisier, executed during the French Revolution for being too severe a tax collector, left behind a list of 23 substances he believed to be elements, based on a trial and error study of their specific chemical reactions; most of these proved to be correct.

Elaborating the Atomic Theory

At the turn of the nineteenth century, chemistry had made great progress, but many problems remained unsolved. For instance, it was known from experimentation that several different compounds could be formed of the same two elements and that one of these compounds might contain more of an element than the other compound. In one type of copper oxide, for example, there are two atoms of oxygen, while in the other type of copper oxide, there is only one atom of oxygen. This constitutes a ratio of 2:1 for the relative amount of oxygen in the first compound to that in the second. But at that time no method existed for determining the exact proportions of one element to the other in each compound. Nor had any rule been discovered that would relate the proportions of an element in one compound of several elements to the proportions of the same element in other compounds containing the same elements.

The French chemist Joseph Proust (1754–1826) solved this problem. Using careful experimentation and the most delicate instruments available, Proust further showed that the elements in copper carbonate existed in the ratio of 5.3 parts of copper to 4 of oxygen and 1 of carbon, and that this ratio was based on the weights of the three elements found in the compound. This led Proust to formulate a chemical law of major importance, the **law of definite proportions**. The law of definite proportions states that when two or more pure elements form a compound, they combine in definite proportions, by weight, of each element. To those who believed that the elements were made up of atoms, the law of definite proportions suggested an interesting corollary; each individual atom of the element in the compound must also have its specific weight.

The law of definite proportions led to the concept that each elemental atom has a definite weight.

DALTON'S WORK   In the early 1800s, John Dalton (1766–1844), an English school teacher, investigated this implication of the law of definite proportions. Dalton found that there was a series of compounds composed of the same two elements in varying proportions. In each of these compounds, the ratio of the first element to the second element, by weight, was different. For example, carbon monoxide had 17 grams of oxygen and 31 grams of carbon; carbon dioxide had 34 grams of oxygen and 31 grams of carbon; carbonate had 51 grams of oxygen and 31 grams of carbon. It struck Dalton as significant that the amount of carbon in each compound was equal. Each compound had the same amount of carbon, but the weight ratio of carbon to oxygen varied; the ratio of oxygen by weight in the second compound to oxygen in the first was 2:1; the oxygen weight ratio of the third compound to the first was 3:1.

From observations such as these, Dalton derived another law of chemistry, known as the **law of multiple proportions**. This law states that in different compounds composed of two or more of the same elements, the ratio by weight of one of the elements in one compound to the weight of the same element in another of the compounds can be expressed as a small whole number.

The observations of Proust and Dalton seemed to indicate that there was always a definite weight ratio between the different elements in a particular compound. Furthermore, there was always a definite ratio between the fraction by weight of an element in one compound and the fraction of the same element in a different compound composed of the same kinds of elements. These findings led Dalton to advance his **atomic theory,** which claimed that the compound particles (later known as molecules) of a substance were composed of small units known as atoms. Each atom had a definite weight, which was characteristic of its element. The weights of the atoms of one element were different from the weights of the atoms of another element. Dalton further stated that these atoms combined, in definite ratios, to form the chemical particles we know as molecules.

The modern definitions of the atom and the molecule are derived from the law of multiple proportions.

What had not been discovered, however, were the true formulas of the compounds Dalton had postulated in his theory. For instance, the gas we know as carbon monoxide was shown by Dalton to consist of a fixed weight ratio of oxygen to carbon. This, Dalton claimed, meant that the gas consisted of one atom of carbon to one atom of oxygen. But might not the true formula of the compound really be a multiple of this? After all, the atomic ratio of 1:1 would still be maintained if the compound consisted of two atoms of carbon and two of oxygen, since by dividing this 2:2 ratio by two, one would still obtain a 1:1 ratio.

Further experiments only added to the confusion. During the eighteenth century, it had been shown that the formation of the gas water vapor always required two volumes of hydrogen and one of oxygen, but that two volumes of water vapor were formed. This might at first seem to be a contradiction of the law of conservation of matter, since there are three volumes of reactants but only two volumes of the product. But the conservation law applies to *mass,* or the total weight of the matter; the units used in this experiment are those of *volume,* or the amount of space the matter takes up. When the two different elements react, they form chemical bonds which bring the atoms closer together; therefore their volume decreases although their mass remains constant. This problem of changing volumes because of the making and breaking of bonds is one of the complexities that impeded the development of chemistry as a science.

Experimentation showed that if 30 liters of hydrogen were used, then 15 liters of oxygen would be required, and this would give 30 liters of vapor. The ratio of elements seemed to indicate that the water molecule consisted of one atom of oxygen and two of hydrogen, giving it the formula H_2O. But this conclusion was contradicted by Dalton's experiments based on the weights, or mass, of hydrogen and oxygen. Dalton had shown that 8 parts by weight of oxygen always combined with one part by weight of hydrogen to give water. Dalton assumed that these weights indicated one atom of hydrogen, with a weight of one, combining with one atom of oxygen, with a weight of 8; he thought the formula for water should be HO.

Other experiments produced conflicting evidence. It had been shown that the reaction producing an oxide of nitrogen consisting of one part by weight of nitrogen to one part by weight of oxygen always required one volume of nitrogen gas and one volume of oxygen gas; this always gave two volumes of this nitrogen oxide gas. How could this be? Was not one atom supposed to combine with one other atom to give one, and not two, molecules of a compound? Even more confusing, in one of the other nitrogen oxide compounds two volumes of nitrogen gas combined with one volume of oxygen to give two volumes of gas. If the formula of the gas were N_2O, as indicated by weight experiments, shouldn't there have been only one volume of gas, since only one molecule was formed?

AVOGADRO'S WORK It was Amadeo Avogadro (1776–1856), an Italian chemist, who explained these puzzles, by a process of educated guesswork. He first postulated that, under identical physical conditions of temperature, pressure, and volume, equal volumes of the gases of different elements contained the same numbers of particles of these elements. This means that if, for instance, one liter of element A contains 1000 atoms of element A, then 1 liter of element B must also contain exactly 1000 atoms of element B. It also means that two liters of element A would contain 2000 atoms of element A, and so forth.

Avogadro also guessed that, in all of the experiments with gases mentioned above, each particle of these gases consisted of two, and not one, atoms. This meant that the particles of the element that were actually reacting to form a new compound consisted of two-atom molecules, and not single atoms, as had previously been supposed. Avogadro's hypothesis clarified matters consider-

Avogadro postulated that one particle of some gases has two atoms, not one.

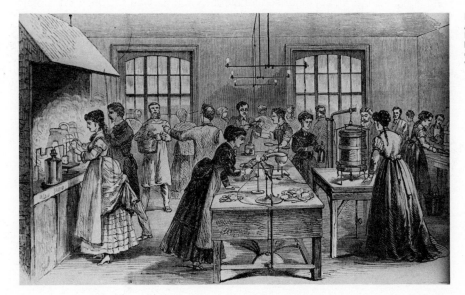

Figure 13-3
An American chemical laboratory in 1869 looked like this. Here, students at the Massachusetts Institute of Technology conduct experiments on various substances with the aid of fire.

ably. For instance, if the particles of hydrogen entering into the formation of water are actually two-atom molecules, and twice as much, by volume, of this hydrogen is needed as oxygen, we will have four parts by weight of hydrogen. If one volume of oxygen contains two-atom molecules of oxygen, we will have two parts of oxygen. This gives us a total of six parts—four of hydrogen and two of oxygen. In the final quantity of water vapor formed—two volumes—we thus have six parts, or three parts to each volume. Since there is twice as much hydrogen as oxygen, the formula for water must be H_2O, and not HO, as Dalton had thought.

Other problems were also explained by Avogadro's hypothesis. The observation that two volumes of hydrogen chloride were formed from one volume of hydrogen and one volume of chlorine was attributable to two-atom molecules of hydrogen and two-atom molecules of chlorine. One volume of hydrogen and one of chlorine totaled two volumes, each containing two parts of one element to give a total of four parts. Since two volumes of hydrogen chloride were formed, and there were four parts of the elements in each of these two volumes, each volume must contain, as had been shown, two parts, or two atoms. One of these atoms was hydrogen, and the other was chlorine.

Extending his hypothesis, Avogadro stated that a given weight of any element or compound always contains exactly the same number of atoms or molecules as the equivalent weight of any other element or compound. This number, called **Avogadro's number**, has been shown to be 6.022×10^{23} atoms or molecules—an extremely large number, but not at all unreasonable considering the minute size of a single atom.

A mole expresses the quantity of atoms or molecules found in a substance.

The equivalent weight is called a **mole,** or the weight, in grams, that corresponds to the atomic weight of an element. Oxygen, the element used as the traditional standard of comparison for atomic weights, was assigned the atomic weight of 16.000; gold then has an atomic weight of 197.2; nitrogen has an atomic weight of 14.008. Thus, one mole of oxygen weighs 16.000 grams; one mole of gold weighs 197.2 grams; one mole of nitrogen weighs 14.008 grams.

We may also speak of compounds in terms of moles. In water, there are two atoms of hydrogen, each having an atomic weight of 1.0080. There is also one atom of oxygen, having an atomic weight of 16.000. Then a molecule of water weighs 1.0080 + 1.0080 + 16.000, or 18.0160 atomic mass units, and one mole of water weighs 18.0160 grams.

It has also been found that one mole of a gas, no matter what its chemical composition, occupies a specific volume under standard conditions of temperature and pressure; this volume is 22.4 liters.

The Classification of Elements

The chief concern of the nineteenth century chemist was the study of the physical and chemical properties of the elements. The knowledge accumulated from these studies began to grow prodigiously and before long, this knowledge became burdensome; the need for some order among the elements became apparent.

The basis of all attempts to achieve order and simplification was the grouping of elements with similar properties. For example, sodium and potassium were both soft and silvery metals; calcium, barium, and strontium could be prepared by similar chemical changes. But such discoveries were isolated and offered little promise of classifying all the known elements into a unifying system.

DOBEREINER'S WORK   The earliest actual attempt at a systematic arrangement of the elements into groups was apparently made when the German chemist Johann Dobereiner (1780–1849), a friend of Goethe, discovered that some of the elements, when arranged by atomic weight in groups of three, seemed to form units, or **triads**, with similar chemical properties, as shown by their reactions with other chemicals. Especially significant was the fact that the elements in a triad formed a sequence; Dobereiner found such a sequence in the elements lithium, sodium, and potassium. These elements were found to resemble each other chemically—they showed some relationship between **reactivity**, or the ease with which they formed bonds with other elements, and their atomic weights. The properties of sodium, the middle member of the group, were seen to be intermediate between those of lithium and potassium. Potassium was found to be the best electrical conductor, sodium was seen to be not quite as good, and lithium was found to be the poorest of the three. Potassium was the densest element of the three, sodium less dense, and lithium the least dense. This sequence continued for every chemical and physical property Dobereiner tested!

When the element bromine was discovered in 1826, Dobereiner suggested that its atomic weight would fall about halfway between the atomic weights of chlorine and iodine. When this was confirmed, he was quick to suggest that this was yet another triad. But Dobereiner was really unable to make any kind of valid generalization, since he could locate fewer than half a dozen triads from among the fifty known elements.

NEWLANDS' LAW OF OCTAVES   In 1864 an English chemist, John Newlands (1838–1898), recognized that when the elements were arranged in the order of increasing atomic weight, every *eighth* element seemed to repeat the properties of the eighth element preceding it; an interval of seven elements separated similar elements. Newlands termed this phenomenon the "law of octaves" intentionally drawing an analogy to the octave in music, and on this basis he arranged the elements in horizontal periods and vertical columns.

Although Newlands had to put many of the elements out of their atomic weight order to make things fit, he believed that he had hit upon something important and was deeply distressed by the fact that his colleagues would not at first take his proposals seriously. One fellow chemist went so far as to suggest face-tiously that Newlands might have been better off listing the elements in alphabetical order. Newlands' paper on the subject was rejected by the editor of the Journal of the Chemical Society as "not adapted for publication in the Society's Journal."

MENDELEYEV'S PERIODIC TABLE OF THE ELEMENTS The culminating achievement of the search for order among the elements came in 1869, when Dmitri Mendeleyev (1834–1907) in Russia and Lothar Meyer (1830–1895) in Germany, each working separately, hit upon a fruitful scheme for classifying and arranging the known elements (leaving gaps in the arrangement that might also serve to predict the properties of elements yet to be discovered). Mendeleyev is generally given credit for its full development.

Mendeleyev's periodic table classified elements according to atomic weight.

It is said that Mendeleyev came upon his ideas one night in a dream. In March of 1869 he presented his first periodic classification table to the Russian Chemical Society. Mendeleyev found that, if the elements are arranged in the order of their increasing atomic weights (starting with the lightest element, hydrogen), they form a table of natural periods and groups in which the elements in any group resemble one another in their properties and chemical reactivity. According to Mendeleyev's table, the first two rows or **periods** of elements contain seven elements each and the next three rows contain seventeen elements each.

Figure 13-4
The Russian chemist Dmitri Ivanovich Mendeleyev developed the periodic law of classification of the elements and predicted properties of elements then unknown.

Mendeleyev's scheme, named the "**periodic table** of the elements," greatly clarified and simplified the comprehension of chemistry. The periodic table integrated what had been just a compendium of facts. The table soon became the basis of a **periodic law**; as Mendeleyev stated it, "The chemical properties of the elements are periodic functions of their atomic weights." In other words, it was seen that chemical properties varied in a regular fashion as atomic weights became larger. The poet Robert K. Duncan expressed this fundamental order in these words: "Just as the pendulum returns again in its swing, just as the moon returns in its orbit, just as the advancing year ever brings the rose of spring, so do the properties of the elements periodically recur as the weights of the atoms rise . . ."

The organization of the elements into families and groups not only simplified the correlation of their properties but also made it possible to predict the properties and behavior of yet undiscovered elements. Being forced to leave gaps in his table in order to achieve the order he desired, Mendeleyev recognized that these gaps were elements to be discovered; more significantly, he was able to hypothesize the properties of missing elements, all of which were eventually found to exist.

The formulation of the periodic table greatly accelerated the search for new elements. Today the table of natural elements has grown to 92. In recent years, the periodic table has provided a useful guide for chemists and physicists who have created artificial elements by nuclear bombardment of existing elements. Thirteen additional radioactive elements have been synthesized and added to the list. These elements are all heavier than uranium and have sparked interest in a further search for the transuranium "heavy" elements.

MOSELEY AND THE ATOMIC NUMBER Eventually it was discovered that the regularity of the table arose from the fact that the chemical properties of the elements were dictated by the electronic structure of the atoms from which they were made. Almost half a century after Mendeleyev first laid down the periodic classification system, an important discovery was made which led to this realization.

In 1914, Henry Moseley (1888–1915), a young Welsh physicist killed in the first world war, used X-rays to determine the number of protons in the nucleus of an atom; this number is called the **atomic number**. When Moseley prepared a periodic table based on atomic numbers, rather than atomic weights, he found that the variations in properties agreed perfectly with the order of the elements in the table.

Moseley's periodic table classified elements according to atomic number.

In Mendeleyev's table, arranged according to atomic weights, there had been slight discrepancies in some cases. For example, potassium precedes argon in Mendeleyev's table; yet, when arranged according to properties, potassium follows argon, and this is in agreement with their atomic numbers, 18 for argon and 19 for potassium. Although Mendeleyev had concluded that the chemical properties of the elements are periodic functions of their atomic weights, today we state the periodic law in slightly different terms; "the chemical properties of the elements are periodic functions of their atomic numbers."

Arrangement of the Periodic Table

The modern periodic table is shown on pages 276–277. It has seven horizontal periods of varying lengths: the first period has only two elements (hydrogen and helium), whereas the sixth period has 32 elements. The seventh period is incomplete. The last element on that period presently is element 106, which has not yet been named, and the period has room for twelve more elements. It is likely that these superheavy elements do no exist in nature, but chemists are trying to synthesize them in the laboratory.

Various versions of the periodic table differ slightly in the number of vertical columns recognized. The major categories are the alkali metals, the alkaline earth metals, the halogens, and the noble gases.

The modern periodic table classifies elements according to atomic structure.

The grouping of elements into metals and nonmetals, shown by the black lines, is not such a neat one; but as a general rule the metals are on the left, nonmetals on the right.

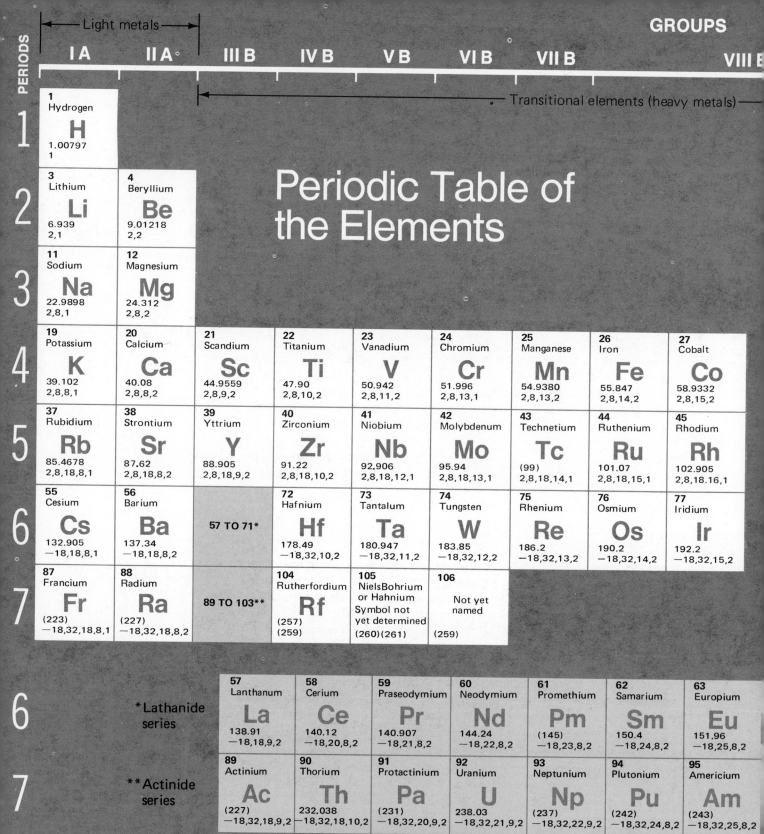

| | I B | II B | III A | IV A | V A | VI A | VII A | VIII A |

Non-metals

2
Helium

He

4.0026
2

KEY

27
Cobalt

Co

58.9332
2,8,15,2

- Atomic number
- Element name
- Element symbol
- Atomic mass (weight)
- Electron configuration

5 Boron **B** 10.81 2,3	**6** Carbon **C** 12.011 2,4	**7** Nitrogen **N** 14.0067 2,5	**8** Oxygen **O** 15.9994 2,6	**9** Fluorine **F** 18.9984 2,7	**10** Neon **Ne** 20.183 2,8
13 Aluminum **Al** 26.9815 2,8,3	**14** Silicon **Si** 28.086 2,8,4	**15** Phosphorus **P** 30.9738 2,8,5	**16** Sulfur **S** 32.064 2,8,6	**17** Chlorine **Cl** 35.453 2,8,7	**18** Argon **Ar** 39.948 2,8,8

28 Nickel **Ni** 58.71 2,8,16,2	**29** Copper **Cu** 63.54 2,8,18,1	**30** Zinc **Zn** 65.37 2,8,18,2	**31** Gallium **Ga** 69.72 2,8,18,3	**32** Germanium **Ge** 72.59 2,8,18,4	**33** Arsenic **As** 74.9216 2,8,18,5	**34** Selenium **Se** 78.96 2,8,18,6	**35** Bromine **Br** 79.909 2,8,18,7	**36** Krypton **Kr** 83.80 2,8,18,8
46 Palladium **Pd** 106.4 2,8,18,18	**47** Silver **Ag** 107.870 2,8,18,18,1	**48** Cadmium **Cd** 112.40 2,8,18,18,2	**49** Indium **In** 114.82 2,8,18,18,3	**50** Tin **Sn** 118.69 2,8,18,18,4	**51** Antimony **Sb** 121.75 2,8,18,18,5	**52** Tellurium **Te** 127.60 2,8,18,18,6	**53** Iodine **I** 126.9044 2,8,18,18,7	**54** Xenon **Xe** 131.30 2,8,18,18,8
78 Platinum **Pt** 195.09 —18,32,17,1	**79** Gold **Au** 196.967 —18,32,18,1	**80** Mercury **Hg** 200.59 —18,32,18,2	**81** Thallium **Tl** 204.37 —18,32,18,3	**82** Lead **Pb** 207.19 —18,32,18,4	**83** Bismuth **Bi** 208.980 —18,32,18,5	**84** Polonium **Po** (210) —18,32,18,6	**85** Astatine **At** (210) —18,32,18,7	**86** Radon **Rn** (222) —18,32,18,8

64 Gadolinium **Gd** 157.25 —18,25,9,2	**65** Terbium **Tb** 158.924 —18,27,8,2	**66** Dysprosium **Dy** 162.50 —18,28,8,2	**67** Holmium **Ho** 164.930 —18,29,8,2	**68** Erbium **Er** 167.26 —18,30,8,2	**69** Thulium **Tm** 168.934 —18,31,8,2	**70** Ytterbium **Yb** 173.04 —18,32,8,2	**71** Lutetium **Lu** 174.97 —18,32,9,2
96 Curium **Cm** (245) —18,32,25,9,2	**97** Berkelium **Bk** (249) —18,32,26,9,2	**98** Californium **Cf** (250) —18,32,28,8,2	**99** Einsteinium **Es** (254) —18,32,29,8,2	**100** Fermium **Fm** (252) —18,32,30,8,2	**101** Mendelevium **Md** (256) —18,32,31,8,2	**102** Nobelium **No** (254) —18,32,32,8,2	**103** Lawrencium **Lw** (257) —18,32,32,9,2

Some elements are radioactive, and the numbers in parentheses denote the mass number of the most stable or common isotope of these elements.

Metals and Nonmetals

The oldest division of the elements is the classification into metals and non-metals. Even before atomic structure was understood, scientists were able to distinguish between the two groups. Metals have a characteristic way of reflecting light which we call luster. They conduct heat and electricity extremely efficiently; they generally have high melting points and a high density and can be drawn into wires or hammered into sheets. Nonmetals, on the other hand lack conductivity; in the solid state they are usually brittle and without metallic luster. Metals are much more numerous than nonmetals; only 20 of the 104 known elements are distinctly nonmetallic.

In chemical behavior also there are fundamental differences between metals and nonmetals. Although metals may show considerable differences even among themselves, certain generalizations may be made. Metals combine more readily with nonmetals than with each other. Nonmetals on the other hand, do show a tendency to combine with one another, although rarely with the ease with which they combine with metals.

Since metals are placed on the left-hand side of the periodic table, there is a gradation, as we progress from left to right, from the metallic into the nonmetallic. As a result, there are certain borderline elements of intermediate character called **metalloids**. These are found on both sides of the diagonal line which separates metals from nonmetals. Metalloids, which show both weak metallic and weak nonmetallic properties, include such elements as boron, germanium, silicon, and arsenic.

The relationship between atomic structure and chemical properties would suggest that the distinction between metals and nonmetals could best be defined in terms of atoms. Metallic elements in the table have atoms which contain one or a few electrons in their outer shell, and the electrons are held rather weakly. This is of central importance in determining their chemical properties. Nonmetals, on the other hand, have several electrons in their outermost shells. These shells are nearly filled and the electrons are held quite firmly.

In general, then, the fewer the number of electrons and the more loosely they are bound, the more pronounced will be the metallic character. As a result, metallic properties increase to the left of the periodic table and down the columns.

On the basis of this generalization, we would expect francium (atomic number 87) to be the most metallic element. Although it is only available in minute quantities due to its radioactivity, experimental evidence has confirmed that it is indeed the strongest metal. Francium has only 1 electron in its outermost shell. But this is true for all metals above francium in Group I of the table. The difference is that the outershell electron of francium is located in the seventh shell, the farthest out from the nucleus and thus the most loosely bound.

Nonmetallic character increases toward the right of the periodic table and up the columns. Fluorine (atomic number 9) is the strongest nonmetal. It contains 7 electrons in its outer shell (remember, it needs only 8 electrons to have a complete shell) and this shell is nearest the nucleus, where positive forces attract the negative electrons very strongly, thus binding them firmly to the atom.

Elements having similar properties fall into the same vertical group or family. Examination of some of the more important chemical groups will help us gain enough understanding of the structure and meaning of the periodic table to predict the elements that will combine together.

Vertical Groups of the Periodic Table

ALKALI METALS The **alkali metals** form Group I of the periodic table. Included are lithium, sodium, potassium, rubidium, cesium, and the radioactive francium. In general, these are all relatively light, chemically reactive metals. They have 1 electron in the outer shell and combine chemically with chlorine to form salts with similar properties. The alkali metals react more or less violently with water, releasing hydrogen gas. Because of their reactivity, these metals must be stored in an inert atmosphere or under oil.

Alkaline earth metals are more active and have higher melting points than alkali metals.

Lithium, soft and silvery in appearance with 1 electron in its outer shell, is used in the manufacture of glass and glazes for porcelain dishes. Sodium is especially useful in the manufacture of organic chemicals, dyes, and tetraethyl lead, a gasoline additive. Sodium is also used in sodium-vapor lamps which provide high-intensity street lighting; their light can be recognized by its rather lurid yellow color. Potassium compounds include those used as plant fertilizers, and also cream of tartar, which has the mysterious ability to keep beaten egg whites stiff; Julia Child always puts a pinch of cream of tartar in her meringues.

ALKALINE EARTH METALS Group II consists of the six **alkaline earth metals**– beryllium, magnesium, calcium, strontium, barium, and radium. They each contain 2 outershell electrons, and as would be expected, are all active metals; they are never found uncombined in nature. In general Group IIa metals are considerably harder than the Group Ia metals, and they are also denser and have higher melting points.

Magnesium, the second most abundant metal in the sea, is also found in many silicate minerals. Magnesium oxide (MgO) is a compound used in medicine (milk of magnesia) to neutralize excess acid in the stomach. Calcium is a silvery metal, used industrially in many lead alloys from which bearings and sheathings for electric cables are made. Calcium is found abundantly as calcium carbonate ($CaCO_3$) in marble, limestone, and chalk. Calcium sulfate ($CaSO_4$) occurs in nature as the mineral gypsum; this substance is converted commercially to plaster of paris. Practical uses of strontium and barium compounds are limited. In combination with nitrates, strontium and barium are used in fireworks to produce red and green fire, respectively. Barium sulfate is used

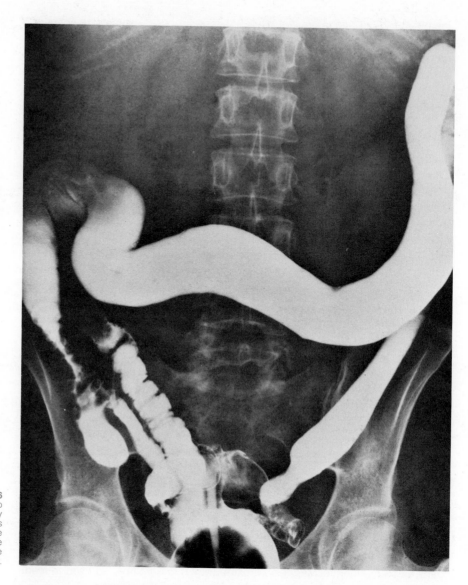

Figure 13-6
Barium sulfate is used by doctors to
diagnose digestive ailments. In the X-ray
photograph shown here, the barium makes
the colon—the large white tube in the
center—stand out. Also clearly visible are
the vertebral column and pelvis of the
patient.

in an interesting way to diagnose medical problems in the digestive tract, such
as ulcers and intestinal obstructions. The patient swallows the barium sulfate
and shortly thereafter is X-rayed; the barium in his stomach and intestines
makes them stand out in high contrast on the X-ray film.

Halogens Group VII elements, called **halogens**, constitute the most strongly nonmetallic
class of elements. The halogens—fluorine, chlorine, bromine, iodine, and asta-

tine—are very reactive chemically, and they form a large number of compounds. Each contains 7 electrons in the outer shell and has a strong tendency to fill this shell by forming chemical bonds. At room temperature, the first two elements, fluorine and chlorine, are gases; bromine is a liquid; and iodine and astatine are solids. Fluorine, the lightest of the halogens and the most reactive of all the elements, is found in minerals such as fluorite (CaF_2) and cryolite (Na_3AlF_6), and in small quantities in sea water and drinking water. Although fluorine is extremely poisonous, fluorides have been found useful in preventing tooth decay and in some communities are added to the supply of drinking water. The successful formulation of a fluoride toothpaste took a long time; because of fluorine's reactivity, the fluorides kept breaking down and forming other compounds with the toothpaste ingredients. The problem was finally solved by combining fluorine with tin, producing the stable compound stannous fluoride.

Chlorine, the greenish-yellow gas with an irritating odor, is an extremely active element. It is useful in killing bacteria, and for that reason it is used as an additive to sterilize drinking water and the water in swimming pools. Bromine, less reactive than chlorine, is found in compounds in sea water and natural salt deposits. The light-sensitive silver bromides are used in photography, to coat the film; light turns the coating dark. Iodine is a crystalline solid, more dense and less reactive than bromine; it is found in sea water and in seaweed. Astatine is extremely radioactive and disintegrates rapidly. It is not found in nature, and little data about it is available.

Halogens form ionic salts when combined with metals.

Noble Gases

The elements in the last group of the periodic table, Group 0, are called inert or **noble gases**. These elements'include helium, neon, argon, krypton, xenon, and radon. Because of the extraordinary stability of their complete outer shell, they are, for the most part, completely unreactive chemically. Each of the noble gases exists in the free state totally uncombined. The densities and atomic weights of these gases increase as we look down the column.

Helium, with 2 electrons, is the first noble gas. Neon, the second, contains an octet of 8 electrons in its filled outer shell; it is used commercially in neon signs. The characteristic glow emitted by these advertising signs results when an electric current is passed through a tube containing neon gas. Argon is used in industry to provide a chemically inert atmosphere for such processes as welding and the manufacture of alloys. Krypton, xenon, and radon comprise the remainder of the inert gases.

Research in the past decade has shown that several compounds of xenon can be synthesized. In 1962 and 1963, the first xenon fluorides were prepared. It would seem understandable that if the inert xenon were to react with any compound it would be fluorine, the most reactive of all the elements. In light of this new evidence, the definition of an inert gas calls for some modification. It might

Figure 13-7
Neon signs light up "The Great White
Way"—Times Square, in New York City. The
light in each sign is produced as an
electrical charge is passed through a glass
tube containing neon gas.

be more accurate to say that the noble gases are unreactive except under special laboratory conditions.

Periodic Properties

Mendeleyev arranged his periodic table on the basis of experimental evidence about the properties of various elements. It was so valuable a predictive tool that he was able to foretell the existence and properties of elements yet to be discovered. But, although Mendeleyev did perceive a connection between the atomic weight and the periodic properties of an element, he did not really know what caused the connection.

The number of electrons in the outer shell of an element determines the periodicity of its properties.

Today it is accepted that the periodicity of properties of the elements is caused by the structure of the outer electron shell—or, to be more specific, by the number of electrons in the outer shell. The noble gases, with full outer shells, are virtually nonreactive. The alkali metals, with a single outershell electron, are highly reactive, readily yielding their electron to some other atom. In fact, it might be said that the science of chemistry really comes down to a glorified form of numerology!

Elements with the same number of outershell electrons have similar properties, but their properties are not identical. The filled shells between the nucleus and the outer, reactive electrons take up space and to some extent shield the nucleus. The more shells there are, the more pronounced is this effect. Thus, lithium and cesium, both alkali metals with 1 electron in the outer shell, have many similar properties. But in lithium there is only one filled shell between the nucleus and the outer electron, whereas in cesium there are five. Because of this difference, the two elements do exhibit slightly different properties.

Ionization Energy

Electrons are held in an atom by the electrostatic force of a positively charged nucleus. It is possible, however, to remove an electron from an atom. **Ionization energy** refers to the amount of energy required to remove the *least* tightly bound electron from an isolated atom. When an electron is removed, the atom remaining has one less electron and a positive electric charge; it is called an **ion.**

An ion is an electrically charged atom.

Metals generally have low ionization energies and lose electrons readily. Conversely a great deal of energy is required to remove an electron from a nonmetallic element. Inert gases have enormously high ionization energies due to the stability of the outershell octet.

The magnitude of the ionization energy depends upon three factors: atomic size; charge on the nucleus; and the screening effects of inner electron shells. With any group, the ionization energies decrease with increasing atomic number. Since metallic character increases with atomic number, it can be inferred that as metallic character increases, ionization energy decreases.

Electron Affinity

Electron affinity is a measure of the energy released by an atom in the formation of a negative ion. It is really the mirror image of the concept of ionization energy. Ionization energy is the energy required to remove an electron from a neutral atom; electron affinity is the energy released when an electron is *added* to a neutral atom. Although electron affinities are difficult to measure, it is known that electron affinity is a good indicator of the strength with which an additional electron will be bound to the atom. Halogens (with 7 outershell electrons) could therefore be expected to have high electron affinities; in these elements, the addition of one more electron would result in a complete and stable outer shell. It could be expected that electron affinities would decrease with increasing atomic number, since the added electron enters shells which are progressively farther from the nucleus.

Valence

Dalton's atomic theory, which stated that atoms combine in definite simple number ratios to form molecules, led quite naturally to further exploration of the way atoms join to form molecules. In 1867, the German chemist Auguste Kekulé (1829–1896) introduced the term **valence** as a measure of the capacity of an atom to combine with other atoms. Hydrogen is used as a standard of comparison; thus an element has a valence of 1, if one atomic weight of the element combines with one atomic weight of hydrogen. Similarly, if one atomic weight of the element combines with two atomic weights of hydrogen, its valence is 2; an example is oxygen, in H_2O. The highest valence observed in naturally occurring compounds is 4, or half the complete octet of outershell electrons. Certain elements, such as sulfur and phosphorous, have multiple valences. For example, when sulfur combines with hydrogen it has a valence of $2(H_2S)$ and when it combines with carbon it has a valence of $1(CS_2)$.

Chemical Bonds

Chemical bonding between atoms occurs when electrons are transferred or shared.

Shortly after the appearance of the valence theory, the English chemist Edward Frankland (1825–1899) proposed the concept of a **chemical bond** to describe the forces holding atoms together. Today we know that there are various kinds of chemical bonds which bind together all matter and dictate its characteristics; the type and strength of bonds depend on the structure of the atoms involved. The most significant factor in the formation of any chemical bond is the number of electrons in the outer shells of each atom. Most bonds are formed by the transfer or sharing of these electrons, which are called **valence electrons.**

Shells and Orbitals

It is the number of electrons orbiting the outermost rings of the nucleus of an atom that determines its physical and chemical properties. These rings are called **atomic shells,** regions containing electrons with approximately the same energy level, orbiting at approximately the same distance from the nucleus. Each shell differs in its level of energy, the shells closest to the nucleus having the lowest energy. As first postulated by Niels Bohr and later confirmed by spectroscopy, there are seven atomic shells, labeled *K, L, M, N, O, P,* and *Q,* the *K* shell being the lowest level energy. The letters correspond to the principal quantum numbers 1, 2, 3 . . . 7, which designate the size of each electron wave, or the energy levels of the elements on the periodic table. The number of electrons at each energy level varies from 2 in the *M* shell to 32 in the *P* shell.

The electrons in each shell do not all have exactly the same amount of energy, and this is the basis for subdividing the shells into subshells. The subshells are labeled *s, p, d,* and *f,* with *s* having the lowest energy and *f* the highest. Each letter designates the shape of an electron wave; *s* stands for sharp, *p* for principal, *d* for diffuse, and *f* for fundamental. The first shell contains 1 subshell (*s*); the second shell contains two subshells (*s* and *p*); the third shell contains three subshells (*s, p,* and *d*); the fourth shell—and all others—have four subshells (*s, p, d,* and *f*).

The electrons in an atomic shell should not be thought of as discrete particles circling the nucleus as the planets circle the sun. Rather, each shell should be thought of as an electron cloud within which there is a high probability that an

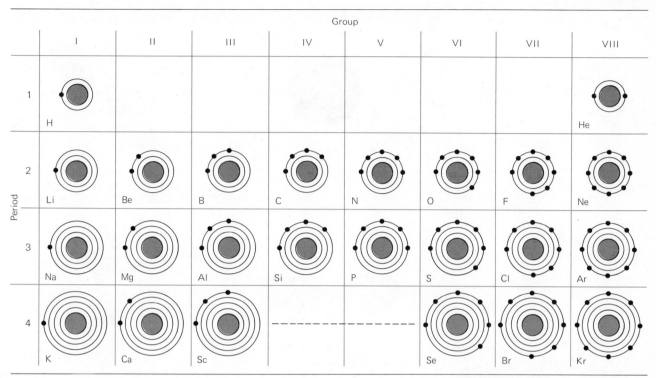

Figure 13-8
This schematic drawing shows the electron configuration of the outer shells from the periodic table of the elements. The broken line in the bottom row indicates the omission of the twelve other elements in this shell.

electron can be located. It is *possible* that an electron orbiting an atomic nucleus within your body might travel across the room; but it is *probable* that it will be located within the area of its subshell. By using complex wave formulas, scientists are able to approximate the size and shape of the area in the shell which probably contains an electron. This area of the shell is known as the **orbital**. Technically, the orbitals are "located" in the subshells.

Each subshell of a shell contains a definite number of orbitals. Thus, the *s* subshell has 1 orbital, the *p* has 3, the *d* has 5, and the *f* subshell has 7 orbitals. According to the Pauli principle of exclusion, each orbital can hold up to two electrons. One must spin in a clockwise direction, the other in a counterclockwise direction.

Physicists have found it useful to indicate graphically the probable three-dimensional shapes of orbitals. What would be the shape of the electron orbital of the hydrogen atom? Hydrogen, as a glance at the periodic table shows, belongs to the first shell. The first shell has only 1 orbital and it is an *s* orbital. The *s* orbital, which in a hydrogen atom holds only one electron, assumes a spherical shape—there is an equal probability that the electron can be located in any direction from the nucleus of the atom.

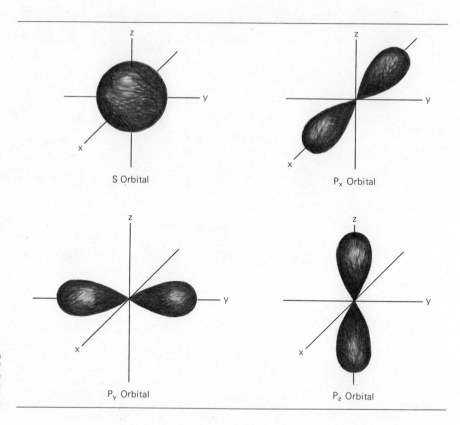

Figure 13-9
The configurations of the two basic kinds of
atomic orbitals—s and p orbitals—are
shown here. It is highly probable that an
electron will be located in the shaded
regions of each of the orbitals.

The second shell contains $1s$ and $3p$ subshells. The s subshell holds 2 electrons; the $3p$ subshells hold 2 electrons each. The s orbital of the second shell (called the $2s$ orbital by combining the quantum number with the letter of the subshell) is also spherical. It differs from the $1s$ orbital above only by the fact that it is larger. The p orbitals, however, are different. Each of the p orbitals assumes the shape of an hourglass or a figure eight, and the three orbitals are perpendicular to one another.

When the atoms bond together, the shape of their orbitals determines the way they will join. The relationship of one atom to another, when bonded, is called the bond angle. Often the orbitals combine to form new shapes, or **hybrid** orbitals.

IONIC BONDS   The most common inorganic chemical bond is the ionic bond. An **ionic bond** is formed when an atom gives up one or more valence electrons to another atom. The donor atom becomes chemically more stable through the loss of its "surplus" outer electron; the receiver atom becomes chemically more stable because it fills up its outer shell. In the process of achieving chemical stability, the atoms become electrically unstable, as ionic bonding leads to the

formation of a positive ion (the donor) and a negative ion (the receiver). Therefore an electrostatic attraction, measurable in coulombs, exists between the
positive and negative ions.

Let us examine a typical example of a compound formed by ionic bonds.
Fluorine, which contains seven electrons in the second shell, becomes a very
stable negative fluoride ion upon the addition of one electron. Similarly, lithium
becomes a stable positive ion upon the loss of its third electron, leaving a filled
first shell. Consequently, the two atoms easily enter into chemical reaction to
become a molecule of lithium fluoride:

$$Li \circ + \circ \ddot{\underset{\cdot\cdot}{F}} \colon \longrightarrow Li + \colon \ddot{\underset{\cdot\cdot}{F}} \colon —$$

Compounds such as LiF, which are formed through electron transfer, are known
as **ionic compounds**. Another example of an ionic compound is sodium chloride, or table salt (NaCl). Upon the loss of an electron, sodium (which has an
atomic number of 11) is left with a filled second shell. As a result, sodium has

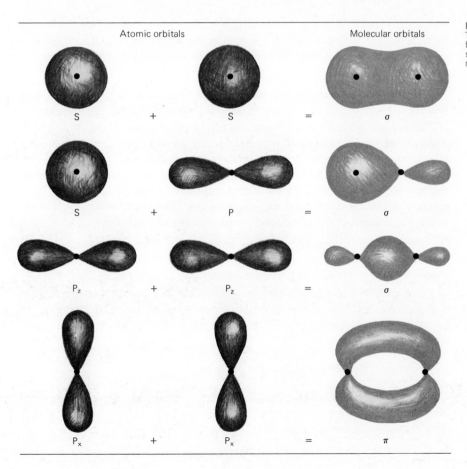

Atomic orbitals		Molecular orbitals
S	+ S	= σ
S	+ P	= σ
P_z	+ P_z	= σ
P_x	+ P_x	= π

Figure 13-10
The s and p atomic orbitals can combine to
form four kinds of molecular orbitals. The
small dot in each figure indicates the
nucleus.

a strong tendency to donate one electron and is considered an active metal. Chloride, on the other hand, needs an additional electron to complete its third shell. We can show this in an electron dot diagram, in which the dots represent the electrons in the outer shell:

$$Na° + .\overset{\circ\circ}{\underset{\circ\circ}{Cl}}: \longrightarrow :\overset{\circ\circ}{Na}: + :\overset{\circ\circ}{\underset{\circ\circ}{Cl}}:^-$$

On the left hand side of the equation, Na is depicted as having only one electron; on the right hand side of the equation, Na^+ is depicted as having eight electrons, because its third shell is empty and its valence electrons are now in the second shell.

Most ionic compounds are a combination of metals, which tend to give up outer electrons, and nonmetals, which grab these electrons to fill their outer shells. As a result of the strong electric nature of the ionic bond, most of these compounds have high melting points and exhibit excellent electrical conduction in solution or molten states.

COVALENT BONDS   There are numerous compounds that cannot be accounted for by ionic bonding. For example, elements from the middle of the table need several electrons to fill their outer shells, but highly charged ions, such as C^{4+}, are very unstable. Consequently, these elements tend to form covalent bonds.

The **covalent bond** consists of a pair of shared electrons, each contributed by one of the bonding atoms. The atoms draw sufficiently close to allow an overlap of shells. Each atom finds itself with an additional electron filling its shell, yet experiences no net charge for it has not really gained or lost an electron. The two atoms are held tightly together by their shared attraction for the same pair of electrons and form a **covalent compound.** Covalent bonds are generally stronger than ionic bonds, as the atoms are more intimately connected. Ionic compounds, on the other hand, can be separated into ions by passing a current through a solution of the compound in the process of electrolysis.

Covalent bonds occur very frequently, even among elements that easily form ionic bonds as well, such as the halogens and hydrogen. Hydrogen, for example, forms ionic bonds with many elements. But when it has no other element to bond with, atoms of hydrogen covalently bond together in pairs to form hydrogen gas (H_2). Each atom then uses the other's single electron to complete its own first shell. Essentially, the two hydrogen atoms are sharing a mutual, filled K shell:

The various halogens which require one more electon to fill their outer shell, form similar simple covalent bonds. Chlorine atoms for example will combine to form molecules of chlorine gas or Cl_2.

Atoms with higher valences will frequently form covalent bonds with more than one atom. Carbon, which needs four electrons to fill its outer shell, can obtain these by covalently bonding with a wide assortment of other atoms. The simplest combination is methane (CH_4) in which carbon forms four single bonds to four hydrogen atoms:

Carbon can also form two **double bonds** with two oxygen atoms, since oxygen has six electrons in its outer shell. In order to fulfill the requirement of the octet rule, each oxygen atom shares two of the four electrons in carbon's outer shell, while carbon obtains an additional two electrons from each oxygen atom:

A **triple bond** is composed of three shared electrons from each atom. Such a bond occurs in nitrogen gas, in which two nitrogen atoms, each of which needs three electrons to fill its outer shell and has three unpaired electrons to share, are connected by a triple bond to form a molecule of N_2:

:N:::N:

Another important factor in covalent bonds is **electronegativity,** the affinity of an atomic nucleus for electrons. As we move across any row of the periodic table, electronegativity increases due to increasing positive charge in the nucleus. As we move down any group, electronegativity decreases, due to the distance of the outer electrons from the nucleus and the shielding effect of inner electrons. In a covalent bond between two atoms of differing electronegativity, the electrons lie closer to the atom with the higher electronegativity. This gives rise to a **dipole moment,** during which each atom has a partial ionic character. The more electronegative atom is said to have a negative dipole and the lesser electronegative atom a positive dipole.

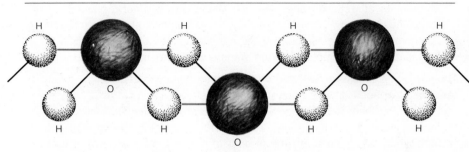

Figure 13-11
Hydrogen bonding in the water molecule depends on the polar orientation and shape of the hydrogen and oxygen atoms. Here, the smaller positive hydrogen atoms in one water molecule bond with the larger negative oxygen atom in the neighboring molecule.

HYDROGEN BONDS   In addition to the major forces represented by ionic and covalent bonds there are several other types of bonds and attractions. Chief among these is the hydrogen bond. Hydrogen has low electronegativity, and therefore when it bonds to more electronegative atoms such as oxygen and flourine, it has an appreciable positive dipole. Consequently, there is often an attraction between a hydrogen atom bonded to a heteroatom, and the unshared electrons or anions of another molecule. This attraction is called a **hydrogen bond**. Though weak in contrast to ionic and covalent bonds, the hydrogen bond is nevertheless very important in aligning molecules.

The water molecule (H_2O) presents a good example of hydrogen bonding. Because the charges are not uniformly distributed, this type of molecule is known as a dipolar molecule. There are more unshared electrons in the area occupied by the oxygen atom than there are in the area occupied by the hydrogen atoms. Thus, one end of the molecule contains two positively-charged hydrogen atoms, while a single negatively-charged oxygen atom occupies the other end. The geometrical arrangement of the molecule is also unbalanced, in that the two hydrogen atoms are not symmetrically arranged around the oxygen atom. The positive charges of the hydrogen atoms cause them to repel one another, with the result that the bond angle formed by the union of the hydrogen atoms with the oxygen atom is about 105°.

Hydrogen bonding occurs when the end of the water molecule containing the positive hydrogen dipole is attracted to the end of another water molecule containing the negative oxygen dipole. The structural formula for hydrogen bonding is:

$$
\begin{array}{ccc}
H & H & H \\
| & | & | \\
\text{---O---H---O---H---O---H}
\end{array}
$$

The hydrogen bonds in this formula are indicated by broken lines. The bond is sometimes known as a bridge because it links the two water molecules.

VAN DER WAALS FORCES   In addition to hydrogen bonds, there are also weak intermolecular attractions called van der Waals forces, after the Dutch physicist Johannes van der Waals (1837–1923). These come into play when two molecules come into very close contact but are not bonded chemically. As a result of this force, the two molecules are transformed into instantaneous dipoles.

The van der Waals force is based on the fact that at any given instant, a temporary shift can occur in the electron charge distribution of two adjoining molecules, causing them to behave like dipolar water molecules. Thus, two nonpolar molecules moving rapidly through their medium may suddenly meet for a brief instant. When this occurs, an abrupt shift in the electric charge of one of the molecules can bring about a corresponding shift in the electric charge in the

Figure 13-12
Van der Waals forces temporarily bind two nonpolar molecules, as this sequence shows. Two nonpolar molecules (a) repel one another. In (b), the van der Waals force causes the two to instantaneously shift their charges, becoming polar molecules which attract one another. After a fraction of a second (c), the two resume their normal nonpolar states and separate.

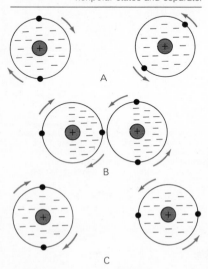

other molecule, causing both molecules to become temporary dipoles. The oppositely-charged ends of the molecules are attracted to one another for perhaps a fraction of a microsecond. Just as instantaneously, an electron shift converts the molecules back to their stable nonpolar condition. The two then repulse one another and continue moving through their medium, eventually forming similar attractions with other molecules.

Glossary

An **anion** is a negative ion.

The **atomic number** is the number of protons in the nucleus; it also indicates the number of electrons in the neutral state.

An **atomic orbital** is a subdivision of an atomic shell and contains a particular type of electron cloud. Each orbital is characterized by a different electron cloud shape.

An **atomic shell** is an energy level which encircles the nucleus at a certain distance and contains electrons.

The **atomic weight** is the combined mass of protons and neutrons, each of which has an atomic mass unit of one.

A **cation** is a positive ion.

A **chemical bond** is the force that holds two or more atoms together to form a discrete unit called a molecule.

A **chemical reaction** occurs when chemical bonds are formed or broken between two or more atoms or molecules.

A **covalent bond** consists of a shared pair of electrons between two atoms.

A **covalent compound** is a substance formed by the combination of two atoms which share a pair of electrons.

A **dipole** is the partical ionic character which results from an atom's electronegativity.

A **double bond** consists of two shared pairs of electrons between two atoms.

Electronegativity is the affinity of an atomic nucleus for electrons.

A **filled shell** indicates that an atomic shell contains its maximum number of electrons and will not form a chemical bond.

A **hydrogen bond** is a weak chemical bridge between a positive hydrogen atom of a polar molecule and the negative atom of a different molecule.

An **ionic bond** is formed by the transfer of an electron from one atom to another. This results in the formation of two oppositely charged ions which exert a powerful electrostatic attraction on each other.

An **ionic compound** is a substance formed through the combination of two or more atoms by ionic bonds.

The **octet rule** states that an atom is particularly stable or inert when its outer shell contains eight electrons.

A **triple bond** consists of three shared pairs of electrons between two atoms.

Valence is the capacity of an atom to combine with another atom.

Valence electrons are the electrons which fill the outer shell of a molecule and are involved in bonding.

Van der Waals forces are weak temporary attractions between two molecules during which the molecules are transformed into instantaneous dipoles.

Exercises

1 If two grams of hydrogen combine with oxygen to give water, how many grams of water will be formed?

2 What is the difference between an atom and a molecule?

3 Fluorine has an atomic number of 9; potassium has an atomic number of 19. Write down the electronic configuration of these two elements. Explain why the compound potassium fluoride (KF) has an ionic bond; which of the two contributes electrons to the other?

4 A container has hydrogen (H_2) gas at 100°C and at a pressure of 2 atmospheres. If the volume of the gas is 1 litre, how many molecules are contained in the enclosure?

5 In transition metals (Fe, Co, Ni, etc.) the 3d orbital is not completely filled, but the 4s orbital is. What is a possible explanation for this discrepancy?

6 What are the valences of sulfur in the following compounds? SO_2, SO_3, H_2SO_4 (sulfuric acid), CuS.

7 Ammonia is formed according to the equation:

$$\tfrac{1}{2}N_2 + \tfrac{2}{3}H_2 = NH_3$$

If 17 grams of ammonia are converted back into nitrogen and hydrogen, how many grams of nitrogen will be liberated? How many atoms of nitrogen?

8 Which of the two has the higher ionization energy; iron or fluorine?

9 In $CuSO_4$, copper has a valence of 2. If one ampere of current is passed for five minutes through a copper sulfate solution, how many atoms of copper will be deposited on the negative electrode?

10 One of the major difficulties in separation of U^{235} from U^{237} is that they have identical chemical properties. Why is this so?

11 Helium has two electrons in its outermost shell, neon has eight. Why then are both of them classified as noble gases?

12 Why is iodine chemically less active than fluorine?

13 Carbon combines with hydrogen to give the following compounds: ethane C_2H_6, ethylene C_2H_6, acetylene C_2H_2. If all the chemical bonds are covalent, how many covalent bonds exist between the carbon atoms in ethane? In ethylene? In acetylene?

14 At ordinary temperature, helium atoms interact with other helium atoms through van der Waals forces. Why do they stay atomic and not form into molecules—He_2, for example?

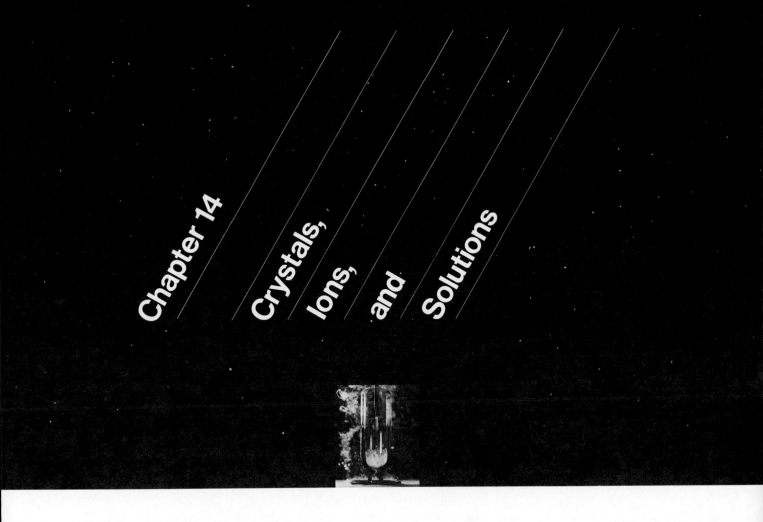

Gases consist of atoms or molecules moving randomly at high speeds but separated by vast distances; these particles frequently collide with one another. As
the temperature of a gas is lowered, kinetic energy is lost and the particles
move more slowly. In the first stages of cooling, the molecules, slowing down,
find it increasingly difficult to escape quickly past each other's forces and attractions, so they begin to collide more often and interact more fully. With further cooling, there is still less space between the molecules, and they enter a
liquid state. Still random and disordered, the molecules then move by gliding
past each other singly or in groups, in the flowing movement of liquids. With additional decreases in temperature, water and nearly all other liquids take a
solid state. The molecules still have some kinetic energy, but no longer enough
to escape the powerful tendencies of atoms to form connections by bonds that

make them more stable. Through this bonding, the solid takes the form of a relatively rigid structure called a **crystal**. Appropriately, the term crystal is taken from the Greek word for ice.

Crystals

The faces of a crystalline solid reflect the regular geometric arrangement of its atoms.

Not every solid forms a crystal. There is a group of solids called the amorphous (shapeless) solids, that have no fixed structure; these include glass, molasses, and tar. Amorphous solids have no fixed melting point, but just sag and soften and start to flow. Most solids, however, have crystalline structures based on regular geometric groups of atoms, with a pattern repeated throughout the body of the solid. These structures are called **lattices** because of their resemblance to the repetitious patterns of crossed strips of wood often used in doors, gates, and window coverings. (Another example of a lattice is the criss-cross pattern often seen in the top crust of fruit pies). The atoms in crystal lattices are held together primarily by covalent and ionic bonds, and also by weaker forces, such as hydrogen bonds and van der Waals forces.

Though the atoms of a solid are in an orderly arrangement and maintain a regular geometric pattern, crystalline solids are not completely rigid. Within a small volume, each of the atoms in a crystal is vibrating continually around a fixed average position.

Ionic Crystals

Solids that are composed of atoms held together by ionic bonds form **ionic crystals**. An ionic bond is formed when atoms either lose or gain electrons to fill their outer shells and thereby become more chemically stable. In this process there is a transfer of one or more electrons between each pair of atoms. This leaves positively charged cations paired with negatively charged anions, held together by the electrostatic force between them. But in achieving this chemical stability, the molecules have sacrificed some degree of electrical stability—they have acquired an electric charge.

Figure 14-1
Although glass atoms have no fixed structure, they are not completely disorganized. Here, each small atom is surrounded at definite distances by three large atoms, forming an equilateral triangle.

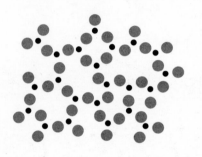

Repulsive electrostatic force between like charges and attractive force between unlike charges create no problem as long as the bond exists between a paired set of ionized atoms. What happens, however, when large numbers of ions are close together, as in a block of salt? In this situation, there are both attractive and repulsive forces acting throughout, and the result is that a certain order emerges. This order maximizes the attractive force between ions of opposite charge and minimizes the repulsive force between unlike ions, so that the group achieves structural stability. This is the basis for the ionic crystal lattice and accounts for its remarkable geometric regularity.

Figure 14-2 shows the crystal structure of NaCl as determined by X-ray crystallography. Sodium cations (Na^+) and chlorine anions (Cl^-) alternate at the corners in a regular pattern like a children's jungle gym. Note that the ions are arranged so that each one is surrounded by six ions of opposite charge. This arrangement can be visualized in another way. The Na⁺ ions are at the corner of one cubic pattern, the Cl⁻ ions at the corners of a second one, and the two cubic patterns are fitted or dovetailed into one another. Thus each ion is seen to be at the center of a face formed by the opposite ions. This pattern maximizes the attractive forces between unlike charged ions, lending stability to the position of each ion and therefore to the whole. Moreover, the distance between like ions is 5.63×10^{-10}m, twice as far as the distance between the nearest unlike ions, so repulsive forces are minimized with this type of geometric structure, called a face-centered cubic. NaCl maximizes the attraction between unlike ions at the expense of the repulsion between like ions.

Other geometric forms also maximize attractive forces and overcome repulsive forces. Another form of ionic crystal is the body-centered cubic, which is the structure of cesium chloride. In this structure (see Figure 14-3) a cesium cation (Cs^+) is found at the center of a cube whose eight corners are chloride anions. The distance between similar ions, about 4.11×10^{-10}m, is greater than the distance between unlike ions. Since electric force is inversely proportional to the square of the distance, this larger distance between like ions greatly diminishes the forces of repulsion relative to the forces of attraction.

The second major type of crystalline solid is the **covalent crystal**. Unlike crystals made with ionic bonds and composed of ionized atoms that will relate electrically with all neighboring ions, the covalently bonded atoms form self-contained units with fixed links to specific pairs of atoms.

The covalent bond usually provides two shared electrons, localized in position between the two bonded atoms, and hence rather strongly directional in the resulting crystal structures. Once the outer shell is filled, a covalently bonded atom is stable and chemically nonreactive (like an inert gas) even though it does carry an electrostatic charge. A simple covalent crystal, such as iodine (I_2), consists of atoms held together in a lattice by van der Waals forces. Since these are relatively weak bonding forces, solid iodine is soft and has a low melting point ($-113°C$).

Because of the strength of covalent bonds, covalent crystals include the hardest substances known. Diamonds, a kind of carbon crystal, are perhaps the most beautiful example of covalent crystals. Their hardness—the greatest known—makes diamonds in such demand for industrial purposes such as grinding, cutting, and polishing other substances that synthetic diamonds are now being produced. This is done by exerting enormous pressure on graphite, another type of carbon crystal, at extremely high temperatures. The effect of the high temperature is to disrupt the crystal pattern of graphite, while the pressure forces its atoms into the much more compact structure of diamond. In this structure, each C atom is surrounded by four adjacent C atoms. The crystal thus assumes the configuration of a tetrahedron or three-dimensional figure with four

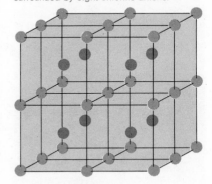

Figure 14-2
Sodium chloride is an example of a face-centered cubic crystal. X-ray crystallography has revealed that the sodium cations and chlorine anions alternate in the patterns shown here.

Covalent Crystals

Figure 14-3
A crystal of cesium chloride has a body-centered cubic structure. The cesium cation, which forms the center of a cube, is surrounded by eight chlorine anions.

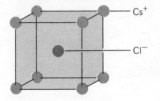

Figure 14-4
The configuration of the faces on this sulfur crystal reflect the ordered repeated arrangements of the atoms which compose it.

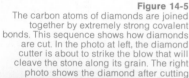

Figure 14-5
The carbon atoms of diamonds are joined together by extremely strong covalent bonds. This sequence shows how diamonds are cut. In the photo at left, the diamond cutter is about to strike the blow that will cleave the stone along its grain. The right photo shows the diamond after cutting.

sides, as shown on page 297. Every C atom shares a pair of electrons with each of the four C atoms surrounding it; thus, every atom in the diamond lattice is bonded to its neighbor in the most rigid fashion possible, fixed at all directions.

With such immensely strong bonds, these purely covalent crystals have high melting points (3500°C for diamond and 1600°C for quartz) and are quite insoluble in normal liquid solvents. Other substances that have a similar crystal lattice are silicon, silicon dioxide or quartz, silicon carbide, and germanium. Covalent solids also include some crystals that are characteristically softer than ionic solids. This is especially true of elements with a valence of 1, which may bond only to one other atom.

SILICATE MINERALS Crystalline solids, like the molecules of which they are made, are not always exclusively covalent or ionic in nature; often a compound contains both covalent and ionic bonds.

There seems to be a wide, even continuous, range of ionic and covalent bonding proportions. For example, NaCl is purely ionic, or 1.00 on a fractional ionic scale from 0 to 1. Cadmium oxide is 0.79; zinc oxide, 0.62; silicon carbide, 0.18; and silicon itself, 0.00.

A special class of compounds with both covalent and ionic bonds in a network

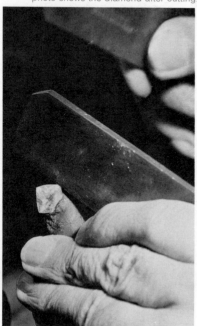

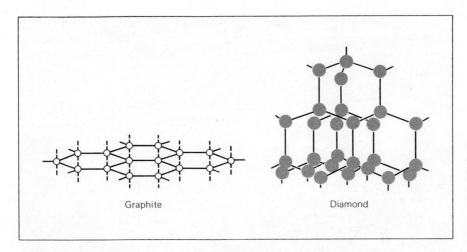

Graphite   Diamond

Figure 14-6
Covalent bonds of different strengths join the
atoms of carbon crystals. In the graphite
crystal at left, weak bonds hold the atoms in
planes of hexagons which can easily slide
over one another. Extremely strong covalent
bonds bind the atoms of the diamond crystal
shown at right into a compact tetrahedron in
which each carbon atom is surrounded by
four other carbon atoms.

structure are the silicates. These include such common examples as feldspar, mica, garnet, asbestos, and topaz, and constitute the most extensive group of minerals.

The basic structure of silicate minerals consists of long chains of silicon tetrahedrons. At the center of each tetrahedron is a silicon atom, which is surrounded by four oxygen atoms at the corners; this forms the prototype building block SiO_4^{-4}. In forming a chain, each corner oxygen may be shared by an adjacent tetrahedron through covalent bonding. These covalently bonded strands may then be held together by positive ions of other elements, such as iron (Fe^+) or manganese (Mn^+). This is the structural basis of asbestos. The tetrahedron building blocks can also be joined by covalent bonds at three corners to produce sheets. These sheets are in turn held together by ionic bonds with positive ions such as K^+, Na^+, and Ca^+. Minerals structured in these two dimensional sheets include mica and talc.

Most covalent crystals are joined by extremely strong bonds.

With such covalent bonding at the corner oxygen atoms, much of the negative charge of the SiO_4^{-4} building block is neutralized. But often there is still enough charge remaining to form ionic bonds between sheets or strands. In sheets of mica, for example, only three of the corner oxygens covalently bond, leaving one negatively charged corner oxygen which then can ionically bond with positive ions such as K^+, If all four corners of oxygens are shared—as in silica (SiO_2)— then the entire structure is covalently bonded in a three dimensional network solid.

Metallic Solids

The third major class of solids are purely **metallic solids,** which consist solely of metal atoms. These seem to have a unique type of structure, for which a new type of bond has been hypothesized.

When a metallic atom bonds to a nonmetallic atom, an ionic bond is formed, in which the metal donates an electron and the nonmetal accepts it. But when metallic atoms bond together, there is a problem. While there are numerous atoms with a tendency to donate valence electrons, there are no atoms available with a matching tendency to accept electrons. More evidence that the bonding structure of metallic substances must be different from the bonding of other solids comes from the unusual properties of metals. They are excellent conductors of electricity and heat, and they can be hammered, drawn into thin wires, pressed into thin plates, or formed into different stable shapes, such as fragile jewelry or high precision gears. These properties are especially apparent in such metals as gold and silver; in contrast, one can only imagine the results of trying to draw table salt into a thin wire!

Covalent crystals do not exhibit these metallic properties, because they have favored bonding directions which will not take arbitrary shapes and relatively fixed electrons which are not free to act as conductors. Ionic solids, which also lack these properties, similarly have a highly ordered structure with electrons moving in fixed shells around very stable ions. It is necessary, therefore, to consider a different type of molecular structure for metallic solids, with a rich supply of freely mobile electrons, called conduction electrons.

Various models of the structure of metallic solids have been proposed, but the most widely accepted hypothesis at this time pictures a lattice of positive charges held or embedded in a fairly uniform **sea of electrons**. According to this model, the valence electrons in the outer shells of the metal atoms are not restricted to fixed average positions surrounding discrete nuclei. Rather the valence electrons are dispersed throughout the metal and shared by the electrostatic attraction of adjacent nuclei. The metallic solid is a sea or gas of electrons, an electron fluid, which engulfs and holds together the positively charged ions nuclei of the metallic atoms. In this manner the metal is joined together as if by one shared great covalent bond. All of the valence electrons are shared by all of the metallic nuclei.

The "sea of electrons" theory attempts to explain the peculiar bonds of metallic solids.

According to this model, the metallic solid also has something of the ionic character. The nuclei of the atoms act as positively charged cations and are held together as a solid by their interaction with negatively charged electrons. However, unlike the bonding in an ionic crystal, the metallic bonding forces are nondirectional, due to the ever shifting attractions between electrons and the charged atomic nuclei. Most metals have closely packed atoms.

With this model, the metallic property of conduction can be understood as arising from the free and random movement of the sea of electrons. A structure in which the outer shells of electrons extend throughout the substance is a highly conductive arrangement; a stream of electrons, forming an electric current, can easily pass through a substance in which there is no strong local attraction to deflect it. This model also helps to explain the metallic properties of ductility and malleability. They are due to the nonpolar and dynamic nature of the bonds. No matter how a metallic solid is shaped, drawn, or compressed, new attractions between nuclei and electrons are immediately formed, holding the solid together in the new shape.

Molecular Crystals

Not all substances form bonded lattices when their molecules are brought together. In particular, many organic substances, like inert gases, are composed of molecules that are so stable they remain chemically independent. They are discrete entities and do not interact to form chemical bonds. And yet there are organic solids and so-called inert gas crystals. How can these stable molecules exist as solids and liquids?

A chemically stable molecule is electrically neutral, with no electric dipole. But this is an *average* description of these nonpolar molecules. Complete neutralization of the positive charge at the nucleus of the atom theoretically occurs when the negatively charged electrons are uniformly distributed. In actual fact, however, the electrons are rarely uniformly distributed. The momentary position of an electron constantly fluctuates. As a result, whenever two nonpolar molecules are sufficiently close, there is a rapidly fluctuating tendency for mutual induction due to the momentary excess or deficit of charge with an instantaneous

Molecular crystals are not linked by chemical bonds.

X-ray Crystallography: Key to the Atomic Structure of Solids

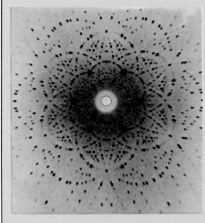

How have scientists deciphered the arrangement of atoms in crystals? Among the ways molecular structure can be determined are microwave and infrared spectroscopy, the indirect evidence from chemical reactions, electron diffraction, and most important, the x-ray diffraction of crystals.

In this process, x-rays are beamed through a crystal and scattered into regular and definite patterns by interaction with the regularly spaced atoms of the crystal. This process was proposed in 1912 by the German physicist Max von Laue (1879–1960) and experimentally demonstrated by Laue, W. Freidrich and P. Knipping in 1913. Scientists had believed earlier that the orderly arrangement of external faces of a large crystal corresponded to the orderly arrangement of its atoms; from this they deduced the atomic structure of a crystal by measuring its faces. Their reasoning was correct, but they had no way to prove it. The process of x-ray diffraction unlocked the structural mysteries of crystalline solids, much as the invention of the microscope two centuries earlier unlocked the mysteries of matter and the cell structures of all life.

The actual mechanism of x-ray diffraction is quite simple. A lead shield containing a tiny slit is placed between the x-ray source and the crystal. The scattered x-rays from the source are emitted in many directions, but they are absorbed by the heavy lead and only a pinpoint beam is allowed through the slit in the shield. When the x-ray beam enters the crystal, it is diffracted by the atoms of the crystal into numerous separate beams. These diffracted beams can produce *constructive* interference when coming from the regularly spaced atoms or ions in the crystal lattice, and in particular from the parallel planes of the lattice structure. The constructively diffracted beams are registered as light spots on a photographic film placed behind the crystal, which is sensitive to x-rays as well as to visible light. Distances between light spots, determined by the wavelength of the x-rays and the angle at which they strike the crystal, correspond to the distance between the atoms in the crystal.

Why must x-rays be used in this process? The answer lies in the very small wavelengths of x-rays. As small as or smaller than the atoms, they are about 0.12×10^{-8} cm at their shortest, and hence they can penetrate the spaces between the atoms, which are about 2×10^{-8} cm wide. Therefore, each atom can scatter the x-radiation visible pattern. Electron waves are too large to penetrate the interatomic spaces, but neutron beams are also useful as an alternative to x-rays. Direct use of an x-ray microscope is impossible because x-rays can only be scattered; they cannot be bent by any available lenslike device.

electric field at one position or another in the outer electron orbits. This weak, unsteady attraction of one atom for the electrons of another is an example of a van der Waals force.

Though much weaker than either covalent or ionic bonds, the van der Waals forces hold molecules of the inert gases together in a class of solids called **molecular crystals**. Similar forces bond many organic crystals. These weak forces operate for atoms or molecules that are close together, where the local electric fields of the fluctuating electron positions can quickly be effective. Therefore these are closely packed crystal lattices. Because of the relative weakness of the van der Waals attractive forces, molecular solids have very low melting and boiling points, and insignificant mechanical strengths. This difference can be seen by comparing an amethyst, which is a quartz with mixed ionic and covalent bonds, with an opal, a quartz with molecular crystals. Although an amethyst can survive rough treatment, an opal that is struck sharply falls apart into powdery fragments, as some wearers of opal rings have discovered with dismay.

Figure 14-8
Ice crystals on a windowpane form intriguing patterns. The frozen water molecules in an ice crystal are held together in a hexagonal configuration by hydrogen bonds.

Hydrogen-bonded Crystals

A quite special crystal bond is provided by neutral hydrogen atoms. With only one electron, H might seem ready for a single covalent bond with one other atom, but the fact an ionic bonding is possible under the right conditions. The electron is captured by another atom in the molecule that forms the crystal, and the proton (which is simply the H nucleus) binds two closely neighboring negative ions. Only two can fit because of the small volume of the proton which, in its electrostatic attraction of the negative ions, requires very close distances. Water provides an important example of the H bond in liquid and solid H_2O. Others are protein molecules, certain polymers such as hydrogen fluorides and formic acid, and the pairing relation of molecules in the organic bases of the DNA molecules that provide the molecular mechanism of genetics.

Solutions and Mixtures

A **solution** is a mixture of two or more different substances that is observed to be homogeneous down to molecular size. Mixtures can be solids, liquids, or gases. Mixtures differ from compounds in that the different atoms or molecules that mix in a solution are not joined by chemical bonds. For example, hydrogen and oxygen when combined chemically form water molecules. However, hydrogen and oxygen can also be mixed together in a gaseous solution which is highly explosive. Similarly, air is a gaseous solution containing various gases such as nitrogen and oxygen.

Mixtures

In practice, mixtures can be distinguished from compounds because the component substances of a mixture can be separated by simple physical means. For example, in work on molecular biology, large molecules can be separated from smaller molecules, in much the same way that you might use a sieve to separate larger particles from smaller ones, by a device called a Sephadex column, which contains tiny hollow spheres. After the larger molecules pass through and are collected at the bottom of the column, the smaller molecules proceed to move down the column.

Mixtures can be separated by magnetism, heating, evaporation, or distillation.

Other methods of separating mixtures include:

1 magnetism to remove ferromagnetic substances.

2 heating, to exploit the different melting points of solids. Bronze, for example, is a mixture of tin and copper. When bronze is heated, the component metals, each of which has a different melting point, will separate under intense heat.

3 evaporation, in which more volatile components vaporize, leaving other elements behind. This method is employed to extract minerals from sea water.

4 distillation, a technique employed to separate two liquids in solutions. This physical method exploits the difference in boiling points of various substances, to separate them into liquid and vapor at a given temperature. The vapor can then be collected and cooled, so that it condenses back to the liquid state of the pure substance. The illegal stills of hillbilly lore are examples of this process, used to distill alcohol.

Liquid Solutions

The most complex solutions are the **liquid solutions**. When two substances are mixed together, chemists conventionally call the one present in greatest amount the **solvent**. In the case of a solid or gas dissolved in a liquid, the latter is always the solvent. Water is the most common solvent, blanketing three-quarters of the earth. The other substances in a liquid solution, those present in lesser amounts, are referred to as **solutes**. Thus, when you drop a tablet of Alka-

Figure 14-9
Three officers of the law stand guard over their capture—a mammoth illegal still seized in a raid on "moonshiners" in Washington, D.C., during the 1920s. The device was used to separate alcohol from fermented materials such as fruits, grains, or sugarcane.

Seltzer in a glass of water, a solution is formed in which water is the solvent, and the salts and bubbles of CO_2 are the solutes.

A solution consists of a solvent and a solute.

SATURATION If we were to keep adding sodium bromide to a liter of water, we would observe that after a while the compound no longer dissolves, no longer remains invisible in the water; rather the sodium bromide will begin to precipitate to the bottom of the flask, where it is easily seen. When a given amount of solvent contains as much solute as it can possibly hold at a given temperature, the solvent is said to be **saturated**. The solubility of a solute is therefore the maximum amount that can be dissolved in a given amount of liquid at a particular temperature. This quantity is usually expressed as grams/liter, or the grams of solute per 100 grams of solvent.

As the temperature increases, the solubility of solids usually also increases. Consequently, if a solution is saturated at a high temperature and then cooled, some of the solid must go out of solution, or precipitate out, falling to the bottom of the container. The amount that remains in solution will equal its solubility for that temperature. **Supersaturation** may occur if the cooling process proceeds very slowly. In a supersaturated solution, the amount of solid in solution exceeds its solubility. However, a supersaturated solution is not stable. If the solution is shaken even slightly, the molecular balance is ruined and much of the solute will precipitate out.

POLAR AND NONPOLAR SOLVENTS Although water is considered the nearly universal solvent, certain substances, such as oil or sulfur, are not soluble in water. Similarly, certain organic solvents—such as ether—will only dissolve organic solutes. The reason for this differential ability to dissolve solutes is that dissolving takes place only when solvent and solute have compatible electric structures. Water, for example, has polar bonds. Even in the liquid state, water molecules are hydrogen-bonded into groups. When a polar substance is mixed with water, similar attractive forces come into play between the water molecules and the solute. But fat molecules and other nonpolar substances are not as attracted to water molecules. Consequently, the attraction of water molecules for each other is greater, and the nonpolar solute is forced out into a separate layer. Nonpolar molecules, such as oils and fats, may be dissolved in nonpolar solvents, such as benzene.

The dissolution of one substance by another depends upon the electric structure of the two.

One exception to the general rule that solvent and solute must have similar polarity is water itself. Water, though it is not itself ionic, dissolves ionic crystals, despite the fact that its hydrogen bonds are weaker than normal ionic bonds in the crystal lattice. However, the combined attraction of several water molecules for a surface ion is strong enough to overcome the internal forces of the crystal lattice. By attacking one layer at a time, water can dissolve many ionic crystals.

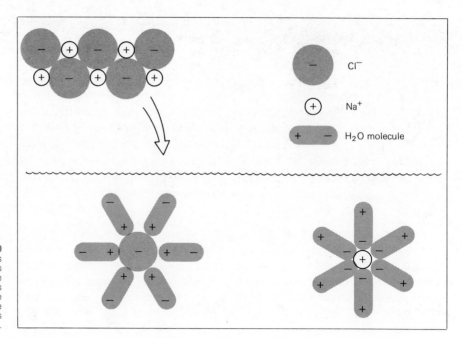

Figure 14-10
The polarity of the water molecule is
responsible for the dissolution of materials
placed in it. Here, the sodium and chlorine
ions are removed from their crystal lattice as
the positive end of the water molecule
attracts the negative chlorine ions, while the
negative end of the water molecule attracts
the positive sodium ions.

Dissociation and Free Ions

When an ionic solid is dissolved, its component ions enter into solution and are said to be free. They are no longer paired and interconnected in a precisely ordered lattice. The separation of an ionic solid by its solvent into free ions is termed **dissociation**. Any ionic solid which dissociates in a solvent is called an **electrolyte**. As a result of their dissociation into *freely moving* charged ions, electrolytes in solution can conduct an electric current. Ions in solution always appear in pairs of cations and anions, originating from the dissociation of a molecule or pair of ions.

An electrolyte conducts electric
current because of the free movement
of its charged ions.

When a molecule dissociates, its ions generally have discrete properties that differ from those of the parent molecule. For example, a common laboratory experiment demonstrates the presence of Cu^{++} (cuprous ion) in a solution by its blue color. Cu, the element copper, on the other hand, has a dull reddish color. Similarly, Cl^- has different properties from both NaCl and Cl_2. The latter is a poisonous greenish yellow gas, whereas Cl^- is colorless.

Electrolytes are divided into three major classes, acids, bases, and salts, based on the behavior of their ions in solution.

ACIDS Our common experience with acids brings to mind a substance which tastes sour and may even burn our skin. Yet the orange juice we drink is also an acid, as is the vinegar we put in our salad dressing.

An **acid** is composed of hydrogen and one or more nonmetallic ions. Upon entering solution, the acid dissociates into hydrogen ions and nonmetal ions. It is possible to write an equation expressing the dissociation of an acid into ions as follows:

$$HCl \longrightarrow H^+ + Cl^-$$

The difference between strong acids which are corrosive and frequently dangerous to handle, and weak acids which we consume in our food, lies in the degree of their dissociation. A strong acid tends to give off hydrogen ions in solution and will dissociate to a larger degree. The reaction is shown graphically in these equations:

Nitric acid $\qquad HNO_3 \rightleftharpoons H^+ + NO_3^-$

Hydrochloric acid $\quad HCl \rightleftharpoons H^+ + Cl^-$

(The longer arrow indicates the direction of the reaction.) Sample weak acid reactions are:

Acetic acid $\qquad HC_2H_3O_2 \rightleftharpoons H^+ + C_2H_3O_2^-$

Carbonic acid $\quad H_2CO_3 \rightleftharpoons 2H^+ + CO_3^-$

Such equations are only a chemical shorthand demonstrating the manner and direction of association. In reality, a strong acid such as HCl only dissociates *in solution* with water. The reaction that actually occurs is:

$$HCl + H_2O \rightleftharpoons H_3O^+ + Cl^-$$

Figure 14-11
Sulfuric acid is used in the manufacture of a great many products, including fertilizers and paper and pulp. The acid wastes sometimes enter water sources near factories and can be a serious source of pollution.

The polar molecules of water tend to attract the hydrogen atoms, which break away from the chloride ion and form hydrogen bonds to oxygen. What results is a positively charged water molecule called a **hydronium ion**, which looks like this:

$$2H_2O \xrightarrow{\text{electrolysis}} 2H_2 + O_2$$

When acid enters a solution there is always an increase of hydronium ions, simply indicated as H^+.

THE PH SCALE The dissociation of pure water has been measured. The concentration of H^+ produced in this manner equals 10^{-7} molar; the concentrations of hydroxyl ions (OH^-) also equals 10^{-7} molar. When acid is introduced into solution with water, the concentration of H^+ increases and the concentration of OH^- decreases. When the added acid is a strong one, such as hydrochloric acid (HCl), there will be an even greater increase in the number of hydronium ions because strong acids have a greater tendency to donate hydrogen ions. In such a case, the concentration of H^+ may rise to 10^{-1} molar, while the concentration of OH^- drops to 10^{-13} molar.

An acid has a high concentration of hydronium ions; a base has a high concentration of hydroxyl ions.

This phenomenon lays the basis for an ingenious method of determining the number of hydronium ions present in a solution, and consequently the strength of various acids. This method simplifies the molar concentrations into a scale of simple numbers from one to fourteen; the scale is called the **pH scale**. Water, with an H^+ concentration of 10^{-7} molar, has a pH of 7, and this is considered

Table 14.1
The pH scale

pH	Condition	Example and pH rating
0	Very acidic	Gastric juice (0.9)
1		
2		Orange juice (2.6–4.4)
3		Vinegar (3.0)
4	Increasingly acidic (H^+ ions stronger)	Tomato juice (4.3)
5		Urine (4.8–7.5)
6		Milk (6.6–6.9)
7	Neutral	Water (7.0) Tears (7.4)
8		Pancreatic juice (7.5–8.0) Baking soda (8.4)
9	Increasingly alkaline (OH^- ions stronger)	
10		Milk of magnesia (10.0)
11		Ammonia (11.9)
12		
13		Sodium hydroxide (13.0)
14	Very alkaline	

the neutral point of the scale. A pH number lower than 7 denotes an acid, and the lower the number the stronger the acid. A solution with a pH above 7 has more hydroxyl ions than hydronium ions.

The magnitude of pH is measured by dipping a strip of specially treated litmus pH paper into the solution. Different amounts of H^+ react with the paper to produce different colors. Each color signifies a certain pH, a pinkish red color indicates an ion pH, whereas a blue indicates a high pH.

BASES Compounds consisting of a metal and a hydroxyl group (OH) are known as **bases**; in solution, they show a pH over 7. Common bases include potassium hydroxide (lye), sodium hydroxide (caustic soda), and ammonium hydroxide. The following equations show some strong and weak bases. For the strong bases the arrows point in the direction of dissociation.

Potassium hydroxide	$KOH \rightleftharpoons K^+ + OH^-$
Sodium hydroxide	$NaOH \rightleftharpoons Na^+ + OH^-$
Calcium hydroxide (lime)	$Ca(OH)_2 \rightleftharpoons Ca^{++} + 2OH^-$
Ammonium hydroxide	$NH_4OH \rightleftharpoons NH_4^+ + OH^-$

Strong bases have properties similar to those of strong acids, except that they have a bitter taste instead of sour. But neither should ever be tasted, as many strong acids and bases are extremely poisonous! The reason strong acids and bases cause burns is that they are so reactive; they will react chemically with the water molecules in various materials, including skin, having much the same effect as steam or a flame.

The strength of a base is determined by its ability to conduct an electric current.

THE BRONSTED-LOWRY THEORY Although acids and bases can be operationally defined by their OH rating, a broader definition of acids and bases has been proposed by Johannes Bronsted and Thomas Lowry. Called the **Bronsted-Lowry theory** of acids and bases, it defines an acid as any compound that donates hydrogen ions in solution; a base is any compound that accepts hydrogen ions. This theory revolutionized the understanding of acids and bases and their behavior.

Water molecules both donate and accept hydrogen ions. Thus, according to the Bronsted-Lowry theory, water is both an acid and a base. Similarly, when a strong acid such as HCl donates hydroxyl ions, it conforms to the Bronsted-Lowry definition. According to the latter, however, the chloride ion is a base; Cl^- is hydrogen ion acceptor in the reverse reaction:

the forward reaction	$HCl^- \longrightarrow H^+ + Cl^-$
the reverse reaction	$H^+ + Cl^- \longrightarrow HCl$

CONJUGATE ACIDS AND BASES All acid-base reactions are **reversible**; that is, they can proceed both ways. In the case of HCl, the long arrow indicates that the forward reaction proceeds more rapidly and completely than the reverse reaction. The latter is represented by a short arrow. This is because HCl is a strong acid and therefore has difficulty holding onto its hydrogen atoms. Conversely, Cl^- is a weak base because it has a weak attraction for hydrogen ions. The negative ion produced by the dissociation of an acid is called its **conjugate base**. A strong acid always has a weak conjugate base.

Weak acids, on the other hand, always have strong conjugate bases. In the case of carbonic acid for example, the reverse reaction

$$2H^+ + CO_3^- \rightleftharpoons H_2CO_3$$

proceeds much more rapidly than the forward reaction. This is because the hydrogen atoms are held relatively tightly by the CO_3^- ion in H_2CO_3. Consequently, according to the Bronsted-Lowry theory, H_2CO_3 is a weak conjugate acid and CO_3^- ion is a strong conjugate base. Conjugate acids and bases always exist as interrelated pairs.

SALTS   The final class of electrolytes are salts. If HCl is added to a solution of sodium hydroxide (NaOH), the solution becomes increasingly less basic and finally neutral. The NaOH, a base, has been neutralized by the HCl, an acid. If the water evaporates, all that is left is table salt, NaCl. What has occurred to the hydronium and hydroxyl ions, and why has a salt precipitated?

When the acid is introduced into the water, H^+ is released. Similarly, the base has dissociated into Na^+ and OH^-. Consequently, as fast as H^+ ions are released into the solution, they unite with the OH^- ions from the base to produce water molecules. As soon as all the OH^- ions previously free in solution have combined with the H^+ ions being introduced to form water, the solution turns neutral. Without free H^+ or OH^- ions there are no acidic or basic properties. The pH is 7, or neutral.

Meanwhile, the Na^+ ions react with the Cl^- ions to form the ionic compound NaCl. The reaction proceeds as follows:

$$H^+Cl^- + Na^+OH^- \longrightarrow H_2O + Na^+ + Cl^- \longrightarrow H_2O + NaCl$$

Compounds formed through the combination of a metal donated by a base and a nonmetal donated by an acid, are called salts. Any salt can be produced by neutralizing a base with the correct acid.

Glossary   An **acid** is an electrolyte composed of hydrogen and one or more nonmetals.

A **base** is an electrolyte composed of hydroxide (OH) and one or more metals or other positively charged ions.

A **body-centered cubic** is an ionic structure in which a cation is formed at the center of a cube whose corners are anions, and which forms a repeating unit in an ionic crystal.

The **Bronsted-Lowry theory** describes an acid as any substance which donates hydrogen ions upon dissociation, and a base as any hydrogen ion acceptor.

A **closest packed structure** is a configuration which metal atoms are assumed to form in metallic solids. In such a structure, each atom is surrounded by twelve neighbors in a form similar to stacked cannonballs or ball bearings in a sack.

A **conjugate base** is the nonmetal ion of an acid formed during dissociations of the acid.

A **covalent crystal** is composed of atoms covalently bonded into network solids, or sheets held together by Van der Waals forces.

A **crystal lattice** is the basic structure of a solid and is formed through the chemical bonding of atoms in repeated patterns.

Dissociation is the separation of an ionic solid into free ions upon entering solution.

An **electrolyte** is any ionic solid which dissociates in solution; the solution will then conduct an electric current.

A **face-centered cubic** is an ionic structure in which cations and anions alternate at the corners and center faces of a repeating cubic pattern.

An **ionic crystal** is a covalently-bonded structure in which every atom is rigidly bonded to its adjacent atoms, usually in a tetrahedral pattern.

Metallic solids are composed of metal atoms. The latter are believed to form a dynamic structure in which ions are held together by the shared valence electrons of the entire solid. Metallic solids are characterized by ductility, malleability, electrical conduction, and heat conduction.

A **mixture** is the common term for a substance consisting of two or more elements or compounds in varying proportions. A solution is a mixture that is homogeneous (has constant properties) down to molecular-size particles.

A **molecular crystal** is composed of discrete molecules held together by weak attractive forces such as hydrogen bonds and van der Waals forces.

The **pH scale** is a logarithmic formula used to measure the acidity of a solution.

A **salt** is the compound that results from the combination of a metal ion and a nonmetal ion upon the neutralization of a base with acid.

The **sea of electrons model** accounts for the unique properties of metals by postulating the existence of free valence electrons dispersed through the metal. These valence electrons collectively bond the positively charged metal nuclei.

Silicate minerals are a major class of minerals composed of covalently bonded chains, sheets or networks held together by or containing ionic bonds. Silicate minerals include feldspar, asbestos and mica.

Solubility is the maximum amount of a solute that can be dissolved in a given solvent at a given temperature. Solubility is usually measured in grams/liter or moles/liter.

The solute is the substance dissolved in a liquid solvent.

A solution is a homogeneous mixture of two or more substances which may be in any state (liquid, gas or solid) and which do not share chemical bonds. The most important class of solutions are liquids.

A solvent refers to the liquid component of a liquid-solid solution or to that component of any solution which is present in the greatest amount.

Supersaturation occurs when the amount of solute in solution exceeds its solubility.

X-ray crystallography is a technique that determines the crystal structure of a solid by beaming x-rays through it and measuring their refracted image positions.

Exercises

1 If a wooden table top is composed of vibrating molecules, why does a book placed on its surface rest there without merging, sinking into, or passing through the table?

2 Which of the following classes of compounds would tend to have higher melting points and why? Diatomic covalent crystals (such as I_2), covalent network solids, ionic solids.

3 Account for the conflicting observation that covalent solids include exceedingly hard substances, such as diamonds, as well as soft solids, such as iodine.

4 Describe the basic chemical structure of the following minerals: asbestos, talc, silica.

5 Why is water considered a polar molecule? What occurs when water freezes into ice? Why is salt used to melt ice on a frozen driveway?

6 Components A, B, and C of a liquid solution have boiling points of 80°C, 100°C, and 115°C respectively. How would you set up a distillation to separate the three liquids? Describe a possible apparatus which would successfully accomplish this task.

7 How many grams of sulfur must be added to three liters of water to form: an 8.1% solution, a 0.15% solution, a 21.0% solution.

8 Write reactions for the dissociations of the following acids: HCl, carbonic acid, nitric acid.

9 Calculate the concentrations of H^+ at the following pH values: pH = 1, pH = 5, pH = 9, pH = 4.3.

10 Arrange the following solutions in order of their increasing acidity (so that the least acidic is at the left): pH = 5, pOH = 8, $[H^+] = 10^{-2}$, $[OH^-] = 10^{-12}$.

11 If a liter of 0.1M HCl has a pH of 1, what will be the new pH after addition of one liter of 0.2M NaOH? (Assume complete dissociation.)

12 Predict the color of a red litmus paper indicator for the following solutions: 1 OM NaOH, 0.2 M HCl, pH = 8, $[OH^-] = 10^{-8}$, $[H_3O^+] = 10^{-2}$.

13 What are the conjugate bases of the following acids? Hydrochloric acid, CH_3CH_2COOH, acetic acid, carbonic acid.

14 What salts are produced by combining the following acids and bases? HCl, KOH; NH_4OH, HCl; $Ca(OH)_2$, H_2CO_3.

15 Write reactions for the formation of the following salts. NaCl, KCl, $NaC_2H_3O_2$, KNO_3.

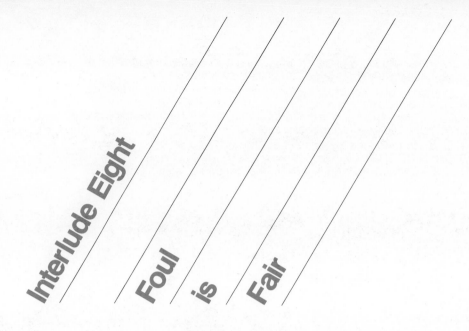

Interlude Eight

Foul is Fair

Fair is foul, and foul is fair:
Hover through the fog and filthy air.

Witches' Chorus from *Macbeth*

Everybody knows auto exhaust fumes can kill. We have all read newspaper accounts of suicides who ran the car engine in a closed garage, and anyone who ever drove through a long tunnel in rush-hour traffic will tell you he came out feeling a little less alive. But it is much harder to determine the cumulative hazard arising from continual breathing of fume-thickened urban air, along with an occasional lungful of undiluted bus exhaust. This was the subject of a recent study by the National Academy of Sciences, which found that in the United States auto pollution alone is directly responsible for 4000 deaths and

4,000,000 lost workdays annually. Furthermore, it turns out that about 20 percent of us—the very young, the aged, people with respiratory or cardiac conditions—are especially vulnerable to air pollution.

What causes the internal combustion engine to pollute? Gasoline is a mixture of volatile hydrocarbons such as pentane (C_5H_{12}) and should burn cleanly to yield only carbon dioxide and water vapor. The chemical reaction for the complete combustion of pentane is:

$$8O_2 + C_5H_{12} \longrightarrow 5CO_2 + 6H_2O$$

The end products, the same as in animal and plant respiration, are normally harmless to life. When gasoline vapor is mixed with air and exploded in a car's engine, though, the results are very different as we can see from the following

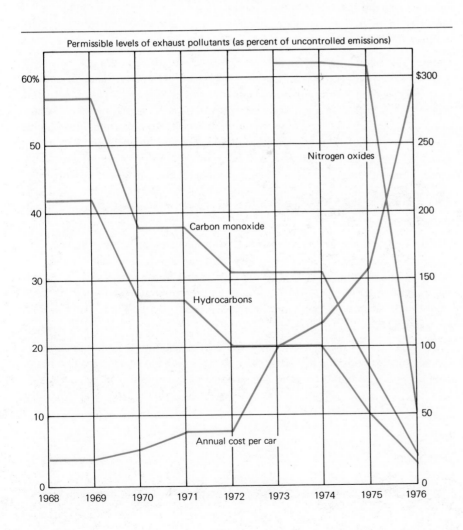

Permissible levels of exhaust pollutants (as percent of uncontrolled emissions)

rather complicated equation:

$$320N_2 + 80O_2 + 10C_5H_{12} \longrightarrow 40CO_2 + 54H_2O + 319N_2 + 8O_2$$
$$+ 2C_2H_6 + 6CO + 2NO_2$$

Here the combustion is incomplete and three pollutants are produced: C_2H_6, or ethane, an unburned hydrocarbon; CO, the poison gas carbon monoxide; and NO_2, nitrogen dioxide gas, also poisonous. The unburned hydrocarbon and carbon monoxide are produced because the mixture of air and gasoline needed for good engine performance (about 15 parts of air for each part gas) is too rich to allow all the fuel in a cylinder to be burned before the exhaust cycle starts. The nitrogen oxide is produced because the nitrogen in air, normally inert, reacts with oxygen at the high engine temperatures and pressures associated with the explosion in the cylinder. The exhaust may also contain pollutants resulting from impurities or additives in the gasoline—chiefly lead compounds emitted as tiny solid particles, a kind of pollution called particulates.

Certain obvious things can be done to reduce the emissions. We can adjust the engines to run on gasoline without lead additives. We can put smaller engines (less horsepower) in our cars, so the total amount of exhaust per car is reduced. We can fiddle with the tuning and modify the design of the engine itself: make it run on a leaner fuel mixture; lower the operating temperature. All of these things have been done, but there is a catch: after a point, whatever is done to further reduce CO and HC production results in formation of more NO_2, and vice versa. Can a new and cleaner type of engine be found? Research and development on this problem are now being conducted by the Environmental Protection Agency's Alternative Automotive Power Systems project (AAPS) and the automobile manufacturers.

And what of the immediate future? A short-term solution is the catalytic converter, a device designed to convert the unwanted components of exhaust into harmless substances by exposing them to appropriate catalysts. This system was approved by the EPA in 1973 and is included on all General Motors 1975 model cars, and on 90 percent of the 1975s produced by Ford, Chrysler, and American Motors.

Since two of the major pollutants (HC and CO) result from incomplete combustion, all that is needed is to induce them to react with the unused oxygen in the exhaust. For our sample pentane reaction this would mean:

$$2C_2H_6 + 7O_2 \longrightarrow 6H_2O + 4CO_2$$

and

$$6CO + 3O_2 \longrightarrow 6CO_2$$

Both these reactions proceed rapidly on a platinum surface. Since both the HC and CO are oxidized in these reactions, they may be eliminated simultaneously in a so-called oxidizing converter.

After that, only NO_2 remains to pollute the exhaust in our sample reaction. This can be broken down by the reaction:

$$2NO_2 \longrightarrow N_2 + 2O_2$$

In this process nitrogen is reduced, again using a platinum surface as catalyst. Because this reaction is optimized at a lower temperature than that for the HC and CO oxidation, a separate reducing converter is used to break up the nitrogen oxides. Even with this dual converter system, some of the NO_2, HC, and CO will get through the system and out the tailpipe. Although not perfect, the system is a substantial improvement.

The chief problem in manufacturing an oxidizing converter is the high cost of platinum, now about $120 an ounce and rising. How can a large enough platinum surface be provided, in a small package, so that most of the pollutants will

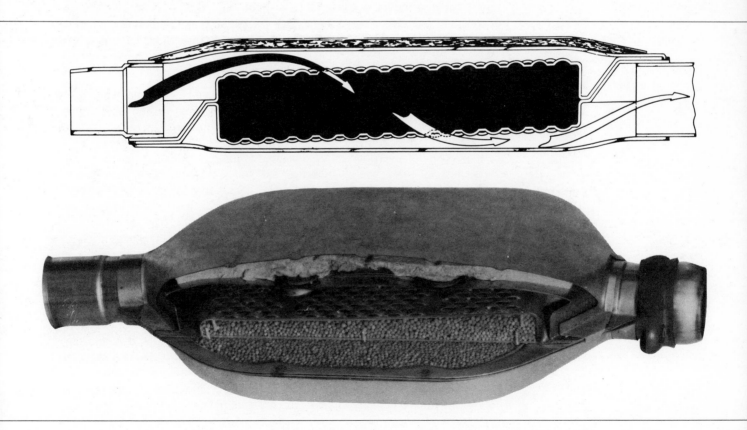

be removed in the short time it takes for the exhaust gases to pass out the exhaust system? Only by making a structure with a very intricate, convoluted surface and then spreading the expensive metal in the thinnest possible layer over the surface. G.M. uses little aluminum oxide pellets, $\frac{1}{8}$" in diameter (about the size of BBs), with a roughed-up surface coated with platinum and palladium; the amount of platinum and palladium used in one of these converters is less than $\frac{1}{10}$ of a troy ounce (about 3 grams). The advantage of using beads is that they form a loose pile, leaving air space for the exhaust gases to pass through. An alternative model used on Fords and Chryslers uses a rigid, honeycomblike structure, with built-in flow passages for the exhaust.

Since the reaction occurring inside the converter is a type of combustion, heat will be liberated. This raises the internal temperature of the devices to 1700° Fahrenheit, so high that expensive stainless steel must be used for the converter shell, and special insulation must be built in. In 1975, the total cost of such a device was about $150, of which about $15 goes for platinum and palladium. It should last about 50,000 miles or 5 years.

What about the reducing converter? The original plan was to pass the exhaust first through a reducing converter to eliminate NO_2, then through an oxidizing converter to remove HC and CO. In this way the oxygen liberated in the breakdown of the nitrogen oxides would be available to react with the HC and CO in the second converter. Unfortunately, the presently available versions of the reduction converter are too unreliable for use in '75 model cars.

The widespread installation of the catalytic converters will also decrease lead particulate pollution. It happens that the lead additives in normal gasoline "poison" the catalyst of the converter; two tankfuls of leaded gas will reduce the efficiency of the catalytic converter by 70 percent. For this reason, cars equipped with these devices are designed to run on nonleaded gasoline only, and a special narrow fuel pump nozzle and small gas tank opening are used to prevent accidental use of leaded gas. Now that lead-free gas is widely available and most new cars will be required to use it, the amount of lead we breathe will probably be gradually reduced to safe levels.

Potential problems with the catalytic converter have been uncovered in preliminary tests, but only large scale production and use will determine if any of these are serious drawbacks. The Academy's Committee on Motor Vehicle Emissions points out some problems: the catalysts are too cold to operate efficiently while the engine is warming up; tend to overheat while coasting downhill; are poisoned by sulfur impurities in gasoline; are subject to severe thermal stress in stop-and-go traffic; and tend to deteriorate if sufficiently overheated. Other investigators fear that small particles of platinum and palladium may emerge from the tailpipe and lodge in human lungs, with unknown consequences. Finally, there are reports that sulfur impurities in gasoline are converted into sulfuric acid mist by the catalytic action.

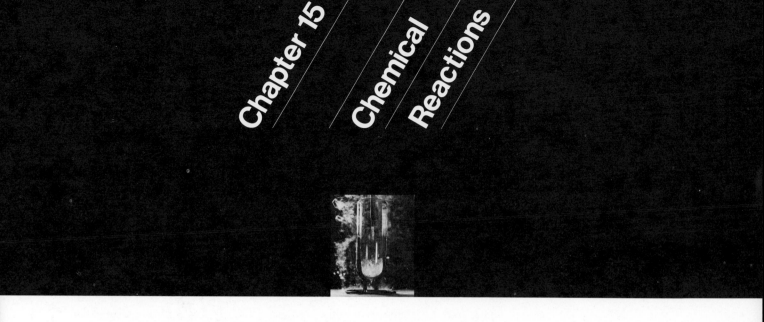

A **chemical reaction** is an event that involves the creation, alteration, or destruction of chemical bonds. The substances involved—they may be either elements or compounds—are altered by the reaction and take on new physical and chemical properties. The most significant aspect of a chemical reaction is that it involves changes in the energy levels of the different reacting materials.

Here is a simple example of a reaction:

nitric oxide + oxygen $\longrightarrow$ nitrogen dioxide

or

$2NO + O_2 \longrightarrow 2NO_2$

or, in generalized form

$$A + B \longrightarrow C$$

In this reaction, A and B are called the **reactants,** or starting materials. C is called the **product,** or the result of the reaction. In the course of the chemical changes involved in a reaction, the reactants disappear and the products come into being.

Most reactions, both in the laboratory and in nature, are not this simple. Reactions frequently occur in several steps, with intermediate molecules which then break up into other compounds. A generalized example of a more complicated reaction is:

$$A + B \longrightarrow C[D] \longrightarrow C + [E] \longrightarrow C + F + G$$

D and E are intermediate molecules formed in reaction and then broken down again as the reaction proceeds, yielding the end products F and G. The overall reaction is written:

$$A + B \longrightarrow C + F + G$$

Chemical reactions involve changes in the energy levels of the reacting materials.

Chemical Energy

The concept of chemical energy is central to the analysis of reactions. Why are molecules said to contain energy? How is this energy stored and how is it used in chemical reactions?

Bond Energy

Due to the bonding arrangement, the electrically charged particles that make up the atoms and molecules take differing energetic states. A state of low stable energy is often characteristic, but in some arrangements a higher potential may be stable. A pebble perched on the top of a cliff has potential energy relative to the ground; a molecular bond permits a certain amount of electrical energy to be stored in the structure of the molecule. This is the molecule's potential energy, ready to be released by the right reaction. The potential energy of the pebble on the cliff may be converted to kinetic energy if it is kicked off the cliff and falls to the ground. In much the same way, the chemical potential energy of gasoline molecules may be released by the process of combustion with oxygen. The chemical energy is thereby converted to heat energy, which may in turn be converted to the kinetic energy of the engine piston that drives the wheels of an automobile.

The chemical bonds of molecules contain potential energy which may be released in a chemical reaction.

Just as the gravitational potential energy of the pebble is stored by its position relative to the ground, so the potential energy of a molecule, which could be released in a reaction, is stored in the attractive energy relations of the chemical bond. In both situations, there is a natural tendency to proceed downward toward a lower potential energy and, given the physical or chemical opportunity, that process will happen. Potential energy is then converted to heat and/or kinetic energy and other lower levels of potential energy.

The lower the potential energy, the more stable are the objects or compounds. Imagine, for example, what happens to a marble on the sloping inside of a cereal bowl. From a position of high potential energy the marble will roll around and around the bowl in a noisy discharge of kinetic energy, until it comes to rest on the bottom, where it has maximum stability. A mark of its great stability is the ease with which it returns to a bottom position if the bowl is moved again. Contrast this stable balance with the delicate balance or unstable equilibrium of a house of cards or a molecule of high explosive.

When molecules with a relatively high and unstable level of energy react by rolling down the potential energy hill to form more stable compounds, energy must be released as heat while the atoms form a new, more stable bond. The **heat of reaction** is the difference in heat content or potential energy between the original reactants and the final products of a chemical reaction. Heat of reaction, usually written as ΔH, is a measure of the heat absorbed or released by a reaction system. The equation for heat of reaction is:

$$H = H_{\text{products}} - H_{\text{reactants}}$$

where H stands for heat.

Once a new bond is formed it contains potential energy of its own, equal to the amount of energy needed for the reaction that formed the bond. In order to break the bond apart, additional energy must be added to it, in the form of heat, or the electrical energy of radiation, or the kinetic energy of bombarding elemental particles. So energy must be added to break bonds, and energy is liberated when new bonds are formed. This can be seen in a common reaction:

$$HCl + NaOH \longrightarrow NaCl + H_2O$$

[hydrochloric acid + sodium hydroxide $\longrightarrow$ sodium chloride + water]

Energy must be put in to break the bonds between H and Cl and Na and OH. But energy is released again when the new bonds are formed between Na and Cl, and H and O. Depending on the individual bond energies involved (which of course are the stored energies that reflect the amount of energy it took to form these bonds initially) the net energy difference in any reaction may be positive or negative.

An overall reaction can be represented as a change in the potential energy, written as ΔE. The equation for potential energy is:

$$\Delta E = E_{\text{products}} - E_{\text{reactants}}$$
$$= \text{sum of bonds formed} - \text{sum of bonds broken}.$$

If more energy is released in the formation of the product bonds than was needed to break the bonds of the reactants, the net energy difference will be negative and the reaction should proceed easily and even spontaneously downward to the state of lower energy (and higher stability) represented by the

Figure 15-1
The Apollo 8 rocket lifts off on December 21, 1968, carrying astronauts Frank Borman, James Lovell, and William Anders on the first manned flight to the moon. The engine of this Saturn V rocket converts chemical energy into heat energy and then into kinetic energy: kerosene and liquid oxygen fuels are shot into the engine's combustion zone; this reaction yields CO_2 and H_2O along with intense heat which thrusts the exhaust gases downward and lifts the rocket off the launching pad.

Exothermic reactions give off energy, usually in the form of heat; endothermic reactions require the addition of energy to proceed.

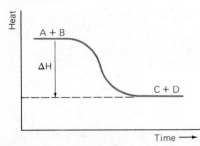

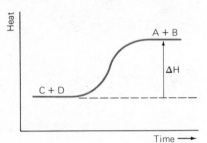

Figure 15-2
The top diagram shows an exothermic reaction; the reactants (A + B) contain more potential energy than the products (C + D) so the reaction produces heat (ΔH), which is released as the products form. The bottom diagram shows an endothermic reaction; here the reactants (C + D) contain less potential energy than the products (A + B), and the reaction requires heat in order to proceed.

Chemical Thermodynamics

products. Such reactions yield excess energy, most often in the form of heat. They are called **exothermic** reactions (from the Greek word for heat, *therme*, and the prefix *ex*, meaning out, as in expel). An example of the exothermic reaction is the formation of water from the elements hydrogen and oxygen, which releases 68.3 kcal of heat energy for each mole of water formed.

$$H_2 + \tfrac{1}{2} O_2 \longrightarrow H_2O + 68.3 \text{ kcal}$$

Water (H_2O) is therefore in a lower potential energy state than the two reactants H_2 and O_2, and so it is more stable. This is the reason there is no spontaneous breakdown of water into its constituent elements; in the language of chemical energy, this would be energetically unfavorable.

A reaction that produces less stable products with higher bond energies requires an energy input, such as heat, in order to proceed. Such a reaction is called an **endothermic reaction** (the prefix *endo* means into). An endothermic reaction proceeds uphill from a lower to a higher potential energy state. There is a natural resistance to this type of reaction and it will not happen spontaneously, any more than a ball will roll uphill. Energy must be put into the system in order to lift the molecules to a higher energy level. For example, water can be broken down into hydrogen and oxygen in the laboratory, using the process of electrolysis, in which an electric current is passed through the water. The electricity provides the needed energy for this endothermic reaction:

$$2H_2O \xdashrightarrow{\text{electrolysis}} 2H_2 + O_2$$

The study of energy transformations is called **thermodynamics**. The field of chemical or molecular thermodynamics looks at the relative energies of molecules and their bonding relations. Thus chemists are able to predict which reactions will proceed readily and which ones will not. There is such a strong natural tendency for all molecules to go toward a position of lower energy that some exothermic reactions are violent explosions. That is the case when sulfur and zinc powder combine with oxygen; an explosion takes place almost immediately.

Some molecules react at room temperature and have enough kinetic energy for bonds to break during ordinary collision. But many exothermic reactions are not self-starting. For example, coal may be stored indefinitely in air without great chemical change, and yet when ignited it reacts vigorously to give off a large amount of heat; the spark provides just the amount of heat needed to start the reaction going.

Although the collision of the molecules of reacting compounds is an obvious prerequisite for reaction, most collisions do not result in reaction. Each colliding molecule must first be brought to a higher energy level, or **transition** state, simply in order to start the reaction process that will result finally in the lower energy of the products. This additional energy required to start the reaction is called the activation energy. **Activation energy** is defined as the minimum energy that is required to make a reaction begin. After activation, the reaction can normally continue without further energy being added. A match is needed to start a fire but not to keep it burning. The energy barrier is usually overcome by the addition of heat, which raises a greater number of molecules to the higher kinetic energy level. The higher the activation energy, the more heat required to initiate reaction.

The notion of activation energy is made clearer by the analogy depicted in Figure 15-3. Here the potential energy of a boy and his sled will be converted to the kinetic energy of motion if he sleds down to the valley below. Even though the boy already has a certain amount of potential energy relative to the valley, the energy transformation cannot take place unless he expends an initial kinetic energy to climb the hill between his present position and the valley below. Once he has invested a certain amount of kinetic energy in getting over the hill, he will more than make up for it as he converts potential into kinetic energy by sledding down to the bottom.

The energy situation in a chemical reaction is similar. Activation energy is needed to achieve bond breaking. The molecular state at the height of activation energy (corresponding to the boy at the top of the hill) is called the **activated complex**. The molecule in this state will have some of its bonds broken and some new bonds formed. Many exothermic reactions require the initial boost of activation energy in order to start a reaction that later becomes self-sustaining due to the heat released. This, of course, is what happens when coal is ignited.

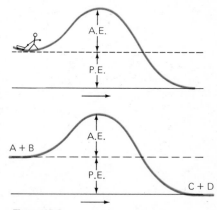

Figure 15-3
In order to realize his potential energy, the boy with the sled must first climb the hill, which represents activation energy. In the same way, the chemical reaction in the bottom diagram requires the addition of activation energy before it can proceed.

Many exothermic reactions require additional energy to begin the reaction process.

Although some reactions, such as the explosion of TNT, occur instantaneously, other reactions may take years to complete; rusting, or the slow oxidation of iron, is an example. What accounts for such great differences in reaction rate?

The study of reaction rates is known as **reaction kinetics.** Chemists have found that the rate at which a reaction proceeds will vary with four principal factors:

1 temperature
2 concentrations of reactants
3 surface areas of reacting solids
4 presence of catalysts

The rate of a reaction is measured in terms of the number of moles of a reactant consumed or the number of moles of product formed per unit volume in a unit of time.

Rate of Reaction

The rate of a reaction is affected by the temperature, concentration, and amount of surface areas of the reactants.

Temperature

For two atoms, ions, or molecules to react, they must come close enough for their mutual forces to affect one another. Therefore, they must achieve a sufficiently high level of kinetic energy to overcome any activation energy barrier. Experimental results show that the higher the activation energy, the slower the reaction. This pattern may be understood in terms of collision probabilities. If the activation energy is very high, the probability is small that two colliding molecules will have sufficient energy to react, so the reaction will be slow, a few molecules at a time. An increase in temperature will cause the molecules of reactants to have increased energy content; the number of collisions between molecules will also increase as temperature rises, just because molecular speeds are increased. So a higher temperature speeds up a reaction in two ways: it increases the total number of collisions per unit time; and it increases the percentage of collisions that lead to reaction.

As a working rule, it has been established that for every 10℃ increase in temperature, the rate of reaction will approximately double. This may seem like a small increase in temperature to produce such a large effect; why should an increase from 260° to 270° double a reaction rate? One clue lies in the fact that the reaction process generally releases heat. When a small amount of external heat is applied and the rate of reaction starts to go up, it leads to an increase in the amount of heat released internally, as potential energy is transformed into

Figure 15-4
Chemical reactions proceed at varying rates of speed. Explosions, such as this TNT blast in a canal-building project, have a very high rate of reaction. Rusting, on the other hand, proceeds so slowly that we cannot see it happening.

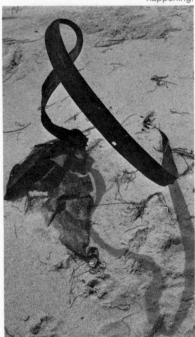

kinetic energy. This increase in internally generated heat then leads to another rise in the reaction heat, causing more heat to be released internally, causing another rise in the reaction rate, and so on in a cycle of self-sustaining increase. This is why so many exothermic reactions proceed with explosive speed once a small amount of external heat—like the spark of a match—is applied.

Since the rate of reaction is directly proportional to the number of successful collisions between molecules in a given time, it would make sense that if more molecules are colliding, more successful collisions will occur in that time. For this reason, **concentration,** or the number of molecules formed in a given volume of the reactants, is an important influence on reaction speed. This relationship is stated in the **law of mass action**: *For a given set of reactants at constant temperature, reaction rate is generally proportional to the concentration of the reactants.* It can also be expressed in the form of an equation:

rate $= K[A][B]...$

where K is an algebraic rate constant, and $[A]$ and $[B]$ are concentrations of the reactants. To demonstrate the effect of concentration on reaction rate, we can insert hypothetical figures into the equation. Suppose that K is equal to 10 and we have a solution composed of 30 parts of A and 20 parts of B.

rate $= 10 \times 30 \times 20 = 6000$

Now increase the concentrations of A and B by just 10 percent, making 33 parts of A and 22 parts of B

rate $= 10 \times 33 \times 22 = 7260$

Clearly, if the concentration of either reactant is increased, the rate at which the reaction proceeds will also increase.

The surface area of solid reactants is an important factor influencing reaction rate. Have you ever wondered why icing recipes call for powdered rather than granulated sugar to be mixed with the liquid? Powdered sugar dissolves more quickly because its surface area is greater. There are more sugar molecules in contact with the liquid, and therefore the chance of successful collisions is greater. Stirring the mixture will also increase the rate of dissolution because more molecules will be exposed for reaction.

In simple terms, any reaction depends on successful collision between molecules of sufficient energy. (By successful, chemists mean that the interacting molecules strike at just the right place and just the right time to cause a break in bonds.) We have seen how the temperature will increase molecular energy

Concentration

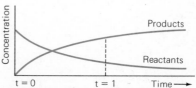

Figure 15-5
A change in the concentration of the reactants and the products occurs as a chemical reaction proceeds. At $t = 0$, before the reaction begins, there are no products; at $t = 1$, the concentration of reactants has decreased and the concentration of products has increased.

Surface Area

Catalysts

and speed of the molecules, and we have also seen how surface area and concentration affect the rate of reaction.

Still another influence—at first somewhat mysterious—comes into play. Experimental chemists have shown that when certain substances are added to reaction mixtures, the reaction is completed in a shorter time, although these substances are not themselves used up or changed in the reaction. For example, bottled hydrogen peroxide on the shelf slowly breaks down into oxygen and water:

$$2H_2O_2 \longrightarrow 2H_2O + O_2$$

If a trace of powdered platinum is added, this reaction is speeded up so much it may become explosive. Platinum and other substances known to hasten or retard chemical reactions are called **catalysts,** and their action—known as catalysis—plays a large role in all fields of chemistry.

Figure 15-6
Crude oil contains a complex mixture of hydrocarbons from which gasoline and other petroleum products are manufactured. The high-octane gasoline stored in these huge tanks is refined from crude oil by means of chemical reactions in which platinum and hydrogen are often used as catalysts.

Catalysts affect reaction rates only; they do not change potential energy differences between initial reactants and ultimate products. But catalysts do seem to lower the required activation energies and as a consequence increase the rate of reaction.

How do catalysts produce this effect? In some reactions, they physically—or electrically—bring reacting molecules closer to each other, so that higher energies are unnecessary for successful interactions.

The catalyst may sometimes work by forming an unstable intermediate compound with one or both of the reactants, allowing them to overcome energy barriers by structurally aligning the molecules in better orientation for collision. This is the mechanism of the action of biological enzymes, complex organic molecules that catalyze reactions in living organisms (enzymes will be discussed more fully in Chapter 17). The irregular shape of enzyme molecules allows for positions on their surfaces which hold the reacting molecules close together. Both types of reacting molecules and the enzyme fit each other like a lock and key, and the reactants are then oriented for successful interaction. So it is the physical configuration of the enzyme molecule that provides the special help to biological reactants.

In some cases, catalysts provide new and faster paths by which a reaction can proceed. For example, the reaction between cerium and thallium ions in solution proceeds very slowly:

$$2Ce^{+4} + Tl^+ \longrightarrow 2Ce^{+3} + Tl^{+3}$$

But the reaction may be speeded up if it is catalyzed by Mn^{++}. The manganese acts by providing new reaction paths, so the single slow reaction is replaced by three faster reactions:

$$Ce^{+4} + Mn^{++} \longrightarrow Ce^{+3} + Mn^{+3}$$
$$Ce^{+4} + Mn^{+3} \longrightarrow Ce^{+3} + Mn^{+4}$$
$$Mn^{+4} + Tl^+ \longrightarrow Tl^{+3} + Mn^{++}$$

As can be seen, the addition of these new paths creates more of the new products Ce^{+3} and Tl^{+3}, so the total reaction rate is increased. The overall reaction remains unchanged and potential energy differences between products and reactants remain the same; the catalytic ions of Mn^{++} are also unchanged.

Some catalysts are highly specific; they will accelerate certain reactions but leave others unchanged. Although there remains much to be discovered about the way certain catalysts work, the process of catalysis has, through an experimental process of trial and error, become very important to many kinds of chemical manufacturing. For example, the catalysts nickel, iron, and platinum are used to accelerate the process by which edible vegetable oils, such as Crisco, are obtained from cottonseed oil. The catalyst aluminum oxide is used

Catalysts speed up or slow down a chemical reaction without affecting the potential energy difference between the reactants and the products.

when water is added to ethylene to form alcohol. Other commonly used catalysts are manganese, vanadium, iron, and oxides of aluminum and chromium.

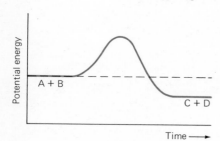

Potential energy

A + B

C + D

Time →

Figure 15-7
In this simple energy diagram, which illustrates the energy exchange in the reaction A + B = C + D, the highest point of the curve represents the activation energy needed to begin the reaction.

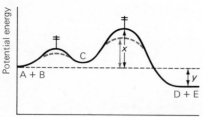

Potential energy

C

A + B

x

y

D + E

Time →

Figure 15-8
This energy diagram is more complicated than Figure 15-7 because the reaction includes an intermediary step. Line x represents the activation energy needed to complete the reaction. Line y indicates the difference between the energy potential of the reactants and the energy potential of the products. The dotted red lines show the effect of a catalyst on this reaction.

The problems associated with activation energy can be more easily understood with the help of an energy diagram. The diagram (see Figure 15-7) is so arranged that the movement from left to right visually represents the reaction mechanism—the continuous change in geometry as the two molecules collide, forcing apart the atoms whose bonds break and bringing together those which are forming new bonds. The potential energy increases until the transition state is reached and then the energy falls off as the transition state complex starts to regroup into the stabler forms of the product molecules.

Reactions which involve intermediates may also be diagrammed on these energy charts. Figure 15-8 is a representation of the reaction:

$$A + B \longrightarrow C \longrightarrow D + E$$

Each step of the reaction involves passage through a transition state.

The highest point on the curve represents the transition state for the rate-controlling step of the reaction: the overall speed of the reaction is only as fast as this step. The greater the energy difference between the reactants at the beginning of the reaction and the reactants of the highest point of the curve (shown in Figure 15-8 by distance X), the slower the reaction. The low points on the curve represent real and observable molecules—the reactants, intermediates, and products. The transition state complexes (high points) do not have full or normal bonds, as some are partially broken and others are partially formed. Their ephemeral existence cannot usually be observed.

Even if the diagram shows more stable products than reactants, the reaction will proceed very slowly if the energy barrier (activation energy) is high and will only proceed at a faster rate if the temperature is increased or a catalyst is used. The red line on Figure 15-8 shows what happens when an appropriate catalyst is added.

Chemical Equilibrium

Almost all chemical reactions can be made reversible since, under the proper conditions, the products of a reaction may interact to produce the original starting materials. For example, ammonia is produced by the reaction of nitrogen gas and hydrogen gas:

$$N_2 + 3H_2 \longrightarrow 2NH_3$$

The reaction is readily reversible when ammonia decomposes:

$$2NH_3 \longrightarrow N_2 + 3H_2$$

In chemical reaction, double arrows are used to show that a reaction is reversible:

$N_2 + 3H_2 \rightleftarrows 2NH_3$

or, in generalized form:

$A + B \leftrightarrow C$

Most reactions exhibit some degree of reversibility, and many chemical changes are known to have forward and reverse reactions occurring simultaneously. This may be theoretically true, but it is often pragmatically impossible, especially if some of the end product escapes from the system, or if the reverse reactions require a catalyst not originally present. For example, the only way to reverse the reaction that takes place when a log is burned in a fireplace is to grow a new tree!

Let us follow the process of forward and reverse reactions in the formation and decomposition of ammonia. As molecules of N_2 and H_2 collide, molecules of NH_3 are formed in increasing numbers. According to the law of mass action, the rate of this forward reaction would be:

forward rate $= K[N_2][H_2]$

Suppose that the molar concentration of each of the reactants is 50, and the value of K is 10. At the very moment that N_2 and H_2 are brought into contact (time $t = 0$) the concentration of the reactants would be at their maximum, so the rate would be:

forward rate $(R_f) = 10 \times 50 \times 50 = 25000$

The formula for the reverse reaction follows the same law; it would be:

reverse rate $(R_r) = K[NH_3]$

At $t = 0$, there would be no molecules of ammonia yet formed, so its rate would be:

$R_r = 10 \times 0 = 0$

What happens next? One second later, at $t = 1$, the concentration of reactants will have dropped, since some of the N_2 and H_2 have been used up. The concentration of product will have increased, since molecules of NH_3 have been formed. Using hypothetical numbers in our formulas, we get:

$R_r = 10 \times 49 \times 47 = 22,930$
$R_r = 10 \times 1 \qquad = \qquad 10$

After a while, the concentration of N_2 and H_2 will steadily decrease, as more NH_3 is formed. There is less N_2 and H_2, so the rate of R_f will taper off. There is more NH_3 available to decompose into N_2 and H_2, so the rate of R_r will pick up.

Almost all chemical reactions can be made reversible; reversible reactions are indicated by double arrows in the equation.

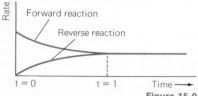

Rate

Forward reaction

Reverse reaction

t = 0   t = 1   Time →

Figure 15-9
When the forward action and the reverse
action in a reversible chemical reaction are
occurring at the same rate, it is said that the
reaction is in equilibrium.

The increase in R_r will produce more of the starting materials N_2 and H_2 and these will go on to react to form more NH_3 which in the meantime has been generally depleted by the reaction forming N_2 and H_2. Eventually N_2 and H_2 will be combining at exactly the same rate as NH_3 is decomposing. When the rates of these two reactions are the same, a state of balance or **chemical equilibrium** is reached.

Perhaps this process will be made clearer if we consider the analogy of a bathroom sink which is stoppered and filling up with water at a steady rate. When the water reaches a certain level near the top of the sink, it will begin to flow out of the safety pipe which is built into most sinks to prevent overflowing. If the rate water comes into the sink from the faucet is exactly equal to the rate it flows out of the safety pipe, then the level of water in the sink will remain the same.

The analogy can be brought one step further if we realize that the sink is in a state of dynamic equilibrium. The water level remains exactly the same but there is fresh water entering as the old water escapes. This is true as well of the reaction $N_2 + 3H_2 \rightarrow 2NH_3$. A state of balance is achieved between the two opposing processes and a dynamic equilibrium is maintained as the results of the forward reaction are constantly being undone by the reverse reaction, and vice versa. At the point of equilibrium, the rates of the forward and reverse reactions will be equal. And during any given period of time, the number of NH_3 molecules decomposing will be equal to the number of NH_3 molecules being formed. The number and concentration of NH_3 molecules will remain constant but the same N_2 and H_2 molecules will sometimes exist as uncombined molecules and sometimes will be combined to form NH_3. At all times the total number of uncombined N_2 molecules, uncombined H_2 molecules, and NH_3 molecules remains the same. The reaction is continuing during equilibrium but there is no net change. The *quantities* of reactants and products are not necessarily equal; it is the *rates* of the two reactions, forward and reverse, that are equal.

Equilibrium Constant

The conditions of concentration obtained at the time the equilibrium condition is reached are expressed by an equation for the equilibrium constant, K_{eq}. The equilibrium constant is a mathematical relationship between reactants and products. The molar concentrations of the reactants is constant, and that relationship is called K_{eq}.

At equilibrium, the rate of the forward reaction is equal to the rate of the reverse reaction:

$$R_f = R_r$$

or

$$K_f[A][B] = K[C]$$

or

$$\frac{K_f}{K_r} = \frac{[C]}{[A][B]} = K_{eq}$$

In working actual problems with this equation, the concentration [A][B] and [C] are expressed in numbers of moles.

The equilibrium constant K_{eq} has a specific numerical value for any given chemical reaction at a particular temperature, and the relationship shows that at equilibrium, the *ratio* of the concentrations of reactants and products will be constant even though the individual concentrations may be different. Because of this relationship, if one concentration is changed, there will have to be a compensation in the concentrations of the other substances. For example, if more N_2 is added to the reaction that produces ammonia, then the concentration of H_2 must decrease and the concentration of NH_3 will increase, in order for the ratio to remain the same. In general terms, then, if the concentration of one or more of the reactants or products is changed then the other concentrations must all change in such a way as to keep the ratio at its original value.

In chemical equilibrium, the rates of the forward and the reverse reactions are equal.

The equilibrium state will continue to exist indefinitely unless it is disturbed by certain external factors. What factors might disturb the equilibrium? We noted that if more of any reactant or product is added to an equilibrium mixture, a disturbance will be created. If, for example, more atoms of substance A are added to the reaction mixture, then the rate of the forward reaction will increase and more of product C will be formed. Product C will be produced faster than it can decompose and thus the concentration of C will increase. In general, changing concentration of one reactant affects the rate of the forward reaction; changing the concentration of one product affects the rate of the reverse reaction. Since the two opposite reactions in the equilibrium are affected differently by changing the concentration of one of the substances, then there will have to be a compensation in the other direction.

Le Chatelier's Principle

This principle was summed up by the French chemist H. L. Le Chatelier (1850–1936) in 1884. **Le Chatelier's principle** states that *if an equilibrium system, or set of reacting substances, is subjected to change, the system itself will shift in such a way as to relieve the disturbance and restore equilibrium.* A shift in the position of equilibrium always counteracts the change imposed on the system. This principle is really another way to state the concept of K_{eq}.

An illustration of Le Chatelier's principle can be seen with the Haber process, used in the industrial manufacture of ammonia. Nitrogen and hydrogen are fed into the system in continuous supply as the ammonia is removed from the system by liquefaction and other methods which do not affect the hydrogen or nitrogen. Withdrawal of the ammonia is the driving force behind the reaction, for as it is removed from the system, the equilibrium shifts to the right, thus

Figure 15-10
Ammonia manufactured in the Haber
process is a major source of the fixed
nitrogen found in chemical fertilizers.

producing more of the product. In chemical equilibria, then, if the concentration of one substance increases, the reaction that reduces the amount of that substance is favored. On the other hand if the concentration of a substance decreases, then the reaction that produces that substance is favored.

Thus, in the Haber process, if the amount of nitrogen were to be suddenly increased, the result would be that more ammonia is suddenly produced. Eventually, as the amount of ammonia builds up, the reverse reaction rate will increase but with different concentrations of the three substances involved. This is so for any reaction: $A + B \leftrightarrow C$

1 The addition of A or B will shift the equilibrium to the right, reducing the concentration of A or B.
2 The removal of C will shift the position of equilibrium to the right, producing more C.

CHANGES IN TEMPERATURE The law of conservation of energy requires that for any chemical equilibrium, if the forward reaction is exothermic, the reverse reaction must be endothermic. Although an increase in temperature will accelerate both reaction rates, the endothermic reaction rate will increase faster. Raising or lowering the temperature (adding or removing heat) in a reaction mixture can be analyzed much the same way as changes in concentration. In general, if the temperature of an exothermic reaction is increased, the position of equilibrium will shift to the left, absorbing heat and lowering the temperature. If we increase the temperature of an endothermic reaction, the position of equilibrium will shift to the right, absorbing heat and lowering the temperature.

What happens if we raise the temperature of the Haber process, an exothermic reaction? This addition of heat to the reaction system would shift the position of equilibrium to the left, favoring the endothermic decomposition of ammonia to hydrogen and nitrogen.

CHANGES IN PRESSURE   Gaseous reactions which take place in closed containers are very dependent on pressure. In a closed system at constant volume and temperature, the pressure will increase when the number of moles of gas increases. Conversely, the pressure will decrease when the number of moles of gas decreases. But pressure may be a cause as well as a result. Thus a change in pressure will produce a change in the equilibrium of a reaction. In our old familiar reaction:

$$N_2 + 3H_3 \longleftrightarrow 2\,NH_3$$

we have one mole of nitrogen gas, three moles of hydrogen gas and two moles of ammonia gas. Thus, there are four moles of gas on the left of the arrows and two moles on the right. What will happen to this reaction (assuming it takes place in a closed system) if the amount of pressure on all gases is increased? Since the number of moles of gas on the left side is greater than the number of moles of gas on the right side, any increase in pressure will disturb the equilibrium of the reaction. If the reaction moved to the left producing more N_2 and H_2, the disequilibrium would be even worse, because there are already more moles of gas on this side of the equation. According to Le Chatelier's principle, equilibrium will be restored when the reaction moves to the right, producing more moles of NH_3, since it is on that side that the lower pressure exists. Thus, an increase in pressure applied to the Haber reaction leads to the production of more ammonia.

Only gases in equilibrium are markedly affected by changes in pressure, because solids and liquids are nearly incompressible. In an equation that involves substances in different states of matter, it is the total number of moles of gas on each side of the arrow that is the determining factor.

Early chemists used the term **oxidation** to mean a chemical change in a compound that resulted in the *gain* of oxygen atoms or molecules. A companion term, **reduction**, referred to a chemical change in a compound that resulted in

A chemical system in equilibrium can be disturbed by changes in the concentration of the reactants or products, by changes in the temperature, or by changes in the pressure.

Oxidation and Reduction Reactions

the *loss* of oxygen. When iron burns in oxygen, for example, it oxidizes and forms iron oxide:

$$2Fe + O_2 + 2H_2O \longrightarrow 2Fe(OH)_2$$

Rusting is also an oxidation process, in which molecules of iron slowly combine with the oxygen in the air to form iron oxide.

Oxidation and reduction reactions are especially important in the field of metallurgy, or the technology of refining metals and preparing them for commercial use. Most metals do not occur in an uncombined state in nature but instead are to be found in metal oxides or ores. To separate out the metal, the metallic oxides must be reduced. This is usually carried out with some form of carbon, as in the reaction that extracts iron ore or hematite:

$$Fe_2O_3 + 3C \longrightarrow 2Fe + 3CO$$
hematite   coke   iron   carbon monoxide

The hematite and the coke are finely ground and then thoroughly mixed. The carbon is oxidized, and the hematite is reduced. Since hematite supplies the oxygen that combines with the carbon, it is called the **oxidizing agent**. The coke, which accepts the oxygen, is called a **reducing agent**. In general, any substance that loses oxygen is reduced and is called an oxidizing agent. Any substance that gains oxygen is oxidized and is a reducing agent.

If we rewrite the hematite equation in terms of the valances of the atoms involved, it looks like this:

$$2Fe^{+++} + 3O^{--} + 3C \longrightarrow 2Fe^0 + 3C^{++} + 3O^{--}$$

We can use the term "oxidation state" in preference to "valence," so that we can discuss reactions that may not necessarily involve ions; oxidation state is not always identical with ionic charge, although in many cases it is the same.

The oxidation state of iron in the hematite is +3. The oxidation state of the uncombined iron is zero. In changing from $2Fe^{+++}$ to $2Fe^0$, the iron gains six electrons.

As carbon changes from $3C^0$ to $3C^{++}$, it loses six electrons. So the substance that was reduced gained electrons, whereas the substance that was oxidized lost electrons.

In oxidation-reduction reactions, one substance is oxidized (loses electrons) while another substance is reduced (gains electrons).

Chemists have decided that the transfer of electrons is more significant in this process than the movement of the oxygen, and therefore they have broadened the definition of oxidation to mean the loss of electrons. Reduction means the gain of electrons. Thus we can speak of an oxidation reaction in which oxygen plays absolutely no part at all! Consider the reaction:

$$2Na + Cl_2 \longrightarrow 2NaCl$$

To show electron transfer, the reaction may be written:

$$2Na + Cl_2 \longrightarrow 2Na^+ + 2Cl^-$$

The sodium loses an electron and changes valence from 0 to +1; it is thus oxidized. Chlorine gains an electron and changes valence from 0 to −1; it is reduced. The loss and gain of the electron is illustrated more graphically in an electron dot diagram:

$$Na\cdot \ + \ \cdot\ddot{\underset{..}{C}}l\!: \ \longrightarrow Na^+ + \ :\!\ddot{\underset{..}{C}}l\!:$$

In any oxidation-reduction reaction, oxidation of one substance is always accompanied by the reduction of another. Oxidation-reduction reactions, commonly abbreviated as **redox** reactions, constitute a major class of chemical reactions and find applications in every field of human endeavor. In medicine, oxidizing agents are used as antiseptics because many infection-causing bacteria cannot grow in the presence of oxygen; hydrogen peroxide (H_2O_2) is one such antiseptic oxidizing agent.

Redox means oxidation-reduction.

Ionic solutions used in chemical batteries furnish a good example of redox reactions. Consider, for example, an ionic solution of copper sulfate, $CuSO_4$, in water. If we immerse a strip of zinc in the solution, we observed the production of copper metal and the generation of zinc ions.

Displacement Reactions

$$Cu^{++} + Zn^0 \longrightarrow Zn^{++} + Cu^0$$

Solid metallic zinc loses electrons as it becomes a positive ion; these electrons are taken up by the positively charged copper, which then becomes neutral in charge. In this way the zinc displaces the copper from its combination with $SO_4^{=}$ ions. Written in nonionic form this becomes:

$$CuSO_4 + Zn^0 \longrightarrow ZnSO_4 + Cu^0$$

Reactions of this type are known as **displacement reactions**. If we examine what happens to the two metals in a displacement reaction, we find that they are involved in a redox reaction.

$$Cu^{++} + 2e^- \longrightarrow Cu^0$$
$$Zn^0 \longrightarrow Zn^{++} + 2e^-$$

The ionic copper gains two electrons in becoming metallic, and the metallic zinc loses two electrons in becoming dissolved ions. In other reactions between copper ions and aluminum and between copper ions and iron, we find that the same sort of thing occurs. Ionic copper gains electrons and is reduced, while the other metal is oxidized. In this respect, the elements zinc, aluminum, and iron have something in common: their ability to displace copper ions in solution. It follows that under similar conditions, copper metal could not displace these ions from solution.

What is it that allows some metals to displace others? Early alchemists who ob-

served the displacement reaction between iron and copper believed that iron
had been transmuted into copper. Today we know that the more chemically or
electrically active metals may displace other less active metals in ionic solution. By performing all the different displacement reactions with all possible
metal-ion pairs, a list can be drawn up according to the order of displacement.
This order is known as the **displacement series,** and it is arranged so that the
strongest displacers are at the top and the weakest are at the bottom. The lower
elements in the series always accept electrons from the metals listed above
them. Thus, iron will always accept electrons from sodium, as the position of
sodium is higher in the table.

Hydrogen is included among the metals in the displacement series, a fact that
is not as unusual as it might seem. After all, reaction between a metal and an
acid is just another displacement reaction. For example:

$$Zn^0 + 2HCl \longrightarrow ZnCl_2 + H_2$$

or

$$Zn^0 + 2H^+ + 2Cl^- \longrightarrow Zn^{++} + 2Cl^- + H_2$$

In this reaction, zinc gives electrons to hydrogen ions, causing the liberation of
free hydrogen gas, while the ionized zinc goes into solution.

The strongest metals in the series, those at the top, react with even small
amounts of H^+ in water. In fact, the reaction of sodium in water is so violent an
explosion occurs.

$$2Na + 2H_2O \longrightarrow 2NaOH + H_2$$

Electrochemical Cells

Scientists reasoned that if displacement reactions were making use of differences in electric potentials, then perhaps these potential differences would
be useful in generating electrical currents. If somehow the opposing processes
of oxidation and reduction could be separated from each other and made to
run in separate compartments connected by an external wire, then perhaps
the difference in electrical potential set up would be great enough to create a
flow of electricity.

In any redox reaction, there are actually two reactions going on at the same
time: the reaction in which electrons are lost and the reaction in which electrons are gained. These **half-reactions** when written separately represent the
reactions that may occur at the two electrodes of a battery or **electrochemical
cell.**

In the apparatus depicted in Figure 15-11 direct contact between ions of one
kind and atoms of another kind is prevented. Two metal electrodes, one zinc
and the other copper, are placed in solutions of their own ions in separate
compartments of **half-cells.** The electrodes are connected by a wire, and a

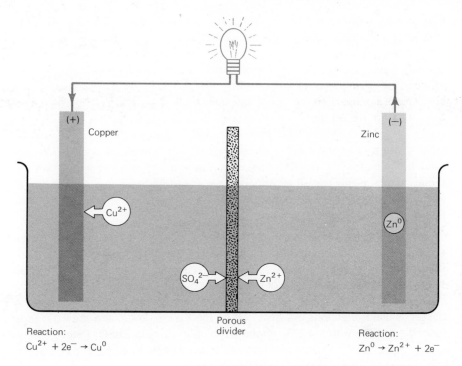

Reaction:
$$Cu^{2+} + 2e^- \rightarrow Cu^0$$

Reaction:
$$Zn^0 \rightarrow Zn^{2+} + 2e^-$$

Figure 15-11
Both of these electrochemical cells convert chemical energy into electrical energy through a redox reaction. In the flashlight battery (below), also called a dry cell, zinc ions form in an oxidation reaction in a paste of aluminum chloride and zinc chloride that provides the electrons for the negative charge at the bottom terminal ($Zn \rightarrow Zn^{++} + 2e^-$). The other half-reaction is a reduction in which the manganese dioxide combines with the electrons released in the zinc reaction ($2\,NH_4^+ + 2\,MnO_2 + 2e^- \rightarrow 2\,NH_3 + H_2O + Mn_2O_3$). The carbon rod is a nonreacting electrode.

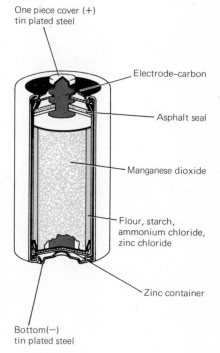

porous divider allows the passage of ions from one compartment to the next. If the zinc metal electrode and the copper ions were in contact in solution the complete reaction would read:

$$Zn^0 + Cu^{++} \longrightarrow Zn^{++} + Cu^0$$

When divided into half reactions, this is the reaction that occurs in a solution of $ZnSO_4$:

1) $Zn^0 \longrightarrow Zn^{++} + 2e^-$

This is the reaction that occurs in a solution of $CuSO_4$:

2) $Cu^{++} + 2e^- \longrightarrow Cu^0$

In the electrochemical cell the transfer of electrons from Zn to Cu^{++} occurs indirectly. At the zinc electrode, zinc atoms lose two electrons and enter the solution as Zn^{++} ions. The electrons travel through the wire to the Cu electrode where they reduce copper ions in solution to copper atoms. The resultant copper atoms will be deposited on the copper electrode.

What happens to the Zn^{++} ions which are left near the zinc electrode? What happens to the SO_4^{--} ions which are left in excess as Cu atoms are formed

Electrical energy is produced from chemical energy in electrochemical cells through a displacement reaction in which the reacting electrons flow through an external circuit.

from ions? If these ions piled up, a positive charge would build up in the zinc half-cell and a negative charge would build up in the copper half-cell. To prevent this, the divider provides a path for ions to flow between compartments. The flow keeps the solutions electrically neutral, and a continuous electric current can be maintained as further electron flow occurs through the wire. This current will persist until the entire zinc electrode goes into solution or all the Cu^{++} ions are plated out.

LEAD STORAGE BATTERY All electrochemical cells make use of oxidation and reduction reactions in which the electron flow is made to occur through an external circuit. In this way, they generate an electric current, and the amount of voltage driving the electron current increases when metals with greater separation on the activity series are used.

Lead storage batteries found in automobiles make use of the same principles employed in the standard electrochemical cell. In this type of battery, both electrodes are immersed in the same ionic solution of sulfuric acid, H_2SO_4. The electrodes are actually plates of metallic lead (Pb) dotted with small deposits of lead dioxide (PbO_2). The H_2SO_4 dissociates into H^+ and So_4^{--} ions. When the battery is used, electrons are given off by the Pb electrode, and the Pb^{++} ions formed combine with the SO_4^{--} ions in solution to yield $PbSO_4$ (lead sulfate). This compound is insoluble and adheres to the electrode surfaces as the reaction proceeds. This reaction is represented by:

$$Pb + SO_4^{--} \longrightarrow PbSO_4 + 2e^-$$

The electrons given up by Pb are furnished to the wires where they are added to PbO_2 in the solution of H^+ and SO_4^{--} to give $PbSO_4$ (lead sulfate) and water:

$$2e^- \; PbO_2 + 4H^+ + SO_4^{--} \longrightarrow PbSO_4 + 2H_2O$$

The reactants on the left side of the arrow are all in a higher energy state than those on the right, so the reaction yields energy.

Lead sulfate is deposited on both electrodes as the reaction continues. As current is drawn from the battery, it is seen that the ions H^+ and SO_4^{--} will gradually diminish and the electrodes will become caked with lead sulfate. When there are not enough reactants to continue the reaction, the battery becomes dead.

An interesting property of the lead storage battery is that it may be recharged by passing a current through in the opposite direction. This causes the two half reactions to proceed in reverse, and it forms Pb again at one electrode and PbO_2 at the other, thus restoring the battery to its original condition. Easy recharging is feasible in the storage battery because the lead sulphate formed in the forward reaction is deposited on the electrodes of the battery and is thus available for use in the reverse, or recharging, reaction.

ELECTROLYSIS Electrochemical cells operate by transforming chemical energy into electrical energy. In electrolysis, the opposite process occurs: an external source of electricity is passed through a liquid or gas, dissolving molecules and releasing free elements, in a chemical reaction that could not take place without this addition of energy.

Electrolysis is another example of a redox reaction, as the electrolysis of molten NaCl illustrates. Sodium chloride contains two ions, sodium (Na^+) and chloride (Cl^-). The Na^+ ions will migrate to the negatively charged cathode when a current is passed through, and the following half-reaction will occur:

$$Na^+ + e^- \longrightarrow Na^0$$

In like manner, the negative chloride ions, Cl^-, will migrate to the positively charged anode:

$$2Cl^- \longrightarrow Cl_2 + 2e^-$$

The net result of this process can be represented as:

$$2NaCl \longrightarrow 2Na + Cl_2$$

This reaction does not occur under normal conditions; it is only when an external voltage is applied that molten sodium chloride can be broken up into its constituent elements.

Electrolysis provides a good demonstration of the way that it has often been possible for experimental scientists to describe a phenomenon in accurate detail without understanding it in the least. Michael Faraday (1791–1867) observed and measured electrolytic effects in thousands of situations, using different compounds in solution, different electrodes, and different currents. He observed that the amount of any element deposited at the electrode was directly related both to the total charge passed through the solution and to the atomic mass and valence of the substance. In 1833, he stated this as the law of electrolysis and expressed it in the formula:

$$\text{mass deposited} = \frac{\text{charge}}{\text{constant}} \times \frac{\text{atomic mass}}{\text{valence}}$$

Faraday was also able to establish experimentally the value of the constant, or that quantity of electricity that will cause the electrolytic deposition of 1 mole of any substance. It is 96,490 coulombs/mole; this unit has been named a faraday.

Faraday's work made the commercial use of electrolysis for plating metals an immediate possibility. Yet since the structure of the atom was unknown to him, he had no idea why his constant held true, or what the nature of the action of the current was. For several generations, Faraday's law was a known but inexplicable fact; only with more recent discoveries about the fine structure of matter have we been able to understand why he achieved such a result.

Glossary   An **activated complex** is the transitional molecular state that occurs in a chemical reaction when reactant bonds are breaking, product bonds are forming, and activation energy is highest.

Activation energy is the energy needed to reach the transition state in a chemical reaction.

Bond energy is the potential energy stored in a molecule in the form of bonds.

A **catalyst** is a substance that speeds up or slows down the rate of a chemical reaction without being used up itself.

Chemical equilibrium occurs when the forward reaction in a reversible reaction is taking place at the same rate as the reverse reaction.

A **displacement reaction** is a type of redox reaction in which one metal plates out as another goes into solution as ions.

A **displacement series** contains an ordering of all displacement reactions according to their relative reactivity.

An **electrochemical cell** is a form of battery in which a redox reaction is used to generate electric potential.

Electrolysis is an electrochemical reaction in which electrical energy is applied to drive the reaction in the opposite direction from the one it would take spontaneously.

An **endothermic reaction** is one that absorbs heat energy.

An **equilibrium constant** is an expression that relates the relative amounts of reactants and products in a given equilibrium reaction.

An **exothermic** reaction is one that gives off heat energy.

A **half-reaction** is the part of the reaction in a redox equation that shows either oxidation or reduction. Two half-reactions combine to form the redox reaction.

Heat of reaction is the heat content of the products of a reaction minus the heat content of the reactants.

Le Chatelier's Principle states that if an equilibrium system, or set of reacting substances, is subjected to change, the system itself will shift in such a way as to relieve the disturbance and restore equilibrium.

Oxidation is the process whereby an atom donates electrons to an oxidizing agent and increases in valence.

The **rate-controlling step** in kinetics is the transition state of a reaction that has the highest activation energy.

Reaction kinetics is the study of the facts affecting reaction rates.

A **redox reaction** is a chemical reaction in which one type of atoms is oxidized (loses electrons) and another type is reduced (gains electrons).

A **reducing agent** is an atom that causes the reduction of another atom and is oxidized in the process.

Reduction is the process whereby an atom gains electrons from a reducing agent.

Thermodynamics is the study of the energy relationships in chemical reactions.

A **transition state** is the point at which the reactants are at their highest energy level in the course of a reaction.

1 Identify the following reactions as either endothermic or exothermic:

    a) $^{C}graphite + 0.45\ kcal \rightleftharpoons {}^{C}diamond$

    b) $N_2 + 3H_2 \rightleftharpoons 2NH_3 + 22\ kcal$

    c) $N_2O_4 \rightleftharpoons 2NO_2 - 14.1\ kcal$

    d) $H_2 + Cl_2 - 44.0\ kcal \rightleftharpoons 2HCl$

2 The thermite reaction produces temperatures of around 3000°C and consequently has been used in bombs. It consists of the two half-reactions given below. Show the total balanced reaction and calculate the heat evolved.

$$2Al + {}^3\!/_2O_2 \longrightarrow Al_2O_3 + 400\ kcal$$
$$Fe_2O_3 \longrightarrow 2Fe + {}^3\!/_2O_2 - 200\ kcal$$

3 Which of the reactions shown in graphs 1 and 2 is most likely to occur? Why?

4 Identify on graph 3:

    a) catalyzed reaction path
    b) uncatalyzed reaction path
    c) transition state
    d) reactant(s)
    e) product(s)

5 Which reaction in graph 4 is the rate-determining step?

6 Draw an energy diagram of the Haber process, labeling all important aspects and chemicals.

7 Pure H_2O ionizes slightly according to the equation:

$$H_2O \rightleftharpoons H^+ + OH^-$$

The equilibrium constant for this ionization is very important in chemistry and has the value 1×10^{-14}. Knowing this, would you expect water to carry an electric current? Why or why not?

8 Consider the equilibrium situation

$$Cl_2 + 2OH^- \rightleftharpoons Cl^- + ClO^- + H_2O$$

Derive an expression for the equilibrium constant, K_{eq}.

9 Write equations for the equilibrium constants of the following reactions and predict whether they go to completion.

Exercises

Graph 1

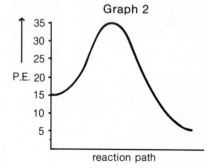

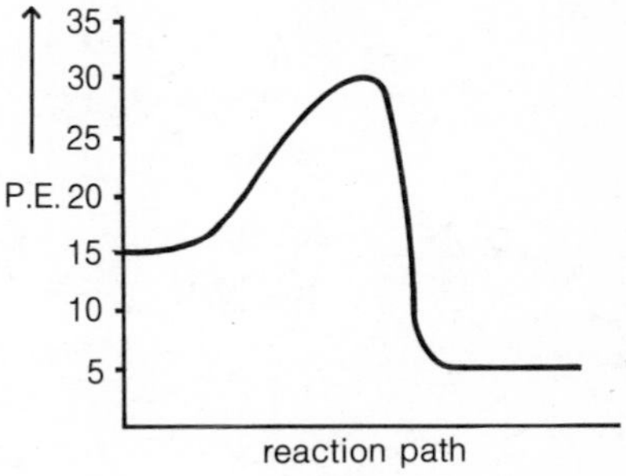

reaction path

Graph 2

reaction path

Graph 3

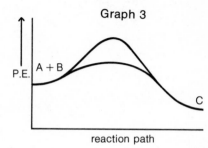

reaction path

Graph 4

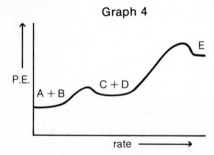

a) $Ag^+ + 2NH_3 \rightleftharpoons Ag(NH_3)_2^+$ $K = 1.7 \times 10^7$
b) $CH_3COOH \rightleftharpoons H^+ + CH_3COO^-$ $K = 1.8 \times 10^{-5}$
c) $AgCl \rightleftharpoons Ag^+ + Cl^-$ $K = 1.7 \times 10^{-10}$
d) $Cu + 2Ag^+ \rightleftharpoons 2Ag + Cu^{++}$ $K = 2 \times 10^{15}$

10 Consider the equilibrium

$$H^+ + OH^+ \rightleftharpoons H_2O$$

The rate constant, an indicator of the rate of reaction, is 1.4×10^{11} for the reaction at 25°C. What will the rate constant be at 35°C? at 45°C?

11 In the unbalanced redox systems below, label the oxidizing and reducing agents.

$$Fe^{++} + MnO_4^- \longrightarrow Mn^{++} + Fe^{3+}$$
$$Cr_2O_7 = +Zn \longrightarrow Zn^{++} \; Cr^{3+}$$
$$Ag + NO_3^- \longrightarrow Ag^+ + NO$$

12 In the electronegative series, can elements below hydrogen donate electrons to elements above it? If so, how? Place these elements in correct order in the series on the basis of the reactions below.

$Na + Fe^{++} \longrightarrow Na^+ + Fe$ spontaneous
$Cr^{3+}Zn \longrightarrow Cr + Zn^{++}$ spontaneous
$Zn^{++} + Fe \longrightarrow Zn + Fe^{++}$ no reaction
$Na + Cr^{3+} \longrightarrow Na^+ + Cr^+$ spontaneous
$Zn + Na^+ \longrightarrow Zn^{++} + Na$ no reaction

13 Mercury batteries used in hearing aids and pacemakers produce energy from the following redox reaction:

$$Zn + HgO \longrightarrow ZnO + Hg$$

How much voltage will the battery produce if the two half-cells involved are

$Zn + 2OH^- \longrightarrow ZnO^{\cdot} + H_2O + 2e^-$ $E = 0.7628V$
$HgO + H_2O + 2e^- \longrightarrow Hg + 2OH^-$ $E = 0.0984V$

14 A glass bulb contains nitrogen dioxide in a state of equilibrium:

$$N_2O_4 + 14.1 \text{ kcal} \rightleftharpoons 2NO_2$$

Using your knowledge of Le Chatelier's Principle, explain the following observations:

a) At 1 atm pressure the bulb is dark brown. As the pressure is increased to 4 atm, the gas mixture turns clear.
b) At 0°C the bulb is colorless, but at 100°C it is dark brown.

15 If a cube of metal 1 cm on a side is dissolved in acid in 1 hour, how long will it take to dissolve the same amount of metal if the cube is cut into smaller cubes each 1mm on a side? (Assume that reaction rate is directly proportional to surface area.)

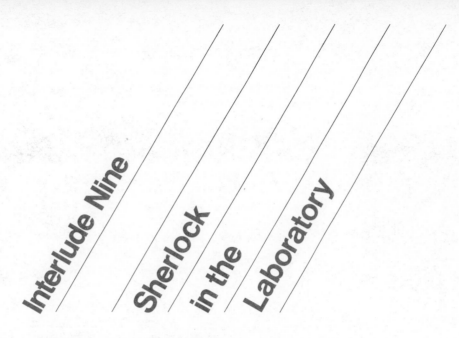

Interlude Nine

Sherlock in the Laboratory

One recent December afternoon, the police picked up Joe Green, a 27-year-old hospital employee, as a suspect in the murder of Edna O'Brien. Ms. O'Brien had been found a few hours earlier, bleeding profusely from bullet wounds in the chest and head. Neighbors had seen Green leaving the victim's house.

At the police station, detectives noticed a blood stain on Green's trousers. They sent the trousers to the crime lab where a forensic scientist—a chemical detective who is a modern-day Sherlock Holmes—tested the blood. After performing a few simple tests to determine that the stain was human blood, the analyst tested it to see if it matched that of the victim, who had type A blood.

For purposes of typing, human blood has two chief components: red blood cells (RBCs) and serum (also called plasma), which is a mixture of proteins, inorganic salts, and water. RBCs contain protein molecules known as antigens; serum contains antibodies that act as defensive agents when foreign antigens are introduced into the blood. Blood grouping is based on the fact that in certain cases, RBCs from one individual agglutinate, or clump together, when added to the blood serum of another individual.

Biochemists have determined that there are two types of antigens—designated A and B—and the blood of an individual is grouped according to the type of antigen it contains. Similarly, there are two types of antibodies—anti-A and anti-B. As the names indicate, antibody A reacts against the antigen in the red blood cells of blood type A, causing the RBCs to clump together. Anti-B, on the other hand, causes the RBCs of type B blood to clump. Thus, a person with type A blood has antigen A in his RBCs and anti-B in his serum. Similarly, B blood types have antigen B and anti-A. AB blood groups have antigen A and B, but no antibodies; O blood has no antigens, but A and B antibodies.

"Holmes was working hard over a chemical investigation."

The blood on Green's trousers was tested by placing a small amount of anti-A serum on one end of a microscopic slide and a drop of anti-B serum on the other. The dried blood from the trousers was scraped off, added to each serum sample, and mixed. When the samples were examined under the microscope, the RBCs from the trousers clumped together when they reacted with anti-A serum. The blood type on the trousers matched that of the victim, and one link in the chain of evidence against Green was forged.

The next step was to determine whether Green had fired a gun recently. If he had, a paraffin test on his hands would detect nitrates from microscopic particles of gunpowder that are ejected from the barrel and trigger housing of the gun by the force of the explosion. Since Green was right-handed, hot paraffin was poured on his right hand to obtain a cast. The heat of the wax opens the pores, forcing out any nitrates that may have been deposited in them. Next, a solution of diphenylamine plus sulfuric acid and water was applied to the inside of the paraffin cast. If nitrates were picked up by the cast, the solution would cause them to appear as blue-violet spots. In this case, the appearance of blue-violet spots on the cast indicated that Green had recently fired a gun.

But the work of the chemical detectives did not stop here. During their search of Green's clothing, they noticed tiny blue specks of what appeared to be paint. The specks, no larger than the head of a pin, were first examined under a stereo microscope—which enables the observer to view specimens in the three dimensions—to determine their physical structure. After determining that the specks were in fact paint, the next step was to identify their exact color, and compare it with a sample of the blue paint from the wall of the room where the body was discovered.

It would seem that the investigator had only to place the two samples in front of him and visually compare them, but this is not so. For one thing, the paint samples on Green's clothes were too tiny. Secondly, the sense of color differs from person to person, and may be influenced by the background against which the samples are being viewed, as well as the quality of the lighting and the angle at which it strikes the samples. Green's freedom hung in the balance, so it was necessary to use a more exact technique known as emission spectography. This identifies the characteristic colors of the light waves emitted by a substance heated until it glows.

In the crime lab, the analyst placed the blue chips under a high flame. He then viewed them through a spectroscope, a tubular device measuring about 75 inches in length with a viewing slit at one end and either a prism or a diffraction grating at the other. The grating disperses the light waves emitted by the burning chips into their own characteristic spectra, or color bands. Because each element has its own spectrum, the nature of the chips—and their true color—can be readily determined. A sample of the paint from the room where Ms. O'Brien body was found was also tested by emission spectrography and found to be identical in color spectra to the specks found on Green's clothes. One more bit of evidence for Green's guilt.

Also found on Green's shoes were small specks of dried mud. They were analyzed by a relatively recent technique, neutron activation analysis (NAA), to determine if the mud had come from the O'Brien's yard. NAA enables investigators to detect minute quantities of trace elements—phosphorous, sulfur, silver, and others—which are present in different quantities in different substances. Unlike emission spectrography, NAA does not change the sample being tested.

The dried mud specks from Green's shoes were placed in a nuclear reactor for about 72 hours. Billions of neutrons struck the mud every second, causing it to become radioactive. The type of radiation emitted by the isotopic trace elements in the mud and their decay rates were detected by a sodium iodide crystal and transmitted to a device which converted this information into numerical quantities on an electronic readout screen. When a sample of the mud from the murder victim's yard was submitted to the same test, it was found that the type and number of trace elements in it matched those in the mud from Green's shoes.

When confronted with this impressive body of scientific evidence, Green broke down and confessed to the murder.

Six

Carbon Chemistry

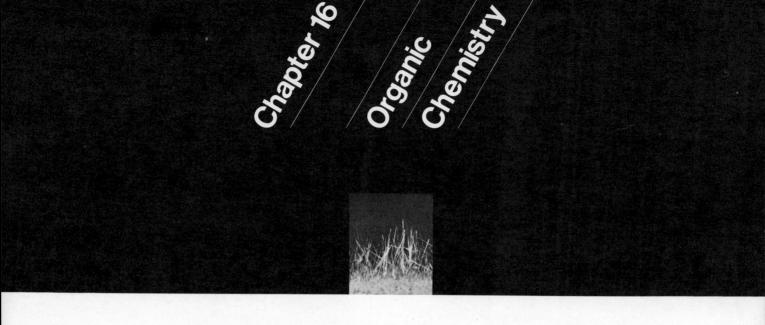

The difference between organic and inorganic compounds is that all organic compounds contain carbon.

Molecules are the submicroscopic particles which comprise all substances. They too have a structure of atoms, and the atoms have a structure of protons, electrons, and other elementary particles; but molecules are the smallest units of a substance that still retain all the qualities of that substance. Even before they were able to see them, chemists understood much about the structure and behavior of the molecules which make up matter. Yet long after they had established many laws regarding the behavior of inorganic substances, such as rocks and minerals, chemists were not sure that these laws applied to other kinds of matter. When the science of chemistry was in its infancy, molecules produced by living things—wood, foodstuffs, and the materials that make up plant and animal body tissues—were considered to be quite different from inorganic materials—rocks, glass, salts, metals—and seemed to be more complex in nature, to have something extra. These substances were thought to possess a "vital force," which seemed untouchable by the science of chemistry.

It was not until 1828 that chemists first realized organic compounds might generally be subject to the same forces that were operating in the realm of inorganic chemistry. In that year Friedrich Wöhler (1800–1882) succeeded in converting ammonium cyanate, an inorganic compound, to urea, a substance appearing naturally in the urine of many animals. Here was direct proof that organic compounds were not necessarily derived from a vital force.

What is it then that sets organic compounds apart from all others? The distinguishing feature of organic compounds was found to be the presence of carbon; the best present definition of organic chemistry is that it is the science of the properties of carbon-containing compounds. Although it has been discovered in the twentieth century that many carbon compounds can be synthesized in the laboratory (plastics, synthetic fibers, many drugs), they are still categorized as organic compounds because their chemical properties are much the same as those of carbon compounds synthesized by living things.

The element carbon is quantitatively one of the less numerous constituents of the earth and its environs; in fact it comprises less than 1 percent of the earth's crust. Over 2 million different organic compounds are known and every year tens of thousands are added to the list! What makes carbon so uniquely able to form new compounds?

The remarkable combining power of carbon is due to the fact that carbon contains six electrons—an inner ring of two and an outer ring of four electrons. Carbon, therefore, needs four more electrons to fill its outer shell, and achieve a complete octet of eight electrons. This is accomplished by covalent bonding, a form of bonding in which atoms share electrons with each other in a kind of mutual deception in which each atom behaves as though it is filling its outer shell, although it is actually only sharing the electrons with its partner. Other elements have similar outer electron shells (Si, Ge, Pb, and so on). But carbon has a smaller size and no inner shells to shield its 8 electrons from high electronegativity.

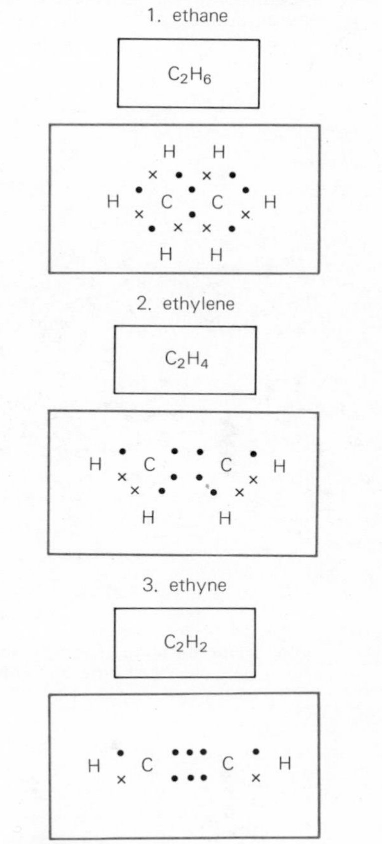

Covalently bonded carbon atoms form chains of straight, branched, or ringed compounds of extremely large numbers and complexity and may even form "double" and "triple" bonds in which sharing occurs with a total of 4 or 6 electrons (2 or 3 from each atom). The carbon-to-carbon bond is very durable, able to withstand breakdown during many chemical reactions, because the outer ring becomes so stable.

The idea that organic molecules are actually three-dimensional structures was not recognized until the late nineteenth century, when van't Hoff and LeBel in-

Figure 16-1
Carbon compounds. In this figure, dots represent carbon electrons, and x's represent hydrogen electrons. In ethane, the carbon-carbon bond is a single bond. A double bond is formed in ethylene with four electrons, two from each atom involved. In ethyne, the triple bond is made up of six electrons.

Simple Carbon Compounds

Carbon atoms have an outer ring of four electrons and a remarkable power to form covalent bonds; there are millions of different carbon-containing compounds.

The carbon-to-carbon bond is very durable because the outer ring is so stable.

dependently proposed the spatial nature of molecules. They suggested that because carbon has four electrons in its outer ring, the compounds it forms will usually assume a tetrahedral arrangement with carbon at the center. An example can be seen with the gas methane (CH_4):

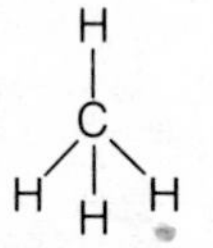

$$
\begin{array}{c}
H \\
| \\
C \\
H \diagup\ | \diagdown H \\
H
\end{array}
$$

Carbon compounds are three-dimensional structures; they usually have a tetrahedral shape with carbon at the center.

The common cleaning fluid carbon tetrachloride, in which the four available electrons react with four chlorine atoms, expresses the geometry of the compound in its name:

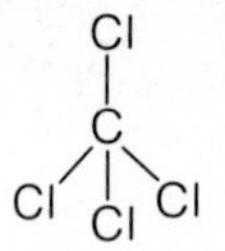

$$
\begin{array}{c}
Cl \\
| \\
C \\
Cl \diagup\ | \diagdown Cl \\
Cl
\end{array}
$$

This simple theory has often been termed the keystone of organic chemistry.

Hydrocarbons—The Methane or Paraffin Series

The simplest way to understand carbon chemistry is to begin with the carbon compounds that have only one other type of atom—hydrogen atoms—attached to them. Hydrogen atoms are the atoms most often found bonded to carbon because they are so commonly available and readily form two-electron bonds; the one electron of hydrogen combines easily with the open electrons of carbon. These carbon-hydrogen compounds are called **hydrocarbons**; they are the principal constituents of petroleum.

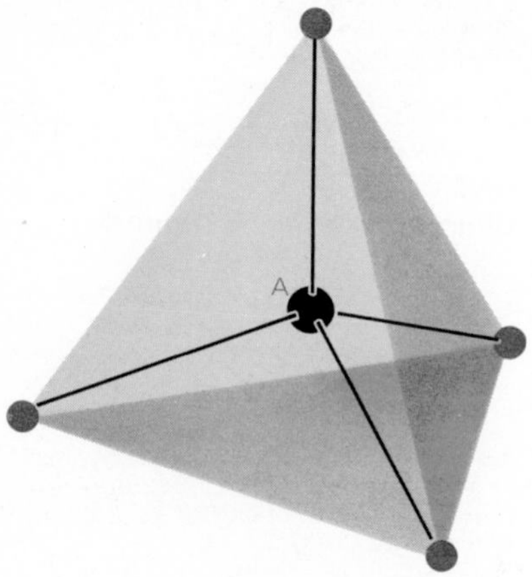

When we use natural gas to cook our food, or a propane container when we camp out, or a butane lighter for a cigarette, we are using hydrocarbons with one to four carbons. These compounds are: methane, CH_4; ethane, C_2H_6; propane, C_3H_8; and butane, C_4H_{10}. Ethane is a small constituent of this group and our most frequent contact with it is when an —OH group is attached to it to form spirits of ethanol, the alcohol of intoxicating beverages. As we can see by simple inspection of their chemical formulas, there is a basic arrangement and order to these hydrocarbons; the number ratio of hydrogen atoms to carbon atoms is two times the number of carbons plus two. This can be expressed mathematically by the formula C_nH_{2n+2}. These are called **saturated** hydrocarbons because each molecule contains the maximum number of hydrogen atoms that its carbon atoms can bond to.

Hydrocarbons have many practical applications, but the pentanes (5-carbons), hexanes (6), heptanes (7), octanes (8), nonanes (9), and decanes (10) form the economically most important group in this series since they are the components of gasoline. Since the pentanes and hexanes are highly volatile, they make gasoline extremely flammable. A combination of pentane and hexane and

Figure 16-2
This three-dimensional drawing of methane (CH_4) shows the tetrahedral arrangement characteristic of organic compounds. The four corners are equidistant from *A*, the location of the carbon atom, and from each other.

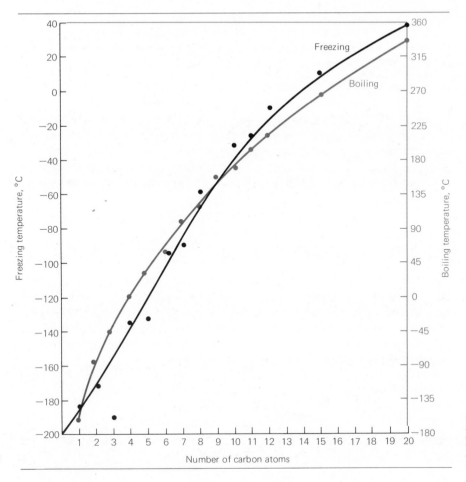

small amounts of some other hydrocarbons forms a colorless liquid used as a cleaning solvent, called petroleum ether or naphtha.

The group containing 11 to 17 atoms forms kerosene, which is heavier and less volatile. The hydrocarbon groups above seventeen carbons form the semisolids such as petroleum jelly, greases, and lubricating oils. The carbon groups above 23 carbons are the constituents of wax.

Since hydrocarbons are practically nonpolar in electrical charge, they are relatively unreactive chemically to the usual acids and bases and most oxidizing agents at common temperatures. Because of their relative unreactivity, they have been termed **paraffins**—from the Latin par (little) and affin (affinity).

All the saturated hydrocarbons have similar physical properties, making it difficult to separate them from each other in petroleum. The melting points, boiling points, and densities increase with carbon number. This fact is taken advantage of when distilling petroleum to separate the different compounds. Because it is

Saturated hydrocarbons contain the maximum possible number of hydrogen atoms that can bond to carbon atoms in that molecule.

Figure 16-4
Many long molecules, such as dodecane ($C_{12}H_{26}$), cannot be used in modern automobile engines because of their poor antiknock characteristics. These long molecules are broken down in a process called cracking into shorter molecules such as olefins (C_6H_{12} and C_5H_{10}), which can be used for motor fuels. Cracking is achieved in fluid catalytic cracking plants such as the one pictured here.

next to impossible to make pure compounds from the petroleum distillate, they are separated into fractions by boiling point. This practice of fractional distillation is a very efficient and quick way of getting the products needed from crude oil, since complete purity is not usually necessary.

Much more gasoline is needed in our economy than any other of these petroleum products. Because the economically important gasoline fractions of crude oil are not very abundant, gasoline must be manufactured from the other products that may be in excess. Larger hydrocarbons, not in as much demand in our economy, are broken down into smaller molecules. This process of cracking (also called pyrolysis) uses high temperatures, high pressure, and chemical catalysts to break the carbon-to-carbon bonds. Kerosene, a C_{12} hydrocarbon, can be cracked into a C_7 and a C_5 compound:

$$C_{12}H_{26} \xrightarrow[\text{catalysts}]{\text{pressure}} C_7H_{16} + C_5H_{10}$$

kerosene   gasoline

The smaller hydrocarbons that are found in natural gas have lower numbers than the gasoline molecules. The gas can be heated in the presence of catalysts which attaches the gas molecules to each other. These large molecules are called **polymers**, from the Latin words for many (poly) and units (mers); the process of assembling the molecules is called **polymerization**. The gases of propane and butane can be polymerized into heptane:

$$C_3H_8 + C_4H_{10} \xrightarrow[\text{catalysts}]{\text{heat}} C_7H_{16}$$

propane butane   heptane

Usually when hydrocarbons burn they completely convert to carbon dioxide and water:

$$C_7H_{16} + 11O_2 \longrightarrow 7CO_2 + 8H_2O$$

In the internal combustion engine, which has a relatively low efficiency rating, not all the hydrocarbons burn completely; the result is the formation of the pollutant carbon monoxide, as well as other compounds:

$$2C_7H_{16} + 15O_2 \longrightarrow 14CO + 16H_2O$$

The other pollutants are the result of the presence of gasoline additives such as tetraethyl lead (in 1922 it was found that the addition of this compound could greatly increase the efficiency of gasoline as a fuel), which during side reactions in the combustion process form many compounds with highly corrosive properties that are dangerous to our health.

Larger Organic Molecules

The hydrocarbons can help us to understand the basic formulations and variations of more complex carbon compounds. In describing them, we have been using a kind of shorthand which expresses the number of carbons and hy-

Table 16-1
Alkane hydrocarbons

Formula	Name	Isomers	Freezing point	Boiling point	Density
CH_4	Methane	—	−184 C	−161 C	0.424
C_2H_6	Ethane	—	−172	−88	0.546
C_3H_8	Propane	—	−190	−45	0.501
C_4H_{10}	Butane	2	−135	−0.5	0.579
C_5H_{12}	Pentane	3	−132	36	0.626
C_6H_{14}	Hexane	5	−94	69	0.659
C_7H_{16}	Heptane	9	−90	98	0.684
C_8H_{18}	Octane	18	−57	125	0.703
C_9H_{20}	Nonane	35	−51	154	0.718
$C_{10}H_{22}$	Decane	75	−32	174	0.730
$C_{11}H_{24}$	Undecane	151	−27	197	0.740

drogens in the compounds; it is called the **molecular formula**. In order to understand the striking variability of the carbon compounds, we must use another format, called the **structural formula**.

The molecular formula for methane is CH_4; its structural formula is:

A molecular formula expresses the number of atoms of each of the elements in a given compound; a structural formula also shows the arrangement of atoms and bonds in the compound.

Although it is only two-dimensional, this way of showing the structure is more specific than the molecular formula, because the same number of C and H atoms often may be combined in different structures.

With the use of structural formulas, it is easy to see that two different structures might exist for the same molecular formula. With methane (CH_4) this obviously cannot occur; there is only one possible way the structure can be drawn. And this is true as well for ethane and propane. But when we reach the level of four carbons in a compound, there is more than one possible arrangement of the carbon atoms. The carbons in butane (C_4H_{10}) can be in a straight chain or may branch as follows:

Isomers

This variation of butane is called isobutane; the continuous chain form of butane is called normal butane or *n*-butane. The greater the number of carbons, the more numerous the variants. These variations of structure for compounds with identical ratios of the various component elements are called structural **isomers**; isomers have the same molecular formula but different structural formulas. The physical and chemical properties of isomers may be quite different. For example, the boiling point of *n*-butane is −0.5°C; the boiling point of isobutane is −11.7°C.

Structural variations of compounds with identical molecular formulas are called structural isomers.

In the five-carbon compound pentane (C_5H_{12}) the number of isomers increases by one. Three isomers of pentane can be formulated:

n-pentane   isopentane   neopentane

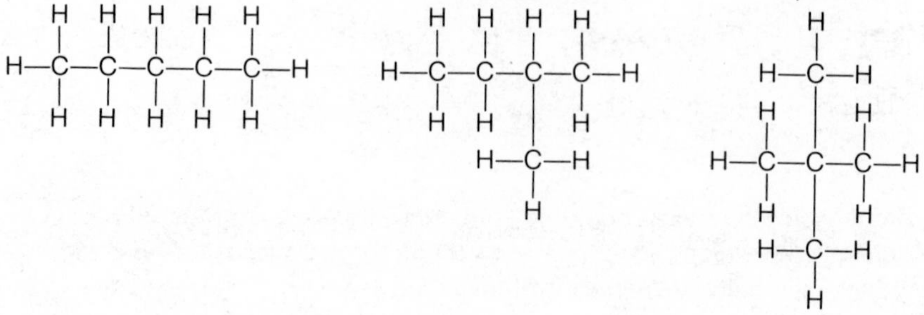

The number of possible isomers increases very rapidly as the carbon number increases. A C_{13} compound can have 813 possible isomers and a C_{20} molecule can have 366,319 theoretical isomers. So far very few of all the possible isomers have been either found in nature or synthesized in the laboratory.

Unsaturated Hydrocarbons

Unsaturated hydrocarbons contain double or triple bonds between the carbon atoms.

The paraffin hydrocarbons have single covalent bonds between carbon and the other atoms. All the bonds are filled with hydrogen and therefore it is said that the paraffin hydrocarbons are saturated with hydrogen. When the available bonds of the carbon atoms are not filled with hydrogen atoms, the compound is said to be **unsaturated**. The open carbon electrons, instead of being attached to hydrogen atoms, are shared within the carbon atoms in the form of double bonds (alkenes) or triple bonds (alkynes). In order to achieve saturation, the covalent bonds to carbon may alternatively be filled by elements other than hydrogen, so long as they have a valence of −1. The halogens chlorine and bromine often substitute for H, yielding ethylene chloride ($C_2H_4Cl_2$) or ethylene bromide ($C_2H_4Br_2$). The ethylene itself without chlorine is the unsaturated C_2H_4, and has the H_{2n} formulation characteristic of unsaturated hydrocarbons (saturated hydrocarbons have the formulation H_{2n+2}). This leaves two electrons—one from each carbon—still available for the chlorine and bromine. If these atoms are not available, then the two carbons share two pairs of electrons in a double bond:

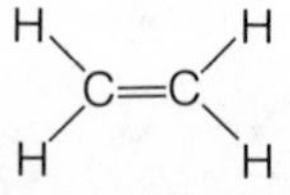

The double bond is signified by two lines rather than one in the structural formula. If the ethylene picked up two hydrogens it would be the paraffin hydrogen ethane, a saturated hydrocarbon:

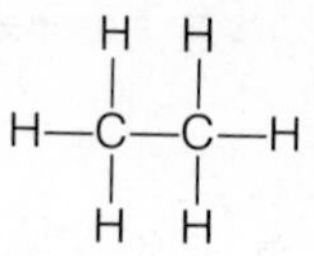

The unsaturated hydrocarbons form a series of double bonded carbon compounds called **alkenes**, also called the ethylene series, since the series begins with ethylene. Unsaturated hydrocarbons can also have triple bonds between carbon atoms. These are the **alkynes**, or acetylene series of compounds, whose first member is acetylene. Acetylene has the structural formula of H—C≡C—H.

Acetylene can be easily and cheaply prepared from calcium carbide (made in an electric furnace from lime and coke) and water:

$$\underset{\text{calcium carbide}}{CaC_2} \;+\; \underset{\text{water}}{2H_2O} \longrightarrow \underset{\text{acetylene}}{CH\equiv CH} \;+\; \underset{\text{calcium hydroxide}}{Ca(OH)_2}$$

Polymerization of the unsaturated hydrocarbons is similar to that of the paraffin series, but also involves the shift of the multiple bonds between different carbons. The unsaturated hydrocarbons may have alternating single and double bonds; these are called conjugated systems. Many conjugated hydrocarbons form the major portion of natural pigments such as carotene, which gives the orange color to carrots and is a precursor of vitamin A.

How Structure Is Determined

The structural arrangement of a molecule can be determined by a deductive process; chemists working with unknown substances break them down chemically to determine their constituent elements and combine them with known substances to determine how they react.

Based on the manner of reactivity and the proportions involved in reactions, chemists can deduce information about the structure of any molecule. This efficient and reliable method of structure determination has been used for more than 100 years. Chemical methods include: degradation, which is the breaking down of undetermined molecules into simpler constituents in order to recognize composition and structure; and reaction with compounds of known structure, which helps to determine information about the unknown structure.

Today's chemist also uses sophisticated electronic devices which employ infrared and other radiation and magnetic resonances in order to deduce the composition and relative positions of the atoms of a molecule.

Physical properties also serve as aids in identification of structure. Analysis of

Figure 16-5
A model polymer chain. [*Omikron*]

Naming the Organic Compounds

In 1892 a congress of chemists adopted a scheme of nomenclature for organic compounds known as the Geneva System. The system is based on the hydrocarbons, a series of compounds with a carbon-to-hydrogen bond. The simplest hydrocarbon compound contains one carbon and is called methane (top right). The four bonds designated by the lines connecting the C and H are called *single covalent bonds.* Carbon, in a hydrocarbon compound, will always have four bonds attached to it.

Methane is a building block to which other carbons atoms can be added by eliminating a hydrogen. One example is a two-carbon system, called ethane (bottom right). Notice that each carbon has four covalent bonds and now a carbon is bonded to another carbon.

Generally, hydrocarbons may be named according to the following characteristics:

1 the number of carbon atoms contained in one straight chain,
2 the number of bonds between each carbon atom (we have mentioned single bonds, but there can also be double and triple bonds),
3 atoms other than H attached to a carbon atom

The left table shows prefixes of the names for carbon chains one through ten.

Prefix	Number of carbon atoms
meth-	1
eth-	2
prop-	3
but-	4
pent-	5
hex-	6
hept-	7
ort-	8
non-	9
dec-	10

Class of compounds	Type of bonding	Suffix
Alkane	single	-ane
Alkene	double	-ene
Alkyne	triple	-yne

Hydrocarbons are grouped into classes according to the types of bonds formed by the carbon atoms. These classes include alkanes (single bonds), alkenes (double bonds), alkynes (triple bonds) and aromatics (hybrids of double bonds). Any of the prefixes from the left table may be combined with the suffixes, -ane, -ene, and -yne, in the right table. Thus, there are hydrocarbons such as methane, ethene, butyne, and so forth. A *saturated* hydrocarbon is one in which each carbon-to-carbon bond is a single covalent bond; such a compound is an alkane. If double or triple covalent bonds exist between carbon atoms, the hydrocarbon is said to be *unsaturated* and, therefore, is either an alkene, an alkyne, or an aromatic.

The aromatic bonding is a hybrid because the double bonding is not fixed and can resonate from carbon to carbon. The name "aromatic" stems from the characteristic pungent odor associated with this type of compound.

The alkane, or paraffin, hydrocarbons are named according to the longest continuous chain of carbons. The general formula for a paraffin is C_nH_{2n+2} when n is the number of carbons in the chain. Propane (top right) is an example.

```
  H   H   H
  |   |   |
H-C - C - C-H
  |   |   |
  H   H   H
```

Additional hydrocarbon groups can be attached to the straight chain by replacing a hydrogen in a process called branching. In the second formula from the top right, for example, the name of the compound is derived as follows: The longest chain is four carbons, so the compound is a butane. But there is also a methyl group on the second carbon counting from left to right. For this reason, the full name of the compound is 2-methyl-butane. As we can see, there are a host of various compounds that could be added to this four-carbon chain, but the parent compound is still butane.

```
          H
          |
        H-C-H
  H       |    H   H
  |       |    |   |
H-C  -   C  - C - C-H
  |       |    |   |
  H       H    H   H
  1       2    3   4
```

A similar procedure is used to name unsaturated hydrocarbons. The general formula for alkenes is C_nH_{2n}. The butane compound shown above would be called a butene if it had a double bond. The double bond could exist between carbon 1 and 2 or carbon 2 and 3. (If it existed between 3 and 4, it would be the same thing as a bond between 1 and 2; one just flips the compound 180°.) The carbon nearest the end of the chain containing the double bond is the key to naming an unsaturated hydrocarbon. The third formula from the top right, for example, would be called 3-methyl-1-butene because the methyl group is on the *third* carbon; and the double bond is between the *first* and second carbon.

```
          H
          |
        H-C-H
  H       |    H   H
  |       |    |   |
H-C  -   C  - C = C
  |       |    |   |
  H       H    H   H
  4       3    2   1
```

The full names of unsaturated hydrocarbons indicate not only the length of the carbon chain, but the location of other types of alkyl groups and bonds attached to it. This is done by placing a number before the hydrocarbon name to indicate where to find the bond, the number being that of the low-numbered carbon of the two carbons involved in the bond. If there is more than one unsaturated bond in the compound, the appropriate prefix is used to indicate the number present. Thus *mono* is used for one, *di* for two, *tri* for three, and so on. Each compound is numbered from the end giving the lowest number. Thus the bottom right formula is called 1, 3 hexa-di-ene. Reading the name from left to right, 1 indicates that there is a double bond between the first and the second carbon; 3 indicates that there is a double bond between the third and the fourth carbon; hexa- indicates that the compound has six carbons; -di- indicates that there are two double bonds in the compound; and -ene indicates that the compound contains double bonds.

```
  H   H   H   H   H   H
  |   |   |   |   |   |
H-C = C - C = C - C - C-H
                    |   |
                    H   H
```

a. Electron dot

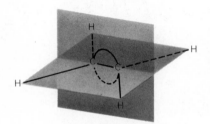

b. Conventional

c. Ball and stick

d. Planar geometry

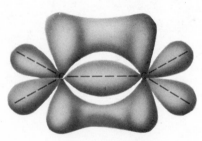

e. Electron–cloud (orbital)

Figure 16-6
Ethylene is the simplest member of the alkene family. The central feature of the ethylene molecule is the double bond between carbon atoms, shown here in the different two- and three-dimensional representations.

the physical properties of a compound such as density, melting point and boiling point helps the chemist to identify the molecular formula.

Let us take ethylene as an example and see how we may determine its molecular structure. The molecular formula of ethylene (C_2H_4) may be determined by spectroscopy or physical property analysis. How can we then determine how the bonds are arranged?

The first step is to calculate all the variant arrangements possible with the molecular formula. There are three possibilities:

a) CH_3CH b) $CH_2—CH_2$ c) $H_2C═CH_2$

Based on our knowledge of carbon chemistry we can clearly see that a and b are not viable possibilities. Carbon must be tetravalent—that is, it must form four bonds in any compound. In possibility a this is not the case, as one carbon atom is attached to just one H and one C. The same is true in b, where both carbon atoms are trivalent. Therefore c must be the correct choice. Here each carbon has 4 attachments: 2 to H atoms and 2 to the other C in the form of a double bond.

What about compounds that might have isomers? If we had the molecular formula of C_2H_6O, how could we determine its structure? We would have to find out what happens when it reacts with some specific inorganic compounds. If we react the compound with sodium, and hydrogen is liberated in the same proportions that the sodium is attached to the compound, we can say that one of the hydrogen atoms is attached differently onto the compound than the other five. Since sodium does not readily remove hydrogen from carbon unless more energy is added to the reaction, two sodiums will react as follows:

$$2Na + 2C_2H_6O \longrightarrow 2C_2H_5ONa \pm H_2.$$

If the hydrogen that comes off is not attached to a carbon, the oxygen must be attached to one of the carbons. The other five hydrogens must be attached to the two carbons. We know that the valence for oxygen is −2. Since one of the electrons is being shared with a carbon, the other must be shared with the remaining sixth hydrogen in the form of an hydroxyl group (—OH). The compound could then be an **alcohol**. If the oxygen was attached to both carbons (positioned between them) then the above reaction could not take place. The latter compound is called an **ether**. The formulas for the two possibilities are:

ethyl alcohol   dimethyl ether

We can prove that the O and the H are together in a hydroxyl group by one further experiment. If the OH group exists in the presumed alcohol, then if we use an acid, which will add a free hydrogen, we should produce water. HCl when it reacts with an alcohol produces water and in this case ethyl chloride, C_2H_5Cl. If this latter reaction occurs we can be certain that ethyl alcohol is the correct alternative. If these reactions do not work out, then we would test for other alternatives. The ether could be tested for by warming with sulfuric acid or aqueous hydrogen iodide; ether does not react with these compounds as readily as does alcohol.

If the carbon compound when dissolved in water gives an acid reaction, then the chemist knows that one end of the molecule contains a carboxyl or organic acid group:

Other end groups such as aldehydes

or ketones ($—C=O$) can also be identified by their reactive characteristics. The organic chemist can draw on a long history of reactions that have been worked out to identify various forms and configurations of organic subunits; from these reactions he can discover the structural formula of very complex molecules. This process may take many years of diligent research, but the successful conclusion is highly rewarding. Today chemicals essential for life, such as vitamins and hormones, can be synthesized industrially.

THREE-DIMENSION STRUCTURE Thus far we have considered structural formulas from only a flat two-dimensional viewpoint. But we know that the molecules actually exist in depth too, which adds the third spatial dimension to their structure, and the carbon chains can be twisted, kinked, and bent without changing their basic configuration. The atoms can vibrate back and forth about an equilibrium position, slightly distorting the symmetry of the molecule. These changes in position can be expressed by using three-dimensional models much like Tinker Toys with spring connections to represent the molecules.

The vibrations of the atoms in a molecule can be detected through a spectroscope. When we pass certain wavelengths of light through molecules in solution, the vibrations of the atoms will selectively absorb the light energy and produce dark (absorption) lines on the instrument viewer. The lines can be compared with standard known configurations, and thus the structure of the molecule can be deduced. For instance, the differences among single, double, or triple

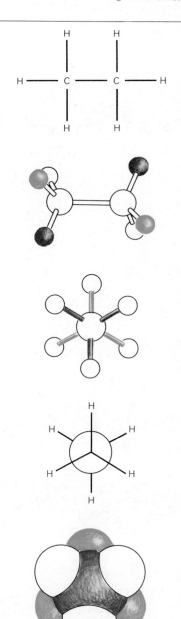

Figure 16-7
Rotation of the bond between carbon atoms in the ethane molecule allows the formation of different conformations of ethane. Above are various models of the staggered conformation. The bottom three are an end-on view.

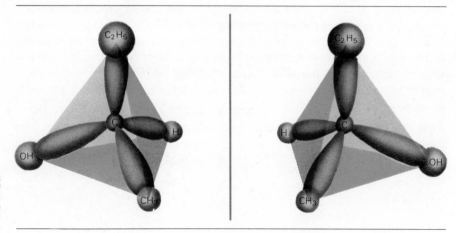

Figure 16-8
These optical isomers are mirror images of the same compound. The physical and chemical properties of the two isomers are the same except that they rotate in opposite directions.

bonding of carbon atoms can be seen, since each state has a different vibrational energy and therefore will absorb slightly different light waves.

Some compounds have identical chemical and physical properties but differ in that they rotate polarized light in opposite directions. These structurally different molecules are called optical isomers; each molecule is the mirror image of the other. When a solution is prepared it can be rendered optically inactive if equal numbers of the two types of molecules are present. The optical isomers are designated as left (*l*) or right (*d*).

Orbital Theory and Carbon Compounds

Up to this point we have been thinking of carbon compounds in terms of the Tinker-Toy model of molecular structure. This model is helpful in understanding the basic makeup of carbon molecules. The structural formulas we have been using also help us to get a sense of the fact that two molecules share electrons in their valence ring. But this model does not explain the reasons for the particular angle of reaction and the behavior of the electrons involved in the covalent bond. To do this, we must add the concept of motion to our model, considering where and how the electrons orbit the atomic nucleus.

The two basic electron orbitals of the covalent bond are the spherical *s* orbital and the figure-eight *p* orbital.

The two basic types of orbitals applicable to the covalent bond are the spherical *s* orbital and the dumbbell-shaped or figure-eight *p* orbital. When the covalent bond is strong, it means that a high amount of energy is necessary to dissociate the bond. This type of bond is designated by an elliptical or sausage shape. Lesser energy states are designated by different shaped loops. Each covalent bond will have a characteristic length. This length is determined by the strength of the repulsion of the two positively charged nuclei, which must be in equilibrium with the negatively charged electrons shared by the two atoms.

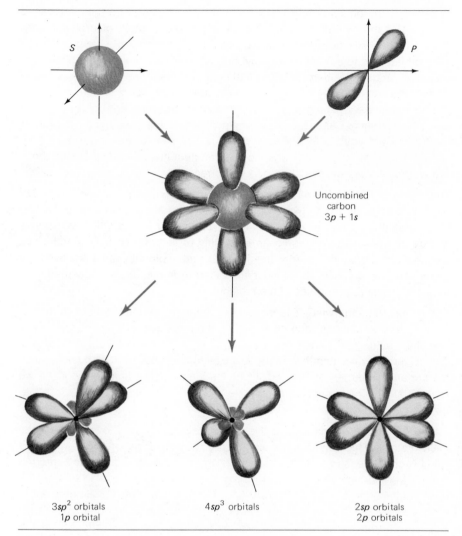

Figure 16-9
An atomic orbital is the region that surrounds the nucleus and contains two electrons. This drawing shows configurations of *s* and *p* orbitals. Uncombined carbon contains electrons in three *p* and *s* orbital with a common origin. These orbital regions may change greatly in shape and directions when the carbon atom combines to form a molecule. The sp^3 hybrid combines all three *p* orbitals and the *s* orbital; the sp^2 hybrid contains two *p* and one *s*, and thus leaves one *p* uncombined; the *sp* hybrid combines one *p* and one *s* and leaves two *p* orbitals intact.

In the methane molecule, carbon forms a tetrahedral molecule with the four bonded hydrogen atoms. The orbitals available on the carbon atom are three *p* orbitals and one *s* orbital. Since each *s* orbital has a spherical shape and each of the *p* orbitals has a figure-eight shape, the carbon atom could not form a symmetrical molecule in a reaction with hydrogen atoms, each of which has a single *s* orbital. Yet the four covalent bonds in the methane molecule are all of equal angles. This condition can only occur if the electrons are functioning in exactly equivalent electron states. The four electrons are therefore considered to act as a combination or hybrid of the one *s* orbital with the three *p* orbitals.

Orbitals in the Alkane Series

These **hybrid** orbitals, or mixtures of orbital types, seem to be the best way to conceive of the symmetry of the bonds within the context of orbital theory. These hybrids are designated as sp^3 orbitals, indicating the contribution of each orbital type to the hybrid. The hybrid concept helps solve other bond configuration problems. Chemists know that unpaired electrons tend to be situated as far away from each other as is physically possible; therefore they can predict the angle, the length, and the strength of any covalent bond by setting up a geometric model. Thus it was hypothesized that the bond strength of methane is 101 kcal/mole; the bond angle is 109.5°; and the length is 12 Angstroms. When technological advances permitted the quantification of these properties, all the guesses proved to be correct.

The sharing of electrons situates the nuclei in a common electron cloud, but electron negativity is not always equally distributed about the two nuclei. Most of the negativity may be on a balanced axis along a line with the positive charge of the nuclei, producing an inequality of electron density which gives the molecule a specific polarity. In such cases the molecule has a positive pole and a negative pole, which together are called a dipole; this polarity in turn affects the physical properties of the molecule.

But it may be otherwise. The bonds in methane, equally distant from one another, exactly cancel each other out, so the molecule exhibits no polarity. If one of the hydrogens is replaced by a chlorine atom, the molecule is polarized. The larger chlorine atom has many more electrons and a larger nucleus; thus the negative charges are pulled in the direction of the chlorine atom. If we replace all the hydrogens with chlorine atoms, the molecule (CCl_4) is equalized again and the dipole is eliminated.

Carbon-to-carbon bonds involve the sp^3 orbitals in an equal overlap between the two atoms. This results in the elliptical shaped bond called the sigma bond (σ); it is characteristic of bonds between s orbitals. Carbon bonded to hydrogen forms a very similar shaped bond, which is also a sigma bond. In ethane, the bond angles of the C—H bonds are 109.5°, as in methane. The C—H bond length is 1.10 Å; the C—C bond, slightly longer, is 1.54 Å. These proportions are consistently carried through the whole alkane series. The bond strength of the C—C bond is 83 kcal/mole.

Orbitals in the Alkene Series   The carbon double bond of the alkenes poses an interesting situation. In order to form bonds with three other atoms (H, H, C) the electrons are arranged in three equivalent hybrid orbitals lying in a single plane. If three equal length lines, arising from a central point between the three orbitals, are drawn out through each orbital, a triangle will be formed:

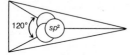

Each orbital is the hybrid of one *s* orbital and two *p* orbitals, forming sp^2 orbitals. The orbital angle is given as 120°. This trigonal arrangement allows the electrons to be as far apart as is possible. Each carbon atom itself lies at the center of the projected triangle formed with its two hydrogens at the base angles and the other carbon at the apex. In ethene, the bonds are sigma bonds similar to the ethane molecule, except that all the bond angles are 120°:

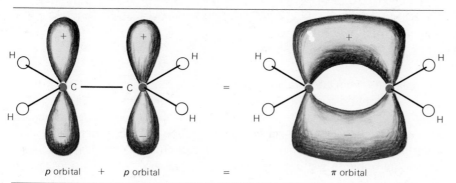

The remaining electron cloud on each carbon atom is in a *p* orbital with a portion of the cloud extending above and below the carbons in the figure-eight arrangement. The *p* orbital is perpendicular to the plane of the sp^2 sigma bonds formed between the C—C and C—H bonds. The *p* orbital clouds on each carbon overlap the other *p* orbital at the widest part of their loop above and below the sigma bond. This double overlap of the *p* orbitals is designated as a pi (π) bond. Since the electrons overlap less in this bond, less energy is needed to break it (40 kcal/mole) than the sigma bond (60 kcal/mole). For the alkene double bond, total dissociation of the bond requires 60 + 40, or 100 kcal/mole, which is greater than the ethane molecule.

Orbital theory may not necessarily agree with other methods of studying these

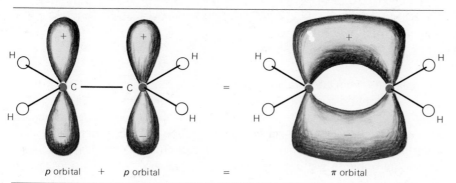

p orbital   +   *p* orbital   = π orbital

Figure 16-10
Overlapping two parallel *p* orbitals creates the familiar electron cloud of the π (pi) bond. Double and triple bonds are π bonds. Thus, sp^2 and *sp* orbitals are characteristic of multiple bonds because these hybrids leave unattached *p* orbitals to overlap and form the π bond needed to make the multiple bond.

molecules. Ethene, according to the orbital theory, has a bond angle of 120°. If we measure the bond angle through a modified electron microscope called an electron diffraction scope, we find that the bond angles are not those predicted by the orbital theory. The ball and stick model suggests yet another difference. These varieties in measurement using differing methods have not been reconciled by organic chemists as yet.

Orbitals in the Alkyne Series

The triple bond hydrocarbons, or alkynes, add another bond to the carbon-to-carbon interaction. These bonds are formed by a hybrid of the *s* orbital with one of the *p* orbitals, producing an *sp* hybrid orbital for the sigma bonds of the C—C and C—H attachments. The remaining two electrons on each carbon are in *p* orbitals. One *p* orbital is in the *y*-axis perpendicular to the carbon atom, as in the alkenes. The other is situated in the *z*-axis, which is perpendicular to the other *p* orbital and the *sp* orbital. The bond distance between carbons is 1.20 Å and the carbon-hydrogen distance is 1.06 Å. The bond angle of the C—C bond is linear (180°). The overlapping *p* orbitals, which form a highly reactive set of electrons, are one of the reasons these compounds are important starting materials for many industrial synthetic processes. The terminal hydrogens arranged linearly from the carbon atoms allow them to act as weak acids which can be removed by strong bases to form acetylide salts. These salts and starting compounds are used in the manufacture of such compounds as vinyl plastics.

Hydrocarbon Derivatives
Halogen Derivatives

One or more atoms of the halogen group, all of which have a valence of −1, can be substituted for hydrogen atoms in the hydrocarbon series to form the halocarbons. The importance of the halogen derivatives lies in the fact that they are used as intermediary steps in synthesizing carbon molecules; they are highly reactive and therefore allow for selective placement of alkyl groups in the construction of the molecule. Some of the simpler ones in common use are carbon tetrachloride (CCl_4), a cleaning fluid; chloroform ($CHCl_3$), an anesthetic; dichlorodifluoromethane or Freon (CCl_2F_2), a refrigerant; DDT ($C_{14}H_9Cl_5$), a chlorinated form of diphenyl ethane; and tetraethyl lead ($(C_2H_5)_4Pb$), an antiknock gasoline additive.

Alcohols

Alcohols differ from the paraffin hydrocarbons by the presence of one hydroxyl group. A methyl group with an attached OH is methanol (CH_3OH); an ethyl group with an attached OH is an ethanol (CH_3CH_2OH) or common grain alcohol. The presence here of one extra carbon makes the difference between a relatively harmless drug and a highly poisonous drug. The shorter chained alcohols are very miscible in water, since water (H_2O) has the —OH group in its own structure. Methanol and ethanol have high dielectric constants—30 and 24 respectively. These are not high enough to dissolve many ionic substances, but they are low enough to allow mixing with many organic compounds, which makes them very good solvents. The alcohols with low carbon numbers are also very volatile at room temperature and are therefore used in rubbing compounds as cooling agents. Their rapid evaporation when applied to the skin draws off heat from the surface of the body. The most common rubbing alcohol is isopropyl alcohol ($(CH_3)_2CHOH$).

The longer-chained alcohol compounds act less and less like water and more and more like oils and waxes. The longer-chained alcohols may not dissolve in water but will readily dissolve in methanol.

A very common hand softener and moisturizer—glycerin, which is a solution of water and glycerol—is an example of an alcohol with more than one —OH group in its structure:

$$
\begin{array}{ccc}
H & H & H \\
| & | & | \\
H-C-C-C-H \\
| & | & | \\
OH & OH & OH
\end{array}
$$

Glycerol is a very viscous liquid because the molecules are cross-bonded together by hydrogen bonds. Some of the common uses of this compound are in toothpastes, creams, lotions, and other cosmetics. Its slow evaporation accounts for its moisturizing features. When the OH groups are replaced by NO_2 groups, the result is the highly explosive compound nitroglycerine.

Organic Acids

Compounds with the carboxyl group COOH are called organic or carboxylic acids. They are usually characterized by a foul or astringent smell. The simplest organic acid is formic acid, which has a single hydrogen atom attached to the carboxyl group; formic acid is responsible for the characteristic smell of ants when they are crushed. The sharp taste and smell of vinegar is from acetic acid (CH_3COOH), which has a methyl group tagged onto the carboxyl:

$$
\begin{array}{cc}
H & O \\
| & \parallel \\
H-C-C-OH \\
| \\
H
\end{array}
$$

Rancid butter contains butryic acid ($CH_3CH_2CH_2COOH$), released by the microbes feeding on the butter. Longer-chain organic acids are important constituents of body fats and derivative compounds such as steroid hormones.

Esters   When alcohols react with organic acids, esters are formed and a water molecule is given off; the H in an OH group is replaced by a regular hydrocarbon
group. Esters are usually strikingly sweet smelling. The odors of ripened fruits
are due to esters in the cells of the fruit being released into the air. Whenever
an acid and an alcohol join to form an ester a molecule of water is formed from
the reaction:

$$H_3C-\overset{\overset{\displaystyle O}{\|}}{C}-OH + H-O-CH_3 \longrightarrow H_3C-\overset{\overset{\displaystyle O}{\|}}{C}-O-CH_3 + H_2O$$

   acetic acid ·   methyl alcohol   methyl acetate   water

The hydrogen for the water molecule comes from the alcohol and the hydroxyl
(—OH) from the acid. The above reaction produces an ester that smells like
bananas and is often referred to as banana oil. These fragrant esters are edible
in small doses in fruits, but in such chemicals as airplane glue they are highly
toxic and debilitating. It is the effect of these compounds on the nervous system
that produces the characteristic high of a glue sniffer. The liver, lungs, and
nasal passages are gradually damaged by persistent use of these toxic chemicals.

Amino Acids   When one H atom in the hydrocarbon group of an organic acid is replaced by an
NH_2 group, the result is an amino acid. One of the simplest amino acids is
glycine:

$$\underset{CH_2COOH}{\overset{NH_2}{|}}$$

When amino acids are linked together, they form proteins, the structural basis of
all living organisms.

Benzene Compounds   Carbon molecules exist not only in linear arrangement but also in rings. The
linear hydrocarbons are often called aliphatic compounds. The simplest ring
compound can be formed from a saturated hydrocarbon of hexane. The two
ends can join together, releasing the terminal hydrogens, to form cyclohexane
(C_6H_{12}). Cyclohexane is a nonplanar (three-dimensional) molecule with some
flexibility in switching between various conformational isomers due to changes
in the energy states of the molecule (see Figure 16-11). Benzene (C_6H_6) and its
pleasant smelling derivatives are called aromatic compounds. Benzene itself is
an unsaturated hydrocarbon ring with alternating double bonds; it is a planar
molecule. There are two structural isomers of benzene:

How does the benzene ring maintain its strong hexagonal shape? Since interatomic distances between single covalent carbon-to-carbon bonds are longer (1.5×10^{-10}m.) than double covalent bond (1.35×10^{-10}m.) the hexagonal figure of the ring might be expected to be distorted. Yet benzene is a very stable compound, stronger than the above resonating structural formula seems to indicate. Since there is no distortion of the ring, the three carbon electrons that are supposedly resonating from one carbon pair to another may be thought of as being equally shared by all the carbons. Therefore, a better formulation of the benzene ring would be:

The delocalized or resonating electrons are shown by the inner circle. These electrons are responsible for one of the characteristic properties of benzene compounds, their differential absorption of ultraviolet light. When several rings combine, brilliant colors are produced. Many of our dyes or pigments are polymers of such ringed compounds.

Benzenelike unsaturated ring compounds, such as toluene, a solvent, and phenol, an antiseptic, are also aromatic compounds.

toluene

phenol

Phenol was the first workable antiseptic, introduced by the English physician Joseph Lister in 1867. An alcohol with a double benzene ring is naptha:

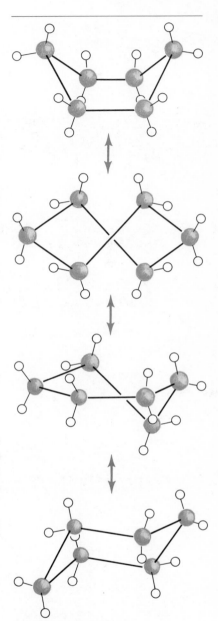

Figure 16-11
The differing configurations of cyclohexane are called conformational isomers, or conformers. The top configuration is called a boat; the second, a twist boat; the third, a half-chair; the bottom, a chair. The chairs are the most stable; the half-chairs, the least stable.

If we add nitrites (NO_2) to three other carbons on the toluene molecule, the ring becomes very unstable and will release a great deal of energy very quickly. This makes it a very potent explosive—trinitrotoluene, or TNT.

TNT

Industrial Organic Chemistry

Simple molecules called monomers can be synthesized into macromolecules called polymers, which are made up of repeating structural units; plastics and other common industrial products are made up of long polymeric chains of simple organic compounds.

In the early decades of this century, chemists discovered how to untangle the long chains of molecules in natural substances and to duplicate these intricate structures in synthetic materials through polymerization. In this process, **monomers**, or simple molecules, are synthesized into a single giant macromolecule, or polymer. Most of the products we take for granted in our daily lives are composed of long polymeric chains of simple organic compounds. Plastic tubing and sheeting, phonograph records, styrofoam, Lucite, Saran wrap, Teflon-coated cookwear, the housings of radio and television sets, even chewing gum, are products of the polymerization of basic compounds into complex compounds such as polyethylene, polyvinyl chloride, or polystyrene. Polymerization has led to the manufacture of products that are generally lighter, more durable, and less expensive than similar goods made of other materials.

The synthetic material that has had the greatest impact on our culture—and which has revolutionized the world's textile industry—is nylon, developed during World War II. Nylon, the first artificial fiber to be successfully synthesized, opened the door for the development of hundreds of man-made textiles and other products that surround us today.

The manufacture of nylon is very complex, involving a huge plant containing many kinds of complicated machinery and hundreds of workers. The synthesis of nylon begins when a nitrogen-containing compound, hexamethylene diamine, and the long-chain carboxylic acid, adipic acid, are united in a long chain, called a polyamide. The union of the two chemicals under high pressure produces a nylon salt, which is then poured into an autoclave, a huge machine similar to a pressure cooker. In the autoclave, the monomers are synthesized by

Figure 16-12
Acrylic fibers are produced by methods similar to those used in the manufacture of nylon. The polymer derived from acrylonitrile (CH_2=CHCN) is dissolved in a solvent to form a viscous solution that is then forced through a spinnerette, shown here. Then the solvent is removed, leaving filaments of the acrylic polymer from which acrylic fibers are made.

superheated steam under high pressure into macromolecules. The viscous polymer is then drained from the autoclave and allowed to cool and harden. Next, it is melt spun. In this process, the solidified polymer is heated again to the melting point and forced through a spinneret, a device with numerous small holes which resembles a shower head. As they emerge from the spinneret, the thin taffy-like threads are cooled and solidified by a jet of cold air, then drawn to about four times their original length. The drawing process causes the polymeric molecules to align themselves so they lie parallel with the long axis of the strand, creating a fiber. Drawing also increases the tensile strength and elasticity of the material.

The strength, elasticity, durability, ease of care, and inexpensiveness of nylon have made it ideal for many applications. It is used in the manufacture of carpets, upholstery, wearing apparel of all kinds, parachutes, sails, and auto tires. Its extensive industrial uses include sewing thread, filters, twine, screens, air springs, and many other items.

Alkanes are the series of saturated carbon compounds whose simplest member is methane (CH_4).

Alkenes are the series of double-bonded carbon compounds whose simplest member is ethylene (C_2H_4).

Alkynes are the series of triple-bonded carbon compounds whose simplest member is acetylene (C_2H_2).

Glossary

Hybrid orbitals are the electron orbitals in a molecule in which orbitals of various types are combined in the bonding.

A hydrocarbon is an organic compound containing only carbon and hydrogen.

Isomers are compounds that have the same molecular formula but a different structural formula.

A molecular formula is a chemical formula that indicates the total number of atoms of each element in a molecule.

A monomer is a chemical compound that can undergo polymerization.

Organic chemistry is the science of the properties of carbon-containing compounds.

A p orbital is an electron orbital that is shaped like a figure eight.

A paraffin is a saturated hydrocarbon.

A polymer is a chemical compound or mixture of compounds formed by polymerization and consisting of repeated structural units.

Polymerization is a chemical reaction in which two or more small molecules combine to form larger molecules that contain repeating structural units of the original molecules.

An s orbital is an electron orbital that has a spherical shape.

A saturated hydrocarbon is an organic compound containing only carbon and hydrogen in which each molecule contains the maximum number of hydrogen atoms that its carbon atoms can bond to.

A structural formula is a chemical formula that shows the arrangement of the atoms and the bonds.

An unsaturated hydrocarbon is an organic compound containing only carbon and hydrogen in which the molecules contain more than one bond between adjacent carbon atoms.

Exercises

1 Can you suggest a reason why the tetrahedral arrangement is the most stable one for carbon molecules?

2 Write out structural formulas for all isomers of C_6H_{14}. How many structural isomers exists? Assign them names on the systematic nomenclature.

3 Describe what is meant by each of the following terms:
   a. isomer
   b. unsaturation
   c. polymer
   d. $C_3H_4Cl_2$

5 Draw the isomers (use structural formulas) of the following organic compound: C_2H_6O.

6 Can you think of one reason why the two structural isomers of butane have markedly different boiling points?

7 Which of the following pairs of organic compounds would one expect to have the higher boiling point?

$CH_3—CH_3$ or $CH_3—CH_2—OH$

$CH_3—CH_2—CH_2—CH_3$ or $CH_3(CH_2)_4CH_3$

$CH_3—CH_2—OH$ or

$$CH_3—\overset{\displaystyle O}{\overset{\|}{C}}—O—CH_2—CH_3$$

$\begin{matrix} CH_2—CH_2 \\ | \qquad | \\ OH \quad\ OH \end{matrix}$ or $CH_3—CH_2—OH$

8 Give an example of a saturated compound and a closely related unsaturated compound. Why are saturated compounds less reactive than unsaturated compounds?

9 Listed below are several organic compounds; characterize each of them as being an alkane, alkene, or alkyne.
a. C_5H_{12}
b. C_7H_{14}
c. $C_{23}H_{46}$
d. C_2H_2
e. C_4H_{10}
f. C_6H_6

10 What is the molecular formula of:
a. an alkane with 5 carbons?
b. an alkane with 37 carbons?
c. an alkene with 9 carbons?
d. an alkyne with 4 carbons?
Draw structural formulas for each example.

11 What is wrong with the following molecular formulas?
a. C_3H_3
b. CH_{30}
c. C_4H_{16}

12 Which of the following compounds are not consistent with the tetravalent nature of carbon?
a. CH_3CH_3
b. $CH=CH_2$
c. $CH\equiv CH$
d. CH_4CH_2
e. $CH_2=CH_2$

13 Experience has shown that the greater the degree of branching that a hydrocarbon possesses, the better fuel it will be. If this is the case, which of the following compounds would one expect to be a better gasoline?
a. 2,2,4-trimethylpentane
b. n-heptane
c. n-butane
d. 2-monomethylnonane

14 What general statement can be made relating hydrocarbon chain length with the boiling point, melting point, and density of a given hydrocarbon?

15 Give an example of:
a. an alcohol.
b. a carboxylic acid.

  c. an aromatic compound.
  d. an ester.

16 Write the structural formulas corresponding to these compounds:
  a. 2,2,3,3-tetramethyl pentane
  b. 2,4-dimethyl heptane
  c. 2-methyl-3-chloropentane
  d. 3-chloropropene
  e. 3-bromo-2,3 dimethyl pentane
  f. butyne
  g. 2,2,4-trimethyl pentane

17 Describe what is meant by the term *optical isomerism* and cite an example.

18 Describe and illustrate what is meant by the terms *s* and *p* orbitals.

19 a. Draw the molecules ethane, ethene, and ethyne showing the hydridized orbital structure and the sigma (σ) and pi(π) bonding.
  b. Describe the geometric properties (i.e., bond length, bond angle, shape of molecule, spare arrangement, etc.) of each type of hydridization and bonding shown above.

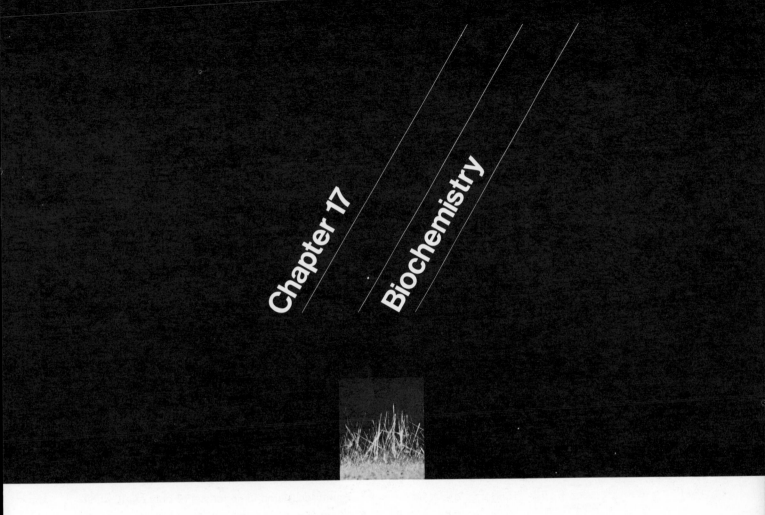

Chapter 17 / Biochemistry

In addition to their importance to industry, carbon compounds are also significant because they are the basic material of life. In the past, the complexity of
carbon compounds found in living organisms made them very difficult to
analyze. However, with advances in methodology and instrumentation, scientists have been able to probe the structure of these vital substances, and to discover some of the chemical reactions that occur at the molecular level in living
organisms. As a result a new science, called biochemistry, was born. **Biochemistry** is the study of carbon compounds found in the structure of living organisms. These compounds are divided into four groups: proteins, carbohydrates,
fats (lipids), and nucleic acids.

Carbohydrates

Americans daily eat a great deal of food which is almost exclusively carbohydrate—bread, cake, spaghetti, candy, potatoes, soft drinks, and sugar. The materials of our environment also contain these carbon compounds. Wooden seats and floors are carbohydrates, as are clothes made of cotton and linen. Carbohydrates are important for our food, clothing, and shelter.

Carbohydrate Synthesis

Mannose

Glucose

Fructose

Galactose

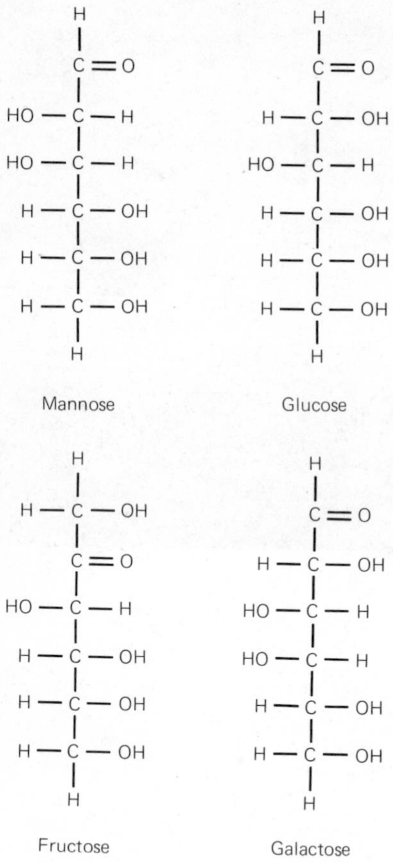

Figure 17-1
Each of these sugar isomers has the same formula—$C_6H_{12}O_6$.

Sugar, starch, and cellulose are forms of carbohydrates.

A **carbohydrate** is a compound of carbon, hydrogen, and oxygen manufactured in the leaves of green plants. From the plant, the carbohydrates are passed along the levels of the food chain (Chapter 21). The plant takes in carbon dioxide and water, and using the sun's energy, breaks down these compounds and then rebonds the carbon, hydrogen, and oxygen atoms to form small chains of carbohydrates. To carry out this process, the plant absorbs the sun's energy through specialized cellular structures that contain chlorophyll, a substance that acts as a necessary catalyst during the reaction. Since the bonding of the carbon atoms is a synthetic process that uses the radiant energy of the sun, the process is known as photosynthesis.

The result of photosynthesis is a 3-carbon compound called phosphoglyceric acid. This compound then combines with itself to form the more familiar sugar **glucose** ($C_6H_{12}O_6$). Glucose is the most important of the sugars because it serves as a source of fuel for all plants and animals. The formula for glucose is the same as certain other sugars, all of them isomers of glucose which form important molecules in the food chain of organisms. These include fructose, galactose, and mannose, which received its name from the Biblical desert food, *manna;* as with all isomers, they differ in molecular structure. The structural formulas for these isomers are shown in Figure 17-1.

Sugars are also known as saccharides. When they consist of only one C_6 unit, simple sugars are called monosaccharides. If two of these sugars are combined, they form a disaccharide. Similarly, the combination of three monosaccharides yields a trisaccharide, and the joining of more than three sugars forms a polysaccharide.

After a plant produces numerous molecules of glucose, these sugars combine into long polymers of from 300 to 1000 monomers with the useful property of storing energy for the future. These are the polysaccharide starch molecules. Starches are a major source of food energy in the roots and grains of plants. Starch is usually insoluble, so it stores well in plant cells, until a starch-using biochemical reaction converts it back into soluble glucose. A similar polysaccharide, chitin, is found in the outer shells of insects, crabs, lobsters, and other shelled animals.

Plants can also build the glucose unit into even longer chains of about 1500 monomers to form cellulose. The walls surrounding plant cells are composed mostly of cellulose, making them extremely tough. When these plant walls are crushed and processed, paper can be made from the wood. Similarly, if the wood is drawn out into threads, it can be woven into cloth.

Carbohydrates are important biochemically because they are the basis for all other organic chemicals used by living organisms. In a way, the cells of plants and animals act as miniature organic chemists, shifting carbons and other atoms of carbohydrates to rearrange them into lipids, proteins, and nucleic acids. To do this, the cells need the energy in the bonds that hold the carbon chains of carbohydrates together. This energy is released through a number of complex steps to separate the carbons that had been put together by the plant during photosynthesis. The process of reversing photosynthesis is an oxidation reaction, which in living matter is called respiration. Respiration, then, is simply the oxidation of glucose to carbon dioxide and water, with release of energy in an exothermic reaction:

Carbohydrates in Metabolism

Carbohydrates are the basic substances from which proteins, lipids, and nucleic acids are made.

$$C_6H_{12}O_6 + 6O_2 \longrightarrow 6CO_2 + 6H_2O + energy$$

Although breathing is often referred to as respiration, it is really only the gas exchange part of this process of energy transfer.

The food we eat supplies the carbon chains for these reactions. After it has been eaten, the food undergoes digestion, which is mainly a process to break down the disaccharides and polysaccharides into monosaccharides; these are then readily absorbed by the cells. During digestion, the long starch chains are broken down into simple sugars by the process of **hydrolysis** (from *hydro,* water + *lysis,* breakdown). This is defined as the addition of water to split the bonds which join 2 or more molecules. In this process, water is inserted at the oxygen bridges which join the starch chains.

Hydrolysis results in the formation of free glucose molecules, which are in turn easily broken down by the cell. The process is aided by other specialized organic compounds called **enzymes.** These are large protein molecules which are known to act as catalysts for all reactions in the cell. During hydrolysis, special starch-digesting enzymes, the amylases, digest starch which

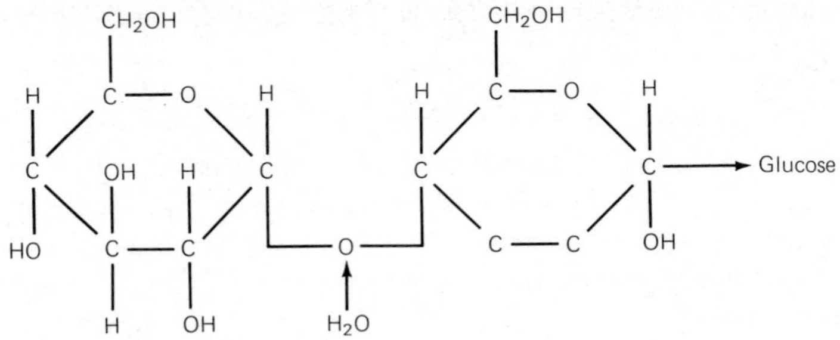

Figure 17-2
During hydrolysis, the polysaccharide starch chain is split into two simple glucose molecules when a water molecule is inserted at the oxygen bridges which join the chain.

helps the insertion of the water molecules at the oxygen bridges. An example of this process is seen in the breakdown of complex sucrose into the simpler glucose and fructose:

$$C_{12}H_{22}O_{11} + H_2O \longrightarrow C_6H_{12}O_6 + C_6H_{12}O_6$$

But this process is not as simple as the reaction equation seems to suggest, and it takes place over several steps. The polysaccharide starches, which are insoluble and too large for ordinary digestive processes, are broken down by the water insertions into simpler disaccharides, which are soluble but still not small enough for absorption. Further hydrolysis breaks these down into monosaccharides which are then absorbed in digestion.

The sugars absorbed into the animal cell as a result of digestion can be stored in the form of the polysaccharide glycogen. This substance is to animal tissues what starch is to plant tissue—a molecule to store energy in its chemical bonds. In man and other animals, glycogen is stored in the liver and in muscle tissues for use as energy during muscle contractions.

Lipids

If an animal or plant takes in more carbohydrate than it needs for carbon chain synthesis and energy production, it synthesizes fatty molecules known as lipids. Fats, oils, waxes, and other compounds such as cholesterol and the steroid hormones (sterols), make up the lipids in living organisms. These compounds, like carbohydrates, contain carbon and hydrogen, but the hydrogen–oxygen ratio is much higher. This can be seen by comparing the formula of a carbohydrate with that of a fat. For example, the formula for a common carbohydrate, glucose, is $C_6H_{12}O_6$; tristearin, which is a fat, has a formula of $C_{57}H_{110}O_6$.

Plants manufacture lipid compounds that are impermeable to water and so help prevent excessive water loss from the surface of the stems and leaves. The cell membrane of all cells in plants and animals is a sandwich affair, with lipid molecules making up the inner layers between two outer layers of protein. Since a membrane is involved in most of the known major chemical reactions in cells, lipids are very important to biochemical functioning. Most of the chemicals produced by organisms to affect membrane activity, such as hormones, are either sterols or proteins.

Lipid Molecules

The fat molecule is composed of two parts: an alcohol—usually glycerol—linked to three fatty acid molecules. These two parts are joined by an ester group in a compact arrangement which makes fat an excellent storage molecule. The structural formulas of the components of the fat molecule and the ester group that joins them are:

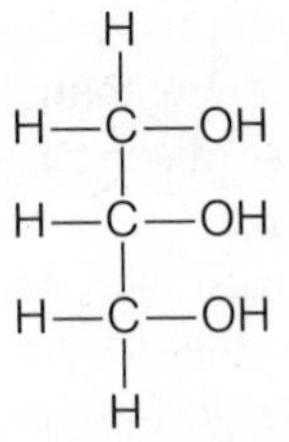

Glycerin

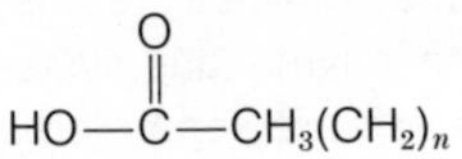

Generalized fatty acid

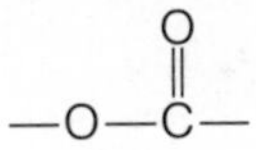

Ester group

During the synthesis of the fat molecule, each carboxyl group of the fatty acid molecules bonds with one of the alcohol groups of the glycerol to form an ester bond. Each ester bond results in the release of a molecule of water. A total of three molecules of water are formed each time a fat molecule is produced.

Fatty acids can be saturated or unsaturated. In **saturated** fatty acids all of the carbon atoms are attached to a hydrogen atom and no more hydrogen can be added. Thus, the structural formula for stearic acid ($C_{17}H_{35}COOH$) is:

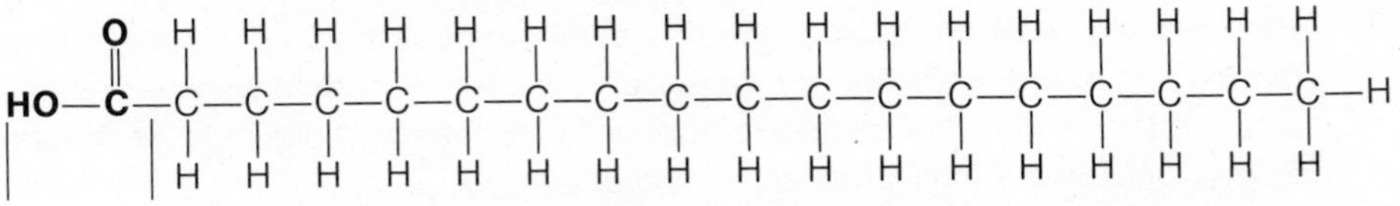

An example of an **unsaturated** fatty acid, or one that could accept an additional H, is linoleic acid, ($C_{17}H_{31}COOH$). Because hydrogen atoms can be attached to four of the carbon atoms in this substance, it is known as a **polyunsaturate**:

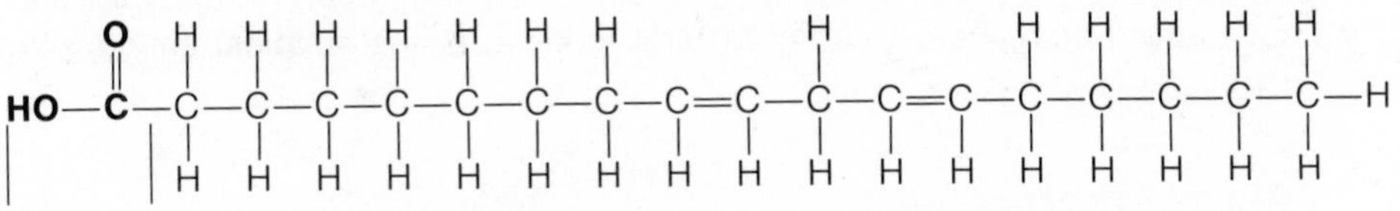

Recently doctors have been advising us to eat polyunsaturated rather than saturated fats. The reason for this is clear from the structural formulas given here: polyunsaturated fats contain fewer hydrogen–carbon bonds and are therefore easier to break down during digestion. Saturated fats, on the other hand, tend to be routed to storage cells because they are difficult to decompose.

A polyunsaturated fat is better for human health because it is more easily decomposed.

During digestion, the fat molecule is broken down by water in the process of hydrolysis. The catalyst, an enzyme known as lipase, splits the fat molecule into

Metabolism of Lipids

glycerol and fatty acids. The glycerol and those fatty acids having fewer than 10 carbon atoms enter the bloodstream; the fatty acids with more than 10 carbons are absorbed into the organism's lymphatic system, later making their way into the blood stream. Fat molecules are separated from one another by bile, a yellowish-green fluid produced by the liver that acts much like soap. Just as soap separates the large lipids that stain our clothes so that they can be dissolved in water, so the bile separates the molecules of fat in the small intestine, readying the molecules for further breakdown.

Since the fat molecule contains many more carbons in a more compact state than the carbohydrate molecule, it is a more stable and useful form of energy for the organism than is the carbohydrate. The carbon chains of the fat molecules that have been broken down can be converted to any other carbon sequence required by the cell. This rearranging of the numerous carbons in lipids produces much more energy than the rearrangement of an equivalent volume of carbohydrate.

Proteins

Amino acids are the building blocks of proteins.

Proteins, the most complex of organic compounds, are very long molecules that contain nitrogen, carbon, hydrogen, and oxygen. In addition, some proteins also contain sulfur. The protein carbon chains contain thousands of carbon atoms. One of the milk proteins, lactoglobulin, for example, has the formula $C_{1864}H_{3012}O_{576}N_{468}S_{21}$ and has a molecular weight of 47,000. The basic monomeric unit of the protein molecule is the **amino acid**, a nitrogenous compound containing an amino group ($-NH_2$) and a carboxyl group ($-COOH$). Each group is located at an opposite end of the molecule. A third important group, known as the R-group, is joined to the carboxyl group. The R-group designates any other atomic group — from a single H to a long alkyl group — and it differs for each amino acid. Here is the structural formula for the simplest amino acid, glycine:

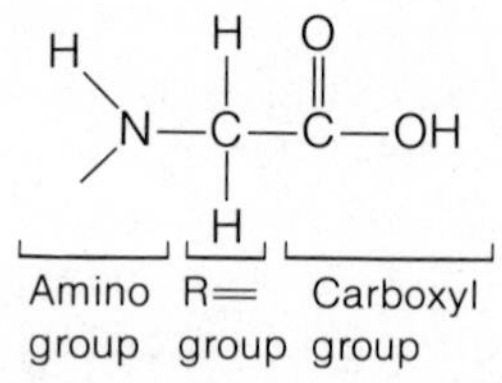

The variations of the R position, which in glycine is occupied by a hydrogen group, change the nature and properties of the amino acid.

Amino acids can join to one another to form proteins by means of the peptide bond. This occurs as the amino group of one amino acid bonds covalently to

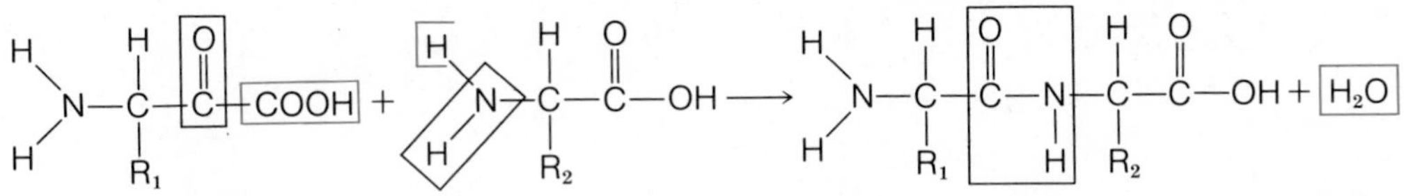

$$H_2N-\overset{H}{\underset{R_1}{C}}-\overset{O}{\underset{}{C}}\boxed{-COOH} + \boxed{H}\overset{}{N}>\overset{H}{\underset{R_2}{C}}-\overset{O}{\underset{}{C}}-OH \longrightarrow H_2N-\overset{H}{\underset{R_1}{C}}-\overset{O}{\underset{}{C}}\boxed{-N-}\overset{H}{\underset{R_2}{C}}-\overset{O}{\underset{}{C}}-OH+\boxed{H_2O}$$

Figure 17-3
During peptide bonding, the carboxyl group
(COOH) of the amino acid molecule at
extreme left joins with the amino group of
the amino acid molecule to its right, and a
molecule of H_2O is liberated. The peptide
bond is shown in the boxed area in the
group to the right of the arrow.

the carboxyl group of its neighbor, resulting in the release of a water molecule. The peptide bonding process is shown in Figure 17-3.

The number of amino acids contained in the peptide formed by the peptide bonds ranges from 2 to 30 or more. Two amino acids bonded by a peptide form a dipeptide. Additional amino acids form longer peptide chains, or polypeptides, and they are the proteins.

Protein Variety

The number of proteins that are known approach 100,000 in the human body alone, and each is a combination of 23 different types of amino acids. This can be best understood by comparing the amino acids with the letters of the alphabet. The words *pat, tap,* and *apt,* for example, contain the same three letters, but the sequence of the letters gives each word its own peculiar meaning. The huge number of different proteins arises from two facts: the 23 types of amino acids can be arranged in an almost astronomical number of combinations, and each acid can be linked end-to-end to other amino acids to form extremely long chains of proteins, some of which are composed of thousands of amino acids. Each protein derives its specific properties and function from the type and sequence of the amino acids comprising it.

Protein Structure

The enormous complexity of the protein molecule is the result of three factors. First, as we have seen, the primary structure of the protein is determined by the amino acid sequence in the polypeptide chain. Some of the chains consist of atoms or monomers joined end-to-end like a train, forming one long molecule. However, the great majority of proteins are spatially arranged in different ways and this is the second reason for their complexity. Many proteins, for example, consist of coiled peptide chains linked by chemical bonds at many parts of the coils. This coiling gives many proteins their secondary structure. Third, an even more complex configuration often results when the coiled polypeptide chains are connected to one another by cross links and then folded in very complex patterns that assume globular shapes.

The type and sequence of amino acids along a polypeptide chain determines the properties of the protein.

A large number of proteins have polypeptide chains coiled together in a ribbon pattern known as the **alpha helix.** (*Helix* means a coil or spiral, and *alpha,* the first letter of the Greek alphabet, is commonly used by scientists to indicate that something is the first of several of its kind.) The amino acids are cross-linked to

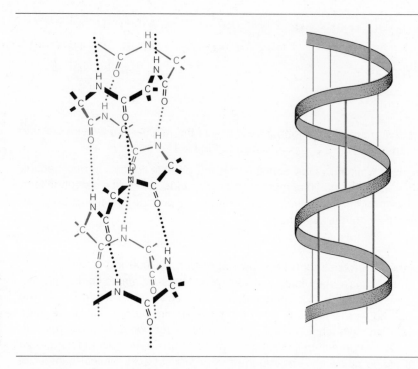

Figure 17-4
The alpha helix of protein consists of a polypeptide backbone, as shown at left. The dotted lines, which connect the C—O with the N—H groups, represent hydrogen bonds which hold the helix together. The figure on the right is a simplified drawing showing the definite helical form of the polypeptide chain.

one another on the helix in two patterns. First, weak hydrogen bonds hold together amino acids positioned above and below each other on the helix. Thus the C=O group of one amino acid is joined to the N—H group of the one above or below it. Each turn of the helix holds 3.7 amino acids along its length. Second, cross links can also occur between molecules of a certain type of amino acid, the cysteines. Since these acids form the beginning and end of each chain, any link between them will result in the folding of the chain in its tertiary structure. Cysteine has a sulfhydral group (—SH) which can join to other sulfhydral groups to form a disulfide bond. These covalent bonds are stronger than the hydrogen bonds that form the helix. When a protein is heated or acids or bases added to it, the hydrogen bonds are broken, changing the structure of the protein irreversibly. This process is called denaturation. Every time we cook an egg, we denature the protein in the albumin, or white, of the egg, causing it to change color and assume a new and rigid shape.

In addition to the alpha helix shape, proteins may also be arranged in a **beta** configuration, which results when two or more uncoiled polypeptide chains lying side by side become attached. In this protein arrangement the adjacent polypeptide chains interact to form a structure like a sheet, sometimes with folds or pleats. The pleating effect is due to the fact that the hydrogen bonds holding the polypeptide chains together shorten the distance between the alternating amino acid groups. Fibers of silk, for example, are made of long filaments of protein in the beta configuration. The filament chains are cross-linked to each other by their hydrogen bonds. This makes the fibers unusually strong and flexible. Wool, in contrast, is made of proteins with an alpha helix structure. When wool fibers are stretched, the hydrogen bonds between the turns of the helix are broken and the fiber extends, forming a beta chain structure. Releasing the tension returns the fiber to its unstretched length and alpha configuration. Collagen, the protein that makes up tendons and other animal tissues, has a different structure. Three polypeptide chains are coiled together in a triple helix, which is then combined with other like strands to build up a strong cord of protein.

Proteins arranged in alpha helixes are often soluble in water, and are active in many chemical reactions that occur in the cell. Proteins with the beta configuration are usually insoluble in water and therefore less active, and hence they usually occur in the structural components of organisms.

A special kind of protein is the enzyme, a complex molecule which helps the cell build up or break down other materials. Enzymes are catalysts, playing key roles in chemical transformations without themselves being changed. Because of this, they can be used and reused many times over, so the organism does not need to expend energy to create new enzymes for the thousands of chemical transformations that continuously occur within its cells.

The Role of Enzymes

An enzyme is a kind of protein that acts as a catalyst.

Enzymes permit chemical transformations in the cell at relatively low temperatures. Thus, enzymes eliminate the need for large inputs of heat or other forms of energy to carry out these reactions.

Each of the thousands of enzymes in the human body is specific in its action; each is a specialist, reacting only with certain molecules. The molecules with which the enzymes react are known as **substrates**. Thus, the enzyme maltase is active only in the transformation of the sugar substrate maltose; it cannot help catalyze reactions involving glucose, lactose, or any other substance. There are some enzymes—trypsin is an example—that can catalyze more than one substrate.

Enzyme specificity is probably due to the great variety in shape of these larger protein molecules. The enzyme surface is like a pattern, evolved to fit perfectly the corresponding surface of the substrate it catalyzes. In combination reactions, the two compounds to be bonded together are aligned in the position and at the distance that maximizes the chance for successful collision. In decomposition reactions, the compound to be broken down is positioned on the enzyme so as to twist or strain the bonds of the molecule at its weakest point. An example of such a reaction can be seen in the enzyme ATP-ase and its substrate, adenosine triphosphate (ATP), an extremely important molecule that contains the high energy phosphate bonds which are the immediate source of energy—the battery, so to speak—for nearly all chemical transformations in all organisms. The ATP molecule consists of adenosine and three phosphate groups, joined by high energy bonds. The molecule is transported to a place in

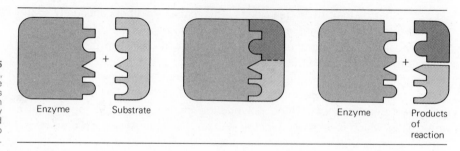

Figure 17-5
Enzymes are specific for certain molecules, as this sequence shows. At left, the opposing surfaces of the enzyme and its substrate molecule make a perfect fit. In center figure, the two combine. When they separate, the enzyme remains unchanged while the substrate has been converted into two molecules.

Enzyme   Substrate   Enzyme   Products of reaction

the cell where energy is needed. Then, to insure its most efficient use, the energy in the molecule is released bit by bit. This takes place as the enzyme ATP-ase breaks the high energy bonds that link the last phosphate group to the ATP molecule, thus releasing a discrete parcel of energy to the cell. The release of the terminal phosphate group leaves a compound called adenosine diphosphate (ADP). This exothermic reaction is then balanced by a subsequent endothermic reaction in which the ADP molecule is converted back into ATP when another enzyme bonds a phosphate group to the ADP molecule.

For almost every organic compound produced within living organisms, there has evolved an enzyme that breaks it down; therefore all the matter on which life depends can be perpetually recycled. Unfortunately, the same cannot be said of most man-made organic compounds, so there is no way that substances such as plastics and pesticides can reenter the natural cycle of decomposition and reuse. DDT will remain unchanged in the soil for decades; plastic detergent bottles buried in a landfill will not decompose. These synthetic compounds therefore present a serious threat to the environment.

COENZYMES AND VITAMINS Enzymes can have helpers known as **coenzymes.** Some coenzymes assist the principal enzyme by joining to it, filling out the surface pattern, and thus creating a better fit with the substrate. Others act as transporters: after the principal enzyme has broken down a compound, the coenzyme bonds to and carries some of the materials produced in the reaction to another site in the cell. An example of a coenzyme is nicotinamide adenine dinucleotide (NAD), which is derived from the vitamin niacine. NAD functions as a transporter of hydrogen that is produced when glucose is oxidized during respiration. The coenzyme carries the hydrogen to another similar hydrogen-transporter coenzyme.

Many other coenzymes are also derived from vitamins, which are complex compounds necessary for the growth and maintenance of the organism. Some plants and animals can synthesize their own vitamins, but human cells can manufacture only vitamin D in useful amounts and must obtain vitamins from food. Certain abnormalities are known to develop in the cell when the enzyme responsible for normal cell activity cannot function for lack of the vitamin coenzyme. If the cell's supply of the vitamin riboflavin is insufficient, for example,

Vitamin	Source	Function	Effect of deficiency
Fat soluble vitamins			
A	leaves, green and yellow vegetables	active in eye pigments, tissue maintenance	night vision affected; cells susceptible to infection
D	fish liver oils fortified milk	controls calcium, phosphorus in bones and teeth	rickets
E	many foods, especially leafy vegetables	unknown	body fails to absorb fat
K	cereal grains, fish	liver synthesis of clotting factors	blood fails to clot
Water soluble vitamins			
B_1 (thiamine)	meat, yeast, cereal grains	coenzyme in cellular respiration	beriberi, heart damage
B_2 (riboflavin)	dairy products	active in cellular respiration	loss of hair, eye damage
niacin	meats, milk	cellular redox reactions	pellagra
B_6 (pyridoxine)	liver, wheat germ, yeast	helps produce energy by breaking down glucose in a reaction series called the Krebs citric acid cycle	convulsions
folic acid	leafy vegetables	transfer of one-carbon fragments in many synthetic pathways	anemia
B_{12}	meats, especially liver	active in metabolism of nucleic acids	pernicious anemia
C	citrus fruits, vegetables	aids in production of collagen	scurvy

Table 17-1
Some vitamins important in human nutrition

certain respiratory functions will be affected. Similarly, if cells do not receive sufficient quantities of vitamin C, or ascorbic acid, enzymes will not be able to produce the collagen that cements connective tissues together. This leads to a debilitating and often fatal disease known as scurvy, characterized by loose teeth, weakened bones, muscle degeneration, and a general feeling of sluggishness.

Nucleic Acids

Although proteins are the most important chemicals in all organisms because they are the basis of all its structural and functional units, they are incapable of maintaining their number and functional patterns by themselves. A relatively small fraction of the cell substance, the nucleic acids, contain the control design, or code, for reproducing proteins, and with them, reproducing the cells too. Nucleic acids are similar to proteins in that both are giant or macromolecules of high molecular weight. Just as the proteins consist of long chains of amino acids, so the nucleic acids contain long chains of substances known as nucleotides. The chemical properties and biological nature of a nucleic acid molecule is determined by the type of nucleotides it contains and by the sequence in which the nucleotides are arranged along its length. However, nucleic acid molecules are much larger than proteins, some having molecular weights in the billions.

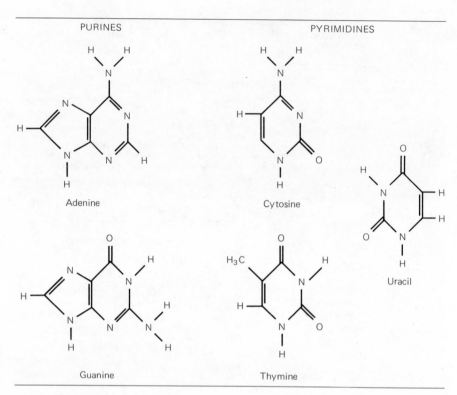

PURINES   PYRIMIDINES

Adenine   Cytosine   Uracil

Guanine   Thymine

Figure 17-6
Nucleic acids contain these five nitrogen bases. Adenine, guanine, cytosine, and thymine are constituents of DNA; RNA contains uracil instead of thyamine.

Each nucleotide is composed of a 5-carbon sugar, a phosphate (—PO_4), and a nitrogen base. The two kinds of nucleic acid are deoxyribonucleic acid, or DNA, with the sugar deoxyribose ($C_5H_{10}O_4$), and ribonucleic acid, or RNA, with the sugar ribose ($C_5H_{10}O_5$). The difference between the two acids arises from the fact that DNA lacks an —OH group. DNA is found in the chromosomes of the cell nucleus, whereas RNA is found both in the nucleus and in the cytoplasm.

DNA is the master molecule that influences the heredity and evolution of all organisms. Its molecular structure resembles a spiral staircase with sugars and phosphates on the outside of the molecule and nitrogen bases linked by hydrogen bonds on the inside.

DNA, which is a macromolecule with a molecular weight of 6 million, is composed of three substances: deoxyribose, phosphoric acid, and two paired nitrogen bases known as pyrimidines and purines. The pyrimidines (cytosine and thymine) are single-ringed structures; the purines (adenine and guanine) are double-ringed. The bases are commonly abbreviated by their initial letters to C, T, A, and G.

DNA governs all the metabolic activities of the cell because it directs the synthesis of necessary cell proteins. It plays a key role in cell reproduction and heredity by reproducing itself and transmitting its characteristics to daughter cells. DNA also provides the mutations necessary for the evolutionary development of the organism. A mutation is a cell that differs in form from its parents in some important and noticeable way. Thus, one or more of the nitrogen bases which comprise the DNA molecule may be mutated, possibly as a result of x-ray radiation or some natural process. When this occurs the mutant cell will not be identical to the cell that produced it, and eventually, in the course of biological evolution, the entire organism will differ from its ancestors.

Credit for unraveling the complex structure of the DNA molecule goes to two biophysicists, James Watson, an American, and Francis Crick, an Englishman. In 1953 these scientists used x-ray crystallography to decipher the structure of the macromolecule. They produced a model of the DNA molecule that showed it to have the shape of a double helix, composed of two intertwined chains of alternating phosphates and sugars connected by the bases A, T, C, and G. In this model, the sugars and phosphates were on the outside of the molecule and the bases were on the inside. The only problem with this configuration was that the bases had to be sequenced in a certain order, and this order was difficult to discover.

The two investigators experimented with various base combinations, finally hitting on the correct configuration, which is based on a system of pairing. They found that adenine and thymine formed one base pair of the molecule, while guanine and cytosine formed the other base pair. The base pairs were attached by hydrogen bonds. Thus, the structure of the DNA molecule resembles a spiral staircase in which the alternating sugars and phosphates form the vertical rails, and the nitrogen bases form the horizontal steps of the structure. The bases can be arranged in any sequence—TA, CG, GC, AT, etc.—but each sequence determines the nature and properties of the particular DNA molecule that they form.

Since the base pairs are fixed—A always pairs with T, and G always pairs with C—it is possible to determine the base sequence of both strands of the molecule even if one of them is missing Thus, if the base sequence on one strand of DNA is G T A A C C G, then the base sequence on the opposite strand must be C A T T G G C.

DNA REPLICATION The Watson—Crick model of the DNA structure explains how DNA makes identical copies of itself in the replication process. Since it transmits hereditary characteristics from one individual to another through the cells of the reproductive process—the sperm and egg of mammals, for example—DNA replication is the mechanism behind the reproduction of all life. DNA begins replication when its two strands become untangled; they unzip just like a zipper as the weak hydrogen bonds connecting them are dissolved. Each separated strand can be thought of as half the staircase with a sequence of unattached bases as its rungs. Free molecular nucleotide base groups floating in the nucleus then attach themselves to those on the half-staircase according to a definite pattern. Thus, a free-floating C base will pair up with an unattached G base on the staircase; a floating T base will attach itself to an A base on the stair, and so on. Since the same process has been occurring with the other half of the DNA, two complete DNA molecules will be formed and they will be exact copies of the original parent molecule. In other words, there are

The Watson—Crick Model of DNA

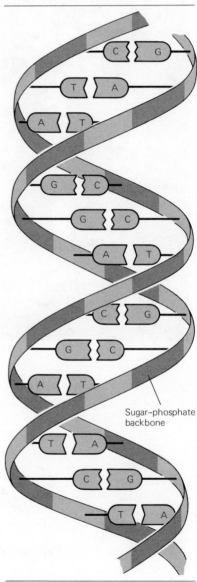

Sugar–phosphate backbone

Figure 17-7
This segment of the DNA molecule shows its chemical structure. Two intertwined strands of alternating phosphates and sugars are joined by nitrogen bases connected by hydrogen bonds.

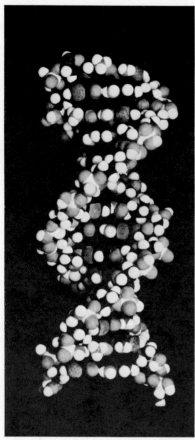

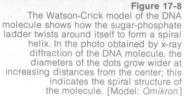

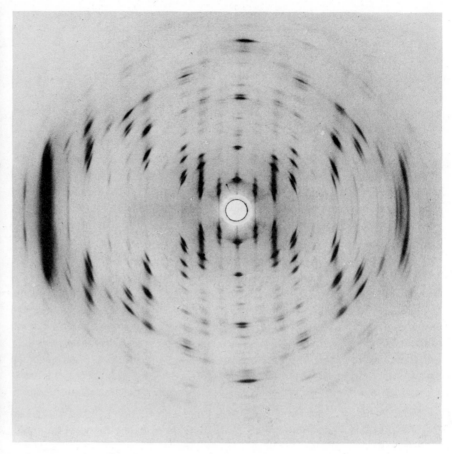

Figure 17-8
The Watson-Crick model of the DNA molecule shows how the sugar-phosphate ladder twists around itself to form a spiral helix. In the photo obtained by x-ray diffraction of the DNA molecule, the diameters of the dots grow wider at increasing distances from the center; this indicates the spiral structure of the molecule. [Model: *Omikron*]

now two identical DNA staircases where only one existed before. This simple process is the basis of the transmission of inherited characteristics in all living organisms.

Structure of RNA

The other important nucleic acid, RNA, differs from DNA in a number of ways and plays a wider role in intracellular reactions. The RNA molecule contains ribose rather than deoxyribose and uracil rather than thymine. Thus, RNA's components are A, U, G, and C. The RNA molecule is shorter than DNA, and has only a single helix. RNA is not involved with the direct production of genetic material; its chief role seems to be that of a go-between, or messenger, in the process of protein synthesis.

Not much is known about the structure of RNA. DNA has a crystalline form which is easily revealed through x-ray crystallography. But because RNA does not have a pure crystalline structure, it is difficult to analyze by x-rays. It is known, however, that RNA is produced by DNA in almost the same way that DNA replicates itself. First, the double strand of DNA uncoils and unzips, leaving the sequence of unattached bases exposed. Ribonucleotides floating about in the nucleus of the cell then attach themselves to the bases on the DNA strands, again in a fixed pattern. Thus, floating G ribonucleotides are joined to C bases on the DNA strand, while floating A nucleotides are joined to T bases. In addition, each A base on the DNA strand picks up a uracil (U) base from the nucleus. At the end of this process, there is a single molecule, half of which is DNA, and half of which is RNA. This unusual molecule exists for a very brief time; it then unzips and the RNA strand separates and leaves. The RNA will contain a sequence of nucleotides which complement that of the DNA from which it was formed. In this way, it obtains from the DNA molecule the sequential code necessary for protein synthesis.

TYPES OF RNA   There are three types of RNA, each of which has a definite function in protein synthesis. Messenger RNA (mRNA) carries the genetic pattern produced by the DNA to the ribosomes where protein molecules are made.

Ribosomal RNA (rRNA), the heaviest variety, is found in the ribosomes, which are small spheroids composed of 70 percent RNA and 30 percent protein and found in the cytoplasm of the cell where protein is manufactured. Although little

Figure 17-9
The DNA molecule reproduces itself in the manner shown here. The molecule at left splits down the center into two separate strands. Nitrogen bases in the cell nucleus then join to the unattached bases on each strand and two new DNA molecules are formed.

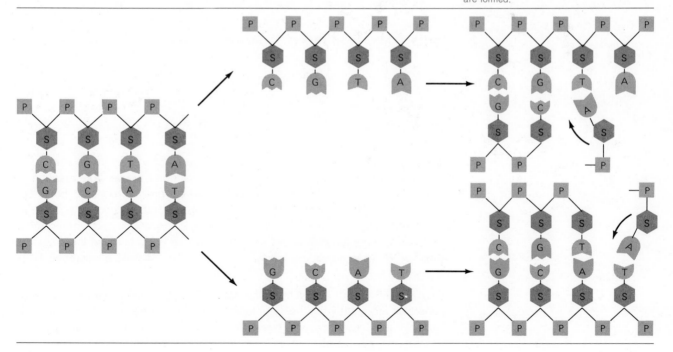

is known about the structure of rRNA, it is thought that part of its structure assumes the configuration of a double helix similar to that of DNA.

Transfer RNA (tRNA) is the smallest variety, and each molecule contains about 80 nucleotides. It is active chiefly during the manufacture of protein, picking up amino acids and transporting them to ribosomes. The molecule is thought to consist of a long strand which doubles back on itself, forming several loops which resemble a cloverleaf. This structure leaves two exposed ends, to one of which the amino acids attach. Since each cell contains about 20 varieties of tRNA, and there are about 23 amino acids, there is roughly one type of tRNA for each type of amino acid.

Protein Synthesis

Self-replication is only one function of the DNA molecule in the cell. This extraordinary molecule is also responsible for protein synthesis. Here the nucleotide sequence of the DNA is a **template**, or blueprint, which is passed on in the form of a code to mRNA and used to assemble the 23 amino acids into thousands of different proteins. Although the composition and structure of the DNA molecule is the same in all organisms, the information contained in its template is different. For this reason, the message transmitted by the DNA in the cell of a dog produces protein that is peculiar only to dogs, not to cats, birds, men, or any other organism. The process by which a single protein is manufactured can be schematized in the following manner:

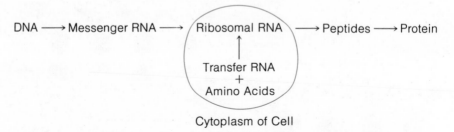

Cytoplasm of Cell

Protein synthesis begins when the DNA molecule transcribes its genetic information into copies of mRNA in the manner described above. This reaction is initiated because the cell requires a particular protein enzyme to build up, break down, or transfer some substance necessary for its functioning. Somehow, the cell is able to send a chemical signal to the DNA about this need. Thus, the primary function of the template in the DNA is to make the enzymes which, as we have noted, are the proteins that govern the cell's metabolic functioning.

The mRNA, tRNA, and rRNA interact at the cell's ribosomes where they manufacture proteins.

While the mRNA is being formed by the DNA on the chromosomes, the tRNA gathers amino acids from the cytoplasm of the cell, in the first of a series of intricate interactions in which codes are used to connect two substances. The

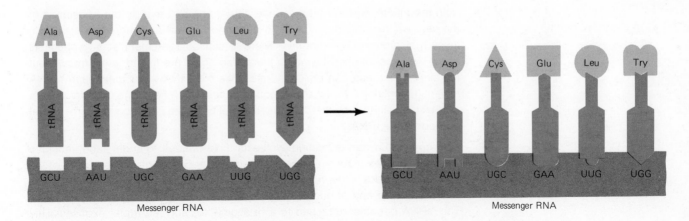

Messenger RNA

Messenger RNA

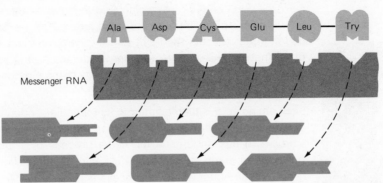

Messenger RNA

Figure 17-10
Protein synthesis occurs within a cell ribosome as amino acids are bonded in a peptide chain. As the top left figure indicates, each molecule of tRNA includes at one end a base sequence that is specific for a particular kind of amino acid and at the other end a base sequence that is complementary to a particular triplet site on the mRNA molecule. The mRNA molecules are located at the ribosomes of the cell. In addition, the sequencing of the molecules of amino acid, tRNA, and mRNA determine which peptide is formed. Top right: the amino acids are bonded to the tRNA, which in turn bonds with the mRNA. It is believed that the rRNA on the surface of the ribosomes facilitates this union. Bottom: the amino acids, joined into a peptide chain, pass into the cytoplasm of the cell, as do the tRNA molecules, which have broken away from the mRNA on the ribosome.

tRNA contains two sets of nitrogen base sequences—one at either end of its structure—arranged in a fixed order. The base sequences at one end of the molecule are arranged in a coded triplet of nucleotides known as a **codon**. (For example, the codon may be CCG or ATT, or TCG, or any of the huge combination of bases.) The codon recognizes, or fits, only one type of amino acid and attaches to it. The arrangement of the bases on the codon is analogous to the insertion of a three-pronged electrical plug into a receptacle socket. If you try to insert the plug into a receptacle having two holes, or if you hold the plug the wrong way—upside down or sideways—you will not be able to make the connection. The metal prongs of the plug and the holes of the receptacle must be of the same size and shape, and they must be aligned exactly. A similar situation exists in the case of the tRNA and the amino acid that is attached.

Both the mRNA and the tRNA–amino acid complex travel to the ribosomes of the cell, where protein synthesis is performed on an assembly line similar to those that produce automobiles. There is still disagreement among chemists about the exact method by which the amino acids are linked to the mRNA, but the overall process can be described. Like the tRNA molecules, each mRNA molecule contains a triplet of nucleotide bases which holds the code for protein synthesis given to it by the DNA. Each triplet of bases corresponds to the triplet carried by the tRNA.

Now comes the part of protein synthesis most difficult to understand, in which the triplets of the tRNA are linked in their proper sequence to the triplets of the mRNA to produce a peptide chain by the peptide bonding of the amino acids at the opposite end of each tRNA molecule positioned along the mRNA code. The mRNA becomes attached to a ribosome, leaving the triplet exposed. This process is thought to be carried out with the aid of the rRNA in the ribosomes.

Next, the tRNA carrying the amino acid recognizes and attaches to complementary codons on the mRNA. The configuration of the bases that make up the codon triplet on both the tRNA and the mRNA molecules determine the exact place that the two will join. In other words, the arrangement of the bases of the two types of RNA is similar to the analogy of the electrical plug and outlet cited earlier.

Each amino acid is then bonded to its particular codon nucleotide with the help of an enzyme. After this union, each amino acid is joined to an adjacent amino acid which has undergone the same process by a peptide bond, forming a polypeptide chain, or protein. The chain detaches from the ribosome and is used by the cell for whatever process is necessary. The new protein then functions in the reactions for which it was made until it deteriorates and is broken down to its constituent amino acids by the cell. The tRNA and the ribosome are then released for use in future protein synthesis.

Thus, at the end of protein synthesis, the code of the DNA molecule, transmitted to mRNA, has been responsible for the proper arrangement of amino acids on a peptide chain that forms the protein. This arrangement then determines the nature and properties of the protein.

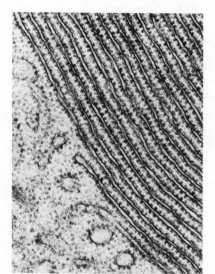

Figure 17-11
The rows of small black dots in this microphotograph of a portion of cell cytoplasm are ribosomes, where protein synthesis takes place.

How did investigators discover the fact that the nucleotides have codes of 3 bases and not 1, 2, or more than 3? It was immediately apparent that words of more than one letter were involved because there are only 4 bases and 23 amino acids. A code containing words of only one letter would mean an alphabet having only 4 letters, and only 4 amino acids would be identified. For the same reason, 2-letter words (bases) would not suffice because $4 \times 4 = 16$. It was obvious then that the code must contain 3 words; thus $4 \times 4 \times 4 = 64$ would provide more than enough different letters to specify the 23 amino acids. Naturally, 4^4, 4^5, and so on were also possible but, given the economy of nature, improbable.

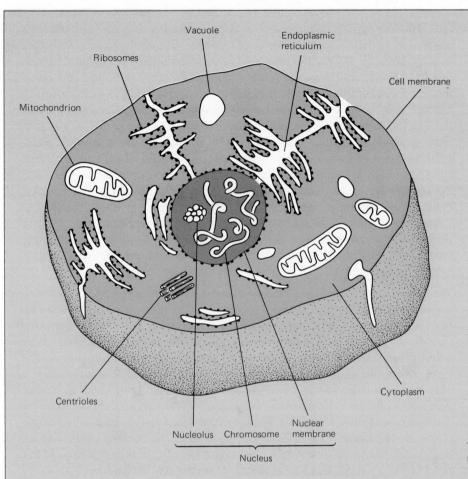

Mitochondrion

Ribosomes

Vacuole

Endoplasmic
reticulum

Cell membrane

Centrioles

Nucleolus   Chromosome

Nuclear
membrane

Nucleus

Cytoplasm

**The Cell as a
Chemical System**

The cell is the basic unit of life. It has the ability to exchange and transform energy, to adjust to the environment, and to grow and reproduce—all the definitive characteristics of life itself. A somatic, or body, cell can have a wide variety of shapes and sizes. It can range in diameter from 0.02 microns (1 micron = 1/1000 of a millimeter) to 3 or 4 millimeters to a gigantic 40 centimeters—which is the size of the single-celled ostrich egg.

The cell is divided into two basic parts—the cell nucleus and the cytoplasm. The cytoplasm, a colloidal suspension of water, carbohydrates, proteins, fats, and other materials, contains a number of small structures known as organelles. Among these are the mitochondria, where respiration occurs; the vacuoles, which in plant cells contain solids or liquids and are surrounded by a membrane; the centrioles, which are active during cell division; and the ribosomes, small spheroid particles containing protein and nucleic acid that play an important role in the manufacture of protein.

The cytoplasm is surrounded by a cell membrane which is semipermeable—it regulates the passage of nutrients and other materials in and out of the cell. The nucleus is in the center of the cell, separated from the cytoplasm by its nuclear membrane. The nucleus is formed from a colloidal material denser then cytoplasm, in which are suspended a number of chromosomes. Each human cell contains 23 pairs—or a total of 46—chromosomes. When a cell divides (in the splitting process called mitosis), the chromosomes also divide. The two new cells, called daughter cells, receive chromosomes that are *nearly* exact copies of those in the parent cell.

Chemotherapy

Chemotherapeutic drugs help control tuberculosis, pneumonia, syphilis, meningitis, and many other diseases; they may also prove effective against cancer.

For centuries, organic compounds that occur in living plants and animals have been used to help fight many kinds of diseases. During the seventeenth century, for example, South American Indians used the dried bark of the cinchona tree to treat malaria. The bark contains quinine, which is still used in modern medicine to treat that disease. The science of pharmacology has catalogued, studied, and compounded these natural materials into the thousands of drugs that are available today. Recently, scientists have used biochemistry to produce synthetic drugs from living organisms. By knowing the shape and metabolic functions of certain organic molecules at the level of the cell, scientists have been able to produce other molecules which lock onto and interfere with the functioning of the natural molecules. As a result, a whole new science, called **chemotherapy**, has been born. Chemotherapeutic drugs, sometimes called wonder drugs, are specifically designed to interfere with the vital metabolic processes of disease-producing microorganisms. For this reason, they are called **antimetabolites**. Chemical substances that are produced by organisms and are detrimental to other organisms are called **antibiotics**.

The emphasis in the science of chemotherapy is on selective toxicity—that is, drugs are designed to be powerful enough to kill or disable the microorganism that invades a host, yet not harm the host. The modern science of chemotherapy was born in 1935 when it was discovered that the sulfonamide drug group could prevent the uncontrolled growth of bacteria without destroying normal cells. When infectious bacteria invade an organism they may overpower its normal defensive mechanisms by growing and reproducing at phenomenal rates. The introduction of a sulfonamide, such as sulfanilamide, actually deceives the bacteria because it resembles para-aminobenzic acid which the microorganisms use to produce their nutrient, folic acid. Since the bacteria cannot use the sulfa compound after ingesting it, they cease growing and multiplying. At this point, the cell's defensive agents step in to control the bacteria.

Figure 17-12
Sulfanilamide is very similar to para-aminobenzic acid, as these structural formulas show. Since infectious bacteria cannot differentiate between the two, they ingest sulfanilamide and stop growing when they are unable to produce their food from it.

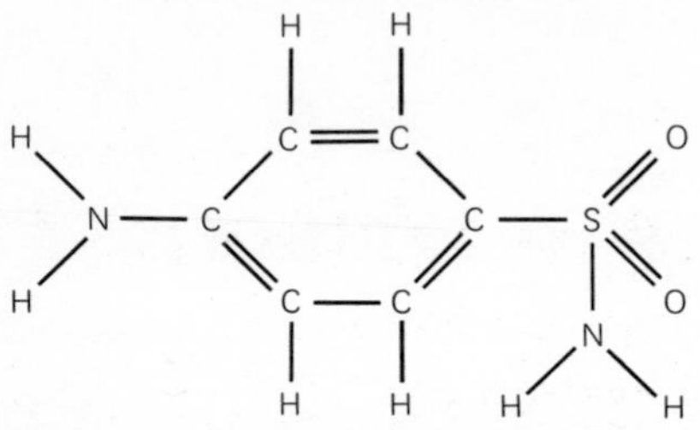

Sulfanilamide

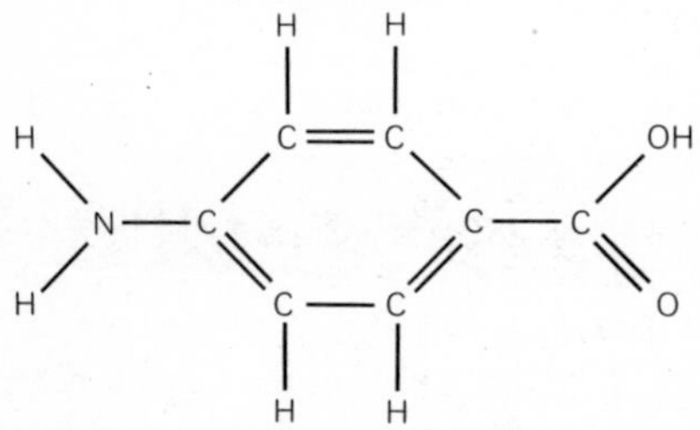

Para–aminobenzoic acid

Some of the more important antibiotics to emerge from chemotherapeutic research are those that destroy a wide range of infectious microorganisms by interfering with certain metabolic activities. These include the penicillins, streptomycin, the tetracyclines, and neomycin. Penicillin, which interferes with the synthesis of the bacterial cell wall, is effective against a number of microorganisms, including those that cause syphilis, meningitis, and certain forms of pneumonia.

Streptomycin acts on bacteria which are responsible for pneumonia, meningitis, and tuberculosis, causing them to manufacture abnormal, nonusable proteins. It does this by joining itself to the ribosomes in the bacterial cell nucleus and causing the mRNA there to incorrectly code the amino acid sequence. The tetracyclines, including aureomycin, and terramycin, among brand names, are effective against typhus, Rocky Mountain spotted fever, one type of pneumonia, and certain eye infections. These drugs bond to the ribosomes in the bacterial cells, preventing the tRNA–amino acid complex from attaching to the ribosomes and so inhibiting protein synthesis.

Another example of chemotherapy is seen in the use of actinomycin in cancer treatment. This drug is an antimetabolite that binds with DNA and blocks RNA transcription. It can be utilized to inhibit cancer because the cancer cell produces RNA at a faster rate than the normal cell. The rapid production rate allows the cancer to pick up the deleterious chemicals faster and in greater amounts than a normal cell, thus killing it before too many normal cells succumb. Unfortunately, the success of actinomycin in curing cancer has been minimal, but its use in nucleic acid research has been most important in working out the patterns of transcription and translation in cells.

Glossary

An **alpha helix** is the spiral pattern which most proteins form when their polypeptide chains coil around one another.

An **amino acid** is a nitrogenous compound containing an amino group and a carboxyl group. It is the basic unit of the protein molecule

An **antibiotic** is a chemical substance produced by one organism, usually a micro-organism, that is detrimental to other organisms.

An **antimetabolite** is a synthetic chemical that interferes with the metabolic functions of a cell.

ATP, adenosine triphosphate, is an organic compound consisting of adenosine and three phosphate groups, two of which are joined by high energy bonds. It is the source of energy for most chemical reactions in all living organisms.

A **beta arrangement** is the ribbonlike shape a protein assumes when uncoiled adjacent polypeptide chains become attached and form smooth or pleated sheets.

Cellulose is a very long-chained polysaccharide which forms the major portion of the cell walls of plants.

Chemotherapy is a branch of biochemistry which produces drugs that selectively control infectious cells by interfering with their metabolic processes.

A **codon** is a sequence of three nucleotide bases, a triplet associated with messenger RNA and transfer RNA.

The **cytoplasm,** which consists of water, carbohydrates, proteins, fats, and other materials, is all of the cell exclusive of the nucleus.

Denaturation is the process by which the physical properties of a protein are irreversibly changed by heat, acid, or alkali.

DNA, **deoxyribonucleic acid,** is a molecule in the cell nucleus consisting of a phosphate group, a deoxyribose sugar ($C_5H_{10}O_4$), and four nitrogen bases. It contains the genetic code for almost all organisms, and is necessary for the cellular reproduction and the genetic evolution of living organisms.

An **enzyme** is a protein that acts as a catalyst to speed up chemical reactions in organisms.

Hydrolysis is the process by which a complex molecule is broken down into simpler molecules by the addition of water.

A **lipid** is a chemical compound which is soluble in fat solvents, but insoluble in water. Lipids include oils, fats, waxes, sterols, and other compounds.

The **nucleus** is that central organelle of a cell which is bordered by a nuclear membrane and contains chromosomes and nucleoli.

A **peptide bond** is the linking of two amino acids which results when the carboxyl group of one amino acid is covalently bonded to the amino group of another.

A **polysaccharide** is a carbohydrate containing numerous joined monosaccharide molecules.

A **ribosome** is a small spherical body in the cell cytoplasm consisting of about 30 percent protein and 70 percent nucleic acid. The ribosomes are the sites of protein synthesis.

RNA, **ribonucleic acid,** is a molecule in the cell consisting of a phosphate group, a ribosome sugar ($C_5H_{10}O_5$), and four nitrogen bases. There are three types of RNA, all of which are active in protein synthesis. Messenger RNA carries the genetic code from the DNA to the ribosomes. Transfer RNA binds with amino acids and joins them to the mRNA at the ribosomes. Ribosomal RNA is thought to help the mRNA attach to the ribosomes.

A **substrate** is the substance for which an enzyme is specific and upon which it acts.

Exercises

1 How is biochemistry different from organic chemistry? Cite some examples to illustrate the difference.
2 Describe a carbohydrate giving the various forms these molecules are found to have.
3 What is hydrolysis? How is it involved in synthesis reactions? Show how the digestion of a polymer involves the synthesis of a water molecule.
4 In what chemical ways are lipids different from carbohydrates? Draw a simple lipid monomer.
5 What type of bond is formed by the molecules that form fat monomers? Indicate these bonds in your diagram for question 4.
6 Using specific molecules, demonstrate the difference among unsaturated, saturated, and polyunsaturated fats.

7 What are the differences among a protein, a lipid, and a carbohydrate?

8 Draw a structural formula of a peptide linkage.

9 What are the functions of proteins in the cells?

10 How many monomers of a protein molecule can there be? How many different kinds of monomers can be used to make up a protein molecule?

11 What are the relationships of a peptide, polypeptide, protein, the alpha helix, and the beta configuration in protein structure?

12 Physiologically active proteins are mostly enzymes. What makes a protein as an enzyme structurally different from a protein that is part of a structure?

13 Describe how an enzyme functions, using diagrams and labels.

14 What two types of nucleic acids are described in the text? In what ways do they differ? How does the difference relate to their function?

15 What chemicals are the basis of the DNA and RNA codes? How are they arranged? How many combinations are possible? How are they paired?

16 Describe the Watson-Crick model of DNA. Draw a portion of the molecule.

17 How do mRNA, tRNA, and rRNA differ?

18 What is chemotherapy? In what ways are chemotherapeutic drugs used in combating disease?

19 You have a cell, deoxyribose, ribose, A, U, T, G, C, all the enzymes you need, all the amino acid you need, and phosphate. Using a cookbook style (list ingredients and amounts of ingredients, then give a step-by-step description of procedures to be followed), tell what you would do to make a protein.

Interlude Ten

Destroying the Ozone Shield

Three-quarters of the way into the twentieth century there looms a possibility that life on earth may come to an end, not only from nuclear wars or mutated viruses, but from the destruction of the ozone barrier that shields us from certain harmful wavelengths in the sun's rays. The sun emits light of all wavelengths. Peak emission occurs at a wavelength of approximately 5000 Å. The amount of radiation emitted falls off rapidly at shorter wavelengths, but there is sufficient radiation short of 3000 Å to destroy life on earth if it were not prevented from reaching the surface by the various layers of the atmosphere.

Beyond the blue-violet end of the visible spectrum is the ultra-violet (uv), and it is this which is so damaging to the delicate balance of chemical processes in biological organisms, mainly by destroying nucleic acids in genetic material. The uv radiation was once helpful to living things, for at the very beginning of life on earth, the uv was the source of energy for conversion of methane, water, and ammonia into the organic chemicals for construction of life forms. Those early elementary forms of living matter were in the organic soup, and were protected from uv destruction by the water of the oceans. Then, as photosynthesis began to convert carbon dioxide (a product of the early fermentation of those organic chemicals) into molecular oxygen, some ozone (O_3) was created in small proportions, along with O_2. Ozone is extremely active chemically, so only those ozone molecules persisted which drifted upward, thereby collecting in the upper atmosphere as a thin, permanent layer of O_3. Curiously, the ozone produced by basic biochemical action protected life down on the earth's surface by its own absorption of uv radiation. This permitted early forms of life to leave the oceans and continue the process of biological evolution on dry land. The ozone protective layer in the stratosphere was thus essential to the development of life on earth.

Oxygen exists in the atmosphere in three forms: single atoms (O), ordinary molecular oxygen (O_2), and ozone (O_3). Oxygen atoms combine readily to form O_2. Oxygen molecules, however, have a structure which absorbs solar radiation in the range of 900–2400 Å, and O_2 then dissociates back into single atoms; these atoms, in turn, combine with other O_2 molecules to form ozone, $O_2 + O \longrightarrow O_3$. Once ozone has been formed, it characteristically absorbs similar solar radiation of 2000–3200 Å, and filters out this portion of the uv. But when such absorption occurs, the ozone dissociates back to ordinary oxygen, and the $O–O_2–O_3$ cycle continues. Ozone is found at about 20–40 km altitude, with maximum concentration at the lower level. Radiation absorption by ozone produces atmospheric heating of the stratosphere, a temperature inversion as high as 270° K. O_3 is unstable, very active, and highly poisonous, and is the cause of the familiar odor following a thunderstorm.

Ozone is also destroyed in other ways. In recent years it has been learned that a principal mechanism of ozone destruction is its reaction with nitric oxide (NO). There is always some nitric oxide in the atmosphere but technological civilization produces vast quantities. The nitric oxide reacts with the ozone; the resulting products are molecular oxygen and nitrogen dioxide, $NO + O_3 \longrightarrow NO_2 + O_2$. The nitrogen dioxide then reacts with any atomic oxygen to form nitric oxide and molecular oxygen, $NO_2 + O \longrightarrow NO + O_2$. The nitric oxide is not used up. These productive and destructive processes maintain the ozone balance in the stratosphere layer between 6.2 mi and 31 mi. Below 6.2 mi the uv penetration is insufficient to cause dissociation of oxygen molecules, and above 31 mi the atmosphere is too thin for any substantial formation of ozone through the reaction described.

The ozone layer has a great influence on the temperature structure of the atmosphere, which gets colder up to the troposphere and then begins to warm up in the stratosphere. This temperature inversion is the result of the absorption of uv radiation by ozone. The chemical stability of the stratosphere makes it vulnerable to contamination: any substance introduced into the layer stays there. Fortunately, most of the substances produced by industry on the surface of the earth never get to the stratosphere; they usually interact with the moisture in the

"By George! I think you've done it!"

air and are washed down in rain storms. In some cases they are absorbed by solid particles in the lower levels of the atmosphere.

Only certain special contaminants reach the stratosphere. Thermonuclear explosions produce oxides of nitrogen in abundance and introduce them into the stratosphere. The atomic bomb tests conducted in the early 1960s produced measurable depletion of the ozone layer. Another danger is from high-flying jet planes. Nitric oxide and other nitrogen oxides are produced in reactions in the high temperatures within jet engines. The new jets are designed to fly well within the lower reaches of the stratosphere. It is there that they emit their combustion products. In 1974 yet another menace to the ozone layer has been recognized: the family of gases used as propellants in aerosol spray cans.

Aerosol sprays require propellants that do not react with the substances being dispensed. Technology has come up with a series of inert, volatile gases, tradenamed Freons, that are satisfactory in this respect. They are compounds of carbon, chlorine, and fluorine. They are nearly inert and almost insoluble in water. This means that they remain for extremely long periods in the atmosphere. Recent experiments have indicated that all the Freon that has ever been released into the atmosphere is still there. What happens to these chlorfluoromethanes, as the chemists call them, when they are dispensed into the atmosphere?

It appears likely that these stable gases slowly find their way into the stratosphere. In the stratosphere they encounter uv radiation and absorb the short wavelengths, <2000 Å, dissociate, and release chlorine (Cl). Chlorine has a remarkable effect on the ozone layer. It combines with ozone to form an oxide and a molecule of oxygen, $Cl + O_3 \longrightarrow ClO + O_2$. The oxide reacts with any atomic hydrogen that happens to be around and reverts back to chlorine, while a molecule of oxygen is formed, $ClO + O \longrightarrow Cl + O_2$. In this respect chlorine acts as a catalyst, just like nitric oxide, but it is many times more effective. It is also not used up in the process and remains in the stratosphere for decades. There is evidence that this sort of ozone depletion has already taken place.

It must be remembered at this point that only a small fraction of the Freon in the lower layers has reached the stratosphere. Ozone depletion has already begun, and it will certainly increase in the next ten years, even if no more Freon is released into the atmosphere. If, on the other hand, these gases continue to be dispersed into the atmosphere, the problem will become increasingly acute for the next hundred years before it can be stabilized. In 1974 two American chemists, F. S. Rowland and M. Molina, estimated that the 1972 rate of atmospheric Freon pollution, if maintained, would reduce the amount of ozone in the protective layer by 10% in a few decades, with a 1% reduction by 1974. Other scientists estimate the current reduction at 0.5–2% and the projected reduction over 20 years at 7–13%. Since it has also been estimated that a 5% reduction in the ozone will allow a 20–25% increase in uv radiation to the earth's surface, the Freon contamination of ozone will be disastrous if it continues.

The depletion of the ozone layer would increase the uv radiation that reaches the surface of the earth. Life would be affected in various ways. The uv rays would promote excessive tanning and burning of the human skin. Sunlight has been established as a primary cause of skin cancer. In addition, uv radiation is blinding. In the intense uv situation, all animals that did not wear protective goggles would suffer from blindness. This would considerably alter their habits and consequently their populations. Marked changes would occur in the plankton population of the sea and crucial links higher in the food chain could be disrupted. Biological adaptation takes thousands of years. Life forms on earth may not be capable of coping with a suddenly altered uv environment. It would be difficult for the human species to survive, even with artificial devices.

It must be remembered that the above analysis is uncertain. The experiments are difficult to perform. The data are also uncertain. It is quite possible that an entirely unknown interaction between solid particles in the atmosphere and the chlorfluoromethanes could alter our pessimistic model. But what if it were true?

A prudent policy would be an immediate stop to production of aerosols with these Freons and a minimization of commercial and military jet flights. Substances that are packaged as aerosol sprays are already being packaged in other ways. This ban on Freon aerosols would not, of course, stop the depletion of ozone. There are already millions of tons of these gases in the troposphere, and they are finding their way up into the stratosphere. What a ban would do is to prevent the situation from becoming unmanageable in about 30 years. Meanwhile, research is urgently needed into various topics such as the air-flow patterns in the stratosphere. Samples from various heights must be analyzed, and the distribution of gases determined. The exact rates of reaction must be studied. Protective measures must be formulated. There is obviously a great deal to do.

Seven

Earth Science

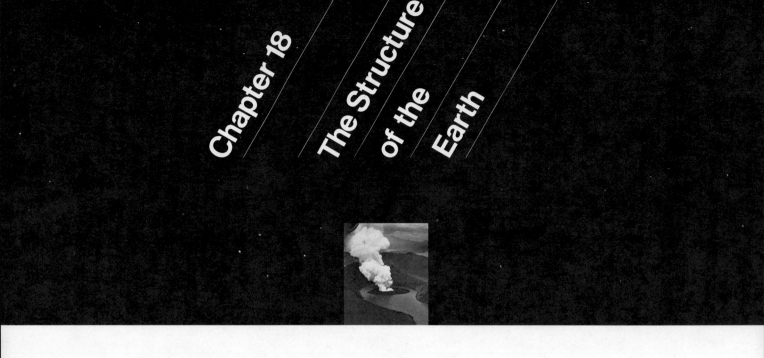

The earth is a dynamic sphere; much of the surface and interior are in a state of constant change. Mountains are continuously being built up and worn down, seas spread and retreat, volcanoes and earthquakes fracture the earth's crust, and rocks form as the result of the forces acting inside and outside the earth. The majority of these changes take place over long periods of time and are detectable only with special instruments. Just as we cannot sense the movement of the earth as it rotates on its axis and revolves around the sun, neither can we notice the usual movements that take place beneath our feet; it takes an earthquake to attract our attention. Yet the same forces and processes that shaped the earth in past eras—forming mountains and canyons, oceans and deserts—are at work today, forming the landscape of tomorrow.

Our planet is not a solid blob of matter but rather a series of concentric, spherical layers. At the surface is the **crust,** a thin rigid rock shell with a topping of sediment, such as gravel, sand, salt, and clay. It ranges in thickness from as little as five kilometers on the ocean floor to as much as 70 kilometers at such continental locations as the Alps. Separating the crust from the second layer is an abrupt break, or discontinuity. This break is named the Mohoroṽićić discontinuity, after its discoverer, a Croatian scientist (1857–1936), and is usually known as the Moho.

Below the crust is the **mantle,** which makes up nearly 80 percent of the earth's volume. It is about 3000 kilometers thick. Nearest the crust the mantle appears to be made up largely of iron and magnesium silicates, probably in the form of the mineral olivine, but near its inner margin, under tremendous pressure, there probably lie iron, magnesium, and silicon oxides—the same materials as olivine, but no longer combined into that mineral because of the effects of heat and high pressure.

The earth is a series of concentric spherical layers.

Figure 18-1
Rock, the basic unit studied by the geologist, is classified as igneous, sedimentary, or metamorphic. Most rocks are aggregates of minerals; they contain clues which help scientists determine when and how they were produced. In this photo, craggy red sandstone forms the defile that is the only entrance to the ancient Nabatean city of Petra in Jordan. Two horsemen and a man on foot make their way through the narrow passageway.

The structure of the mantle appears to be complex. Just below the crust it seems to be quite rigid, and this section, together with the crust, is commonly called the **lithosphere**. Below the lithosphere the mantle, because of high heat, becomes more plastic, and over long periods of time it will slowly flow in response to pressure. This region is called the **asthenosphere**. Below the level of the asthenosphere, where the weight of overlying matter develops enormous pressure, the mantle once again becomes rigid.

The earth's mantle makes up 80 percent of its volume.

Under the mantle is the molten **outer core**, with temperatures higher than the melting point of iron at those pressures, and a composition which geologists believe to be primarily iron and nickel. The core has a solid center, or **inner core**, mostly iron. The whole core is about 3400 kilometers in radius.

Composition of the Crust

Although the crust is only a small proportion of the total volume of the earth, it is important to study for two reasons. Since it is right under our feet, it is most accessible for study. More significantly, it has been created by the same forces that created the interior of the earth. If we know something about the way the surface is modified by underlying forces, we can predict the way those forces will affect the earth's mantle and core.

Ninety-six percent of the rocks found in the earth's crust are silicates.

Of the 106 chemical elements now known, 92 are found in the crust. Yet just eight of them make up the bulk of the earth's surface—98.5 percent by mass, 99.9 percent by total number of atoms. The eight elements and their contributing percentages by mass are: oxygen, 46.6 percent; silicon, 27.7 percent; aluminum, 8.1 percent; iron, 5 percent; calcium, 3.6 percent; sodium, 2.8 percent; potassium, 2.6 percent; and magnesium, 2.1 percent. Surprisingly, a number of metals that are very familiar to us and essential to modern technology—copper, tin, zinc, nickel, lead, and silver—are just traces, rare compared to the other elements. They are grouped with nitrogen and the halogens in a miscellaneous category that totals 1.4 percent of the earth's crust. And stranger still, carbon, nitrogen, and hydrogen, vital to all living things, comprise less than 2 percent of the total. (These percentages apply to the crust alone; for the earth as a whole, iron is by far the most prevalent element.)

Silicon and Silicates

Silicon, which comprises a little more than one quarter of the earth's crust, is a most important mineral element; it has the same prominent function in the inorganic world that carbon has in the organic. Silicon is never seen in a free state because it is chemically more active than carbon; it may combine with oxygen alone to form silicon dioxide, or quartz. But practically all of the earth's silicon combines with oxygen along with various metals to form a mineral group known as silicates. The various silicate minerals are a large group, comprising about 96 percent of the rocks found in the earth's crust.

A

B

C

Figure 18-2
Examples of mineral cleavage. Cleavage is the tendency of a mineral to break along certain planes determined by its ionic or atomic structure. *A* shows how a piece of phlogopite, which has one perfect cleavage, is easily separated into sheets. *B* shows a mass of galena whose cubic structure is quite evident. *C* shows a fragment of quartz which has no cleavage.

Inorganic chemists have been impressed by the complexity of silicon's chemistry; in fact, little is known about the atomic structure of many individual silicates. The overall patterns of silicon crystals, however, have been determined by the use of X-rays. Like its close chemical relative carbon, silicon can combine with other atoms by means of its four outer electrons. Thus, silicon joins with oxygen to form orderly, two- or three-dimensional chains or lattices of silicon and oxygen atoms, in structures called **crystals**. In their simplest forms, these combinations usually form a tetrahedron or four-pointed shape, with the silicon atom in the middle and an oxygen atom at each corner; the atoms are linked by strong bonds. In their more complex patterns, the tetrahedrons unite to form chains, sheets, or three-dimensional structures of silicon and oxygen atoms with metal ions separating the two.

The manner in which the tetrahedrons are linked determines the physical properties of the silicate they have formed. Mica, for example, is composed of silicate tetrahedral layers, arranged one above the other. Corresponding to this molecular structure is mica's physical property of separating into thin sheetlike flakes. Because of the nature of silicon—it is neither metallic nor nonmetallic—it is not limited to behaving as one or the other and can form many different sorts of compounds, each having the possibility of more than one crystalline structure. Some common examples of silicates include the feldspars (comprised of silicon, oxygen, and aluminum), the micas, garnet, and topaz.

Minerals

A mineral is a naturally occurring solid inorganic substance with a characteristic specific atomic structure and fixed chemical components. Although a few minerals are single elements, the majority are chemical compounds. Most minerals are found in rocks, and for the most part, a rock is either a single mineral or an aggregate of minerals. Like everything in nature, there are exceptions to this statement: for example, coals and the volcanic glasses such as obsidian, pumice, and scoria contain no crystalline minerals at all, yet they are considered rocks.

Of the more than 2000 known minerals, only a handful are abundant in rocks that make up the earth's crust. Also fortunate for the beginner is the fact that most minerals are easily recognizable and can be identified by a practiced eye. During examination, one would note such obvious characteristics as the color, hardness, density, and form of the crystals that comprise the mineral. Hardness can be operationally defined by the scratch test, which consists of rubbing one mineral against another. A harder mineral, such as quartz or feldspar, will scratch a softer one, such as calcite or gypsum.

After noting the obvious physical characteristics of the mineral, one should then examine its two other important traits: crystal shape and cleavage. We have seen that it is the internal atomic structure of silica that makes it a crystal. Like silica, most other minerals are crystalline solids that exhibit definite crystal shapes and forms as a result of their atomic structures. Since crystals of different minerals have different shapes, it is usually quite easy to identify the mineral by visual examination. However, because of varying growth conditions, perfect crystals are relatively rare, so identification sometimes requires close examination.

Cleavage, which is also determined by atomic structure, is the tendency of a mineral to split along a smooth plane. Minerals have different cleavage patterns. When struck with a hammer some minerals—halite, for example—break in the same shape as their crystals. Others, like quartz, have no cleavage and shatter into irregular fragments when struck. Mica, on the other hand, has perfect cleavage: it can easily be separated into thin sheets. Cleavage is particularly important in diamond cutting; the diamond cutter must be able to recognize the cleavage lines of a gem and predict the way it will split before he strikes it. A cleavage error can mean the difference between a large perfect gem worth a million dollars and a handful of small stones worth only a fraction of that figure.

Common Minerals Silicates, the most widespread of the rock-forming minerals, include feldspar, mica, clay minerals, and the ferromagnesian minerals. Feldspar, the most widely distributed of the silicate minerals, constitutes more than half the ingredients of the rocks of the earth's crust. It is characterized by two good cleavages and is usually white or colorless. Mica, which can be recognized as easily scratched, with thinly layered flakes in many rocks, consists of two varieties, biotite (black) and muscovite (white or colorless). Clay minerals are very soft materials that are the chief constituents of clay and are principally silicates of hydrogen and aluminum. The ferromagnesian minerals are a group of

Mineral	Composition	Characteristics	Where found
Orthoclase	Potassium aluminum silicate	Pink, white, or gray color; transparent to translucent; two good cleavages; moderately hard.	Colorado, Nevada
Albite	Sodium aluminum silicate	White, yellow, or gray; transparent to translucent; two good cleavages; moderately hard.	New England, Virginia
Quartz	Silicon dioxide	Colorless, white, or various tints; transparent to translucent; no cleavage; hard.	Throughout United States
Magnetite	Ferric oxide	Black metallic luster; no cleavage; moderately hard.	Eastern states, Arkansas, Utah
Hematite	Ferrous oxide	Red or black; translucent; no cleavage; hard.	Missouri, Michigan
Olivine	Magnesium silicate or Iron silicate	Green, gray, or brown; transparent to translucent; no cleavage; hard.	North Carolina, Arizona
Muscovite	Hydrous potassium aluminum silicate	White, yellow, or colorless; translucent to transparent; one perfect cleavage; soft.	New England, North Carolina, Colorado
Biotite	Black mica	Dark brown or black; opaque to translucent; one perfect cleavage; soft.	New England
Kaolinite	Hydrous aluminum silicate	White; opaque; no cleavage; very soft.	Throughout northern United States
Augite	Mafic silicate	Black with transparent edges; two perfect cleavages; moderately hard.	Throughout United States
Calcite	Calcium carbonate	Colorless, white, or various tints; transparent to translucent; three good cleavages; soft.	Throughout United States; especially California, Missouri, Kansas
Pyrite	Iron sulfide	Light yellow (known as "fool's gold"); no cleavage; hard.	Colorado, Utah, Pennsylvania, and other areas
Galena	Lead sulfide	Lead-gray; three good cleavages; very soft.	Missouri, Kansas, Oklahoma
Gypsum	Hydrous calcium sulfate	Colorless, white; fluorescent or phosphorescent; one perfect and two imperfect cleavages; very soft.	Widespread, but especially New York, Michigan, Texas, California
Halite	Sodium chloride	Colorless or white; transparent; three perfect cleavages; very soft.	New York, Michigan, Ohio, Louisiana

Table 18-1
Some common minerals

silicates of magnesium and iron with colors shading from dark green to black. The most common mineral of this group is olivine which looks like greenish-gray sugar and is an important part of many of the heavier rocks of the earth, particularly those of the mantle.

A common carbonate mineral is calcite. Although chemically the same as common chalk—calcium carbonate—in its crystal structure calcite resembles quartz, to which it has no chemical relation at all. It differs from quartz in cleaving perfectly.

Quartz is a silicon compound, although not a silicate. In this form, silicon acts like a metal and oxidizes, forming six-sided crystals that can be water-clear or cloudy. Colored forms are sometimes used for jewelry—amethyst is an example. Despite its regular, often large, crystals, quartz does not cleave; it tends to shatter into curved fragments like broken glass (also a silicon compound).

Rocks and Their Classification

Atoms make up elements; elements make up minerals; minerals make up rocks. Rocks are the basic study of the geologist, for they constitute the fundamental landforms of the crust. Near the surface of the crust, the solid, underlying mass of bedrock is blanketed with loose fragments collectively called regolith, the product of breakdown of the bedrock by atmospheric agents. This breakdown process eventually results in the formation of soil. Classified according to their origin, there are three principal categories of rock: **igneous** (from the Latin word

The three principal categories of rock are igneous, sedimentary, and metamorphic.

for fire), that has solidified from molten material; **sedimentary**, composed of fragments of other rocks that have eroded away and been deposited by wind or precipitated from water and then compressed and compacted as the sedimentary layer gets thicker; and **metamorphic**, which is any rock that has been changed in form by the heat and pressure of overlying rocks or by the heat of nearby volcanic activity.

Igneous Rocks

The texture of an igneous rock tells how and where it was formed.

Most of the earth's crust is composed of fire-formed igneous rock. Common examples are granite, rough-textured and slowly cooled; obsidian, with a black glassy texture due to rapid cooling; and a dark mixture of feldspars and ferromagnesians called basalt, which may be either slowly or rapidly cooled. The chief ingredients of igneous rocks are minerals that contain silicon, such as mica, quartz, and the ferromagnesian minerals.

Igneous rocks are formed from molten material deep in the earth. When it is beneath the earth's surface, this liquid rocky material is called **magma**; if it pours out on the surface, it is called **lava**. The term lava also refers to the surface igneous rock that results after the hot lava cools and solidifies.

Figure 18-3
Volcanic lava fields in the Valley of 10,000 Smokes, Alaska. Lava, an example of extrusive vulcanism, is molten rock that was formed deep in the earth. After it flows or erupts at the surface, the lava cools and solidifies to produce patterns similar to those in this figure. In some parts of the northwestern United States, lava fields extend over 200,000 square miles, often reaching 3000 feet in thickness.

Magma is really a molten mineral "soup" consisting of silicates, oxides, and sulfides. It contains large volumes of dissolved gases held in solution by pressure. Volcanoes and lava streams are the result of the eruption of this fluid mass through the crust of the earth. After a volcano erupts, the lava cools and solidifies to form **extrusive** rock. **Intrusive** rock, the more abundant type, is solidified magma that never reached the surface; it cooled and solidified somewhere beneath the earth's crust and is seen only when erosion breaks down and carries away the overlying rocks.

Volcanic mountains, such as Mauna Loa (13,680 feet high), are built up of extrusive rock. Intrusive forms, since they are formed deep below the surface, are not as readily recognized for what they are. Yet some of our most familiar mountain ranges—the Appalachians, for example—are hard igneous intrusions, or great sheets of granite, that have been exposed after years of erosion have worn away the softer covering rock. The texture of igneous rock is very important, for it tells geologists something of how and where the rock was formed. A coarse, grainy appearance in a rock means that the mineral crystals are large, a sign that the rock solidified deep in the earth's crust after very slow cooling. Smaller crystals and a fine-grained appearance, on the other hand, indicate that the rock cooled quickly somewhere on the surface.

These rocks may begin as eroded and weathered fragments of larger rock masses. Washed or wind-blown into lakes and streams, they settle out of suspension—larger fragments first, then finer material. According to their sizes, they form sediments ranging from coarse-to-fine gravels, sands, clays, and lime muds. With the passage of time, these soft sediments are transformed by a process called consolidation, or lithification, into hard rock. During consolidation, compaction, usually in the form of pressure from other deposits, squeezes the water out of the clay and lime muds and transforms them into rock; in the process of cementation, groundwater percolates through the gravels and precipitates calcite, silica, or limonite, which cement the fine gravel or sand particles together. After such consolidation, the sands become sandstone, the gravels become conglomerates, the clay muds shales, and the lime muds limestones.

Another mechanism for the formation of sedimentary rock is chemical deposition from minerals in water solution or from calcium carbonate in organic marine life. Chalk is formed in this way; an example is the famous White Cliffs of Dover, England, which were formed by chemical deposits from marine shells.

Although they comprise only 5 percent of the earth's crust, sedimentary rocks and sediments blanket 75 percent of the continents and much more of the ocean floor. They are important to us because they are the sources of fossil fuels like petroleum and coal, of common metals such as iron, and of the stone

Sedimentary Rock

Sedimentary rocks often contain fossils, which are keys to earth history.

Figure 18-4
An excellent example of sedimentary rock is this formation at Punakaiki on the western coast of Finland. Remarkable for the uniform thickness of each layer, these sediments are known as the "pancake rocks" because of their flattened upper surfaces. They have been eroded by wave action of the sea, which has long since receded. Sedimentary rocks, which have been laid down layer upon layer over millions of years, cover about three-fourths of the earth's surface.
They often contain fossils which help geologists determine the age of the rocks and the earth processes responsible for their formation. The smaller photo, showing mud with ripple marks caused by water, illustrates how sedimentary layers form on the surface of earth materials.

used in building construction. Sedimentary rocks are typically found in strata, or layers, and often contain fossils; they are important aids to the study of earth history and biological evolution. In addition, other characteristics of sedimentary rocks such as composition, color, and position, provide the geologist with information about the climate, environment, and geologic changes that have taken place in the past. Perhaps the most famous display of sedimentary rock layers in the world is the Grand Canyon of the Colorado River. Its dramatic slopes and cliffs were formed by 10 million years of river erosion through layer upon layer of sedimentary rock that required a billion years for deposition and formation.

The three basic chemical types of sedimentary rock are sandstone, shale, and limestone. Sandstone is composed of tiny sand grains "cemented" together by silica, calcite, or iron oxides; it is permeable to water. Shale is composed of densely compacted clay or mud; it is impermeable to water. Limestone is a fine-grained rock composed mainly of calcite (calcium carbonate).

Metamorphic rocks, which comprise 15 percent of the earth's crust, are formed when sedimentary and igneous rocks are subjected to tremendous heat and pressure several miles deep in the earth, but above the melting zone. Another cause of metamorphosis is the heat of intrusive magma, acting upon adjacent rocks. Evidence of these actions upon such rock generally indicates that it was buried deeply and uplifted by the dynamic action of the earth. When a metamorphic rock is being formed, two kinds of changes occur: first, an increase in pressure causes shattering, fracture, and tremendous frictional heating that may help alter the rock; second, the great heat and especially the pressure cause the rock to become plastic—to flow ever so slightly and to "deform" over a period of many thousands of years. This is followed by re-establishment of the rock's crystalline structure, a process known as recrystallization. This may simply add to the size of the mineral grains present, or it may produce totally new

Metamorphic Rock

Metamorphic rocks are formed from igneous and sedimentary rocks.

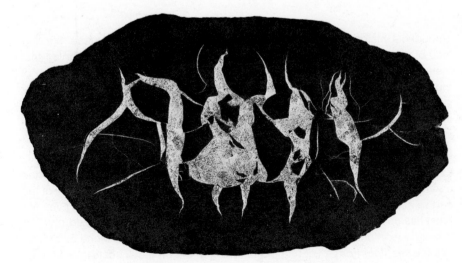

Figure 18-5
Example of a concretion, a mass frequently found embedded in sedimentary rocks. This particular concretion, known as a septaria, is a piece of clay limestone which has been cracked as a result of the movement of adjacent rocks. The cracks were later filled in with calcite, forming the extraordinary pattern shown here. Other septaria have been discovered with similar unusual configurations, some resembling fossil turtle shells.

minerals. Metamorphic rocks are likely to be very old.

Metamorphic rocks are classified into two broad groups: foliated and non-foliated. When crystals of some minerals—mica and chlorite are the most common—are present in rocks being subjected to heat and pressure, they tend to spread sideways, with their granules forming parallel bands like the icing between the layers of a cake. The presence of such flat parallel mineral streaks is called foliation. The minerals that form such layers must possess flat or long grains with closer atomic spacing; thus only minerals whose crystals form sheets, such as mica, hornblende, and chlorite, can form foliated rock. Foliation also causes the rock to cleave parallel to the layers when it is struck.

The most common foliated rocks include slate, made from shale and unstable clay minerals and often thinner than paper; schist, formed from shale under high temperature or from small-grained igneous rocks with layers of varying thickness; and gneiss, formed under high temperature and pressure from almost any other kind of rock, with layers about a millimeter thick.

Unfoliated metamorphic rocks are produced when a rock is formed by heat alone without unidirectional pressure, or more often when the minerals making up the rock do not possess flat or long grains. Examples of this type of rock include marble, which is a crystalline calcareous rock formed by the metamorphosis of calcium carbonate (limestone) and quartzite, which may be composed of sandstone cemented by quartz or partially fused.

It is important to note that these classifications of rock are arbitrary ones that the human mind imposes on nature. In reality, rocks don't always fit the neat pigeonholes geologists have created for them; often they fall between one category and another or simply fail to fit any of the categories.

Composition of the Earth's Interior

Throughout history, man has been curious about what lay underneath the surface of the earth. People who had been in underground caverns observed that the air there seemed to be cooler. This led to the speculation that the center of the earth was a huge cake of ice; this theory was compellingly and poetically stated by Dante in medieval times. Another source of visual evidence regarding the earth's interior was the eruption of volcanoes. The frightening spectacle of volcanic power led many ancient scholars to believe that the earth's core was a place of eternal fire, either molten lava or sulfur; they called it Hell.

Even today, man continues his attempt to obtain firsthand knowledge of the earth's interior. For example, geologists have tried determining the nature of the earth's composition by drilling into the ocean floor. To date, however, even the crust has not been completely penetrated. The most widely publicized of these drilling operations was Project Mohole, an attempt to reach the Moho by drilling into the sea floor from a floating platform. When the project's costs rose from an expected $5 million to $100 million, due to the inadequacy of drilling tools in

The most widely publicized attempt to drill to the earth's interior was Project Mohole.

Figure 18-6
Nighttime operations during the Mohole
Project. Shown here are the Cuss
I, the Mohole drilling vessel, and a worker
scraping mud for analysis from the deep
core pipe after a successful drill. In its initial
stages the drilling, which took place in the
Pacific Ocean off the Coast of Mexico,
penetrated 12,000 feet of sea water and
about 500 feet into the sea bed.

the face of the hardness of the rock at lower levels, the United States Congress
stopped further funding. However, knowledge gained from Mohole's attempt to
drill below the ocean floor will assist those making future attempts.

Seismic Waves

Modern geology's most useful tool in fathoming the depths of the earth's interior
has been the earthquake, a geologic disturbance most often caused when
enormous sections of the earth's crust suddenly give way to some built-up
stress and slip against one another. When this occurs, **seismic waves** (shock
waves) are released, causing tremors which travel outward in all directions
from the center of the disturbance. These waves can be detected at stations
over most of the earth's surface. Using instruments called seismographs, which
measure and record these seismic waves, geologists have been able to determine that an earthquake produces three kinds of waves: primary (P) waves,
which are longitudinal waves similar to sound waves and which travel fastest;
secondary (S) waves, which are transverse waves; and long (L) waves, which
travel with a motion similar to that of water waves and are by far the most intense and damaging. Measurement of seismic waves has helped provide useful
information; for example, Mohorovičić inferred the presence of the Moho from a
sharp increase in the speed of seismic waves at that depth.

The most useful tool in measuring the
earth's depth is the earthquake.

Temperature of the Interior

Current evidence regarding the temperature of the earth's interior suggests that
the core is certainly not a giant ice cube; and although it is assuredly hot, it is
not a fire either. It is true that the temperature is lower just under the surface

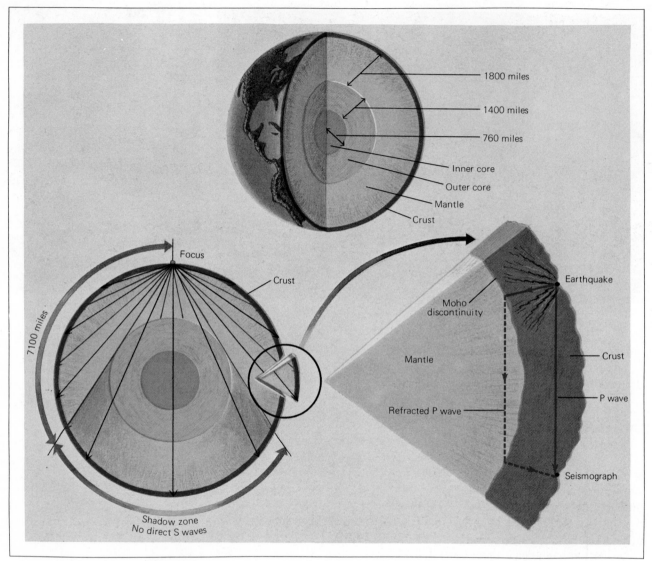

Figure 18-7
Seismic waves generated by earthquakes have shed much light on the structure of the earth. At top is a cross-section of the earth. The diagram at the lower left shows seismic waves passing through the earth after being produced at earthquake's focus. Only the P waves can travel through the liquid core, but they are refracted as they do so. Here, the core has refracted the P waves (shown in dark red) to either side of it, causing the "shadow zone." The lower right diagram shows a shallow earthquake from which P and S waves travel outward and downward, the first directly through the crust to a seismograph, and the second (dashed lines) refracted at the Moho, first inward and then outward back to the crust and on to the seismograph. The different arrival times of the two sets of waves indicate the thickness of the crust.

than it is on top of it; this cooling effect is due to the lack of solar radiation, which heats up the surface. But deep drilling has confirmed that the deeper one goes into the earth, the warmer it gets. Since this warming takes place at the rate of about 95°F per mile of depth, and the distance to the center of the earth is about 4000 miles, such a progression would yield a theoretical temperature in the center of about 400,000°F. However, because the heat of the earth is largely due to energy released by radioactive decay, and since it is known that radioactive materials are most abundant at the crust, diminishing toward the interior, the rate at which the earth heats up can be assumed to slow down toward the center.

On the basis of information obtained about the core's composition by seismic waves, geologists assume that the temperature at the core must be at least as high as the melting point of iron (2795°F at the surface). But because the inner core is thought to be solid, it is likely that the temperature is not high enough to melt iron under conditions of intense pressure (the pressure at 5000 kilometers raises iron's melting point to 6737°F). These hypotheses establish the range of possible temperatures, certainly ruling out 400,000°F, since at that temperature the inner core too would be molten. A more realistic estimate of the temperature at the earth's center is about 5800°F.

In addition to being like a huge furnace, the center of the earth also produces magnetic effects on the surface and in the space around the earth, just as if it were a giant magnet. The earth's power to attract iron needles is one of the oldest observed and recorded geological facts. And the effect of the geomagnetic field upon incoming cosmic rays is also readily observed with modern in-

Earth's Magnetic Field

Figure 18-8
The Van Allen Radiation Belt is simulated in this NASA laboratory. The belt, discovered by orbiting American satellites, is composed of several layers of rapidly moving charged particles trapped by the earth's magnetic field. The belt is composed mostly of electrons, but protons can be found in the inner layers. The layers are found from 2,000 to 12,000 miles above the earth's surface. Clearly visible in this photo are two distinct layers to the right of the model of the earth.

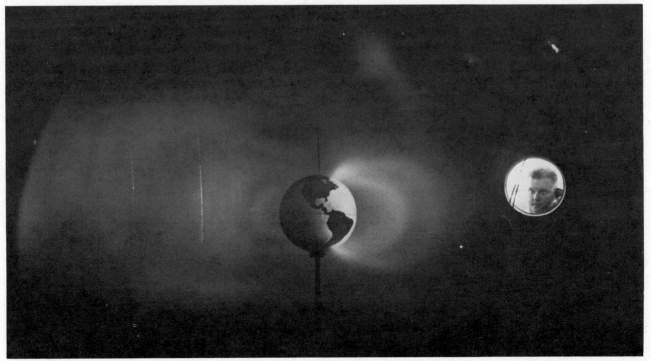

struments. Although it is easy to observe geomagnetism, it is not at all easy to explain it.

In order to envision the effects of geomagnetism, it may help to imagine a giant bar magnet 200 miles long lying in the center of the earth. (We know, of course, that no such magnet could really exist, since iron loses its magnetic properties above a temperature 1480°F.) The points at which the magnetic axis would intersect the earth's surface are the magnetic poles; at present this axis is tilted by 11° from the earth's rotational axis.

Since the compass is quite an old tool, geologists are able to consult records of compass readings in various cities for hundreds of years. These show an interesting irregularity. In 1576, a compass in London pointed to magnetic north at 8° east of true north; in 1823, it was 24° west; today it is about 8° west. The rapidity of these changes suggests that they must in turn be caused by something capable of relatively rapid change. This would seem to rule out the mantle as the source of geomagnetism; since it is for the most part solid, it is capable of only very slow change. It is most likely the molten outer core in which geomagnetism is generated.

Geomagnetism is most likely generated in the earth's molten outer core.

The cause of the earth's magnetism is not understood with any certainty, although it is clear that the magnetic forces we observe at the surface are due in some way to the interior conditions. Since there is no good evidence for the existence of a huge bar magnet at the center of the earth, it seems most likely that geomagnetism arises from dynamic actions in the fluid core. If we can assume the presence of electrically charged material in the fluid, then the large scale motions of such matter act as electrical currents, eddying and circulating, under the convection effect due to different temperatures at the different depths. Such convection typically is a kind of loop flow, like convection heating of air in a room. The varying electrical currents produce magnetic fields. Even though the electrical currents themselves may be feeble the magnetic fields produced could be gigantic, in view of the sheer size and volume of the loops involved; it has been calculated that a velocity of flow of 0.03 cm/sec is enough to give rise to magnetic fields of the size and strength observed on the surface. According to this theory, the fluid core acts as though it were an electric dynamo; this is far more plausible for us to postulate than a gigantic bar magnet at the solid center, because this moving fluid dynamo can account for the history of shifting magnetic north and south poles, and for the paleological evidence for great shifts and reversals of earth magnetism.

Dating the Earth's History   For many centuries, men had no scientific basis for determining the age of the earth. There were many explanations for fossils, with some people holding that they were merely chance resemblances to living forms—coincidences on a grand scale! It was not until the early nineteenth century that Georges Cuvier in

France and William Smith in England published findings that suggested the earth was considerably older than was commonly believed. Cuvier (called the father of paleontology, the study of fossils), in digging through deeper layers, observed animal fossils there that were far less similar to living animals than the fossils he found in upper levels. This observation is the basis of the principle of stratigraphy. Stated most simply, stratigraphy assumes that the earth's crust has been laid down in layers; if they are undisturbed, the lowest strata are the oldest. Such a principle may seem simple and obvious to us, but its value in establishing chronology was not appreciated until the early to mid-nineteenth century.

Cuvier supported a theory of the earth's history known as Catastrophism, which postulated that several times in geologic history, major worldwide catastrophes wiped out all life forms of a given epoch; each catastrophe was then followed by the special new and divine creation of new forms of life. This seemed to him the only way to account for the observable fact that fossil plants and animals are unlike living things found in the world today. Opposing this hypothesis were James Hutton (1726–1797) and Charles Lyell (1797–1875), who proposed the theory of Uniformitarianism. This states that the same dynamic forces of change that are active in the earth today were active in the past as well. In other words, the present is the key to the past. This theory that ancient rock can be interpreted in the light of today's geological processes is widely accepted by modern earth scientists. One of the most important corollaries of Uniformitarianism is that it postulates long stretches of unbroken time, during which natural forces, acting upon both the physical environment and the living things within it, could produce gradual change. Charles Darwin took Lyell's book (*Principles of Geology,* published in 1830) to read during his voyage on the Beagle. Lyell's suggestion of the earth's long history was the background against which Darwin organized his theory of evolution.

There are two approaches to the problem of dating materials in the earth's crust. One is to establish the relationship of past events; this method, called **relative dating,** sets up a reliable chronological sequence of events but cannot necessarily determine the actual time at which any single event occurred. The precision of actual dates comes from the analysis of radioactive materials contained in the specimen to be dated; this approach is called **absolute dating.**

The principal task of geology is to unravel the long, tangled history of the earth, discovering how and when the various landforms and oceans developed the way they did. To accomplish this, the geologist studies rocks. Every rock is like a page of a book: it tells part of a story to one who can read its meaning. In addition to rocks, the geologist also studies fossils, if they are present.

Relative dating of rocks and fossils depends on three important features. One is superposition. The **law of superposition** states that in an undisturbed

Relative Dating of the Past

Three important features in the relative dating of rocks are superposition, unconformity, and correlation.

Fossils

A fossil is any trace or impression of an organism preserved in any of a number of ways. Fossilization usually involves the hard parts of an organism: bones, teeth, shells, and the woody tissues of plants are the most readily fossilized materials. The soft parts of an organism are fossilized only very rarely.

An organism may be preserved without becoming a fossil in the "petrified," or turned to stone, sense. It may be frozen whole in ice, trapped in amber, or preserved in tarpits—anywhere the chemical environment prevents the growth of decay-producing bacteria. By far the majority of fossils, however, are found embedded in the earth's crust as part of rock deposits. A part of an organism typically is deposited in soft sediment, such as mud, silt, or sand. With the passage of time these materials gradually harden into sedimentary rocks, encasing the organic remains. The internal cavities of the organism and its bones are then filled with mineral deposits from the sediment. Finally the external shell or

the walls of the skeletal bones may decay and are replaced eventually by calcium carbonate or silica. In some rocks, traces of the original organic material are still present in the fossilized remains.

Marine animals are the most commonly fossilized creatures. Their corpses accumulate in shallow sea or river bottoms away from wave and tidal action. In this still water, the organisms are more quickly covered and completely enclosed by soft marine sediments that eventually harden into shale and limestone. Absence of wave action also means that the remains are not broken up and that oxygen, which enters solution more readily in turbulent water, will not be present in great quantities; without oxygen, most decay-causing bacteria cannot exist.

Perhaps the most interesting revelation of the fossil records is that organisms tend to become more complex over time so that the simplest fossils are likely to be found in the oldest strata. This fact has two implications: for geology, it means that the age of the stratum can usually be approximated by the structure of the fossils embedded in it; for biology, it supports Darwin's theory that evolution consists of the development of living things from the primitive to the more highly evolved.

The accompanying illustrations reveal some aspects of fossilization. In the top photo is the skeleton of a camel showing deposition, the first stage of fossilization. After centuries of sedimentation and calcification, the camel will become completely fossilized like the fish in the bottom photo. The drawings show common paleozoic fossils.

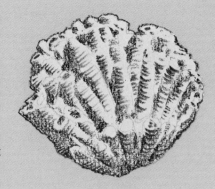

Rugose coral

Compound coral

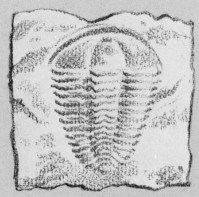

Trilobite

Gastropod

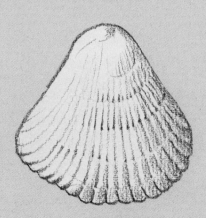

Brachiopod

sequence of strata, the older strata lie below the younger ones. Theoretically, this is quite a simple statement; in practice, however, it is often difficult to ascertain the order of strata because dynamic processes, such as mountain building or erosion, may upset their natural order. This can be seen dramatically in parts of the Sierra Nevada mountain range.

A second feature geologists look for in trying to establish relative dates is **unconformity,** which is said to exist when older layers of rock are separated from younger ones by a surface (of older rock) that has undergone erosion. For example, suppose a many-layered seabed of sedimentary rock is thrust above sea level and tilted by forces in the earth's crust. The exposed portion will erode, revealing a cross-section of its layers (remember, it tilted when it was upthrust, so its layers are now more or less on edge). Later the area sinks again, or the ocean rises, and sediments once more begin to deposit on the eroded surface, eventually forming rock strata that lie at an angle to the tilted, once-exposed strata below. This is called angular unconformity. If the underlying strata are not tilted, so that the new deposits upon worn-away surfaces are parallel to the older layers of rock, the unconformity is called a disconformity. Unconformities of both types are useful because they help reveal the approximate dates of earth movements and the way that land and sea areas were distributed in the past.

William Smith (he was nicknamed "Strata") pioneered the development of the third tool, **correlation.** This technique matches either the sequences of layers of rock or their fossils, or both. Since life forms have generally tended to emerge, flourish, and die off within a relatively limited geological time, one can assume that sandstone in Montana and sandstone in New Jersey that both bear the same marine invertebrate fossils were laid down as sediments on the site of former coastlines at about the same time. And if there are fossils, but the succession of strata show correlated periods of wetness, with heavy deposition of sediments, and dryness, shown by salt deposits from drying-up oceans, it is a clue that the two series of layerings are histories of the same phenomena, no matter how far apart they may be geographically, and thus may represent the same excerpt from the past. If the age of one can be determined, that of the other is also known.

Absolute Dating In 1947, the American chemist Willard Libby, following an idea earlier proposed by Boltwood, demonstrated a means of determining the absolute dates of many natural and manmade objects from the past. His method depends on the fact that in an atomic sense, nearly every rock is a clock. Most rocks contain tiny amounts of radioactive elements; because scientists know that certain radioactive material decays at constant, known rates, independent of external conditions, it is fairly easy to determine the age of any given rock.

Years ago (millions)	Era	Period	Epoch	Biologic events	Geologic events
.01 2 10 25		Quaternary Tertiary	Recent Pleistocene Pliocene Miocene	First man, large mammals Flowering plants prevail Grasses abundant; spread of ungulates, other grazers	Time since last glaciation Great Ice Age and interglacials Formation of Alps, Himalayas, Andes
38 55 65 135	Cenozoic	 Cretaceous	Oligocene Eocene Paleocene	Apes, elephants appear Primitive horses, camels appear First primates First flowering plants; dinosaurs die out	 U.S. undergoes extensive volcanic activity Rocky mountains formed in U.S.
185 225 280 320 340	Mesozoic	Jurassic Triassic Permian Pennsylvanian Mississippian		First birds; dinosaurs dominate Dinosaurs, primitive mammals appear Rise of reptiles; large insects abundant Large flowering plants Large amphibians; fish; extensive forests	Sierra Nevada mountains formed in U.S. Volcanic activity in northeastern U.S. Appalachian mountains begin formation Coal-forming swamps
405 430 500 600	Paleozoic	Devonian Silurian Ordovician Cambrian		First forests; amphibians, fish abound First land plants, air-breathing animals First vertebrates Marine-shelled animals	Caledonian revolution (Caledonian mountains formed in Wales) Ongoing mountain folding and vulcanism on many parts of earth
3500 to 4500	Precambrian	Divisions not firmly established		Few fossils	Oldest dated rocks; formation of continental cores

Table 18-2
Geologic time

Suppose that a geologist is trying to date a rock that is found to contain uranium-238, a common radioactive material. Uranium-238 decays in steps, changing from one radioactive element to another, finally stabilizing as an isotope of lead (Pb-206). Uranium-238 has a half-life of 4500 million years—that is, half of it decays in that span of time; half of the remainder decays in another 4500 million years, and so on. By determining the ratio of the weights of the lead end product and the undecayed U-238 still in the rock, the geologist can compute the age of the rock. Other radioactive materials used to date rocks include U-235, potassium-40, and rubidium-87. Fossils are usually dated by measuring the specimen's ratio of radioactive carbon-14 to carbon-12, the product of radioactive decay. Carbon dating is also widely used to date ancient tools, pottery, and even traces of grain that were found still stuck to a Stone Age grinding dish.

A date obtained in this way can sometimes be difficult to interpret, because it may indicate any of several ages for the rock. For example, it may tell the time lapse since the rock last recrystallized rather than the time when it first solidified. In the case of sedimentary or metamorphic rocks, the date obtained may be that of the older rock from which the new rock was formed. In spite of these

Figure 18-9
Moonlight over the Grand Canyon of the
Colorado. This majestic river valley, which
dates back over 200 million years, is one
mile deep and twelve miles wide. The
sedimentary rocks in the foreground were
laid down during the Paleozoic Era. Beneath
these are older metamorphic rocks dating
back to the Precambrian.

limitations, radioactive dating has established many widely accepted dates. For example, it has indicated that although the limestone on the rim of the Grand Canyon is a mere 225 million years old, the granite in the inner gorge is at least 1000–2000 million years old. Paleontologists have established that the first hominid (humanlike) fossils are about 5.5 million years old.

Continental Drift

In studying the history of the earth's crust, geologists have made some discoveries that posed puzzling problems. For example, fossils of earthworms belonging to the same species have been found on both sides of the Atlantic.

The worm itself was not mobile, and it certainly could not have been carried so far by the action of wind or water. On the other hand, it is also unlikely that two similar worms would have evolved by coincidence; evolutionary probability is greatly against it. How then did this species become distributed across the ocean barrier?

An even more baffling problem came with the discovery that rocks can be fossil magnets. These fossil magnets are actually ancient rocks containing iron oxides; they were magnetized millions of years ago when they were formed with their poles oriented parallel to the earth's magnetic field as it existed then. Today their poles are still pointing toward that ancient magnetic north. They have been unaffected by any succeeding changes in magnetic north because they were "locked" into these polarizations when the lava or sediment of which they were a part solidified. Because of this, we can today determine variations that have taken place in the location of magnetic north since the rocks were deposited.

Two startling facts have arisen from the study of fossil magnets. First, the magnetic poles have reversed themselves in the past 75 million years as many as 170 times—the North Pole becoming the South Pole and vice-versa. More important, all the continents have, in the process of time, changed their positions radically, both with respect to the earth's geomagnetic field and also with respect to each other. For example, it was recently discovered that about 600 million years ago the magnetic north pole was apparently off the coast of California! Although this phenomenon has been labeled "polar wandering," geologists are sure that it is not the poles that are wandering but the land masses, thus changing their relationship with the polar axes.

600 million years ago, the magnetic north pole was off the coast of California.

The theory advanced to explain these observations is the theory of **continental drift,** in which the continents are assumed to be ponderously floating on the surface of the earth's mantle like great icebergs grinding against one another in some frigid sea.

Theories of Continental Formation

Early in this century, the German meteorologist Alfred Wegener noticed that the east coast of South America and the west coast of Africa appeared to fit together like a jigsaw puzzle. He suggested that at one time these two land masses, as well as all the other continents, were united in a supercontinent he called Pangaea. Wegener was hooted at by the conventional thinkers of his day, especially when he went on to state that this supercontinent first broke into two pieces, and that these pieces then divided to form the continents as we know them today. His ideas went largely unheeded even in the face of the evidence he pointed out. Wegener showed that many fossil and mineral samples from areas he said had once been joined were similar, even though their localities are now separated by oceans.

In the middle decades of this century, geophysicists and geochemists produced a revolution in knowledge about the structure of the earth. Theories and techniques developed during studies of the magnetic nature of rocks beneath the ocean yielded information about many areas of the earth; this information threw new light on Wegener's theory, which, in revised form, has gained general acceptance.

It is now thought that there was never one supercontinent, but two great land areas: a northern one called Laurasia, made up approximately of what is today North America, Greenland, Europe, and the great bulk of Asia; and a southern continent, Gondwanaland, encompassed what is now South America, Africa, parts of southern Asia, Antarctica, and Australia. The two land masses were joined where the Straits of Gibraltar are today; a long, narrow sea called the Tethys separated them along a line where now the eastern Mediterranean, the Persian Gulf, and northern India lie.

It is thought that there were once two great land masses called Laurasia and Gondwanaland.

How did the continents get where they are? It is believed that they "drift"—move slowly over the upper layers of the earth. This process takes millions of years and depends on three factors: the "floating" nature of the earth's lithosphere, which is broken into huge masses called plates; the plasticity of the immediately underlying asthenosphere; and a propulsive force— perhaps provided by upwelling of magma along certain oceanic ridges, pushing the plates slowly apart, aided by the circulation of material in the asthenosphere.

Spreading of the Ocean Floor

One reason scientists were slow to accept the theory of continental drift was the question of how the enormous continental chunks could bulldoze their way along, plowing through the solid ocean floor. The solution to this problem came in the mid-1960s with the discovery by two American geologists, Harry Heiss and Robert Dietz, that the sea floor is spreading. This conclusion was reached by linking together a number of ocean studies involving such phenomena as deep sea-core sampling, rock magnetism, age of lava flows, measurements of convection currents beneath the ocean floor, and also the location of underwater earthquakes.

One method of studying the ocean floor is the technique of echo sounding. Using this method, which involves transmitting sound waves to the ocean floor and timing their return, geologists have been able to determine the ocean's depth and, to some extent, the structure of its floor. A second method, sea-core drilling, obtains samples of actual seabed for study. Determination of the earth's magnetic field is achieved by towing a magnetometer behind a ship; this device measures the direction and strength of the magnetism in rocks beneath the floor.

Such studies have revealed a number of interesting facts about the ocean floor. First, the rocks of which they are made seem to be much younger than typical continental rocks. The oldest continental rocks have been dated at about 3500

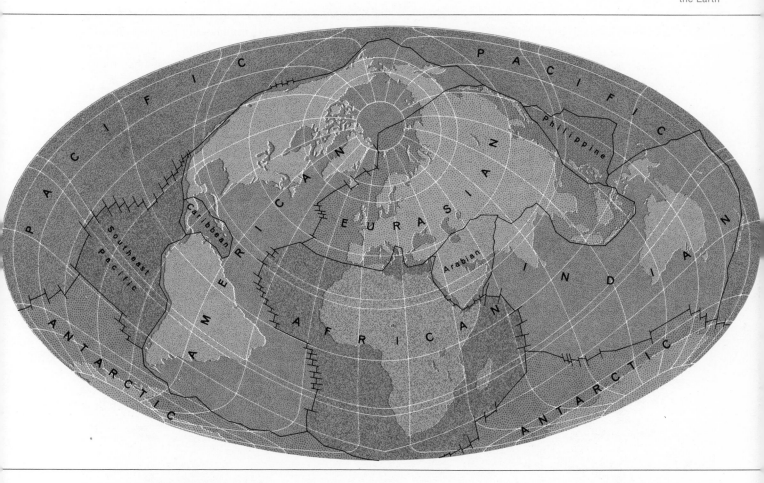

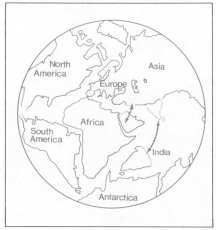

Figure 18-10
According to current plate tectonics theory, the map of the earth may have looked like the first drawing (left) about 200 million years ago. The land masses were combined into two supercontinents. In the north was Laurasia, composed of modern-day North America, Europe, Greenland, and most of Asia. To the south was Gondwanaland, consisting of South America, India, Antarctica, Australia, and Africa. The supercontinents were separated by the Tethys Sea. The second drawing at left shows the continents as they are today. The earth is thought to be divided into the huge tectonic plates shown above. The divisions between the plates are the scenes of frequent diastrophic activity including earthquakes, volcanoes, and mountain building.

million years, while oceanic rocks seem to be no older than 135 to 150 million years at the oldest. Second, a system of ridges (uplifts) and trenches (troughs) traverse the ocean floors. Third, and most importantly for the theory of continental drift, the rock material that lies on either side of the ridges has repeatedly reversed magnetic direction over time. Thus, strips of crust containing north-oriented material alternate with those containing south-oriented material and lie parallel to the ridges.

Based on these observations, geologists have postulated that the sea floor is actually spreading. The way this happens is quite simple and is shown in Figure 18-11. The ridges, which contain rifts parallel to their directions, may be thought of as underwater volcanoes that continuously ooze forth magma. This material rises vertically from inside the ridge and spreads out on either side of it, pushing away already formed lithosphere and taking its place. More freshly flowed magma then pushes it further away from the fissure. This spreading movement takes place at a rate of from two inches to two feet per year. The farther away from a ridge the magma-formed lithosphere is, the older it is. However, a 1974 diving expedition (see the interlude on Project Famous) has raised some doubts about whether the magma is pushing the plates apart. The ex-

Figure 18-11
Sea floor spreading at mid-ocean ridges may be the mechanism behind continental drift. Molten rock in the asthenosphere rises by convection through the fissure at the lithosphere. As it reaches the lithosphere, the magma cools, solidifies, and forms crustal plates containing underwater volcanoes and mountain ridges similar to those shown here. Continuous outpouring of the magma apparently forces apart the plates on either side of the fissure, moving them at the rate of about 1 per year.

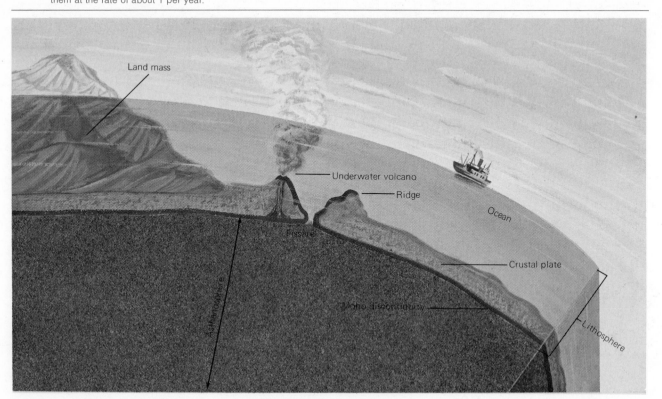

pedition found that the strata on either side of the rift had not been compressed, as they would have been if they were being pushed. Instead, the scientists think the plates are being pulled apart by forces acting in other parts of the sea floor.

While the lithosphere is spreading and slowly moving out from its place of creation (the ocean ridges), the continents, which are merely those parts of the lithosphere that project above sea level, move about 1 centimeter per year. This explanation of the phenomenon of continental drift is called the theory of **plate tectonics**; tectonics is from a classical Greek word meaning carpentry, or building, and refers to the creation of the moving lithospheric plates.

But what of these plates, formed by the ejection of molten magma from the ocean ridges? Where are they located? How do they move? Figure 18-10 shows that there is a worldwide mosaic of six principal plates (a number of smaller ones are not shown for the sake of clarity). The largest plate transports the entire Pacific Ocean, while one of the smallest moves an area about the size of Texas.

The plates, which are in constant motion, are also builders of mountains and causers of earthquakes. For example, when two plates collide in a great grinding crunch, the leading edge of one plate passes over or under the edge of its neighboring plate; the result of this collision is certainly an earthquake and possibly a mountain. It is thought that the slow collision some 40 million years ago of the plate carrying India with the plate carrying Asia produced the Himalayas which are, consequently, made up of material originally on the Tethys seabed. The plate which passes under another plate usually breaks off and descends into the asthenosphere, where it melts and becomes part of this layer. Usually, the continental mass is not dragged down with it because of its buoyancy. Where the lithosphere has carried part of the continental crust with it down into the asthenosphere, the crust melts, then rises, and solidifies to poke through openings in the lithosphere forming island arcs, such as those in the eastern Pacific—Japan, the Philippines, and Indonesia—and, to the north, the Aleutian chain.

These arcs consist of extrusive rock in the form of volcanoes above the surface and massive slabs of intrusive rock below it. Some geologists have theorized that the continents were formed in this manner; after rising to the surface, the island arcs gradually increased in size, finally becoming the continents as they exist today. Structures that resemble island arcs, for example, have been discovered deep below the surface of the North American continent.

Earthquakes can also occur when two plates separated by a break in the earth's crust slide past one another. Along such a break in the earth's crust can be a chain of volcanoes or mountains, a rift in the ocean floor, deep oceanic trenches or a fault—such as the San Andreas fault in Southern California. The volcanic activity associated with plates sliding past, over, or under one another

Figure 18-12
Part of the San Andreas Fault near San Francisco. The entire fault is about 600 miles long, extending from the Gulf of California in the south to a point in the Pacific Ocean just north of San Francisco. Geologists think the fault separates two huge crustal plates—the Pacific Plate west of the fault may be moving northwesterly relative to the American plate to the east. The slow movement of the plates may be the cause of the many earthquakes that have occurred in this area, including the one that devastated San Francisco in 1906.

is thought to be caused in large part by the great heat generated by the enormous friction of the plates' moving contact.

Other structures sometimes present near the boundaries of moving tectonic plates are inactive volcanoes. Common to the Pacific Ocean, these ancient structures rose above the ocean's surface in the past, but have since sunk beneath it. Although they show wave-action erosion from their existence above the ocean's surface, they are now underwater. The lithosphere pushed these volcanoes into deeper and deeper water as it spread out from its formative ridges. As the volcanoes slowly sank, waves eroded them from bottom to top.

Finally, and most importantly, how do the continental plates move? Two theories have been proposed to explain their movement. Many geologists think the answer may be the creation of convection currents, caused by the rising of hot, less dense matter and the sinking of cool, dense matter in the asthenosphere, where temperatures are hotter at lower levels than at higher ones. Thus, the friction of convection currents under each plate may be drawing that plate along with those currents. Such currents, if visualized as traveling in a closed system with circular motion, could be responsible for moving the plates across the earth's surface. However, the physics of such a convection flow seems to be contrary to the way in which such eddies normally behave in laboratory studies.

Some geologists believe the plates may be moving at least partly as a result of the pressure produced by the movement of new lithosphere being formed at ocean ridges. According to this theory, the rock that is continuously being formed by the ocean ridges pushes against adjoining plates and moves them. It is possible that the cause of the motion may be a combination of the two forces: convection and outward push. The fact is that the plates do move, and that movement provides a challenging and fruitful opportunity for future research.

Glossary

Absolute dating is a method of determining the date of the formation of a rock or a fossil by means of a chemical analysis of the radioactive elements contained in the sample.

The **asthenosphere** is a portion of the earth's mantle just below the lithosphere, believed to be plastic rock, ranging in thickness from 100 to 400 kilometers. It is believed that the lithosphere floats on the asthenosphere.

Cleavage is the tendency of a mineral to split along a smooth plane.

Continental drift is the movement of giant land masses over the earth's surface.

Continental plates are formed by cracks in the lithosphere which divide it into six major and a number of minor plates. These plates move very slowly, accounting for continental drift.

Correlation is a geological means for dating the past. This principle enables geologists to correlate the ages of rocks located in different places on the earth by matching strata of the same age.

Crust is the earth's outermost layer. Most of the earth is covered with solid rock called bedrock. Seventy-five percent of the earth's crust is composed of oxygen and silicon.

Extrusive rock is rock formed after the eruption of a volcano. The lava cools and solidifies, and this igneous rock is called extrusive. Formed from magma deep in the earth, intrusive rock is magma which never reaches the surface.

Fossils are traces or impressions of an organism preserved as part of rock deposits. Important in geological dating.

Igneous rock is rock that has solidified from molten material. (From Latin word for fire.) Common examples are granite, basalt, and obsidian.

The **lithosphere** is the crust and upper portion of the earth's mantle thought to be a rigid rock mass about 50 kilometers thick.

Magma is a molten mineral mixture of silicates, oxides, and sulfides, containing a large amount of dissolved gases held in solution by pressure. Formed deep in the earth.

The mantle of the earth is composed largely of rocks primarily with a large iron content, making up about 80 percent of the earth's volume.

Metamorphic rock is rock that has been modified or metamorphosed by pressure deep within the earth.

Minerals are naturally occurring inorganic substances with specific atomic structures, chemical components, and physical properties. Most minerals are chemical compounds but a few are elements.

Mohorovićic discontinuity is a section from 5 to 40 miles thick separating the earth's crust from the mantle. Discovered by an analysis of seismic waves by A. Mohorovićic, a Croatian, in 1909.

Relative dating is an approach to dating materials in the earth's crust by establishing their relationships to past events, setting up a reliable chronological sequence of events.

Sedimentary rock is formed by fragments of older rocks, broken down by wind and water and transformed by consolidation into hard rock.

Seismic waves are waves produced by earthquakes; primary (P) waves similar to sound waves, which travel the fastest; secondary (S) waves, which are transverse waves; and long (L) waves similar to ocean waves, which cause the most damage.

Stratigraphy is the study of the history of the earth by an examination of rock strata and the fossils found in equivalent strata in different parts of the world. The underlying principle is that the lowest strata are the oldest ones.

The law of **superposition** states that in an undisturbed sequence of strata, the older strata lie under the younger ones.

Unconformity is an important relative dating tool of geology. Means that between two beds of rock, erosion surfaces separate the younger strata from the older. Unconformities reveal the approximate dates of earth movements as well as the distribution of land and sea areas in the past.

Uniformitarianism is a hypothesis developed independently by James Hutton (1726–97) and Charles Lyell (1797–1875) that says ancient rocks can be interpreted in the light of today's geological processes. A basic geological theory.

1 Show how the atomic bonding and structure of quartz accounts for its useful properties of hardness and crystallinity.

2 What is the difference between rocks and minerals? Why are the rocks rather than minerals the basic study units of geology?

3 What simple tests for classifying rock specimens are available to a geologist working in the field? Which two tests would distinguish a specimen of calcite from a specimen of quartz with very similar crystal structure and color?

4 An amateur geologist is collecting specimens of basalt near the blasting site for a new building in downtown Manila. What are the two forms of igneous rock he may find and how would he distinguish between them?

5 Why do sedimentary rocks blanket 75 percent of the continents and even more of the ocean floor, despite the fact that they comprise only 5 percent of the earth's crust?

6 What is the relationship between the theory of Uniformitarianism and the Darwinian theory of evolution? Can Cuvier's theory of catastrophism be similarly reconciled with Darwinism? Explain.

7 Discuss the relative advantages and disadvantages of using (a) superposition (b) unconformities (c) correlations as techniques for dating rocks.

8 Suppose that on an archeological dig in North Africa, fossil remains of shellfish and trilobites are unearthed from a layer of rock 25 feet below the surface. Meanwhile, in Tangiers, 125 miles south of the dig, a construction crew comes across fossilized fish skeletons while blasting a rock layer 100 feet down. Which layer of rock is older and why?

9 In order definitively to solve the preceding problem, suppose that the curator of the Tangiers Museum of Anthropology resorts to radioactive dating. If the half life of the radioactive material he is using for dating is 170 million years and the ratio of undecayed isotope to lead is 1:7 in a fish skeleton fossil, compute the age of the fossil.

10 What do P and S waves tell us about the earth's crust, mantle and core? How was the liquid nature of the latter deduced?

11 How does paleomagnetism support the theory of "continental drift"?

Exercises

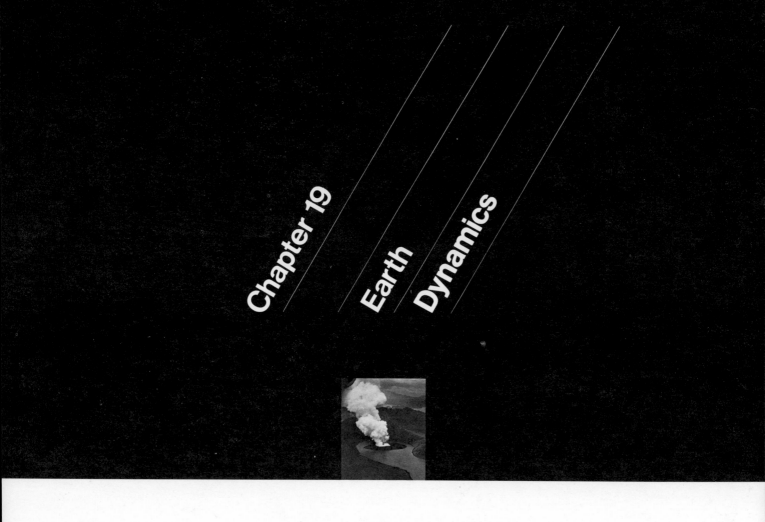

Chapter 19 Earth Dynamics

The structural features of the earth are constantly being changed by vulcanism, diastrophism, and erosion.

To most of us, the earth seems unchanging. Its basic structural features, the mountain ranges and oceans, appear to stand as eternal witnesses to its stability. But these mountains and seas also give mute evidence that over the vast stretches of geologic time, the earth has changed radically. In fact the earth is continually being altered by geologic forces; the primary forces for change are vulcanism, diastrophism, and erosion. Vulcanism, the movement of molten rock beneath the earth, is responsible for volcanoes. Diastrophism refers to the vertical and horizontal movements of the crust that typically result in mountains. Erosion includes the actions of all the destructive external processes—streams, glaciers, and wind—that wear down land masses.

In 1883, the explosion of the island of Krakatoa in the East Indies literally blew that island to pieces. More than 5 cubic miles of rock were hurled into the air, some of it as high as 17 miles. Thirty-six thousand people were killed by the accompanying tidal wave alone. The explosion was heard 2500 miles away, and the sound waves affected air pressure all over the world. This startling natural explosion, in which the earth actually blew its top off, was a terrifying example of a volcanic eruption. Such eruptions are among the most dynamic ways by which the earth restructures itself.

Vulcanism is defined as the movement of molten rock. A volcano is an example of extrusive vulcanism, the external movement of magma, or molten rock, through and over the earth's surface. Volcanoes are widely distributed around the earth, both on the continents and on the sea floor. At one time or another in the earth's history, they must have been present on virtually every part of the earth's crust, since evidence of volcanic activity is found in all major rock formations. Today, however, live volcanoes are found only in certain areas, the most important of which is the **circle of fire**, a belt of volcanoes which rings the Pacific Ocean. (See Figure 19-1.) This belt runs north from New Zealand, through Indonesia, the Philippines and Japan, and across the Aleutian Archipelago of Alaska. It continues south down the west coast of the Americas, in the mountains of the Cascade range in California, the volcanoes of Central America, and those of the Andes Mountains which extend along the entire thousand-mile west coast of South America.

Although geologists do not yet fully understand either the processes of vulcanism or the formation of volcanoes, they have been able to determine the resulting structure of volcanoes from direct observation and indirect evidence. A volcano is typically a cone-shaped hill or mountain ranging in height from 20 feet to 5 miles, surrounding an opening (called a vent) in the earth's surface. When the volcano is active, hot gases, molten rock, and solid fragments are ejected through the vent, which is at the center of a depression called a crater.

Volcanic eruptions, some of them the most spectacular of all earth-forming processes, occur in many patterns. They range in intensity from small lava trickles to huge sprays of liquid fire with masses of rock and clouds of dust; from muted bangs to rumbling, air-rending explosions heard thousands of miles away. What are the forces that determine the intensity of volcanic eruptions? Deep below the vent of the volcano lies the magmatic reservoir, a large deposit of magma which is under pressure because of the weight of rock above it, with temperatures that may reach 1200°C. In addition to molten minerals, magma contains a number of gases—chiefly water vapor, carbon dioxide, and sulfur vapors—that remain in solution in the liquid magma as long as they are under pressure. The expansive force of the magmatic gases through the action of heat drives the magma to the surface; this is the volcanic explosion. As the magma

Vulcanism: The Release of Heat Energy

Extrusive igneous activity causes the deposition of molten rock on the earth's surface. This process may be violent, as in volcanic eruptions, or gradual, as in the formation of lava flows.

Structure of the Volcano

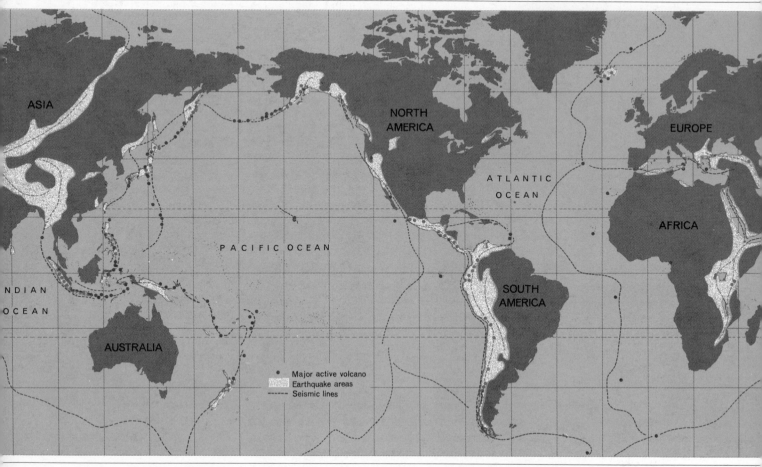

ASIA

NORTH
AMERICA

EUROPE

ATLANTIC
OCEAN

PACIFIC OCEAN

AFRICA

INDIAN
OCEAN

SOUTH
AMERICA

AUSTRALIA

- Major active volcano
  Earthquake areas
- - - - Seismic lines

Figure 19-1

Volcanic activity is distributed widely
throughout the world, as this map shows. The
"circle of fire" which borders the Pacific
Ocean is indicated by the dark line.
Volcanoes, earthquakes, and mountain
building all occur in the same regions, a fact
which current geolophysical theory
attributes to the movement of the tectonic
plates in the earth's lithosphere.

rises, pressure decreases, causing most of the gases to be released. These gases, suddenly free of high pressure, often freeze in the cool air, forming a foamlike rocky solid called pumice.

The intensity of an eruption is determined by the viscosity of the magma and the amount of gases present. The viscosity depends in great part upon the amount of silicates in the magma. Magma with low silicate content is relatively fluid and easy flowing so the gases in it can expand, bubble up through the liquid, and escape early. But the sluggish viscosity of high silicate magma traps the gases, causing them to build up pressure. A volcanic eruption will be highly explosive if the magma is sufficiently dense with silicates. Conversely, magma with little silica moves to the surface with silent or relatively quiet eruptions.

Pyroclastic Debris

When a volcano erupts, one or more explosions thrust out a huge cloud composed of released gases, water vapor, dust, and solids of various sizes. The

Figure 19-2
The Danakil Depression in Ethiopia is all that remains of a former volcano. Some time ago, the cone of the volcano collapsed or exploded, leaving this depression at the top, called a caldera or crater. Rain water has accumulated in the depression forming a crater lake.

Figure 19-3
This section of earth shows examples of extrusive and intrusive activity. The cross-section of a volcano is in the foreground. At the bottom of the volcano is the magmatic reservior. Magma, which is forced up the conduit, erupts from the vent as lava, pyroclastic fragments, water vapor, and gases.

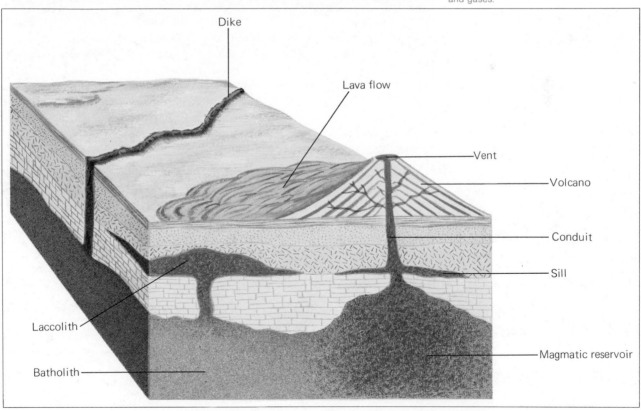

Dike

Lava flow

Vent

Volcano

Conduit

Sill

Laccolith

Batholith

Magmatic reservoir

solid rocks blown from the vent are called **pyroclastic debris.** Pyroclastic (the word means "fire-formed") fragments and debris range in size from blocks weighing two tons to small chunks, large cinders, and fine dust or ash. Ash is formed when the lava, emerging as a stream or spray, suddenly solidifies as it meets the cool surface air.

Volcanoes erupt in a number of patterns. The two active volcanoes on the island of Hawaii, for example, are nonexplosive, ejecting low silica magma and hot fluid lava. In contrast, Mt. Pelée on Martinique in the West Indies erupted in 1902 with no lava discharge at all. Instead, an extremely hot cloud of volcanic ash and glowing gases swept down the slope of the volcano at speeds up to 100 miles per hour, burning to death all but one of the 30,000 inhabitants of the city of St. Pierre. (The survivor was a prisoner in a thick stone-walled prison cell; he doubtless concluded that crime does pay.) Mt. Stromboli, in the Mediterranean Sea north of Sicily, erupts every ten or fifteen minutes with muted explosions. Many volcanoes erupt once soon after they are born, then are never heard from again. Others, after lying dormant for hundreds of years, suddenly explode with great violence.

Recently geologists have been able to study for the first time the birth and development of a new volcano. This one, known as Paricutín, originated in February, 1943, near the town of San Juan, Mexico (about 200 miles west of Mexico City). The volcano began as a small crack in the earth from which steam fizzed with the characteristic volcanic sulfur odor, followed by a brief series of explosions. Explosions, some accompanied by pyroclastic debris, continued throughout the night. By morning, a cone of cinders and ash some 25 feet high had accumulated. Shortly afterward, the volcano spewed forth lava which covered the town of San Juan. (Since there was ample warning, there were no casualties.) By 1949, the volcano, which for five years had been continuously exploding and hurling pyroclastic debris in all directions, had grown to 2000 feet. By 1952, Paricutín was dormant.

In many predominantly volcanic areas, true volcanoes are absent. Present instead are other phenomena known as volcanic plains or plateaus, which are huge sheets of lava that have oozed out of many separate vents and solidified. Similarly, lava may seep through surface fissures many miles long. The hot lava is a genuine liquid, and it flows as such. In other areas, volcanoes that have become almost extinct in recent years have left traces in the form of solidified lava flows. Other traces of volcanic activity, in which magma under the earth is pushing toward the surface, include hot springs, geysers, and steam vents.

Intrusive Activity   Not all vulcanism is explosive. There are masses of magma beneath the earth that solidified before reaching the surface, in the process of intrusive vulcanism. In effect they are jagged solids that are pressured from below to cut and force their way upward.

Intrusive igneous rocks are known as **plutons,** after Pluto, Roman god of the underworld. Generally they have coarser grains than rocks from volcanic eruptions, because the insulating effect of overlying rock causes them to cool very slowly. Intrusive forms are given different names based on their shape and reaction to the rocks that enclose them.

Dikes are igneous forms that, under great pressure cut fractures or fissures across previously formed metamorphic, sedimentary, or igneous rocks. Dikes vary greatly in size. One of the largest is in Africa, the Great Dike in Rhodesia, which is 300 miles long and 5 miles wide.

Sills or intrusive sheets, are solidified lava flows that originally forced their way between and parallel to older layers of rock. In many cases, a sill branches into adjoining rock layers in the form of a small tongue, thus becoming a kind of dike. A familiar example of a large sill is the one forming the Hudson Palisades just west of New York City, known as the Palisades Sill. More than 1000 feet thick in places, it extends 40 miles up the Hudson River from Staten Island to Haverstraw, New York. Originally, it may have been many miles wide. A **laccolith** differs from a sill, in that at points it has lifted into overlying strata, forming a domelike structure. A **batholith,** a huge mass of coarse-grained granite, is the intrusive rock that forms the core of many mountain chains. Batholiths project into the earth, often as much as 6 miles. Some of these large structures measure more than 100 miles wide, 1000 miles long and 4 to 6 miles deep.

Intrusive igneous activity produces rock formations below the earth's surface, such as dikes, sills, laccoliths, and batholiths.

The second of the processes that shape the face of the earth is diastrophism (from the Greek for distortion). **Diastrophism** includes all large-scale movements in which one part of the earth's crust changes position relative to another due to internal pressures. These giant earth movements are best explained by the theory of plate tectonics.

Rocks of the earth's crust subjected to internal pressure by compressional forces, such as the movement of one plate over another, will either fold or fault. When a rock **folds,** it becomes deformed but it does not actually break. A **fault,** on the other hand, is an actual break. It is a fracture in the earth's crust, and it is accompanied by movement of the rocks on either side of the break.

Faults are found in many places in the crust. They vary in length from short faults extending less than a mile to gigantic ones, such as the San Andreas Fault in California, that span hundreds of miles. Displacements vary from less than an inch to 45 feet, which is the largest known. The actual fracture in the earth is known as the fault plane.

Diastrophism

Diastrophism alters the earth's surface through movement of the crust. Rocks may fold or break (fault) from regional pressure or stress.

Folds and Faults

The Greatest Eruption

For many years the eruption of Krakatoa was thought to be the most powerful volcanic explosion mankind has known. However, recent archeological diggings on the Greek island of Thera (also known as Santorini) have discovered that an explosion of far greater power took place there about 1500 BC. The extent of the eruption, and the difficulty of digging out the ruins, is evident from the fact that volcanic ash 150 to 180 feet thick now covers most of Thera. The area where the archeological expedition, headed by the Greek scientist Spirodon Marinatas, began to dig near the modern village of Akroteri had three things to recommend it. First, the layer of ash there was only 30 to 35 feet thick. Second, rains had cut deep ravines down through the ash, so that it was possible to stand at the bottom of a ravine and trace geological history in the layers which were uncovered. Third, some impressive pieces of pottery, clearly from the same time and style as the nearby ancient civilization Crete, had been found in that area.

The bottom layer of one of the ravines at Akroteri is lava from the earliest time of the emergence of the volcanic mountain. On this is a 3-foot thick layer of soil, rich in decomposed vegetation from a long time of growth and decay, uninterrupted by any volcanic activity. It was this rich earth which was farmed by the civilization destroyed by the eruption. Next above the rich earth is 10 to 15 feet of sharp pink pumice stones, and then one or two inches of sandy earth with no traces of civilization, probably from a period in which there was an interruption of volcanic activity but when all human life had already left Thera. Above are 35 feet of white volcanic ash, containing several subordinate layers of small lava stones. On the very top is a thin layer of soil; Thera has been a farming island again for many centuries.

Thera today has a crescent shape, but when it was first settled (around 3000 BC) it was a cone-shaped island—a single, round mountainous mass projecting from the Mediterranean Sea. The volcanic eruption destroyed an area of 83 square km, compared with the 23 square km which exploded and sank at Krakatoa. How much more awful must have been the eruption at Thera! Some scientists have estimated the energy to have been four to five times as great as that released at Krakatoa, with an initial great tidal wave height of 650 feet. Most of the marble and lava mountain collapsed and sank into the vast abyss of the volcanic crater, which is even now 1300 feet deep.

Scalding pumice showers must have fallen from the skies for days, creating a long night which may have been the basis for the legendary triple night of Hercules. Whole towns were swept into the sea. Since so great a portion of circular Thera disappeared beneath the waves, this dreadful event is one plausible origin for the legend of Atlantis. Ever since the German archeological explorer Heinrich Schliemann found the buried but real cities of Troy and Mycenae by taking Homer and the Greek myths to be literally true stories, scientists have paid closer attention to such folk tales.

In the lovely and ghostly town which now is being delicately liberated from the volcanic suffocation of centuries, there is evidence of people who decorated their walls with fresco paintings of great charm, depicting daily life, flowers, sea creatures—especially dolphins and octopi—and birds in flight. One testament to the awesome power of the eruption at Thera is this: the frescoes most beautifully show the meetings of swallows in mid-air. But no swallows have been seen at Thera since.

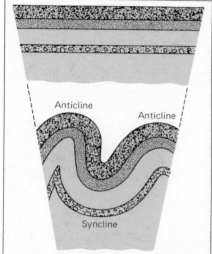

Figure 19-4
These sandstone strata in Anza-Borrego Desert State Park, California, exhibit remarkable folding. The beds toward the bottom of the photo have been compressed from either side into an anticline (upward fold) by powerful internal forces.

Faults, which occur in several varieties, are classified according to the type of movement that has taken place along their surfaces. When the crust along the fracture appears to have slipped down relative to the crust below, the fault is called a normal fault. A normal fault lengthens the surface area previously occupied by all the earth involved in the faulting. If the crust has moved up, the result is a reverse fault, caused by the compression forces which press the crust from either side, forcing one part to ride higher over the other. A reverse fault decreases the surface area of the rock involved in the faulting. Generally, normal faults have a steeper incline than do reverse faults. If the incline of the reverse fault is nearly horizontal, it is called a thrust fault. The third type of fault is the strike-slip, or lateral fault, in which the motion of the crust along the fracture is horizontal rather than vertical.

When the earth's crust has been displaced vertically along a normal or reverse fault, where the crust on one side of the fracture is higher than the other, a cliff may be produced. Often the top of the cliff is eaten away by erosion, thus reducing it to the same height as the lower lying crust on the opposite side.

Earthquakes

An earthquake is a natural movement of the earth that results from the faulting of rock beneath the surface. The best way to understand the dynamics of a major earthquake is to imagine the site to be a large rectangular mass of rock. As a result of the diastrophic forces acting upon it, part of the rock strains to move in one direction, while the other part strains to move in the opposite direction. Movement may be up-and-down, as well as side-to-side. Usually, this stress on the rock is a start-stop process over a long period of time. Part of the rock will be acted upon, causing it to shift slightly, perhaps one-half inch or so. Then, after a period of several years, the opposite end of the rock will be similarly moved.

Because it is slightly elastic, the rock can be stretched or compressed up to a point without breaking. But when its elastic limit is surpassed, the straining rock suddenly snaps, causing an earthquake. (The same thing happens to a rubber band: when it is stretched beyond its elastic limit, it breaks.)

When rock fractures in this way, seismic waves carrying kinetic energy are sent out in all directions. There are three kinds of seismic waves:

1   *P* (Primary) waves, longitudinal like sound waves, are push waves. The wave particles are pushed back and forth parallel to the wave's path.
2   *S* (Secondary) waves, transverse like rope waves, are shear or sideways-twisting waves. They move sideways, perpendicular to the wave's path.
3   *L* (Long Period) or surface waves, are transverse but move like water waves. They ripple out from the center of the disturbance, traveling in all directions.

The *L* waves are the most intense and most damaging. These move slowly and with greater amplitude, which means they transmit more of the earthquake's energy than the other waves. Since *L* waves travel only on the surface of the earth, they are stronger than the *P* and *S* waves, which dissipate much of their energy as they move through the earth's interior.

By timing the speed differential between the *P* and *S* waves, scientists at seismic stations can work backward to pinpoint the **epicenter,** or point on the surface immediately above the focus (which ranges in depth from 35 to 435 miles) of the quake. The epicenter receives most of the shock of the quake as it opens upward.

Although the forces responsible for earthquakes are not completely identified, it is thought that lithospheric plates cause the earth to tremble as they move very slowly past one another. The immediately observable effect of plate movement is the sudden shifting of masses of the earth's crust along faults, resulting in an earthquake.

The force of different earthquakes may be compared by measuring the displacement along a visual surface break, but these are not good enough evidence of the various possible deep earth movements. A more precise comparison computes the total energy released in an earthquake by measuring outward movement of the earth's crust. Thus, the intensity of an earthquake can be measured on a seismograph according to a magnitude scale devised in 1935 by the American geologist C. F. Richter, called the **Richter Scale**. The Richter scale has no maximum figure, but earthquakes that record amplitudes of seven and above are considered catastrophic. The San Francisco earthquake of 1906 is thought to have measured 8.2 on the Richter scale. This quake, which only lasted 40 seconds, destroyed a large part of the city and killed 700 people.

A seismograph measures the amount of energy released in an earthquake.

Each year, approximately one million earthquakes occur all over the world. Fortunately, only a handful of these—ten to fifteen—register amplitudes of seven or above on the scale. The great majority of these quakes register below two and are relatively unnoticed.

Mountain Building

Mountains, which are elevated areas of the crust, result from the action of varied and complex geologic forces. They are usually classified into three groups: erosion mountains, accumulation mountains, and diastrophic mountains. **Erosion mountains**, in which part of the elevated surface has been eroded away leaving a mountain, are examplified by the Catskills in New York State. The volcano, the best example of an **accumulation mountain**, is formed when solidified lava and other rock spewed out pile up around the vent forming a cone. The third type, the **diastrophic mountain**, is formed by complex earth processes such as warping, folding, and faulting. This is the most important type, since the major mountain belts of the world, including the Alps, Pyrenees, Himalayas, Andes, Appalachians, and the Rockies, are all diastrophic.

How is a diastrophic mountain actually formed? Geologists can trace the facts of the formation of a mountain chain, such as the Appalachians, from birth through the present day, but no one is certain of the forces that brought them to their present configurations. It is known that the Appalachians and all other folded mountains undergo at least three formative stages.

GEOSYNCLINAL STAGE   The first stage of the Appalachians, called the geosynclinal stage, began about 500 million years ago with a shallow depressed area of sea bordered by a series of volcanic islands. Sediments were washed and driven by wind from the volcanic regions into the depression. The sediments included materials such as sand, small pebbles, clay, and silt. The huge trough into which they settled is known as a **geosyncline**. As sediment-laden streams emptied into the trough, it was covered by a shallow sea. Later, as more and more solid sediment was deposited, the sea was displaced. The sea repeatedly advanced and retreated for millions of years. Eventually, the sea

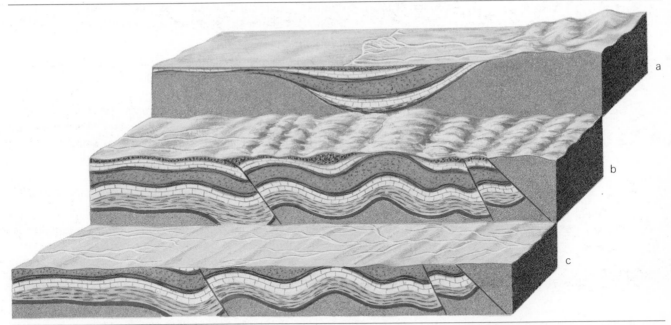

Figure 19-5
The Appalachian Mountains were 500
million years in the making. This figure
shows the stages of their development. A
geosyncline (a) was formed druing the
Paleozoic Era as sediments from adjacent
highlands washed down into it. Internal
forces (b) compressed the strata, leaving
faults and folds toward the end of the
Paleozoic. Stream erosion (c) peneplained
the uplifted landforms. Today, after having
experienced 3½ cycles of uplift and erosion,
the Appalachians appear quite different.

was permanently displaced by increasing sedimentary deposits that finally
rose some 40,000 feet from the deepening center of the geosyncline. The
huge mass of sediment forced the trough deeper into the earth's crust (perhaps even as far as the mantle). Magma began to intrude into the sediments,
forcing them to expand and fold.

THE DEFORMATIONAL STAGE The second stage was the **deformational stage**.
The beds of sedimentary rock lying in the geosyncline were subjected to
tremendous compressional forces from either side, causing them to fault, tilt,
metamorphose, and fold. Rock strata folding upward resulted in structures
known as **anticlines**; downward folding produced **synclines**. In time, the entire
deformed structure, consisting of folded, faulted, and metamorphosed rock, intruded by huge igneous granite batholiths, was heaved far above the surrounding terrain, and a new mountain range was born.

THE EROSION AND UPLIFT STAGE As soon as the mountain rose above sea level,
it was simultaneously subjected to the forces of erosion and isostatic uplift.
Isostatic uplift is the balancing process by which a mountain is forced to rise
like a see-saw as sediments washed from it are deposited onto the connected
adjoining lowlands, forcing them to sink. **Isostasy** means equal weight, and it is
best understood by examining an illustration of Archimedes' principle of the flo-

tation of solids in liquids. Figure 19-6 shows three blocks of wood floating in water; the taller blocks float higher but also farther downward than the shorter blocks.

Using this principle, the British astronomer George Airy postulated a theory of the roots of mountains, his hypothesis of isostasy, in 1855. Airy was attempting to understand why a great crustal mass of heavy material such as a mountain, a plateau, or a continent did not collapse and crash through the underlying layer of earth, the asthenosphere. According to Airy, the mountain, like the wooden blocks, is of lower density than the substrate, or the material below it. Thus, a mountain (like the taller wooden block) floats higher upon the surface, but it also sinks deeper into the heavier fluid (asthenosphere) below it. The mountain has a root which penetrates the lower layer and displaces it, compensating for its greater height.

How does isostasy relate to mountain building? According to the principle of isostasy, there is a continuous state of equilibrium, or balance, between different segments of the earth's crust. Portions of the earth rise and fall as processes such as erosion and deposition shift material from one place to another. For example, in the case of mountains that are continuously subjected to erosion, the rocks and other sediment being eroded fall into adjacent valleys and oceans. This creates an imbalance, in that the mountain is becoming lighter while the adjacent area becomes heavier. Over time, this imbalance creates tremendous pressure beneath the mountain and the adjoining areas; equilibrium is again restored when the pressure causes the adjacent overburdened area to sink and the mountain roots to rise again. Let us return to our

Figure 19-6
The principle of isostasy helps explain why uplifted landforms do not sink into the strata underlying them. At left, wooden blocks of different densities and lengths float in water at different depths. Analogous to these floating blocks are the landforms shown at right. The hot plastic rock of the mantle is like the water in that it allows landforms of varying densities and lengths to "float" in it at varying depths.

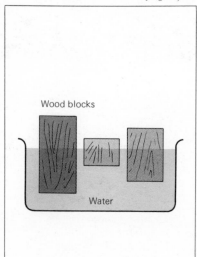

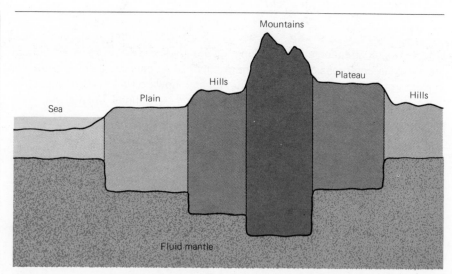

experiment and think of the mountain as a large wooden block, any adjoining area as a small wooden block, and the substratum as the water. If we were to slice a piece from the taller block and place it on a smaller block, the shorter block would sink and the taller block would rise somewhat because weight had been taken from it. The same principle is involved in mountain building in particular and isostasy in general.

Erosion, acting over a long period of time will cancel the effect of isostatic uplift, reducing the mountain's height to a relatively flat surface called peneplain. (A **peneplain** is defined as a land surface worn down by erosion to a flat or gently rolling plain). Eons later, the crust may again be lifted—as were the Appalachians—to be acted upon anew by erosion. Indeed, the Appalachians are now in the midst of their fourth uplift-erosion cycle. Having been uplifted in the recent geologic past, they are currently being levelled. In time all the mountains existing today will have been worn down, and many may then be rejuvenated again.

Geologists are quite certain that all of the world's complex folded mountain chains arose in the manner described above. However, they are not sure what caused the compressional forces that deform the rocks. Immense amounts of energy are needed to push strata of land into folds and to thrust them up thousands of feet above sea level. Attempts to explain these forces have produced several interesting theories.

Causes of Mountain Building

One of the earlier, the thermal contraction theory, is based on the belief that the interior of the earth, once a molten mass, is gradually cooling and contracting. The crust on the surface, having become too large, begins to shrink like the skin of an old apple. As it shrinks, it leaves wrinkles—folded mountains—on the surface. Although it seems to give a plausible model, this theory is now doubted because recent studies show that the earth's temperature appears to be stable (or even rising) from the presence of heat-generating radioactive materials in the rocks of the crust and upper mantle. Further doubt is thrown on the thermal contraction theory, which pictures the earth's interior as a mostly molten mass, by the fact that *S* waves of an earthquake (which are transverse waves and therefore can travel only through solids) have been measured to depths of 1800 miles beneath the surface.

Another theory, the convection current hypothesis, states that mountains are formed primarily as the result of the activity of numerous convection cells generated in the mantle of the earth. This theory requires two conditions: heat source deep in the mantle capable of initiating the convection flow—presumed to be radioactive metals in massive quantity; and a plastic mantle capable of very slow motion, perhaps 2 or 3 centimeters a year. According to the convection theory, the deepest part of the mantle, heated by radioactive materials, slowly rises by convection till it meets the bottom strata of the cool and rigid crust. Here being stopped from further rise, it spreads laterally, until thin layers begin to cool. Once cooled, this layer of mantle, as in any case of convection, slowly sinks toward the depths, dragging with it a portion of the crust that has to some extent fused with it. This downward drag forms a geosyncline in the crust. The high temperatures in the depth of the mantle then cause the sediments in the lower portions of the downcoming geosyncline to melt and even to expand. This expansion can only be upward and so it forces the upper strata in the trough to fold and rise. Next, the molten lower sediments begin to rise into the spaces within the overlying folded strata. As they rise, they cool and solidify into the granite intrusions that form the cores of great folded mountain chains.

According to this theory, the mechanism behind mountain building is not one giant convection cell but a number of smaller cells spread throughout the mantle at varying depths, the uppermost of which lie immediately below the crust. These upper cells vary in thickness from 10 to 300 or 400 miles. This theory has the merit of explaining how a force powerful enough to warp the crust is generated and transmitted, without pointing to the crust as the origin.

A more recent theory, still being developed, involves plate tectonics. According to this theory, diastrophic mountains have been formed as a result of plate movements. Some geologists argue that as each lithospheric plate slowly crunches its way across the earth, its leading edge acts like the blade of a giant snow plow, pushing sediments before it and causing the folds and faults that eventually become mountains. In this way the plate carrying North America and South America moved westward, plowing up mounds of sediment in its path, finally stopping when it met the Pacific plate, which carries the Pacific Ocean. At the meeting line, the mountain chains that run up the Pacific coast of the Americas were formed; the same mechanism produced the Himalayas when the Indian plate met the Asian plate.

According to some other theorists, geosynclines are formed along the edges of continents and then crushed into folded mountains when one plate collides with another. The Appalachians, for example, may have been created in this way when the North American Plate and the African Plate collided eons ago, then drifted apart to their present positions. This theory is supported by the fact that all seven continents have mountain ranges on their seaward edges.

Erosion

Erosion changes the earth's surface by wearing away rock formations and carrying the debris to other locations.

The third process responsible for shaping the earth is **erosion**, a destructive process through which mountains and other uplifted landforms are gradually worn down. Over the long ages of geologic time, the forces of vulcanism and diastrophism have been engaged in a game of give and take with the forces of erosion. Vulcanism and diastrophism thrust up huge masses of rock, and erosion whittles and wears them down and carries them away. Both these processes occur simultaneously, but one or the other usually dominates.

Before erosive agents can act on a rock, it must be prepared by weathering. In the process of **weathering**, atmospheric forces—wind, water, and gases—mechanically and chemically disintegrate and decompose rocks.

Mechanical weathering is simply the breaking of rock into smaller fragments. It occurs when rock is subjected to intense heat or cold. Heat can affect a rock suddenly, during a forest fire or lightning strike, or gradually through exposure to the sun's rays for a long period of time. During the day, the heat of the sun causes the rock to expand; during the night, the cold air causes it to contract. Such rapid temperature alterations eventually cause the rock to crack and break. Water from rain or snow seeps into the pores and cracks of the rock, freezes, and in the form of ice expands to a far greater volume. As it expands, the frozen water exerts great pressure and acts like a wedge, cracking rock that is less hard or already weakened by fractures and decay. This process, called frost wedging, is also seen on road surfaces when water seeps into the surface, freezes, expands, and cracks, causing potholes and cracks.

Chemical weathering, or decomposition, is a more complex process in which the chemical structure of the rock is changed. It is caused especially by hydration and oxidation. In the process of hydration, water combines with the other elements that comprise the rock. During oxidation, oxygen and water combine with the various minerals in the rock, particularly ferromagnesians, and break them down to ferric acids. Atmospheric moisture and acid from decaying vegetable matter join with oxygen to erode these iron compounds. Igneous and metamorphic rocks are particularly susceptible to this type of weathering, for they were formed by heat and pressure beneath the earth. As long as they remain in their original stable environment, they will not undergo chemical weathering. However, when exposed to oxidation and other atmospheric forces, the rocks become unstable and fracture.

Another type of erosionary change is mass movement, in which gravity moves weathered rocks downhill. Typically, mass wasting is found in mountainous areas with steep cliffs and slopes. The force that triggers the downhill movement of great chunks of rock may be an earthquake, high saturation of the area by rain or snow, frost wedging, animal activity, or such human activities as construction blasting.

A landslide is a rapid form of mass movement that occurs when great masses of land tumble down a slope, often burying the area below with tons of rock. One of the largest landslides in recent history took place in 1957 in Madison Canyon, Montana. An earthquake sent 35 million cubic yards of rock, soil, and trees rumbling down the side of the canyon at speeds up to 60 mph. The canyon below was filled with debris to a height of nearly 200 feet.

The rockslide is a particularly destructive type of landslide. Boulders and

Figure 19-7
Mature soil profile. Soil, the product of
weathering, is a mixture of inorganic and
organic material. It contains rock fragments
and clays as well as live and decaying plant
and animal residues. The latter elements,
known as humus, give the upper layers of
the soil their dark brown color. Shown here is
a vertical cut through the soil revealing its
layers. The *A* layer, or topsoil, contains rich
deposits of humus. The *B* layer, also called
the subsoil, is high in clay and iron oxide.
The *C* layer, composed of loose and partially
weathered rock, is the source of soil for the
overlying layers. Beneath the *C* layer is the
bedrock, the solid unweathered rock that
comprises the earth's crust.

Topsoil

Subsoil

Partially weathered rock

Bedrock

smaller rocks, rock fragments, and soil tumble down a slope at great speeds, often with an accumulation of debris at the base called a talus. Eventually, the talus rocks and smaller materials are further weathered and carried to streams by rainwater.

In contrast to the fairly rapid local movements, there is an important kind of slow mass movement called creep. This phenomenon, which involves the slow movement of rock and soil down a slope, occurs over long periods of time and is imperceptible to the eye. Typically, creep is caused by chemical weathering of the slope. Its effects are visibly obvious in the tilt of fence posts, telephone poles, and gravestones lying on a slope.

Just as active as the forces of wind and chemical erosion, but less obvious, are the effects of groundwater, the water that seeps into the pores and cracks of the soil and rocks below the earth's surface. The primary sources of groundwater are rainfall and melting snow. The amount of groundwater in the earth at any given time depends on a variety of factors, including land slope, vegetation, porosity of the rocks and soil, and the rate of precipitation.

At a depth varying from 5 to 500 feet beneath the surface is the water table, lying between a lower region of the earth saturated with water and a higher, unsaturated zone. The table is not really flat as the name implies; generally, it follows the contours of the land above it. The height of the water table at any given location fluctuates. After periods of heavy precipitation, it rises; after droughts, it falls.

Groundwater typically seeps slowly downward and sideways, discharging into streams and lakes. Its rate of movement depends on the porosity of the material through which it passes. Water travels faster through coarse material such as sand or gravel than through fine material such as clay which is at times nearly nonporous. Water may return to the surface in a number of ways—through seeps, wells, or springs. By far the most important groundwater discharges are springs, formed where the water table meets the surface, allowing the groundwater to flow out.

The powerful erosional force of groundwater acting upon soluble rocks, such as limestone and sandstone, forms caves and sinkholes. Particles of these rocks dissolve in the groundwater, so it contains minerals such as calcium, magnesium, and iron ores. Limestone dissolved in water yields carbonic acid, a compound which increases the water's ability to further erode limestone, causing underground caves containing stalactites and stalagmites. A stalactite results when water dripping from the cave ceiling precipitates a calcium carbonate deposit in the shape of an icicle. A stalagmite is built up from the cave floor by drops from the same ceiling source. Both forms are also called dripstones. After thousands of years, the two dripstones may meet and form a massive pillar.

Figure 19-8
These dripstones at Carlsbad Caverns, New Mexico, were formed as groundwater gradually ate through limestone. The stalagmites, stalactites, and pillars shown here required thousands of years to form.

Carlsbad Caverns in New Mexico contains remarkable examples of such dripstone forms. Limestone is also the sea's principal source of salt. After groundwater erodes the limestone, the dissolved salts are transported into streams, which carry them to the sea.

The most common features of limestone regions are sinkholes, depressions caused by seepage of water into areas where rocks are riddled with cracks and fractures. These holes, which vary in depth from 1 to 100 feet, range in circumference from a few yards to several acres.

Water that is not absorbed into the ground may become a stream. Running water, one of the most powerful of all erosional agents, is the force most responsible for the shape and appearance of the earth's surface; water constantly erodes and lowers the land to sea level. Probably the most famous example of the remarkable power of running water is seen in the Grand Canyon in Arizona. This huge gash in the earth—4 to 20 miles wide and 1 mile deep—was cut by the Colorado River, which carries more than 28,000 tons of rock, soil, and other debris an hour.

How do streams get started? Streams are formed when rainwater, under gravity, runs off land surfaces in large sheets and then into small channels in the irregular surface which eventually form gullies. After repeated rains, some gullies deepen enough to become streams.

As a stream develops, it erodes the earth in a number of ways. For example, headwater erosion occurs at the head of a gully when the high velocity of the stream simply jets it through the lower layer of the gully, causing the top layers to crumple and tumble into the stream. Downward erosion takes place when the stream cuts a groove in the earth, forming the stream bed. In the process of lateral erosion, the stream widens its channels by undercutting the shoulders of the land on either side of it. The length, breadth, and depth of the stream channel are determined chiefly by its angle of slope and the volume and velocity of the flowing water. As the water courses forward, it scrapes and digs up the bottom of its bed using the sand and gravel it accumulates as it moves. Slamming this material against its banks and crumpling them gradually, the stream increases in volume and velocity as it rushes forward. Eventually, a V-shaped channel results as the stream knifes deeper into the earth's crust.

HOW VALLEYS ARE FORMED What happens to the face of the earth when powerful running water acts on it for a long period of time? This question is best answered by examining the development of a stream valley over the three periods of its life: youth, maturity, and old age.

The stream begins as a long narrow ribbon of running water that has cut a deep V-shaped channel in the crust. The stream is straight or gently curving and the crust on either side of it is flat or rolling. Swamps and lakes or waterfalls and rapids are usually located along its length. During this stage, the stream cuts into the earth as deeply as possible, down to the base level, which is the level of the water into which the stream flows. (Usually, the base level is sea level). As it reaches the base level, the gushing water loses some of its gradient and hence its power. At this point, the stream begins to erode laterally, causing the bottom of the steep V-shaped channel to become slightly rounded. Continued lateral erosion gives the valley a flat broad floor, called a flood plain.

As the stream ages and becomes more sluggish, it pursues a sinuous, twisting course across the flood plain, widening the plain by eroding the sides of the

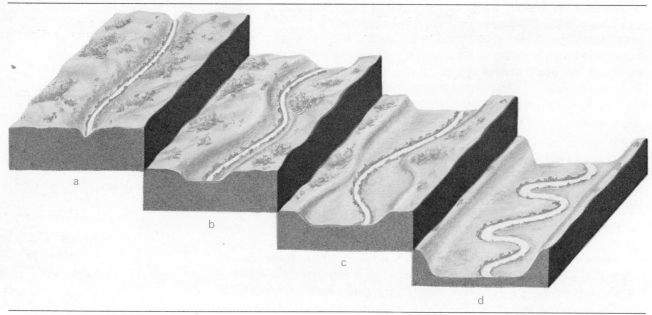

Figure 19-9
This diagram shows the stages in the development of a river valley. In its youthful stage (a), the valley is V-shaped with a straight, narrow stream cutting through it. During early maturity (b), the valley is gradually widened as the stream begins to erode it laterally. In the mature stage (c), the valley floor is greatly widened. In old age (d), the valley has been eroded to broad peneplains by the stream.

valley. From the air, this configuration resembles the sinuous body of a snake and is called a meander. One of the loops of the meander may be cut off from the rest of the stream as the stream channel shifts position, forming an oxbow lake.

During the mature stage of the valley's development, tributaries of the main stream develop, creating their own valleys, and the entire region eventually becomes criss-crossed with a pattern resembling the branches of a tree. During this phase, weathering and mass wasting have been at work eroding the slopes of the higher landforms. This material is deposited into the many creeks and brooks which carry it to the tributaries. These then carry the debris to the main stream, which deposits it in the sea. Gradually, the tributaries undergo the same processes as the parent stream, till they too widen their valleys with flat, broad flood plains, eroding the land that divides them. In the stage of old age, meandering rivers with broad flood plains erode the land between them to low rounded contours, or peneplains.

Such a description is of course idealized. In reality, the formation of river valleys fails to match exactly the sequence presented here. For example, in some areas, uplifted landforms called monadnocks resist stream erosion and remain standing principally because they are of hard rock. Examples of these structures, known as mesas and buttes, are found in the American West.

STREAM DEPOSITION: ALLUVIAL FANS AND DELTAS In addition to eroding and transporting debris, streams also deposit it. Two examples of stream-deposited debris, called **alluvium**, are the alluvial fan and the river delta. When a stream traveling at high speed down a mountain slope suddenly loses its velocity at the foot of the gradient, it drops its alluvium. Because the debris spreads out over the flatter land in the shape of a fan, the deposit is known as an alluvial fan. Examples of these are found at Death Valley, California. Similarly, a river delta may be formed when a fast-flowing river dumps its sediment as it empties into a body of standing water. Such deposits, named after the triangular shape of the Greek letter Δ, are found at the mouths of the Nile, the Irrawaddy, and the Mississippi Rivers. But deltas are not formed at all river mouths. Alluvium will accumulate only if the river deposits it faster than the currents can remove it.

Another type of meeting place between river and sea is the estuary. An **estuary** is a body of water, with fluctuating boundaries, in which the fresh water of a river mixes with the salt water of a sea. Estuarial environments are important food-producing areas in which a number of small plants and animals thrive, making them rich feeding grounds for larger fish and all water birds. The Chesapeake Bay is a typical estuary formed when rising seas flooded an old river valley. The picturesque Norwegian fjords are also estuaries, with floors carved out of bedrock by tongues of glaciers during the ice age. A third type of estuary, created by a fault just off the coastline, is exemplified by Tomales Bay on the California coast. Many estuaries are eventually destroyed by accumulated alluvium, which silts up the estuary bottom at an average rate of 2 meters every 1000 years.

Glaciers

A **glacier**, defined as a mass of slowly flowing ice, forms in regions where the annual mean temperature is below freezing and more snow accumulates in winter than melts in summer. Geologists classify glaciers into two groups. One is the mountain glacier, a giant tongue of ice that flows down from high peaks. These are found in the Alps, the Western United States, and in other areas. As they slide, these glaciers modify existing V-shaped valleys into more rounded U-shapes. The second type of glacier, the ice cap, is a huge mass of ice up to 12,000 feet thick that covers thousands of square miles of level area. Also called continental glaciers because of the large amounts of land they cover, these ice masses extend over great parts of Greenland, northern Siberia, and Antarctica.

Glaciers grow by the process of accretion. As the snow piles up, the lower layers are transformed into ice by pressure from overlying layers. When the ice thickens to approximately 150 feet, it begins to flow under its own weight, The forces responsible for glacier flow are not understood in detail but a number of factors are involved. These include the sliding of ice over rock, deformation of ice crystals, the melting under high pressure and re-freezing of ice, and

Figure 19-10
The Gorner glacier in Switzerland is an example of a valley glacier. The moraines to the front and sides of the two peaks are compacted deposits of rock, soil, and other debris accumulated by the glacier as it plowed its way across the earth over a long period of time.

cracking of the ice inside the glacier. The rate of glacial movement varies, depending upon the angle of slope, ice thickness, and the atmospheric temperature. Some glaciers move less than an inch a day. In rare cases, a glacier may move as much as 100 feet a day. A glacier will move forward as long as its rate of advance exceeds its rate of melting. When the ice mass melts faster than it advances, glacial retreat begins.

In its forward movement, a glacier crunches its way over rocks, carving the bed with them just as a stream does. The glacier fractures and pulverizes the smaller rocks, while it picks up the boulders and other large rocks in its path. From the scratches left on the bedrock by a moving ice mass, geologists are able to determine the direction in which the ice moved thousands of years ago. The glacier also gouges holes in the valley floor. After it has passed or melted, these fill in with water and become lakes.

Mountain glaciers and ice caps leave different erosive effects. The former chisel the land into uneven and angular shapes. Ice caps level the land by building up lower areas with deposits from higher areas. As they slide over the surface, glaciers leave a number of erosional features in their tracks. A till, for example, deposited directly by ice, consists of unstratified clay, sand, gravel, and boulders. The most important depositions of mountain glaciers are moraines—long, uneven, crescent-shaped ridges of till deposited along the glacier's edge. A striking example of a moraine in the United States is the huge ridge of sediment left in the last ice age by a glacier that slid south from Canada, severely modifying all the land over which it passed. This gigantic moraine stretches from Long Island in the east all the way to the Rockies in the west.

In arid regions where there is an abundance of sand or very dry soil, the wind can be an important agent of erosion. Wind erodes the earth in two ways: deflation and abrasion. In the process of deflation, the wind sweeps up soil and may carry it for miles as a dust storm. One of the clearest examples of a dust storm

Wind and Wave Erosion

Figure 19-11
Wind erosion produces many unique effects. At first glance, this unusual cascade of sand might easily be mistaken for a water fountain. The sand is being driven by the wind over layers of stratified silt. In other areas, the wind has chiselled rock into spectacular shapes, some resembling the prows of ships, the bodies of men, even entire cathedrals.

Figure 19-12
This remarkable rock, which would not look out of place in the modern sculpture gallery of a museum, is the result of water erosion. Found on a beach near the ocean, it has been polished smooth by waves carrying sand and small pebbles such as those shown here. The pebbles were probably eroded from rocks elsewhere along the shore.

was the one that occurred in 1934 on the American Great Plains. The winds were able to erode this land because years of poor farming techniques had removed much of the natural vegetation and topsoil. Some 300 million tons of soil were blown out of the Plains by Canadian winds, ruining thousands of acres of farmland and forming the great Dust Bowl. During its flight, the wind may scoop out large areas of soil. When the wind deposits a small portion of its load against boulders or vegetation, a mound of sand called a dune may result.

Abrasion produces more spectacular effects. In this process, the wind smashes sand particles and tiny pebbles against objects, literally sandblasting them. Abrasions have polished and pitted boulders, telephone poles, fence posts, and of course the Egyptian pyramids. Pedestal rocks are produced when the wind-blown materials eat away at the bottom of a rock form over a long period, leaving the rock thicker at top than at bottom.

Wind is also the force behind erosion caused by ocean waves and currents. Waves wear away the land by abrasion. During storms, for example, powerful waves smash pebbles and large rocks against the beach and adjoining land-forms. Over time, the uplifted land is eroded, often leaving behind a wave-cut cliff and an accompanying terrace, in which the land is levelled off just below the water level by wave action.

All of the sedimentary debris created by erosion is ultimately transported to and deposited in the sea. It has been estimated that the earth's rivers dump about 20 billion tons of rock debris into the seas every year. As a stream enters the sea, it dumps its load of sediments. Ocean currents then carry the debris out to sea, depositing it at varying distances from the shore. In the process, the coarse materials are sorted out from the fine materials. The velocity and turbulence of the water determine how far it will transport the material it carries. Usually, heavy sediments are carried out to sea by powerful currents located in the centers of breaking waves. Typically, the sediments which contain heavy minerals such as gold, garnet, and magnetite, are dropped to the ocean bottom closer to the shore and become gravel. Local turbidity currents, which keep the finer sediments in a muddy solution by creating swirling eddies of high velocity, flow down slopes under the influence of gravity and deposit the finer debris far out at sea. Thus, light materials, such as quartz and feldspar, are dropped further out from the coast and become clay and mud.

Sedimentation

Figure 19-13
This diagram shows a cross-section of the sedimentary structures formed at the continental edges and on the sea floor.

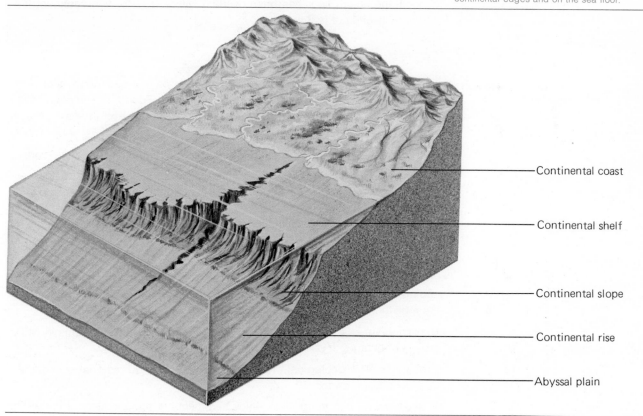

Continental coast

Continental shelf

Continental slope

Continental rise

Abyssal plain

In some areas, currents which flow parallel to the shore deposit layers of sand and other sediments across the mouths of bays. These submerged ridges which parallel the shore are called bars. In some cases, the bars may grow until they protrude above the surface, forming land projections called spits.

The automatic sorting out process is responsible for the shape and thickness of the suboceanic crust, which slopes downward, gradually levelling out at the ocean floor. As a result, immediately offshore is the **continental shelf**, a huge terrace formed by the action of waves on sedimentary deposits. This relatively flat apron of sediment ranges in width up to 800 miles, with an average width of 30 miles. The extreme seaward edge of the shelf is characterized by a sharp 3000 to 10,000 foot incline, called the **continental slope**. Sediments coursing down the slope cause its bottom to fan out (like an alluvial fan) forming the **continental rise**, which extends hundreds of miles from shore at places. The deep end of the rise meets the sea floor, which is marked by **abyssal plains**, very flat surfaces covered with fine sediments.

Sedimentation on the ocean floor builds up slowly, perhaps at the rate of one-half inch of sediment every 50,000 years. Over time, the basic basaltic rock beneath the ocean floor has been overlain with a layer of fine sediment from 100 to thousands of feet thick.

Glossary

The **abyssal plains** are flat surfaces composed of fine sediments, that occur where the continental rise meets the sea floor. Found at varying depths throughout the world, they lie 17,000 feet below sea level off the east coast of North America.

Batholiths are huge masses of intrusive rock that form the granite core of mountain chains.

The **continental shelf** is a huge terrace formed by the action of waves on sedimentary deposits. The extreme seaward edge of the shelf is characterized by a sharp 3000 to 10,000 foot incline, called the **continental slope**. Sediments coursing down the slope cause its bottom to spread out like an alluvial fan forming the **continental rise**.

The **deformational stage** of mountain building is the stage in which beds of sedimentary rock are subjected to tremendous compressional forces causing the rock to fault, tilt, metamorphose, and fold.

Diastrophism is a geological process which includes all movements in which one part of the earth changes position relative to another.

Erosin is a slow geological process in which weathered material is collected, transported and deposited.

A **geosyncline** is a trough into which sediments of rocks collect. It is the first stage in the formation of a diastrophic mountain.

A **glacier** is a large mass of slow moving ice. A mountain glacier is a tongue of ice that flows down from high peaks. An ice cap or continental glacier is a huge mass of ice up to 12,000 feet thick that covers thousands of square miles of level area.

Groundwater is water that seeps into the pores and cracks of the soil and rocks below the surface of the earth.

The **theory of isostasy** states that there is a continuous state of balance between different segments at the earth's crust, as heavier crustal areas sink and lighter ones rise.

A **peneplain** is a relatively flat land surface created by the forces of erosion over a long period of time.

Pyroclastic debris is the solid igneous material thrown from an active volcano.

The **Richter scale** is a magnitude scale which indicates the amount of energy expended by an earthquake.

Vulcanism is the geological process in which magma is generated, transported, and emplaced beneath or above the surface of the earth.

Exercises

1 In 1902 Mt. Pelée erupted, razing a city and killing 38,000 people on the Caribbean island of Martinique. What factors may have led to such an intense volcanic eruption?

2 Despite catastrophic eruptions such as Mt. Pelée, volcanism is generally considered to be more beneficial than harmful by many geologists. Discuss the useful consequences, including one that pertains to the current energy crisis.

3 Distinguish between the three major types of faults and discuss their relationship to the phenomenon of earthquakes.

4 How does the theory of plate tectonics explain (a) volcanic mountains, (b) earthquakes, (c) diastrophic mountains (d) Continental drift?

5 As a youth, Leonardo da Vinci was puzzled when he discovered rocks containing images of ocean shells in the mountains of his native province. Today we know that Leonardo stumbled across fossils. What other evidence exists that the land now forming mountains ranges was once below sea-level? How is this latter fact accounted for by the general model of mountain building?

6 If marine fossils indicate that certain areas have been uplifted what evidence exists that other land masses are sinking?

7 Why does a hot glass crack when cold water is poured into it? What analogous process takes place in nature and where is it most likely to occur?

8 What types of mass movement are most likely to occur in a dry, hilly region with sparse vegetation, subject to harsh winters?

9 What formative factors are common to both soil and sand? How does soil differ from sand? Why doesn't sand support a nitrogen cycle?

Interlude Eleven

A 10,000-Foot

Dive to the

Sea Bottom

It resembles a scene from a Jules Verne adventure. A tiny submarine gently noses its way down through inky waters. Inside, two men sit in a cramped circular space about the size of the interior of a Volkswagen minibus, intently peering from portholes. Although the ship's spotlights are able to penetrate only a few dozen feet, the light is enough to reveal to the crew a strange new world. The eerie, surrealistic landscape contains high cliffs and a sharply-angled stepped wall intruded with large volcanic dikes. From the walls dangle large puffballs of lava resembling giant cocoons; entire fields of structures that look like hugh mushrooms with eyes sit on the valley floor. Continuous landslides send rocks tumbling down the walls. The only sounds are the dull drumming of the engines and a strange metallic ping that seems to be bounding off the outside of the sub. As the craft settles on the sea bottom, the pilot signals to a

vessel above them at the surface that they have touched bottom. The little ship
has just descended through nearly two miles—10,000 feet of ice-cold ocean, 50
times deeper than any submarine has ever been. The pressure, an incredible 2
tons per square inch, is enough to crush an ordinary submarine and turn a
human body to pulp.

What is this vessel, where did it come from, and what is it looking for? The ship
is the *Alvin*, the United States Navy's principal deep-diving submarine,
operated by the Woods Hole Oceanographic Institution. The *Alvin* is the American participant in Project Famous—an acronym of French-American Mid-
Ocean Undersea Study—the most ambitious manned, deep-ocean exploration

ever attempted. The goal of the project, which began in 1971 and ended in the summer of 1974, was the exploration of a portion of the earth's most extensive surface feature. This is the 10,000-mile long mid-Atlantic ridge, an underwater chain of volcanic mountains and valleys that runs pole to pole in the Eastern hemisphere, where the earth's crust is created by volcanic eruptions beneath the surface. Scientists in the project sought to add to their knowledge of the mechanism behind the spreading of the sea floor. They also wanted to determine the conditions responsible for the formation of metallic ores, and the nature of earthquakes that occur at and near the ridge.

Alvin's two French counterparts in this historic voyage of discovery were the French vessels *Archimede* and *Cyana*. The three submersible ships probed a major portion of the mid-Altantic Ridge about 200 miles southwest of the Azores. Under these waters, between latitudes 35–37°N and longitudes 33–36°W, is a sprawling volcanic mountain range averaging about 7200 feet in height and 900 miles in width. Splitting the ridge, along its north-south axis, is a rift valley which cuts the Altantic floor from the Arctic Ocean almost to Antarctica. The rift valley is also the boundary between the North American tectonic plate and the African plate.

Through the long rift, molten rock pours out like lava from the cone of a volcano. The lava cools and solidifies into the plates that move the earth's continents and seas around. As the lava emerges, small tremors, known as microearthquakes, shake the rift zone; in some areas, as many as a dozen tiny quakes may occur every hour. By directly observing the rift's features from the deep-diving ships, the scientists hoped to learn what causes the earth's giant tectonic plates to move from either side of the ridge at the rate of an inch or so every year.

The most important geological discovery from Project Famous concerns the mechanism behind the spreading of the sea floor and continental drift. Previously, it was thought that the huge tectonic plates carrying the continents were pushed up and apart at the rift valley as lava erupted from beneath the sea floor. But geologists from the *Alvin* found that the ocean floor, on either side of the fissure, had not been compressed as it would have been if it were being pushed apart at the rift. In addition, they found numerous fissures paralleling the median rift; these were restricted to an area that extended some 500 feet outward from either side of the median rift. These facts suggested to the investigators that the ocean floor is not being *pushed* apart—rather, it is being *stretched* or pulled apart from either side of the rift by unknown forces operating in other parts of the sea. A negative discovery, since the exact mechanism behind spreading of the sea floor remains a mystery!

The *Alvin,* whose crew made this important discovery, is technically called a submersible, the general name for any maneuverable deep-diving, pressure-sealed, manned capsule equipped with viewing portholes. The 23-foot long submarine has a range of 15 to 50 miles and a cruising speed of about 1½ knots; but in an emergency, she can make up to 3 knots. The ship can stay underwater for about 7 or 8 hours before surfacing for battery recharging. To withstand the tremendous undersea pressure, she is equipped with a pressure sphere of special two-inch thick titanium, which enables her to descend to depths of 12,000 feet. The craft has three propellers—two small ones on either side, which lift the ship, and a huge one astern, which drives her.

What did the sea floor look like from inside *Alvin*? The dive began as the submersible was hydraulically lowered on a platform from between the twin pontoons of the catamaran *Lulu*, which is *Alvin's* mother ship. *Lulu* has been described as a kind of floating mission control, because the ship maintains telephone and sonar contact with *Alvin* at all times. Once the sub was in the water, her crew—a pilot and a scientist-observer—climbed into her personnel sphere, the only part of the ship not flooded with ballast water. After her seals were checked by divers, the ship began its descent.

When she reached the rift valley, the men at the portholes saw an eerie, forbidding topography full of crags and unusual formations. Among the fantastic configurations protruding from the valley walls and floor were volcanic pillows, bulbous structures that form after lava has been squeezed like toothpaste from fissures in the valley walls; 5-foot high formations with arms that appeared like cacti; and gracefully curving structures that reminded one observer of the necks and heads of swans.

During its explorations, the *Alvin* collected hundreds of pounds of rock samples, plus corals, crabs, and samples of seawater and sediment. From inside the ship, scientists were able to operate a variety of special tools, including drills and hammers, which are attached to the outside of the hull. The *Alvin* was also equipped with a manipulative arm ending in a movable, crablike claw which enabled it to latch onto samples and place them in a rotatable basket mounted on its bow. Numerous photos were taken by television and still cameras mounted outside the ship, and by cameras held by the crew. During one dive across the central part of the valley, the ship actually descended into one of the fissures, drove around it, and stopped at the bottom to gather more rock samples.

A few miles to the north of the *Alvin,* the highly-maneuverable French submersible *Cyana* gathered samples from a fracture zone that lies perpendicular to the rift valley. At one of the transverse canyons, the crew of the *Cyana* made the important discovery that undersea geysers often erupted hot water rich in dissolved iron, magnesium, and copper ores. This finding was significant because it supports the theory that the eruptions are the source of much of the world's metallic ore deposits.

The other French craft, the bathyscaphe *Archimede,* explored still another section of the rift valley floor to the south of the *Alvin.* The *Archimede,* at 70 feet the largest of the three submersibles, cannot be maneuvered as well as the other two ships, but she can penetrate greater depths than the *Alvin* and the *Cyana* — she had previously descended 31,000 feet to a deep sea trench off Japan. A bathyscaphe is basically an underwater elevator in the form of a balloon consisting of an inflatable bag and a personnel cabin. Thousands of gallons of gasoline — or some other lighter-than-water liquid — are used as ballast. Tons of iron shotgun pellets are used as weights to make the lumbering craft dive; when her crew wants to resurface, they simply dump the pellets on the sea floor. The gasoline ballast enables the vessel to ascend to the surface, where the gasoline is vented into the sea.

The combined French-American costs of Project Famous over a four-year period were about $9 million. The three made a total of 50 dives, staying underwater for a cumulative total of nine days.

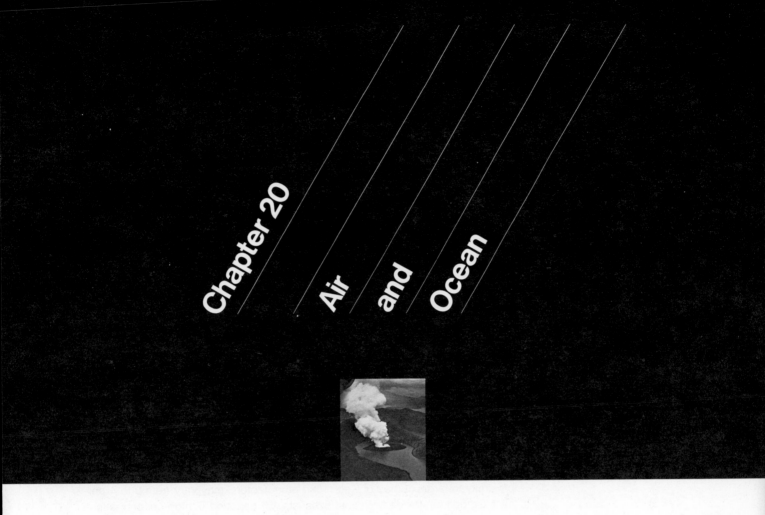

Chapter 20

Air and Ocean

The solid earth, the atmosphere, and the oceans are related in an intricate system of exchange that makes life on this planet possible. Oxygen, nitrogen, carbon dioxide, water, and other materials are continually moving among the three parts of the system in naturally occurring cycles, supplemented by the artificial input of social production and consumption. What is the nature of these relationships, what is their importance for life, and what are the processes responsible for their continuation? This chapter will attempt to answer these questions by exploring the elements of air, the dynamics of weather and climate, and the composition and properties of the oceans.

Atmosphere

Our planet is surrounded by an ocean of air called the **atmosphere**, which contains a mixture of gases extending thousands of miles into space. Indispensable for life itself, the atmosphere also determines the type of clothes we wear, the food we grow and eat, and the limits within which we live; some go so far as to say it shapes our culture, since it is the guiding medium for radio and television waves. It protects us from the dangerous ultraviolet rays of the sun by day, and prevents the accumulated heat from escaping into space at night. Without this natural insulator, the earth would be like the moon—devoid of life, a scorching oven by day, a freezing icebox by night. The atmosphere includes the clouds, rain, snow, winds; it is responsible for the colors of the sky. It produces the running water that slices across the earth's surface, sculpting and altering it through time.

Chemical Composition

Although the atmosphere contains a mixture of at least 11 different gases, only two of these—nitrogen and oxygen—make up more than 99 percent of its volume near sea level. Table 20-1 gives the average composition of dry atmospheric air below 15 miles. Depending upon altitude, latitude, time, and season, (as well as man-made conditions) the composition of the atmosphere may vary slightly from the figures in the table, but the ratio of the major gases in the atmosphere tends to remain constant because of such processes as the nitrogen cycle and the oxygen-carbon dioxide cycle (see Chapter 21 for a further discussion of these cycles).

The amount of water vapor, which depends upon atmospheric temperature, proximity to large bodies of water, and other factors, may vary from 0 to 4 percent. Carbon monoxide from auto exhausts will be found in high concentrations around cities and expressways, but will be relatively low in rural areas. The solid particles found in the air include salt particles from the sea, microscopic animals and plants, lava from volcanic eruptions, grains of wind-blown sand, pollen, and dust from weathered rock. Dust particles are responsible for the beautiful red colors of the sky at sunrise and sunset; they also provide nuclei, or centers, around which water vapor condenses to produce rain or snow.

Table 20-1
Composition of the atmosphere

Gas	Symbol	Percent
Nitrogen	N_2	78
Oxygen	O_2	21
Argon	Ar	0.9
Carbon dioxide	CO_2	0.03
Neon	Ne	0.0018
Methane	CH_4	0.0002
Nitrous oxide	N_2O	0.0005
Hydrogen	H_2	0.0005
Helium	He	0.00052

Divisions of the Atmosphere

The pressure and temperature of the atmosphere vary with altitude; pressure decreases as altitude increases. At sea level, the pressure of the atmosphere is 14.7 lbs/in^2 (1013 millibars); 3.5 miles above sea level, it is decreased by about one-half. But the rate of pressure decrease is not constant, and above 20 miles the rate of decrease is very small. Atmospheric pressure is measured with a device called a barometer (see Chapter 3).

It is the variation in temperature that provides the basis for the division of the atmosphere into four major layers: troposphere, stratosphere, thermosphere, and exosphere. These man-made divisions are purely arbitrary; indeed, there is much discussion among scientists about the boundaries between the layers and even their exact number.

TROPOSPHERE The **troposphere** extends about 10 miles above sea level at the equator, and about 3 to 5 miles at the poles; it is usually thicker in summer than in winter. The temperature of this layer decreases as altitude increases, at the rate of about 3.5°F per 1000 feet. Known as the lapse rate, this temperature drop continues to a height of 10 miles, where it reaches −70°F at the tropopause, or the zone separating the troposphere from the layer above.

STRATOSPHERE The **stratosphere** extends from 10 to 50 miles above the earth's surface. Above the tropopause, the temperature increases from −70°F to 85°F at a height of about 30 miles; above this height it again begins to decrease quickly. The average temperature of the stratosphere is approximately − 67°F.

An important characteristic of the stratosphere is the presence of ozone (O_3), with three atoms of oxygen per molecule instead of the usual two atoms found in molecules of atmospheric oxygen. Ozone is produced by the energizing action of ultraviolet light on O_2, and this process absorbs most of the ultraviolet radiation emitted by the sun. If it were not for this chemical, life would not exist on the surface of the earth—plant and animal tissues would be destroyed by the sun's ultraviolet rays.

Ozone is produced in two stages by these reactions: O_2 + ultraviolet gamma ray $\longrightarrow$ 2 O, and $O + O_2 \longrightarrow O_3$.

Most scientists refer to the upper reaches of the stratosphere, between 35 and 50 miles up, as the **mesosphere**. Here, at an altitude of about 35 miles, temperature begins falling quickly, to about −170°F at 50 miles.

THERMOSPHERE Above the mesosphere, beginning about 50 miles above the surface and extending to a height of approximately 400 miles, is the **thermosphere**. The air in this layer is thin (the molecules of atmospheric gases are spread out, far apart from one another) and it absorbs a great deal of the sun's ultraviolet and X-ray radiation, causing temperatures to rise as high as 2000°C. Pressure is zero. But for such rarefied gases, the normal parameters of temperature and pressure mean little; the significant fact is the high energy level of the individual molecules due to the extremely hot solar radiation in this layer of the atmosphere.

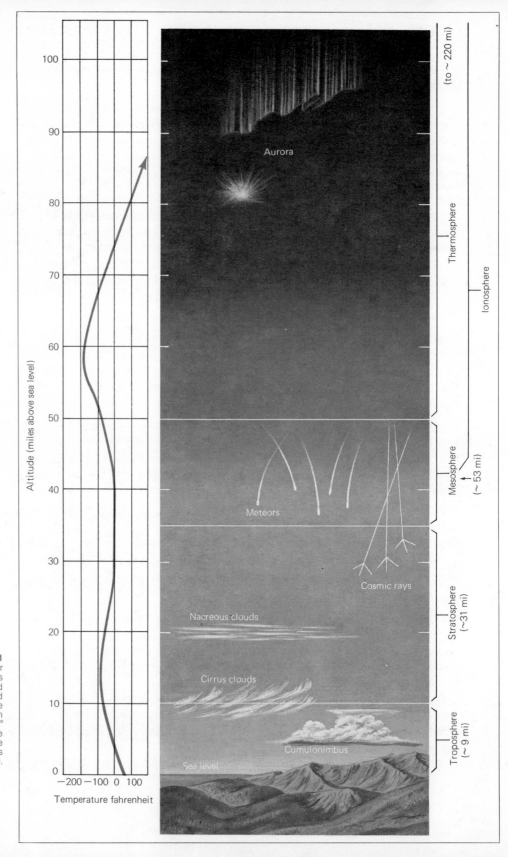

Figure 20-1
The earth's atmosphere consists of a number of layers. Variations in temperature, as shown in the left panel, provide one method to define the layers. Heights and boundaries, as well as temperatures of the layers, vary greatly with latitude and with daily, seasonal, and solar cycles, from ±25° near the surface to as much as ±900° in the high thermosphere, above which the atmospheric temperature is constant (isothermal).

IONOSPHERE The lower part of the thermosphere is often referred to as the **ionosphere**, because it contains ions produced by the action of the sun's ultraviolet and X-rays on atoms of nitrogen and oxygen. The ions create an electric field which prevents the passage of certain types of waves, such as radio waves.

The ionosphere is subdivided too and contains four distinct layers, designated simply D, E, F_1, and F_2. Recent tests indicate the possibility of a fifth, or G, layer. The density of each layer declines with increase in altitude, while ionization increases. Thus, the D layer, a belt about 10 miles in radius, contains oxygen (O_2), nitrogen (N_2), plus traces of other elements, as well as ionic particles of nitric oxide (NO^+), molecular oxygen (O_2^+), and nitrogen (N_2^+). The E layer, which starts about 60 miles above the earth and extends to 90 miles, contains O_2^+ and NO^+. The F_1 layer (from 90 to 150 miles up), consists of atomic oxygen (O) and molecular nitrogen (N_2), plus the ions of oxygen and hydrogen. During the night, when there is no solar radiation to cause ionization, the ions and free electrons in the D, E, and F_1 layers recombine to form neutral atoms or molecules; this process occurs most rapidly in the D layer where the ions are nearest to one another. The F_2 layer (about 150 miles high) consists of atomic oxygen (O) and the ions of oxygen, hydrogen, and helium; it persists even during the night because the ions there are so few and far between they have little chance to recombine.

The layers of the ionosphere reflect radio waves differently, due to their chemical differences. Ionic response to electromagnetic waves will vary, just as materials of different colors will reflect or absorb different parts of the visible portion of the spectrum. The D layer tends to absorb radio waves, whereas the E, F_1, and F_2 layers reflect radio waves back to earth. This property of the atmosphere was discovered in 1901 by Guglielmo Marconi (1874–1937), who found he was able to transmit radio signals from one point on the earth's surface to another, despite the curvature of the earth; the signals do go straight out into the atmosphere but they are reflected back, at an angle, by the middle layers of the ionosphere. Radio transmission is strongest in the early part of the night, when the absorptive D layer is rapidly dissipating. As the night wears on, radio reception becomes increasingly difficult, because the reflective E and F layers have also disappeared.

EXOSPHERE The fifth atmospheric layer is the **exosphere**. This region begins about 400 miles above the surface and extends to an altitude of some 600 miles. This layer is composed only of isolated atomic oxygen, ionized oxygen, and hydrogen atoms.

Some scientists now recognize another layer, beginning above the exosphere and extending 30,000 to 40,000 miles above the earth's surface, called the **magnetosphere**. The only existing matter in this layer is free elementary par-

The middle layers of the ionosphere reflect radio waves back to earth, facilitating worldwide communication.

ticles—mainly electrons and protons governed by the earth's geomagnetic field. These particles are able to retain their electrical charges because they can travel vast distances without colliding; only in the lower regions where the particles are present in greater density do they combine to form electrically neutral atoms.

Heat Energy of the Atmosphere

The earth's air is heated principally by the sun, which emits radiant energy in the form of short electromagnetic waves. Striking the earth's atmosphere, this form of energy is known as **insolation**, an abbreviation for *in*coming *sol*ar radia*tion*. Every day, the outer atmosphere of the earth receives about one-two billionth of the total radiation emitted by the sun—about 700 calories/cm². Of this, approximately 34 percent is reflected back into space by clouds and dust; 19 percent is absorbed by ozone, water vapor, and water in the clouds; and 47 percent reaches the ground. On any given day, the effective amount of solar radiation received by the earth's surface averages about 330 cal/cm². This is a remarkably small amount of heat—about enough to melt a small ice cube. Scientists have obtained these figures by monitoring earth-orbiting satellites. These vehicles contain instruments able to measure solar insolation as well as radiation reflected both from the atmosphere and the earth's surface.

Insolation received at the earth's surface averages about 300 cal/cm² per day.

Surprisingly, the lower atmosphere (including the earth's surface) is not heated directly by the sun's rays, but by heat reradiated from the earth. This refers to the average temperature. Obviously, an individual standing in the sun will receive heat directly from the sun's rays as well as heat reradiated from the soil, rocks, and other environmental features. The troposphere screens out the dangerous long wavelengths and permits the shorter wavelengths to pass through to the surface, where they are absorbed and reradiated back to the troposphere. But the reradiating waves are longer and tend to become trapped, because molecules of water vapor and CO_2 in the troposphere absorb them, causing the temperature to rise. Thus the lower part of the atmosphere acts as a kind of florist's greenhouse: like the glass of a greenhouse, the atmospheric gas allows the short waves of solar radiation to reach the earth, but prevents the long wavelengths from reradiating back into space. This phenomenon of a natural radiation trap is known, naturally enough, as the greenhouse effect; it provides the protection that allows earth's relatively frail creatures to thrive.

The atmosphere warms the earth by trapping the sun's reradiated long wavelengths.

Air pollution—CO, CO_2, SO_2, and radioactive substances—appears to be interfering with the balance of the greenhouse effect. In the United States alone, it has been estimated that 164 million metric tons of pollutants are poured into the air every year; about half of this is produced by auto emissions. Moreover, coal, oil, and wood are burned to heat and power homes and factories, and the CO_2 thus produced has been accumulating in the atmosphere since the Industrial Revolution. CO_2 atmospheric accumulation has grown from about 295 parts per million in 1860 to about 320 parts per million in 1970, an increase of about 10 percent. The more CO_2 and particulates that accumulate in the air, the more efficient an insulator the lower atmosphere becomes.

Is it mere coincidence that the temperature of the earth has been changing in the last century? According to Weather Bureau statistics, the average temperature of the earth began to rise about 1880. Between 1920 and 1940, average world temperatures increased by about 0.4°C. Since 1940, however, temperatures have fallen by about 0.2°C.

Scientists have advanced two opposing theories to account for these temperature variations. The first states that the increase in temperature is due to the increase in CO_2. Large quantities of this chemical added to the atmosphere increase its effectiveness as an insulator, and thereby cause increased temperatures. The opposite theory holds that the increase in CO_2 and particulates in the atmosphere has had the effect of lowering world temperatures by blocking insolation. Although there is some possibility that the two effects will cancel each other out, most investigators seem to think one or the other will prevail. There is agreement among scientists that industrial technology can effect worldwide temperatures, but there is no absolute proof to confirm either of these two theories.

Variations in Air Temperature

Everyone knows that certain regions of the earth are hotter or colder than others. The Sahara Desert, for example, can become unbearably hot, whereas the polar regions are often unbearably cold. The variation in temperature occurs because the earth is tilted on its axis by $23\frac{1}{2}°$ in relation to the plane of its orbit around the sun; uneven heating of the earth's surface causes differences in temperature which lead to differences in air pressure and eventually to winds. The sun's rays strike the equator vertically and directly, concentrating heat in a small area; but the rays strike the polar regions obliquely, covering a larger area with the same amount of heat. Most of the energy of the vertical rays reaches the earth because they pass through a thin section of the atmosphere. The oblique rays, on the other hand, must travel through a much thicker section of the atmosphere, where much of their radiation is absorbed or reflected back into space.

The nature of the surface is another important factor in temperature variation. Areas covered by snow, for example, reflect rather than absorb the sun's rays; areas with dark-colored soils are usually warmer than those with light-colored soils because the darker colored chemical absorbs the heat.

Land surfaces both warm up and cool down faster than water surfaces. Water reflects the sun's rays and therefore absorbs less heat during the day. Since water has an intrinsically higher heat-retaining capacity than almost all soils,

Figure 20-2
Low-lying fair weather cumulus clouds form over a desert area in New Mexico. The clouds are produced as warm air, heated by the earth's surface, rises and condenses into water vapor. The left portion of the cloud, which resembles the head and jaws of a crocodile, is being separated from the main formation to the right.

rocks, and other land surfaces (high specific heat), the night-time atmosphere near large, deep bodies of water is warmer than that over land. Obviously, an area with heavy cloud cover will be cooler by day than a region with sun. Desert regions experience great extremes in temperature from day to night. In these arid areas where the air is very dry, the sun heats the earth during the day, often to temperatures exceeding 90°F. At night, the heat escapes into space because there is little moisture in the air to absorb it, and temperatures fall to below freezing. Other factors that determine the air temperature of an area include the length of the day and atmospheric condition.

Weather and Climate

The most obvious effect of the atmosphere on our daily lives is weather. Stated most simply, **weather** is the combination of temperature, pressure, humidity, cloud cover, wind, and precipitation that is experienced by a region at any given time. Weather is distinguished from climate solely on the basis of time: **climate** may be defined as the long-term weather of a region. The branch of physical science that studies both weather and climate is **meteorology**. The practitioner of this science is the meteorologist, commonly known as the weatherman. (Meteorology is the subject of the Interlude following this chapter.)

The amount of moisture in the air at any given time varies from day-to-day and from one area of the earth to another depending on air temperature and pressure. The amount of water vapor in a unit volume of air—usually expressed in grams/m³—is known as the **absolute humidity**. The maximum amount of water vapor the air can hold at any given temperature is limited. When the air cannot hold any more water than it does, it is said to be saturated. Unsaturated air, on the other hand, is capable of holding more water vapor than it actually does.

More useful for meteorologists is a measure of humidity that reflects not just mere water content but also other environmental conditions, such as temperature and pressure. Thus, they use a term more familiar to us from the daily weather report—relative humidity. For example, if a cubic meter of air holds four grams of water vapor under conditions in which it could hold eight grams at complete saturation, the air is half saturated and has a relative humidity of 50 percent. **Relative humidity** is defined as the ratio between the amount of water vapor actually present in the air and the maximum amount it could hold at that temperature; that maximum is of course a function of temperature and pressure. The relative humidity ratio is usually expressed as a percentage.

If the temperature increases and the amount of water vapor in the air remains constant, the relative humidity will decrease; conversely, if the temperature is lowered while the water vapor remains constant, the relative humidity will

Figure 20-3
Alto-cumulus clouds, towering up to 40,000 feet, bring a thunderstorm to Miami, Florida. The halo effect in the upper left is caused by a veil of ice crystals that reflect the sun's rays. The various mushroom-shaped forms that appear throughout the upper reaches of the cloud are known as anvils.

Figure 20-4
The billows clouds at the left of the photo are
being pushed by a dark cumulus formation.
The billows are actually a group of
individual cloudlets which combine to form
a single large cloud.

increase. The temperature at which the relative humidity increases to 100 percent, causing moisture to condense upon dust particles or cool surfaces, is known as the **dew point**. If the temperature drops below freezing, the water vapor will condense as ice or frost.

Above the earth's surface, rising warm air containing water vapor may be cooled by expansion to the dew point. Condensation occurs wherever there are dust particles or other solid matter to act as nuclei, resulting in the formation of extremely fine droplets of water; this makes a cloud. Figures 20-2 through 20-5 illustrate several different kinds of clouds. When the cooling of moist air occurs at ground level, there is a fog, which is nothing more than a cloud lying close enough to the earth to interfere with ground-level visibility. When fog combines with smoke and other pollutants, it is known as smog.

RAIN Formation of droplets around nuclei of dust particles or salt crystals, and then the collision of those droplets causing them to gather together into big drops, are mainly a tropical effect that yields warm rains falling from high cumulus clouds whose temperature is above freezing. In temperate climates the nuclei are thought to be ice crystals at the cold upper limits of rain clouds.

Figure 20-5
Cumulus and cirrus clouds move over a lake during sunset at Anchorage, Alaska. The thick dark clouds in the foreground and far background are cumulus formations. The wispy light and dark clouds in the center are cirrus.

Figure 20-6
Fog is formed when vapor is condensed into fine water particles as warm air masses are cooled on contact with cool bodies of water or earth. Here, fog settles over a coastal inlet.

In this theory, first worked out by the Swedish meteorologist Tor Bergeron in 1935, super-cooled water vapor condenses on the ice crystals like frost on ice-cold metal. The crystals enlarge rapidly to form snowflakes. If the air is cold enough, snow falls to the ground; or the crystals may pass through a warm layer, then freeze again as small icy particles, and land as sleet; or, more usually, the snow and ice crystals melt on their way down, remain liquid, and fall as rain.

Air in Motion: Wind

Winds transport heat and moisture over the earth's surface, causing local weather conditions. Winds are the result of convection currents caused by differences in surface temperatures at the earth, which in turn creates differences in air pressure. When air heated by the earth's surface rises, leaving behind a low pressure area, cooler air descends from elsewhere and immediately rushes in to equalize the pressure and buoy up the rising layer of warm air. When the cool air in turn is heated and rises, more cool air rushes in to take its place, and the cycle begins over again. Winds at the earth's surface are the horizontal movements of air during the surface phase of the convection loop.

Wind is the horizontal phase of a convection loop.

If the earth were heated uniformly at the equator, did not rotate, and were completely covered with either land or water, the pattern of convection currents over the surface would be a blessedly simple one. In the northern hemisphere, warm air would rise from the equator, creating a low pressure area; cool polar air would then rush in and buoy up the equatorial warm air. This means that the only wind in the northern hemisphere would be a steady north wind. The pattern would reverse itself in the southern hemisphere where a steady south wind would blow continually.

However, as is so often the case, ideal conditions do not prevail. The earth rotates, is unevenly heated, exhibits great surface variation, and therefore has very complex wind patterns. The rotation of the earth affects wind direction through the **Coriolis effect,** named after G.G. Coriolis, the French engineer who discovered it. This effect can best be understood by a simple example. Imagine you are standing at the North Pole and decide to fire a rocket at a target near the equator. After you fire the missile, it will appear to swerve to your right because it goes in its inertial path in a straight line, while the earth beneath it is rotating toward the east. Ultimately it misses the target, which is moving with the moving earth, by a good distance. The rocket's change in direction is only apparent; with respect to the rotating earth, it has shifted its course, but it follows an inertial path and hence with respect to the fixed stars it has not changed direction at all—except for the vertical downward acceleration due to gravity.

The Coriolis effect deflects north-south winds from their straight line paths.

The Coriolis phenomenon has the same effect on atmospheric circulation as it has on the imaginary rocket—it causes winds going due north or south to veer from their straight-line paths. Thus, in the northern hemisphere, a wind traveling north or south swerves to the right of its course; conversely, in the southern hemisphere, a north or south wind will swerve to the left of its course.

In 1957, a powerful spiral of wind some 300 to 500 miles in diameter ripped across the Louisiana coast killing 500 people and causing millions of dollars of property damage. In 1970, a similar disaster took 300,000 lives when it struck East Pakistan. Meteorologists call such monster windstorms tropical cyclones—intense low pressure systems of air with raging winds that spiral upward and inward in a giant funnel. They are called hurricanes by inhabitants of the Caribbean and North America, typhoons by those living in the Western Pacific, and willy-willies by Australians.

These revolving windstorms are one of nature's most destructive forces. The tropical cyclone releases tremendous energy—the equivalent of about 500 billion kilowatts of electricity, enough to power the whole United States for an entire year. Although its internal winds range in velocity from 75 to 200 mph, it travels at only about 12 to 20 mph. For several days, its violent winds strip away everything in its path, battering buildings to pieces and capsizing ships at sea. Because they seldom travel inland, hurricanes do most of their damage in coastal regions. What the winds do not rip apart, the accompanying floods and 40-foot waves carry away. Flooding results when sea water, driven by high winds, piles up in bays and other coastal inlets and raises sea level as much as 10 to 15 feet. In addition, intense rain accompanies the cyclone. For example, Baguio in the Phillipines received 88 inches of rain in a single day during one storm.

Although the dynamics of their formation are not fully understood, it is known that these storms arise only in the tropical oceans between 8° and 15° on either side of the equator. Here, the pressure is low, surface temperature is high—about 80°F—and the Coriolis effect prevails. It is thought that the Coriolis effect plays a key role in the formation of these windstorms, by sending trade winds into a spin around a low pressure area. Once in motion, the storm is driven by heat reradiated from the surface at tropical oceans. The intense upward spiraling winds at the center of the storm suck up great quantities of moist warm air from tropical seas as the cyclone passes over them. As the air travels up the central zone, the water vapor in it condenses, releasing latent heat. In this way, the central zone of the cyclone is always kept warmer (as much as 10°C) than the periphery. The large temperature difference between the two regions of the storm creates a sharp pressure gradient—often up to 40 mb; it is this gradient that is responsible for the powerful winds. At the very center of the cyclone is an area of windless calm known as the eye.

Generally, cyclones are born during the late summer and early autumn in the trade wind belt. They slowly move westward for a few days, then shift to a northerly course. A storm dies as soon as it moves inland because its supply of moist warm air is cut off.

Another form of windstorm similar to the tropical cyclone is the tornado (from the Spanish *tronada,* for thunderstorm). A tornado, which is a funnel of rapidly spiraling air about 250 yards in diameter, carries winds that range between 100 to 500 mph. These storms usually arise in conjunction with cold fronts. Although it is more powerful than a tropical cyclone, the tornado does less extensive damage because it is so shortlived, typically less than ten minutes.

A 500 Billion Kilowatt Powerhouse

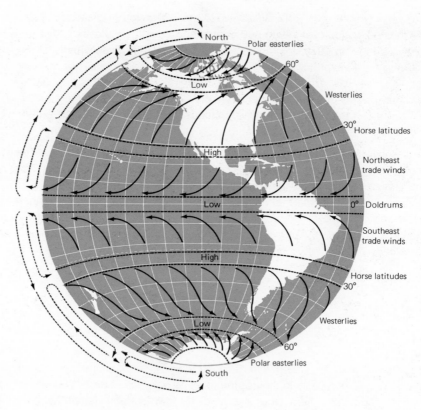

North
Polar easterlies
60°
Low
Westerlies
30°
Horse latitudes
High
Northeast
trade winds
Low
0° Doldrums
Southeast
trade winds
High
Horse latitudes
30°
Low
Westerlies
60°
Polar easterlies
South

Figure 20-7
The earth's general atmospheric circulation
is due to the fact that it rotates and is
unevenly heated. Encircling the earth are 5
belts of alternating high and low pressure
zones which are responsible for wind
patterns. In addition to the winds indicated
here, the jet stream, a narrow ribbon of air,
travels in an easterly direction at altitudes of
30,000 to 40,000 feet with speeds up to 300
miles an hour.

The understanding of convection currents and the Coriolis effect is vital to
meteorology because of the roles they play in creating weather changes and in
influencing the major wind systems of the general atmospheric circulation. An
example can be seen in the twin phenomena of cyclones and anticyclones.
These are gigantic swirls of winds ranging in size from several hundred to
thousands of miles long. A **cyclone** forms where relatively thin warm air heated
by the earth's surface reaches a high enough temperature and suddenly rises,
leaving behind a low pressure area at the surface. Surrounding surface air then
rushes in from all directions to the center of the low pressure zone, spiraling
upward in an elliptical or circular funnel. An **anticyclone**, on the other hand, is
a descending column of cool air in which pressure is increased at the center,
causing the air to spiral outward in all directions. In the northern hemisphere,
the Coriolis effect forces the upward spiral of cyclonic air into a counter-
clockwise swirl, whereas it forces the downward spiral in an anticyclone in a
clockwise direction. In the southern hemisphere, these wind directions are
reversed.

Cyclones are low pressure systems,
anticyclones are high pressure
systems.

As a result of the Coriolis effect and global convection currents, three convection cells, or pressure belts of alternating high and low pressure, circle the earth parallel to the equator. The first of these cells extends from the equator to 30°N and S latitudes; the second lies between 30° and 60°N, and between 30° and 60°S; and the third is found between 60°N and the north pole, and 60°S and the south pole.

At the equator is a low pressure belt created by hot rising air called the doldrums; the pressure at this zone averages 29.9 inches of mercury. Because the air is rising, there is little horizontal movement. The winds are so light that many an early sailing ship often found itself becalmed for weeks in this hot, humid region.

The first convection cell is located between 30°N and 30°S on either side of the equator. It includes the hot air rising from the equator and the cooler winds that rush in toward the equator from the north and south to fill the low pressure zone created by the doldrums. These winds, known as the trade winds (from an old meaning of the word "trade"—direction or course), are light and steady. Because they are deflected by the Coriolis effect, the winds coming toward the equator from the north are called the northeast trade winds; the corresponding winds south of the equator are called the southeast trade winds.

At about latitude 30°, both north and south, are the horse latitudes, subtropical high pressure areas where air accumulates at the earth's surface. There is very little wind in this zone. The area gets its unusual name from the fact that the crews of sailing ships, often becalmed here, were forced to jettison horses in order to conserve food and water.

In the second convection cell, located in the middle latitudes between 30° and 60° (north and south), blow the prevailing westerlies. The air that accumulates at the horse latitudes is the source of these winds. As more and more air gathers in this high pressure zone, the air nearest the surface is squeezed out by that piling up above. Most of the outrushing air travels toward the equator, becoming the tropical easterlies. The rest, which moves northeast, constitutes the westerlies.

In the southern hemisphere, in the latitudes of 40–50°, the prevailing westerlies are known as the "roaring forties." The winds are particularly strong in this area, often reaching gale force of 47–54 mph because there are no land forms to obstruct them as there are in the corresponding latitudes to the north.

In the third cell, from latitude 60°N and S toward the poles, blow the polar northeasterlies (also called the easterlies) which move in a southwesterly direction in the northern hemisphere, and in a northwesterly direction in the southern latitudes. They come from an outward flow of cold polar air.

Another major wind current was discovered by high-flying warplanes during World War II. Called the **jetstream**, this narrow flow of air travels in the tropo-

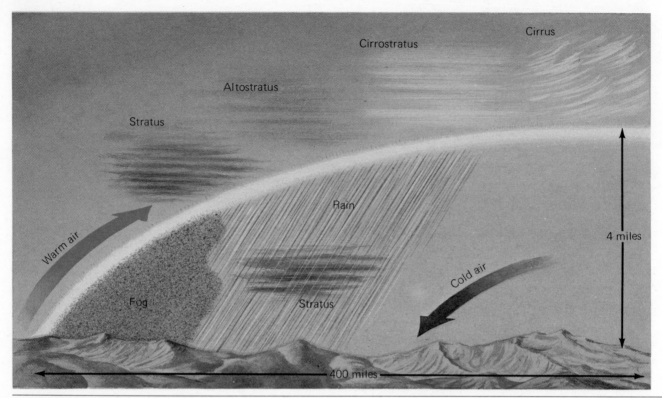

Figure 20-8
These diagrams show the temperature,
precipitation, cloud, and wind conditions
that accompany a warm front (above) and a
cold front (below). The slope of the front
differs sharply in each case, causing
precipitation of varying duration and
intensity.

sphere at speeds varying from 100 to 300 mph, and altitudes of approximately 30,000 to 45,000 feet. Jet streams are typically three or four miles thick. Wind velocities are highest at the center of the jet, decreasing sharply toward its perimeters. In some areas, the jet stream circles the entire earth in one long belt; in other areas it forms a discontinuous belt with hundreds of miles between each segment. Shifts in jet stream flow positions undoubtedly influence weather at the earth's surface—for example, it is known that a jet stream can produce a cyclone situation, but the mechanism of this effect has yet to be completely explained.

Air Masses and Fronts

The daily weather is caused by large volumes of moving air which are of relatively the same temperature and humidity at any given altitude. These huge accumulations of air are called **air masses**. Extending at times to several thousand miles in length, an air mass accumulates and grows because it remains over the same area of the earth a long period of time. Indeed, it gets its temperature and moisture content from the surface it overlies.

The air masses that move over the United States and other parts of the northern hemisphere can be of several types. The moist cold air of the maritime polar air mass that sweeps down on the United States originates over the northern Pacific and northeast Atlantic Oceans. Dry cold continental polar air moves down from Canada over much of the northcentral and northeast United States. The dry warm air of the continental tropical air mass originating in the northern parts of Mexico affects the southwestern part of the United States.

What happens when two air masses of differing temperatures—such as a continental polar and a maritime tropical air mass—collide? The area of contact is a frontal surface, and the line along which the surface meets the ground is called a **front**. The formation of the front produces a change in the weather, bringing precipitation or a storm. The two huge air masses create storm waves with a frontal zone up to 100 miles or more wide, producing alternating hot and cold temperatures and wet and dry periods of weather.

When the masses meet, the lighter warm air rises above the heavier cold air. If a warm air mass displaces cold air, a warm front is formed; conversely, a cold

A weather front is a border between air masses of different temperatures.

Figure 20-9
Oases in the desert often provide travelers with water and shade. Most of the world's great deserts are located in the belts of the trade winds and horse latitudes. The rate of evaporation in desert areas exceeds precipitation, causing hot, dry winds. Deserts cover about 20 percent of the earth's land mass.

front results when a cold air mass displaces warm air. The boundary formed by an advancing cold front is typically twice as steep as that formed along an advancing warm front, because the cold air wedges itself under and displaces the warm air very quickly; the warm air is forced to move rapidly up a steep slope of cold air. Since the warm air is thrust up vigorously and irregularly, it cools very rapidly and disturbs the atmosphere. As a result, a cold front brings sudden weather changes with temperature drops, squalls, and heavy concentrated precipitation thunderstorms. Along an advancing warm front, weather changes are not as sudden. The warm air, gently and steadily pushed up the slope by the cold air, cools slowly as it rises. At lower temperatures, the air can hold less moisture so it condenses, resulting in steady drizzles and showers.

Climate   **Climate** is the average weather in a particular region on a long-term basis—approximately 20 or 30 years. The specific factors that influence the climate of a given area include temperature, wind, pressure, and precipitation. Other more

Climate	Latitude	Outstanding characteristics	Inches annual rainfall	Temperature degrees Fahrenheit	Example
1. Tropical		No cold season			
a. Rainforest	10°N–10°S	Always wet	70–100	75–80	Congo Basin
b. Tropical savanna	5°–25°N and S	Wet and dry seasons	40–70	75–80	South Africa
2 Dry		Evaporation exceeds precipitation; scarce vegetation			
a. Tropical desert and steppe	15°–35°N and S	Arid, low precipitation	10	Annual average above 64	Sahara
b. Mid latitude desert and steppe	35°–50°N and S	Semi-arid	10–25	Annual average below 64	Southwest U.S.
3. Humid mesothermal		Warm, humid with mild winters			
a. Humid temperate	20°–35°N and S	Hot summers; cool winters	30–60	Summer average 75–80	Southeast U.S.
b. Mediterranean	30°–45°N and S	Wet winters; dry summers	15–30	Summer average 75–80	S. California
c. Marine	40°–60°N and S	Extensive cloudiness	30–100	Summer average 60–70	Northwest U.S.
4. Humid microthermal		Strong seasonal contrasts			
a. Humid continental	35°–60°N	Cold winters; hot summers	20–40	−30 in winter; 100+ in summer	North Central U.S.
b. Subarctic	50°–70°N	Long, cold winters; short, hot summers	10–25	Winter average 15 to −60	Alaska
5. Polar		No warm season; average temperature in warmest month below 50°F			
a. Tundra	North of 55°N; south of 50°S	Subsoil permanently frozen; high humidity; cold winters; cool summers	10–15	Average monthly temperature below 50; warm month temperature range 32–50	Lands bordering Arctic Sea
b. Ice cap	North and south poles	Ground permanently frozen; average summer month temperature below freezing	1–5 (snow)	Averages −30 to −60 in winter	Greenland; Antarctica

Table 20-2
Köppen climatological classification system

general variables are the presence of mountains or bodies of water, height above sea level, distance from the ocean, ocean currents, and prevailing wind patterns.

Using one or more of these factors, earth scientists have attempted to categorize world climates, but all such classifications have their limitations. The system most widely used today is known as the Köppen classification system, named after the Austrian scientist Wladimir Köppen who devised it in 1918. Based on temperature and precipitation, the Köppen system divides the climates of the world into five major types (see Table 20-2).

It is important to note that, stable though they may seem to us, climates are subject to change over a long period of time. Scientists have determined this by examining fossil plants and animals and the tracks left by prehistoric glaciers. It is well known, for example, that 20,000 years ago during most of the recent ice age, much of North America and Western Europe was covered by glaciers. Indeed, some investigators think we are currently in a period between glaciations to be followed by another ice age.

Many theories have been advanced to account for world climate fluctuations. One attempts to explain recent temperature increase (within the past 150 years) by stating that temperatures have risen as a result of the high concentration of

CO_2 emitted into the atmosphere since the Industrial Revolution. On a much larger time scale, some scientists say world climate changes are caused by variations in the amount of radiation emitted by the sun over geologic time. Still another theory was advanced by the Yugoslavian scientist M. Milankovitch early in this century. According to him, climates have been fluctuating as a result of changes in the orbit of the earth relative to the sun.

Most investigators would concede, however, that the answer to this question probably involves a combination of these and other factors (such as changes of the earth's surface features over time due to uplift and erosion and the location of the polar ice caps). The answer may come from future research.

The Oceans

Most of the world is ocean: 71 percent of the earth's area is covered by more than 317 million cubic miles of salt water. The oceans are vital to man's existence: in addition to being a source of food—some 55 million tons of seafood are harvested from the seas every year—the ocean floor contains vast deposits of gas, oil, and minerals. The oceans also play a central part in the hydrologic cycle and in the oxygen-carbon dioxide cycle. Moreover, they are an important regulator of world climate. Finally, the huge amount of carbon dioxide which is dissolved in the seas—60 times more than is dissolved in the atmosphere—helps reduce the carbon dioxide in the atmosphere.

The depths of the world's oceans vary. The average depth of the Atlantic, for example, reaches 3926 meters, the depth of the Pacific averages 4283 meters, and the Indian Ocean has an average depth of 3962 meters. The deepest point in the oceans is found near the Marianas in the Pacific, where a depth of 11 kilometers (about 7 miles) has been recorded.

Pressure in the oceans increases with depth; temperature decreases with depth.

Ocean temperatures vary with depth. Surface temperature of water near the equator, for example, is about 80°F, while half a mile beneath the surface the temperature falls about 40°F. Temperature on the ocean floor at the equator and other low latitude zones may be as low as 35°F. The lowest recorded temperature of ocean water is 28°F, recorded near the polar regions. (The freezing point of sea water lies between 28° and 31°F.) Pressure too varies with depth. Since a cubic foot of sea water weighs 64 pounds, the pressure at a depth of 1000 feet is 64,000 lbs/ft²; at 35,000 feet, the pressure is about 2,200,000 lbs/ft².

Sea water is a complex mixture of many materials: water; dissolved gases, such as oxygen, nitrogen, carbon dioxide; dissolved inorganic salts and their ions, such as chloride, sulfate, bicarbonate, bromide, and fluoride. Such salts, produced by chemical and mechanical weathering of rocks on land, are transported to the sea by rivers and streams. Salt remains in the sea because the process of evaporation removes only the water molecules. The sea contains about 56 billion tons of dissolved mineral salts, 80% common salt.

Also present in the ocean are huge chunks of ice in the form of icebergs, which are fragments of glaciers, and ice islands, large masses of frozen sea water which form in the arctic seas. Icebergs, which have reached diameters exceeding several hundred feet, can take as long as ten years to melt. Other large ice structures are the polar ice caps, huge masses of ice covering Antarctica and Greenland. It has been estimated that the melting of the ice caps would cause the level of the earth's seas to rise by approximately 500 feet.

Table 20-3
Salt composition of seawater. The salts make up 3.5% of total seawater volume.

Salt Ions	Symbol	Percent
Sodium	Na^+	31
Chlorine	Cl^-	55
Sulfates	SO_4^{--}	7.7
Carbonates	HCO_3^-	0.4
Potassium	K^+	1.1
Calcium	Ca^{++}	1.2
Magnesium	Mg^{++}	3.7

By absorbing, transferring, and releasing the sun's heat, the oceans affect the climate of the land masses immediately adjacent to them, limiting extremes of temperature. Regions bordering oceans are neither as cold in winter nor as hot in summer as inland areas.

Ocean Currents

More importantly, large-scale ocean currents, driven by winds, transport heat in complex patterns around the world. At the equator, for example, the equatorial current moves from east to west in the Atlantic, Pacific, and Indian Oceans, impelled by the northeast and southeast trade winds. In the northern hemisphere, the Coriolis effect deflects these to the right, while in the southern hemisphere, it deflects them to the left, creating huge eddies on both sides of the equator in the Atlantic and Pacific Oceans. In the northern hemisphere, the Atlantic whirl travels northwest till it reaches the Gulf of Mexico. When it sweeps out of the Gulf, it is known as the Florida current; moving up the east coast, it becomes the well-known Gulf Stream, a 50-mile wide, 100-foot deep flow that travels at speeds of two to six miles per hour. Carrying its warm, 80° water across the Atlantic past Great Britain and Scandinavia, the Gulf Stream finally empties into the Arctic Ocean.

There are many other large-scale ocean currents driven by one or more of the world's great wind systems. In cases where the winds blow towards shore, the currents are an important variable in the climates of neighboring land masses.

In addition to the driving forces of the winds, there are three other reasons for the movement of ocean surface currents:

1 because of gravity and buoyancy, war waters in the equatorial zones and elsewhere expand, become less dense, and rise, and the cold waters in the polar zones contract, become denser, and sink;
2 subsurface pressure variations, which cause water in a high pressure zone to move to a low pressure zone;
3 the Coriolis effect, which deflects moving currents to the right in the northern hemisphere and to the left in the southern hemisphere.

While the relatively shallow currents travel across the ocean's surface, complex submarine currents may be traveling some distance below. For example, as the South Equatorial Current flows westward, the Pacific Equatorial Current moves

Winds, gravity, pressure variations, and the Coriolis effect move ocean currents.

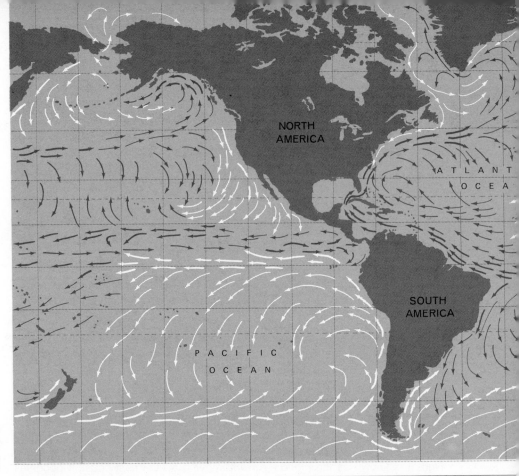

Figure 20-10
The major ocean currents help determine
world climate. Wind is the principal driving
force of the currents, but gravity, pressure
differences, and the Coriolis effect also play
important roles in their movement.

several hundred feet below in the opposite direction. Oceanographers have
discovered that there are two principal sources of deep water currents, one at
the North Pole, the other at the South Pole. At the North Pole, when sea water
turns to ice, only the water freezes, leaving the salt in solution; hence, the water
is denser. The dense water sinks in places at depths up to two miles, and flows
southward toward the equator. Similarly, at the South Pole, the dense water
sinks to great depths and flows north toward the equator. The cold water is important for life in the ocean depths, for it carries large amounts of oxygen with it.

Life in the Ocean

The ocean is home for a wide variety of plant and animal life. Since suboceanic
plants, like land plants, depend on photosynthesis for their existence, they are
not found below the depth that sunlight penetrates, or about 200–300 feet.
Where there is sufficient sunlight, carbon dioxide, and nutrient salts, phytoplankton—minute plants, most of which are algae—thrive. Phytoplankton constitute the basic diet of zooplankton, tiny animals such as protozoa and small
shrimp. Both phytoplankton and zooplankton are consumed by fish and other
marine animals.

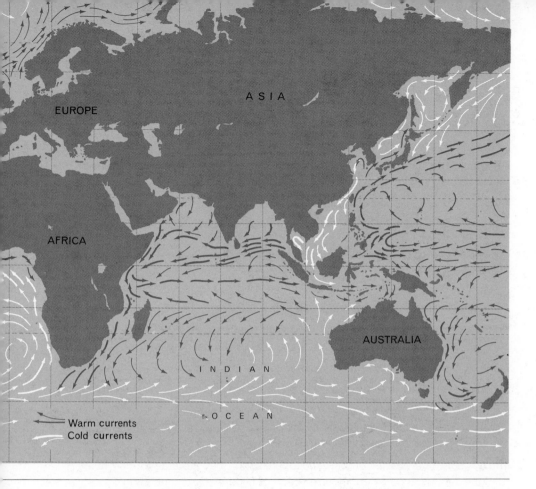

Warm currents
Cold currents

Plant and animal life are more abundant in cold waters than in tropical waters. In the tropical zones, the warm surface water does not mix with the colder water lying beneath it. When phytoplankton, which consume nutrients at the surface, die, they sink into the deeper cold water. Since this represents a drain of nutrients from the surface, very few animals are found there because of the absence of food. In cold water, fish and other sea life abound, because nutrients are plentiful as a result of the rapid and continuous recycling of nutrients.

The ocean basin and the continental land masses adjacent to it are separated by a continuum of sedimentary structures. The Continental Shelf, Continental Slope, and Continental Rise, which gradually phase into the abyssal plains, constitute more than one-half of the earth's surface. These freezing dark flats are covered by sediments transported from land by ocean currents, as well as oceanic sediments, clays, and oozes that represent the remains of sea animals and plants. The thickness of the sediments on the plains is 150–9000 ft.

The ocean basins are very deep—more than five times as deep as the continental land masses are high. For example, the average depth of the Pacific

Topography of the Ocean Depths

Sedimentary structures at ocean depths are similar to those on land.

Ocean is about 14,000 feet, whereas the average land surface extends only about 2700 feet above sea level. Through echo-sounding and other techniques, oceanographers have discovered that the ocean floor contains midocean ridges, mountains, valleys, and canyons which are just as high, wide, and deep as their counterparts on the land. For example, the island arcs of the Pacific Ocean are really huge volcanic mountains, some of which extend up to 30,000 feet above the ocean floor.

Glossary

Absolute humidity is the amount of water vapor in a unit volume of air, usually expressed in grams/m³.

An **air mass** is a large volume of moving air at any given altitude and of relatively the same temperature and humidity.

The **atmosphere** is the shell of air which surrounds the earth, extending thousands of miles into space. It is composed principally of nitrogen and oxygen.

Climate is the long-term weather of a region.

The **Coriolis effect** is a phenomenon affecting wind direction, causing north-south winds to veer from the straight line paths.

The **dew-point** is the temperature at which the relative humidity increases to 100 percent, causing moisture to condense.

The **exosphere** is a layer of the atmosphere which begins at about 300 miles from the surface and extends to an altitude of 600 miles from the earth.

The **greenhouse effect** is a phenomenon in which heat is reradiated from the earth and retained by the lower layer of the atmosphere.

The **ionosphere** is a zone of the atmosphere that begins at the mesosphere and extends into space. It contains ions produced by the action of the sun's ultraviolet and x-rays on gas molecules.

Insolation is radiant energy from the sun in the form of short electromagnetic waves. The word is an abbreviation for *in*coming *solar* radi*ation*.

A **jet stream** is a narrow flow of air that travels in the upper troposphere at speeds varying from 100 to 300 mph at altitudes from 30 to 40 thousand feet.

The **mesosphere** comprises the upper reaches of the stratosphere, between 35 to 50 miles from the earth's surface.

Meteorology is the branch of physical science which studies weather and climate.

The **pressure gradient** is defined in meteorology as the rate at which the atmospheric pressure varies between different areas.

Relative humidity is the ratio between the amount of water vapor in the air and the maximum amount that it could hold at a given temperature, expressed as a percentage.

The **stratosphere** is a layer of the atmosphere extending from roughly 10 to 50 miles above the earth's surface. It contains ozone, which absorbs most of the ultraviolet rays of the sun.

The **thermosphere** is a layer of the atmosphere extending from about 50 miles above the earth's surface to a distance of approximately 300 miles.

The **troposphere** is the layer or zone of atmosphere closest to the earth. It comprises 75 percent of the total atmosphere.

Weather is the combined temperature, atmospheric pressure, humidity, cloud cover, and precipitation experienced by a given region at any given time.

Exercises

1 What are the major pollutants and sources of pollution in our atmosphere? Discuss some of the potential dangers of such pollution. What is the "greenhouse effect"?

2 Why does the troposphere extend two to three times as far over the equator as it does over the poles?

3 What is the primary importance to man of the following atmospheric layers: (a) troposphere (b) stratosphere (c) ionosphere?

4 The amount of solar radiation emitted by the sun is such that, were it to reach the earth, it would equal 700 cal/in². Approximately how much cal/in² are (a) absorbed by the ozone and water vapor, (b) reflected by clouds and dust, and (c) received by the earth?

5 Why do tropical countries experience very warm days and nights? What accounts for the extreme temperature drop which accompanies nightfall in the Sahara Desert?

6 What determines relative humidity and how is it related to cloud formation, rain, snow, and other forms of precipitation?

7 What accounts for the direction and velocity of the Trade Winds 30° north and south of the equator?

8 Why does the formation of a cold front generally cause sudden storms and heavy precipitation?

9 Trace the movements of the Gulf Stream from its origin at the equator, and show how it affects the climate along the coasts.

10 Why is the seashore frequently more comfortable during the summer than inland areas?

11 The Caribbean attracts skindivers from all over the world due to its brilliant coral reefs, warmth, and excellent visibility. What accounts for the clarity of tropical waters as compared with the murkiness of colder seas?

Interlude Twelve
Sunny with Patches of Doubt

When the residents of Kansas City left their homes for work or school on the morning of September 15, 1974, most were carrying umbrellas and wearing boots and raincoats, because the weatherman had predicted showers for the day. But not one drop of rain fell that day. How did the weatherman make such a mistake?

To answer this question, we must visit a typical National Weather Service office. Here meteorologists work around the clock with maps and machines, trying to glimpse the future. On a typical day, a visitor might see a man poring over sheaves of maps, marking them with technical symbols. A woman might be checking the chart at a console where local air, temperature, pressure, humidity, and wind direction are being recorded. Another woman might be peering into a radar scope, checking the echoes from a large advancing rain storm.

Like others in thousands of weather stations throughout the world, these meteorologists are observing and recording weather elements in order to construct what is called a synoptic weather map, or one that contains all the factors that contribute to the weather of a region at a given period of time. The first entries on the map come from widely scattered observers who note the weather conditions occurring in their areas, reporting at four designated times daily: 6 A.M., noon, 6 P.M., and midnight. Temperature, air pressure, humidity, wind speed and direction, and precipitation are recorded by instruments. Clouds are observed and categorized visually, and their height is determined by radar, plane reports, or other means. Phenomena such as rain and snow, fog, dust storms, sandstorms, squalls, or tornados are noted.

All this information is first entered on local maps, then put into an international code devised by the World Meteorological Organization. The WMO links weather stations in countries all over the world, including the United States, the USSR, and China. The coded data are teletyped at 400 words per minute to a national central collecting station such as the U.S. Weather Services' National Meteorological Center (NMC) in Suitland, Maryland. The NMC, the world's largest weather center, receives reports from thousands of American weather stations, from six weather ships in the Atlantic and Pacific Oceans, and from five stations in the Arctic. In addition, data from orbiting satellites, aircraft, military installations, and foreign meteorological centers pour into the center constantly.

In all, the center receives some 43,500 reports every day. Each report is broken down into bits of individual data, which are entered on punch cards and fed into computers. In about an hour and a half, an analysis is completed. Then coded forecasts are distributed by teletype to weather stations in this country and abroad. In about three minutes, the NMC's computer-plotter draws a synoptic weather map showing the weather conditions for the entire Northern Hemisphere.

Given this incredible capability, how could a forecast possibly be incorrect? Let's look at the weather maps (see next spread) to try to determine what went wrong in the Kansas City forecast. The lower map shows the predicted weather for the country, and the upper map shows the actual weather. The first things the meteorologist looks for are pressure centers and frontal systems, because as these move they are the primary determiners of weather. On the lower map, you will notice that a low pressure system was expected to form over Oklahoma City, bringing rain to that area, including Kansas City. Such a low pressure system usually brings precipitation or cloudiness because of the way it is formed; less dense air warmed by the earth's surface travels upward and inward in a counterclockwise spiral. As the air rises, it expands and cools to the point where the moisture in it condenses, eventually falling as precipitation. Why didn't the low pressure system over Phoenix cause rain or cloudiness in that area? Because the dry desert air around Phoenix holds very little moisture. It was this low that was expected to move toward Kansas City. However, it did not, and so other factors influenced that city's weather.

The first of these was the high pressure system centered around Chattanooga, Tennessee, shown in the upper map. The high pressure zone was formed when a descending column of cooler air spiralled outward in a clockwise direction over an irregularly shaped area about 800 miles in diameter. As the cooler air traveled down toward the earth's surface, the pressure on it increased, raising its temperature and causing the dry cloudless conditions in Chattanooga.

If the high were the only feature in the picture, Kansas City would have experienced sunny skies too. However, as the upper map shows, a frontal system moving down from Canada and carrying cold, dry, polar air was draped over the Great Lakes area like a necklace. At its southernmost extremity, the front burrowed under the warm air of the high pressure system over Chattanooga, forcing it to rise. This in turn caused the moisture in the air to condense, bringing clouds and/or precipitation. In this case, it caused clouds to form over Kansas City and St. Louis. If the low over Phoenix had moved on Kansas City as predicted, it would have interacted with the cold front, and the residents of that community would have needed their umbrellas and raincoats.

Because of the tremendous number of variables involved, the meteorologist's job is a difficult one. Even with the aid of high-speed computers, the tasks of calculating known factors and trying to estimate unknown ones is quite demanding. For this reason, forecasts are provided in terms of probability. The weatherman, like the habitual card player, is a kind of professional gambler, interested in the percentages. But he works at a greater disadvantage than the gambler—if there is more than a 50 percent probability of an event, he must predict it. Thus, if there is a 55 percent chance that it will rain, and a 45 percent chance that it will not, the weatherman will predict rain (although he usually does tell the public what the odds are, so we can all share the risk).

However, the modern meteorological center has a number of tools to help overcome the odds. In addition to the thermometer, barometer, and anemometer (which measures wind velocity), meteorologists also employ a variety of techniques for remote recording of weather in the upper atmosphere, where conditions heavily influence those at the earth's surface. One of these techniques is radar. Because raindrops reflect radar waves, a rotating radar scanning system can be employed to detect rain patterns up to 100 miles away in any direction.

Conditions in the upper atmosphere are also monitored by radiosondes, which are meteorological sounding balloons. These small balloons, filled with hydrogen or helium, are capable of rising 100,000 feet into the atmosphere. Instruments to record temperature, barometric pressure, and humidity are combined with a small radio transmitter in a compact unit weighing about two pounds. At designated times, the radio automatically transmits readings back to receivers on the ground. Even further up are the orbiting weather satellites of the Tiros and Nimbus series. These satellites orbit the earth every 110 minutes at a height of about 900 miles. A television camera aimed at the earth takes pictures of cloud formations and violent storms.

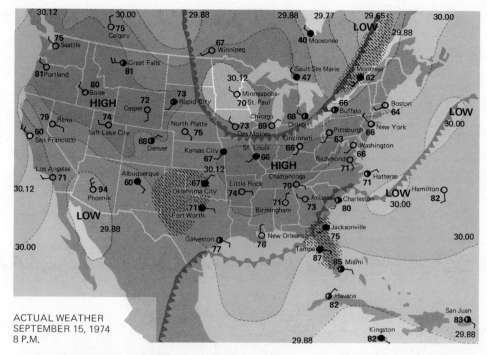

ACTUAL WEATHER
SEPTEMBER 15, 1974
8 P.M.

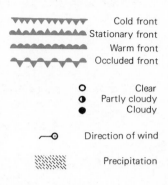

Cold front
Stationary front
Warm front
Occluded front

○ Clear
◐ Partly cloudy
● Cloudy

⌐○ Direction of wind

▨ Precipitation

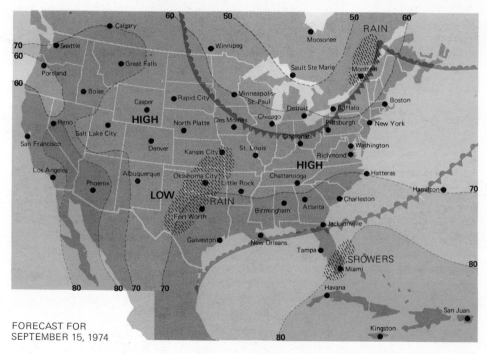

FORECAST FOR
SEPTEMBER 15, 1974

The upper map shows the actual weather in the continental United States on September 15, 1974. The number next to the city name indicates the temperature reported by the weather station in that city. A cold front, such as the one along Des Moines and Detroit on this map, is a boundary between cold air and warmer air. The colder air pushes under the warm air like a wedge. Cold fronts usually move south and east. A warm front, such as the one east of Great Falls and Rapid City on this map, is a boundary between warm air and a retreating wedge of colder air over which the warm air is forced as it advances. Warm fronts usually move north and east. An occluded front, such as the one along New Orleans and Atlanta on this map, is a line along which warm air has been lifted by two opposing wedges of colder air. Often occluded front conditions bring precipitation, and rain was reported in New Orleans on September 15, 1974. The red dotted lines on the top map are isobar lines, which show the boundaries of areas of equal barometric pressure. Barometer readings are given in inches on the top map. The areas of equal barometric pressure, shaded in red, form air-flow patterns.

The lower map shows the U. S. Weather Service forecast for the weather shown in the upper map. The red dotted lines on the bottom map are isotherms; they indicate the boundaries of temperature zones in the forecast. The predicted temperatures are given around the edges of the lower map. The forecast shown was made at 8 P.M. on September 14, 1974.

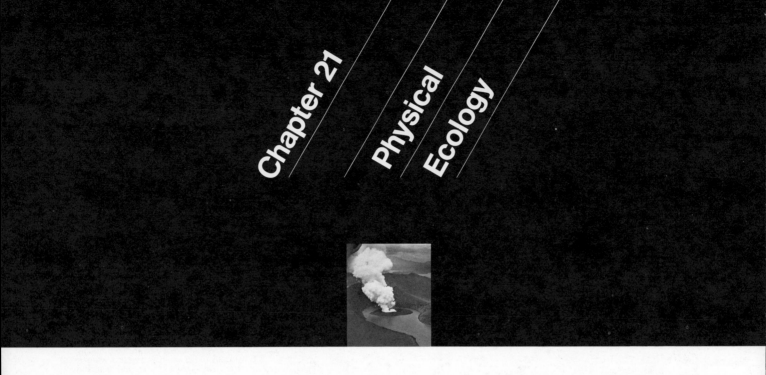

Chapter 21 Physical Ecology

In the early 1960s, a huge dam to control the perennial floodwaters of the Nile River in their passage into the lower Nile valley was built at Aswan, Egypt. The dam is yielding economic benefits for the people of that ancient country by controlling floods, increasing the acreage of cultivable land, and producing electricity. But all of these benefits are substantially offset by numerous undesirable side effects created by the project. The dam has prevented the natural fertilization that occurs when floodwaters deposit silt from the highlands down onto the farming land, with the result that synthetic fertilizers must be imported as chemical substitutes. An extensive sardine industry in the Mediterranean Sea off the Nile's mouth has been destroyed because the dam traps some of the essential nutrients that the fish feed on. The Egyptian coast has been eroded seriously because the restorative flow of silt into the Nile delta has been halted. The incidence of schistosomiasis, a debilitating disease carried by water snails, has been enormously increased because the snails thrive in the new irrigation canals. The salinity of the water stored in Lake Nasser, held back by the dam, has been increased by extended evaporation, lowering crop yield. In addition, the water level of the lake is far lower than expected, due both to the annual evaporation from a large water body in this desert climate, and to the extraordinary drain of dammed-up water into the vast sands of the desert around Aswan and the Nile River.

In short, the dam is an ecological failure. Many of the harmful environmental byproducts could have been minimized, or even avoided, had this structure been planned on fully ecological principles. **Ecology** is defined as the study of the interactions among organisms and their total environment. All of the components which are involved in these interactions form an ecological system, or **ecosystem**. Such a system may be balanced and self-preserving, or it may be disturbed and self-destructive. Since ecology seeks to determine the structural arrangements and functional relationships of ecosystems, it has become an increasingly important analytic tool in helping science to understand, perhaps even to solve, the environmental problems produced by industrial progress within the last two centuries.

Ecology is the branch of the natural sciences that recognizes the systematic nature of everything that occurs on earth—physical, chemical, biological, social, and personal. It considers our planet, or any given part of it that can be isolated, as a completely interrelated energy system in which a change in any part causes reciprocal changes in other parts.

All of the organisms and environments of the earth are interrelated so that a change in one part of the system causes reciprocal changes in other parts.

Principles of Ecology

The science of ecology has two general principles concerning matter and energy. According to the first, energy flows in a one-way direction through the components of any natural ecosystem. Thus, energy flows from the sun to plants, and then from plants to the animals that feed on them.

According to the second principle, matter continuously cycles through the components of a natural ecosystem. For example, matter given off as waste by one organism is consumed as food by another. The same material may then be used and reused by a number of organisms in the ecosystem, being transformed each time it is transferred. Both processes are governed by the laws of thermodynamics (see Chapter 5). The most fruitful way to study the systematic quality of an ecosystem is to examine the ways energy and matter interact as they move through the energy-flow paths and the matter cycles of the system.

Two general principles govern a natural ecosystem: energy flows through the system in a one-way direction, and matter continually cycles through the components of the system.

Energy Transformations in an Ecosystem

Where does energy originate in an ecosystem? How is it transformed? What becomes of it? The ultimate source of all energy in the solar system is the sun.

Ecology As Systems Analysis

Ecology is the study of nature in systems. A system consists of two or more elements related by their interaction in such a way that a change in one of them causes a change in the others. Every natural system has a number of essential processes and components: these include material bodies, an input to the system through which matter and energy are introduced, an output through which matter and energy are released, and energy transformations which result from the interaction of the material bodies and their parts with energy processes.

As energy is transported or transformed in different parts of a system, it interacts with matter through energy pathways. In photosynthesis, for example, energy enters the pathway from outside the system in the form of light from the sun; the sunlight interacts with chlorophyll, a chemical in the leaf which catalyzes the transformation of CO_2 and water from the air into carbohydrates and oxygen; then the surplus energy leaves the pathway to be stored as chemical energy in the plant.

Another characteristic of energy systems is that the rate of energy input equals the rate of energy output, provided energy is not stored in, or released from, a reservoir within the system. This conservative energy budget, which accounts for all the energy within a system, is based on the law of the conservation of energy: energy can neither be created nor destroyed, but it can be transformed. The energy budget is really a kind of bookkeeping that records and tallies energy input, output, and storage.

There are two kinds of systems: open and closed (or isolated). An open system exchanges matter and energy with its surroundings and maintains dynamic equilibrium so that the rates at which matter and energy enter, are stored within, and leave the system remain the same. In a closed system, matter and energy are exchanged among the parts of the system, but neither matter nor energy enters or leaves the system. It is difficult to think of a natural example of such a system because it would have to be completely isolated from all outside forces and influences and, if it continued to exist, self-sustaining like a fictional perpetual motion machine. A terrarium, for example, appears to be a closed system; however, because it receives its energy from the sun, it is actually an open system.

For purposes of analysis, open but relatively isolated natural systems are often treated as if they were closed when models are constructed for understanding them. The biogeochemical cycle discussed in this chapter, for example, is treated as if it were a closed system. Perhaps the only genuinely closed system is the whole universe since it contains all the matter and energy that exist. But doesn't that depend on one's definition of *universe*? If the universe is infinitely extended rather than finite, it can hardly be called closed. What *can* safely be said about all the bodies and all the systems within the universe—galaxies, solar systems, ecosystems, local plant and animal communities, and everything else down to the microorganisms that teem in a temporary puddle—is that they are open systems involved in the continuous exchange, transformation, or storage of matter and energy.

This star, which radiates immense amounts of energy throughout the full spectrum of electromagnetic waves, provides the earth with light and heat. According to the second law of thermodynamics, it is physically impossible to convert a given amount of heat energy entirely into work. As a result, when energy is transferred and transformed from one organism or physical entity in an ecosystem to another, some heat energy is lost in the process. Similarly, during the transformations which convert one form of energy to another through the metabolic functions within individual organisms, considerable energy is lost in heating the plant or animal. Thus, as energy travels through an entire ecosystem or a single organism it is moving in one direction, and the useful energy decreases with each transformation.

Almost all of the energy in the solar system ultimately comes from the sun.

ENERGY AND THE FOOD CHAIN The one-way flow of energy through an ecosystem can be seen by examining the pathways it takes as it passes from one organism to another. A natural ecosystem has two components: abiotic and

Figure 21-1
In this example of the food chain, energy from the sun is transferred to an autotroph, the plant, and then through a series of heterotrophs, beginning with the bacteria that decompose the plant.

Fish–eating bird

Water animals

Water insects

Small anthropods and zooplankton

Single celled plants

Bacterial decomposers acting on organic matter

biotic. The **abiotic** components are the nonliving parts of the environment such as rock, air, and water. The **biotic** components, which are the living parts, are divided into two primary groups: autotrophs ("self feeders") that make their own food, and heterotrophs ("other feeders") that must consume energy-containing food by eating plants or other animals. Autotrophs are mostly green plants; they use solar energy during photosynthesis to manufacture their own food from simpler inorganic substances—mainly water and carbon dioxide. The products of photosynthesis are rich storehouses of energy, quite different from the H_2O and CO_2. Heterotrophs, which cannot produce their own food, feed on autotrophs. They transform the autotroph's organic substances—containing mostly carbon compounds—into usable energy through a long series of chemical reactions.

A food chain is the path that energy takes from autotroph to final heterotroph. It is a sequence by which organisms obtain energy. Because the relationship between eaters and eaten is complex and systematic rather than simple, the food chain is often referred to as the food web. All food chains are composed of three major levels of organisms—producers, consumers, and decomposers. The producers are the autotrophs. The consumers are all those organisms (called herbivores) which must obtain their food from the plant producers, and the carnivores which feed on herbivores or each other. The decomposers include bacteria, fungi, and yeast; these organisms, the microconsumers in the web, obtain energy by decomposing organic remains and wastes. Food chains are broken down into feeding links called **trophic levels**. All the organisms at the same level consume the organisms of a lower trophic level.

As energy is transferred by the movement of matter from one trophic level to another, some of it is converted into energy usable by the organism and some of it is dissipated into the environment as heat. The actual amount of energy dissipated at each level has been estimated by ecologists, and it is quite high. Nature is not a very efficient converter of energy. For example, plants living in temperate regions absorb and utilize only about 1 percent of the solar energy that reaches them. Herbivores feeding on the plants receive only about 5 to 20 percent of the energy which had been absorbed by the plants; carnivores eating the herbivores receive about 5 to 20 percent of the energy consumed by the herbivores.

We can see how this pyramid of energy works by assigning a numerical value to the solar energy received by plants—say 3000 calories. Of the initial 3000 calories of solar energy, about 30 calories would be converted into food energy by the plant. Then herbivores feeding on the plants would receive about 3 calories; carnivores feeding on the herbivores would receive 0.30 calories. At the end of the chain, the animals feeding on the carnivores receive about 10 percent of the 0.30 calories, or 0.030 calories.

These rough figures give concrete evidence that the second law of thermodynamics imposes a practical constraint upon natural processes. Energy is in-

Living things get their energy in a complex and systematic sequence of interrelationships called a food chain.

Figure 21-2
The lion, a carnivore, obtains his food from herbivores like the wildebeest, which feeds on plains grasses.

deed flowing one way through the ecosystem, and the amount lost at each successive trophic level ranges from 99 percent for plants to 80 to 95 percent for the rest of the chain. The huge amount of energy dissipated through any food chain necessitates very short chains containing three or four, and certainly no more than five, links because the increasing scale of energy used up through eating would be literally impossible to sustain. Obviously, the organisms closest to the beginning of the food chain receive proportionately more energy.

There is a lesson here for man. If he ate at the bottom of the food chain — closer to the primary sources of energy — more people could be supported on less land or water than is currently the case. If we became vegetarians and obtained our food energy from soybeans, rice, wheat, plankton, seaweed, and other similar plants, we would have many more usable calories available.

Mankind has tried to increase the energy production efficiency of plant food-making processes by agricultural techniques such as irrigation, fertilization, and use of pesticides. But in so doing, he has all too often introduced other waste, other dissipations (in thermodynamic terms, more entropy) into the natural system by polluting the soil, water, and atmosphere and by interfering with the natural nutrient cycles of the system.

Because usable energy is lost as it is passed along the food chain, organisms close to the beginning of the chain get proportionately more energy.

In addition to the flow of energy, the organisms of an ecosystem also require a certain number of chemical materials for their existence and upkeep. These materials are continuously circulating among the biotic and abiotic parts of the environment in pathways known as **biogeochemical cycles**, so-called because the materials involved are biological, geological, and chemical. During biogeochemical cycles, the same natural materials are used and reused by the various systems interacting in the cycle.

Just as the flow of energy through the ecosystem is governed by the second law of thermodynamics, so the cyclic flow paths of matter are governed by the great conservation law of energy, the first law of thermodynamics, and by the law of the conservation of matter. Matter can neither be created nor destroyed, but it can be transferred and transformed. These laws govern the hydrologic (water) cycle, the nitrogen cycle, and the carbon cycle.

Material Cycles

THE HYDROLOGIC CYCLE Water is both the most important and the most abundant inorganic compound in the biosphere: three-quarters of the earth's surface is covered by this liquid. It is a storage place for biologically important elements — carbon, oxygen, nitrogen, sulphur, and phosphorus — which it carries in solution. Water is the universal environmental medium for life.

The processes involved in the transfer of water from the sea to the land and back again form the **hydrologic cycle**. Water can exist as a gas (water vapor), a liquid, or a solid. As a gas, it is the most physically active component of the

Water, which is the universal environmental medium for life, is being continually cycled through the sea, the air, and the land.

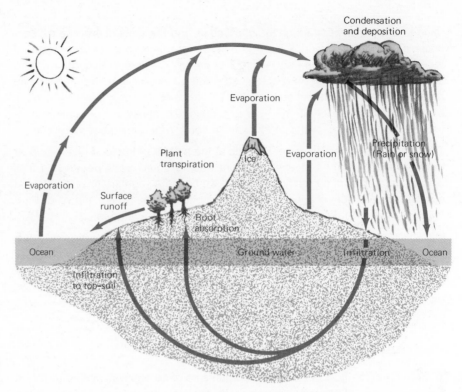

Figure 21-3
Water is circulated through the earth's seas,
land, and air in the hydrologic cycle.

atmosphere. As a liquid and a solid, it forms the hydrosphere—an irregular and discontinuous global layer of water, which includes the world's oceans and ice caps, rivers and glaciers, lakes and aquifers (underground water layers).

In the first stage of the hydrologic cycle, water is evaporated into the atmosphere from reservoirs of water, snow, or ice in the hydrosphere. Since the oceans contain almost 98 percent of the water, most of the atmosphere's water vapor comes from evaporation of this great reservoir. A comparatively small amount of water vapor comes from evaporation of water in lakes, rivers, glaciers, snow fields, soil, and vegetation.

Vapor is carried over the continent by moving air masses. If the vapor is cooled to its dew point, it condenses around any available particle, such as a speck of dust, into visible water droplets which form clouds or fog. Under favorable conditions, the tiny droplets grow large enough to fall to the earth as rain or other forms of precipitation.

About two-thirds of the precipitation which reaches the land surface is quickly returned to the atmosphere by renewed evaporation from ponds, lakes, rivers, soil, and vegetation. The remainder ultimately returns to the ocean as liquid through surface runoff or ground water flow.

A droplet of water can complete its trip through the hydrologic cycle quickly in a few hours, or slowly in a few centuries. Imagine rain falling on a tropical rain forest, with its characteristic dense vegetation. Much of the rain is held as droplets on leaves and plant stems; these droplets will probably soon return directly to the atmosphere by evaporation. If rainfall continues, water will eventually drip down to reach the soil, and infiltration, or movement of water into the soil, will begin. Like the atmosphere, the soil can become saturated as it chemically or biologically absorbs water; when this occurs, pools or puddles will form on the surface. At the same time, water also percolates down through the soil and eventually reaches the aquifers, where it accumulates.

Most of the water that falls on a rain forest is intercepted by the vegetation and is ready to be evaporated again. The route that these droplets follow, as they are intercepted by the plant leaves and stems and eventually return to the atmosphere, is a simple one that takes only a few hours. Similarly, the majority of water in any part of the earth's hydrologic cycle follows a simple path: water evaporated from the ocean falls back on the ocean as precipitation. On the other hand, a very small fraction of the rain droplets which fell on the forest may be stored for hundreds of years in aquifers. In other parts of the earth, water may be stored for long periods on the surface of a glacier.

THE NITROGEN CYCLE Nitrogen is an important part of many complex organic

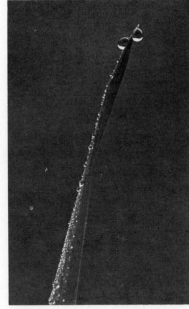

Figure 21-4
This water droplet will be evaporated by the sun and return to the atmosphere from which it has come.

Figure 21-5
We obtain the proteins essential for body maintenance and growth as part of the nitrogen cycle.

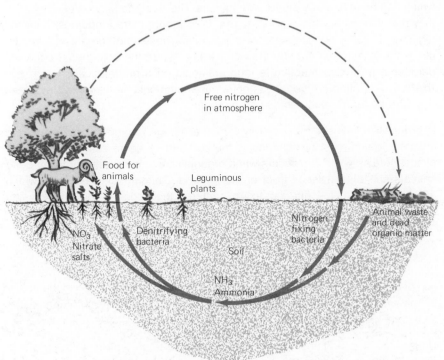

Nitrogen is an important part of many complex organic molecules including proteins, nucleic acids, enzymes, and vitamins.

molecules. Proteins, the essential body chemicals whose building blocks are amino acids, are nitrogen compounds; nucleic acids, enzymes, and vitamins also contain nitrogen. This element can even be used as an indicator of the quality of nutrition in an area. As we saw earlier, animal protein is high in the food chain and energy-expensive to produce, so only relatively wealthy societies can afford a diet rich in animal protein; a low rate of consumption of nitrogen-containing food reflects impoverished conditions.

The indirect source of all of our usable nitrogen is plants. Animals, in their role as secondary consumers, eat plants and obtain nitrogen in various compounds from them. People eat protein-containing plants, such as soybeans, or they may move further up the food chain and eat such animal byproducts as meat, milk, and eggs, which are high in protein content.

But why not use nitrogen directly? After all, four-fifths of the earth's nitrogen is in the form of the chemically inert nitrogen gas that constitutes 79 percent of the atmosphere. The major part of the remainder can be found in the soil's humus; the remaining fraction is contained in organic compounds of living matter.

But the evolution of most organisms took a different chemical turn. They depend upon the simplest forms of life, and the autotrophs—the green plants—are no exception. Gaseous nitrogen cannot be used directly by most organisms. Only certain bacteria and the blue-green algae can convert the inert gas into a form which can be utilized by plants and animals. The bacteria may be free-living, or they may live within the root hairs of certain plants called legumes; common legumes include peas, beans, and peanuts. These microorganisms absorb atmospheric nitrogen through their cell walls. Inside the cell, the nitrogen undergoes a reduction reaction to form ammonia, in a process called **nitrogen fixation**. This simple nitrogenous compound, ammonia, is then used by green plants to build up amino acids and proteins.

In order to maintain a balanced system, the nitrogen component of plant and animal proteins must be returned to the soil and, to some extent, back to the atmosphere as well. This occurs when organisms excrete wastes or when dead animals and plants decay in or on the ground. Decomposing bacteria, which are different from the nitrogen-fixing bacteria, then convert the proteins into ammonia and ammonium salts. Although some plants can use ammonia directly, it is more often oxidized by yet other bacteria in the soil into nitrites and nitrates. Nitrates are the most common form of nitrogen used by plants.

Some of the nitrogen is eventually lost as it returns to the atmosphere by the action of denitrifying bacteria, which is another type of bacteria that has the ability to free nitrogen. Nitrogen is also lost from the soil when nitrates and ammonium salts form solutions with water.

Under natural conditions, the amount of nitrogen in the soil remains constant.

However, human actions have upset the balance, simply by not returning much of the used nitrogen to the soil. We bury organic garbage in landfills as a foundation for buildings: we dump human nitrogenous wastes, chemically treated or not, into rivers and oceans.

The problem of keeping nitrates in the soil used for farming has been worsened by modern agricultural machinery. For example, farmers used to harvest corn by picking the ripe ears off the plants; in the fall, the corn plant died and the farmer plowed it under the soil, where its decay by decomposing bacteria released its nitrogenous molecules back into the soil. Today almost all corn is harvested by a machine that pulls up the whole plant; at the processing plant another machine separates the edible part from the rest of the plant. The result is that the field has no plant nitrogen, and the town where the processing plant is located has a serious garbage disposal problem.

To compensate for this loss, artificial fertilizers are synthesized to provide the soil with nitrogen. In the process of nitrogen fixation, molecules of H_2 and N_2 are combined catalytically to form ammonia. After a series of chemical reactions, ammonium nitrate (NH_4NO_3), a widely used form of nitrogen fertilizer, is obtained, and spread on the fields.

Human intervention has disturbed the natural balance of the nitrogen cycle.

THE CARBON CYCLE   The element carbon is proportionately one of the less

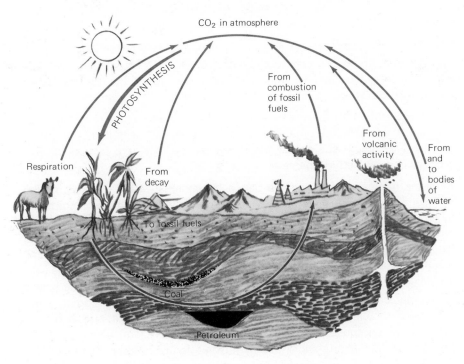

Figure 21-6
Fossil fuels are produced as part of the carbon cycle.

CO$_2$ in atmosphere

PHOTOSYNTHESIS

From combustion of fossil fuels

From volcanic activity

From and to bodies of water

Respiration

From decay

To fossil fuels

Coal

Petroleum

numerous constituents of the earth and its environs. It comprises less than 1 percent of the earth's crust but its sheer bulk seems large. There are about 8×10^{16} metric tons of carbon in the form of carbonates, such as limestone and marble. The atmosphere contains about 6×10^{11} metric tons of carbon in the form of carbon dioxide. The oceans contain approximately 5×10^{23} metric tons of readily available carbon in the form of recyclable detritus and living organisms. Additional carbon can be found in petroleum, coal, peat, and graphite. All of this carbon will eventually pass through a living organism—either frequently, as in the case of atmospheric and oceanic carbon, or slowly, as with the carbon contained in rock. The estimated turnover rate of carbon through the normal carbon cycle on the earth is 200 to 400 years, which means that every available carbon atom will pass through a living organism at least once in that time. But there is a second, very slow, carbon cycle, a kind of epicycle, that locks carbon into the sedimentary rocks.

The balance of carbon in the biosphere is maintained by photosynthesis, respiration, and decay.

Control of the balance of carbon in the biosphere is maintained by three natural processes: photosynthesis, respiration, and decay. During **photosynthesis**, carbon dioxide from the atmosphere is used by green plants to construct the organic molecules of carbohydrates. This is a reduction reaction in which solar energy is used both to break and form bonds. CO_2 and water combine to form a simple sugar, such as glucose, and oxygen. The equation can be written:

$$H_2O + CO_2 \xrightarrow[\text{chlorophyll}]{\text{light}} C_6H_{12}O_6 + O_2$$

Although some of the synthesized sugar will be used by the plant in respiration, much will be stored in the form of complex sugars or starches. These carbon compounds are then passed along the food chain, first to the herbivores and then to the carnivores.

During **respiration**, which takes place in both plants and animals, organic molecules are oxidized. The organism takes in oxygen and uses it to break down glucose into simple carbon compounds, releasing CO_2 as a byproduct. This process is explained in Chapter 17.

As in the nitrogen cycle, bacteria play a fundamental role in maintaining the cyclical carbon flow. Through the process of decay, bacteria and fungi return much carbon to the atmosphere by reducing the carbon-bearing molecules of dead plants and animals. However, some of the carbon is trapped by the decomposing processes of fossilization, producing not only the occasional preserved fossil, but, more importantly, the world's supply of fossil fuels—oil, gas, coal, and peat. It is believed that the latter two fuels are produced when microorganisms act on decaying organic matter to form highly carboniferous peat. Over the millenia, the peat is overlain with strata of sand, clay, and other materials which squeeze out the gases and eventually form bituminous coal. This, in turn, is further compressed by tectonic forces to form anthracite coal.

We return the carbon to the cycle when we burn the fuel. But burning the fossil fuel is a very rapid loss, a drastic interruption, to the carbon cycle.

Power and Pollution

Since the Industrial Revolution, man has been interfering with the energy flows and material cycles of the natural environment. Advances in technology have brought increased demands not only for material goods, but also for more power because industrialized nations require large energy inputs from fossil fuels and other energy resources.

Industrialization disrupts the energy flow and biogeochemical cycles of the earth's ecosystem.

Because of the law of the conservation of matter, the matter in the earth's resources is not really being used up. Instead, it is converted into other forms of matter—such as ash, steam, and gases—which are treated as wastes. Since these residues are not recycled and used as energy sources, the earth's supply of fossil fuels is rapidly diminishing. In turn, increased energy is required to retrieve new fossil fuels and to provide more power as industrialism spreads. In this way, the earth's fossil fuels and other natural resources are decreasing, while the demand for energy to locate, mine, and process them is increasing.

Kinds of Pollution

The immediate result of industrialized society's need for energy is **pollution**, the unfavorable alteration of the physical environment. Pollution is the product of mankind's social and technological systems—the result of goods produced, used, and thrown away. It grows as a consequence of increased population and the desire for a higher standard of living by economically wealthy nations.

Although pollution of any kind is undesirable, pollution of the atmosphere is the most harmful to everything on earth. Hundreds of millions of tons of pollutants are vented into the atmosphere every year by automobiles, airplanes, and industries. Air pollution is believed to be a primary factor in such diseases as lung cancer, emphysema, and bronchitis. Indeed, scientists have estimated that individuals who spend a day in the streets of New York City inhale the same amount of pollutants as they would by smoking almost two packs of cigarettes. Moreover, air pollution causes billions of dollars of damage to crops, clothing, and buildings every year.

The world's rivers and oceans have also become polluted as large quantities of oil, pesticides, herbicides, metals, detergents, and other materials are dumped into them. This degradation, combined with overfishing, has led to the disappearance of many marine species. In addition, pollution of the ocean may be killing the algae which produce over one-half of the earth's oxygen, seriously endangering future life on earth. The earth's waters are also being polluted thermally—that is, by the addition of heat. Hot water discharged by industries

into streams seriously injures or destroys plants and animals dwelling in the water and along the stream bed.

Substances used in agriculture, such as fertilizers and pesticides, have placed an additional burden on the environment. Nitrogen and phosphate fertilizers interfere with the natural cycle of the soil by destroying the bacteria in it. They also seep into waterways where they create demands on the oxygen supply. Pesticides, such as DDT, are poisonous to all life; most of them are forms of chlorinated hydrocarbons whose complex molecular bonds degrade very slowly, resulting in a large chemical residue in the environment. Quantities of fertilizers and pesticides remain in the foods we eat and have harmful effects on cells and body tissues. It has been estimated that the total amount of DDT in the atmosphere may be about one billion pounds, and each person in the United States may hold about one-tenth of a gram of this poison in his fatty tissues.

Noise, which can be defined as unwanted or disordered sound, can also be a form of pollution. Noise pollution can cause hearing loss and lead to hypertension and psychological disorders; it may also be a factor in heart disease and stomach ulcers. The World Health Organization has estimated that the noises produced by industries alone lead to absenteeism, accidents, and compensation claims—all of which cost American industry about $4 billion a year. Moreover, the noise levels in the business district of the typical American city are equal to those produced by industry. Noise from jet aircraft has shattered windows, caused rockslides, and undoubtedly affected the hearing of men and animals alike. It is not surprising, then, that some 20 million Americans suffer from some form of temporary or permanent hearing loss.

Light, an unexpected source of pollution, is produced by bright street lights and advertising signs in urban areas. Light waves, which are scattered by the dust and particles in polluted air, prevent astronomers from studying stars by blocking out their light.

Radioactive pollutants, in the form of iodine-131, strontium-90, cesium-137, and krypton-85, have been discharged into the atmosphere as á result of the testing of nuclear weapons. Moreover, the disposal of radioactive wastes from nuclear power plants has become a problem. The two methods currently utilized—burial in deep mines and burial at sea—have generated widespread controversy because of the possibility of eventual leakage and because the long lives of the materials may make them a threat to future generations. Other sources of radiation include medical X-rays for diagnostic and therapeutic purposes; what were once thought to be safe doses of these can cause genetic damage.

Finally, solid wastes produced by civilization are a major source of pollution. Every day, tons of paper, food, glass, metals, wood, plastic, rubber, cloth, and other materials must be disposed of by municipalities around the world. This is

Industrialized societies require increasing amounts of energy and produce increasing amounts of pollution.

done by incinerating the garbage, by using it as landfill, or by dumping it at sea. Some cities have devised means of separating their garbage and recycling the materials. Others burn the waste materials at high temperatures in an airless container and then use the gases produced by this process as fuel to heat and air-condition buildings.

What Technology is Doing

Scientists and engineers have attempted to deal with some of the problems created by industrialized nations. They have created devices which control the gases emitted by autos and factories. They have discovered new ways to obtain power which do not deplete fossil fuels, and they have found ways to limit population growth by birth control. In addition, modern technology has attempted to deal systematically with one of our most vital problems—the treatment and disposal of sewage.

Sewage Treatment

Sewage, or wastewater, includes the wastes carried by water from residences and industrial plants as well as groundwater, surface water, and water from storm drains. It may contain human wastes, food wastes, detergents, and industrial chemicals. In the United States, each person uses an average of about 150 gallons of water every day for drinking, bathing, and washing away wastes. Of this, 100 gallons end up as sewage water. Thus, in a city of 1 million inhabitants, sewage water averages about 100 million gallons a day.

One example of attempts to solve the problems of pollution through technology is the design of sewage treatment facilities.

By any chemical or geological standards this is a considerable amount of water, and the chances of its polluting fresh water streams and rivers is quite high. For example, some 1400 towns and cities in the United States discharge their sewage into streams without treating it. In biochemical terms, many other towns process wastewater inadequately; and as many as 300,000 industrial plants dump toxic complex wastes into the sewage systems of cities that do not possess the facilities to treat the wastes adequately. A dramatic example of this ecological failure occurred in Cleveland in 1969. During the summer of that year, the Cuyahoga River caught fire because it had become clogged with flammable oils and other chemical wastes. Naturally, there were no fish and few microorganisms in this flaming sewer because the river had a very high **biochemical oxygen demand (BOD)**, which would cause the fish to smother. The BOD is the amount of oxygen required by microorganisms to decompose organic wastes in the water.

In recent years, good progress has been made in sewage treatment, which basically severs the carbon-carbon and carbon-hydrogen bonds in organic wastes, breaking the more complex molecules down into carbon dioxide and water. Sewage processing is divided into three stages: primary or mechanical treatment; secondary or biochemical treatment; and tertiary or chemical treatment.

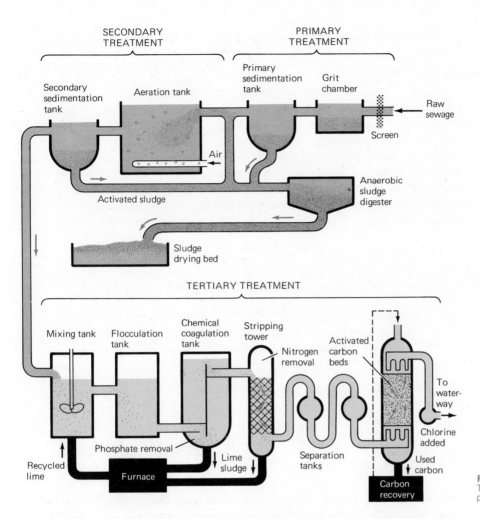

Figure 21-8
The most thorough sewage treatment process involves three stages, shown here.

PRIMARY TREATMENT In the primary stage, wastewaters are transported to a processing center where they are mechanically screened. Large objects, such as rocks, bottles, and cans are screened out, permitting the water containing the finer matter to enter a grit tank where materials such as sand and gravel settle to the bottom. The remaining suspended organic material continues to a third tank, or sedimentation tank, where it is allowed to settle, forming a thick viscous sludge. Primary treatment removes about 60 percent of the suspended solids, 30 percent of the BOD, 20 percent of the nitrogen, and 10 percent of the phosphorus compounds.

SECONDARY TREATMENT About 70 percent of American cities carry their sewage processing a step further to secondary treatment. This stage uses aerobic microorganisms to decompose organic wastes. (Aerobic microorganisms require oxygen to live; anaerobic organisms do not.) The basic idea behind secondary treatment is to remove the fine particles of solid wastes that are in colloidal suspension.

One type of secondary treatment is the activated sludge process. Sewage is forced into an aeration tank where air is introduced, causing the water to bubble. At the same time, large quantities of aerobic microorganisms are pumped into the tank, and they too are suspended with the colloidal organic materials. After four to eight hours, the microorganisms and the colloidal wastes coagulate to form a substance known as biological floc. The suspended colloidal wastes contain carbohydrates, proteins, and fats, plus various inorganic materials. In a typical reaction, the microbes act on carbohydrates to produce pyruvic acid, which is then oxidized, as in the following reactions:

$$C_6H_{12}O_6 \xrightarrow{\text{bacteria}} 2CH_3\overset{\overset{\displaystyle O}{\|}}{C}COOH + 4H$$

$$2CH_3\overset{\overset{\displaystyle O}{\|}}{C}COOH + 4H + 6O_2 \xrightarrow{\text{bacteria}} 6CO_2 + 6H_2O$$

The amino acids in protein are also broken down:

$$R-\overset{\overset{\displaystyle NH_2}{|}}{\underset{\underset{\displaystyle H}{|}}{C}}-COOH + O_2 \xrightarrow{\text{bacteria}} R-COOH + CO_2 + NH_3$$

The floc resulting from these reactions is pumped into a sedimention tank where it settles, leaving a further clarified liquid, called effluent, which may then be emptied into streams. A large part of the biological floc is recycled to the aeration tank so it can process new sewage entering there.

Secondary treatment removes about 95 percent of the BOD, 90 percent of the suspended organic matter, 50 percent of the nitrogen, and 30 percent of the phosphorus that remains after primary treatment. Many other inorganic materials are not removed by secondary treatment; these include salts, insecticides, acids, and other toxic pollutants The treatment also converts the nitrogen and phosphorus into minerals. These minerals are used as nutrients by algae in waterways, which causes the algae to undergo a terrific population explosion. Decomposition of the large algae population creates more BOD in the stream, removing oxygen needed by other organisms. This runaway process, called eutrophication, has led to the destruction of a number of large ecosystems, including all of Lake Erie.

TERTIARY TREATMENT A few communities carry sewage treatment a step further to tertiary treatment. The first phase is the process of coagulation, which uses chemical flocs to agglomerate (cause to group together) the organic colloids remaining after secondary treatment. Because colloids carry negative charges, positive salts of metals such as aluminum or iron are introduced to coagulate them. Aluminum sulfate is added to the sewage, the materials are agitated, and the following reaction occurs:

$$Al_2(SO_4)_3 + 3Ca(HCO_3)_2 \longrightarrow 2Al(OH)_3 + 3CaSO_4 + 6CO_2$$

During this reaction the aluminum sulfate reduces the normal repulsive electrostatic forces that exist between colloidal particles and the aluminum salts. The electrostatic layer surrounding each negatively charged colloidal particle is weakened as a positively charged aluminum ion becomes adsorbed on its surface. Adsorption is the process by which the ions adhere *only* to the surface of the solid colloidal particles. Eventually, the layer weakens to the point where van der Waals forces between the two become 1000 times stronger than the normal electrostatic forces separating the two; at this point, coagulation occurs. As the colloidal particles combine with one another and increase in size, chemical flocs are formed. The floc is then pumped to a sedimentation tank.

The second step in the tertiary treatment is filtration. During this phase, the liquid is pumped through beds of granulated activated carbon. Approximately 90 percent of the phosphate is trapped within the carbon, which can be treated and used again. The liquid is then disinfected with chlorine and released into waterways.

Figure 21-9
This geothermal-electric plant in Wairakei, New Zealand, harnesses natural steam from geysers and converts it into electrical power.

Sludge Disposal   A critical problem after sewage treatment is sludge disposal, which can be more expensive than sewage treatment itself. Sludge accumulates in enormous quantities—as much as 35,000 tons per day in a city of 1 million inhabitants. Some cities dry and then incinerate their sludge, thus increasing air pollution. Others dump it 30 or 40 miles out at sea. Some communities deposit sludge on the land in either open dumps or sanitary landfills. An open dump is an isolated area where garbage is dumped and left uncovered. Sanitary landfilling is a bit healthier; sludge is dumped into deep trenches in designated areas and covered with earth.

Another method is composting, in which the organic materials in the sludge are decomposed by aerobic microorganisms for 5 to 10 weeks after being deposited in beds. The dried compost, which contains nitrogen, phosphoric acid, and potash, is used as a fertilizer or as a soil strengthener.

One of the most efficient methods of sludge reduction is anaerobic digestion. In this process, the sludge is pumped into an airless reservoir containing anaerobic microbes which reduce it to methane and other gases. Proteins in the sludge are reduced to amino acids, polysaccharides are degraded to sugars and then to pyruvic acid, and the fats are broken down to glycerol and fatty acids. The microbes then act on the acids to produce the gases methane, hydrogen, carbon dioxide, and hydrogen sulfide. The methane gas can be recycled to serve as fuel to heat the reservoir and the small amount of sludge which has settled in the tank can be used as soil builder.

Recycling Solid Wastes   Recently, scientists have rediscovered the fact that no waste accumulates in nature and all materials are perpetually used and reused in natural processes. Following the model of nature, a few American communities have devised ways of converting today's junk into tomorrow's resources.

In Franklin, Ohio, for example, a recycling plant accepts 50 tons of unsorted household garbage daily and turns it into reusable paper, glass, and metals. First, the metals are magnetically removed and sent to a nearby steel company. Next, the glass, dirt, and sand are separated from the mixture by centrifuge. The remainder of the garbage, consisting mostly of paper, is converted into paper pulp, which is then reused.

The city of Baltimore is currently building a pyrolysis system to convert its garbage into fuel. Pyrolysis is the process by which solid wastes are incinerated at very high temperatures in an airless container. The gases so produced are purified, then fed into a boiler where they are converted to steam. As a result of this process, five million pounds of steam will be produced every day; the steam will be used to help heat and air-condition buildings in Baltimore's business district. These plants are outstanding examples of the intelligent recycling of wastes.

An **abiotic component** is a nonliving part of the environment.

An **aerobic microorganism** is a microscopic organism which requires oxygen to live. An **anaerobic** microorganism can live without oxygen.

An **autotroph** is an organism that produces its own food. Most green plants, which produce food by photosynthesis, are autotrophs.

Biochemical oxygen demand (BOD) is the amount of oxygen required by microorganisms to decompose organic wastes in water.

The **biogeochemical cycle** is the circular path taken by materials through the biological, geological, and chemical components of the environment.

A **biotic component** is a living part of the environment.

Ecology is that branch of physical science that studies the interactions among organisms and their total environment.

A **heterotroph** is an organism that consumes producers. Herbivores, carnivores, omnivores, and decomposers are heterotrophs.

Photosynthesis is the process by which the chlorophyll cells of green plants use solar energy to convert carbon dioxide and water into carbohydrates and release oxygen.

Respiration is the process by which an organism takes in oxygen, uses it to break down fuel molecules for use as energy, and releases carbon dioxide.

A **system** is two or more bodies united by interaction so that a change in one of the bodies causes a change in the others. An open system exchanges matter and energy with its environment. A closed system is one in which matter and energy are conserved, exchanged, and stored by and among the parts of the system.

A **trophic level** is a link in the food chain of an ecosystem.

1   Why can it be said that energy is transferred but matter is recycled in the ecosystem?
2   What is a trophic level? What determines the trophic level of an organism? Explain how the size of the total population of a given trophic level is related to that level's energy potential.
3   Draw labeled diagrams of the hydrologic cycle, the nitrogen cycle, and the carbon cycle. Indicate places in each of these cycles where human beings have interfered with them.
4   The cultivation of leguminous plants increases the fertility of the soil. Why?
5   Most industrial wastes ultimately end up on the ocean floor. What effects do these wastes have on the ecosystem that exists there?
6   Assume that a nuclear power plant, a sewage disposal plant, or a garbage landfill has been proposed for your neighborhood. Devise a checklist of points to use in the defense of one such project. Devise a checklist of items to oppose the same project.
7   You are a living organism and have a position in the ecosystem. Classify yourself on the basis of food habits, energy transfer, biogeochemical cycling, and integration into the ecosystem in which you live.

Eight

Beyond the Earth

The science of astronomy is thousands of years old—probably older than recorded history. Certainly the earliest records show man's interest in and knowledge of the activity of the sun, moon, and stars. By the time of the Babylonians, a detailed history of hundreds of years of observation was available to students of astronomy. The flaw in these early observations was that they were not systematic. There was a great deal of data about the moon, about the sun, and about the brightest of the constellations; but these were just bits of separate data. It took hundreds of years until these data were assembled within the framework of a system, and then it took hundreds more years until everyone agreed about what that system was.

Our planet earth, the sun, the moon, and the other planets that we sometimes see looking like stars in the morning or evening sky, are indeed part of a system, and that system can be analyzed in exactly the same way that we analyze a system of reacting chemicals, or a system of moving mechanical parts. Our solar system—an open system, of course—has fixed components and processes; its inputs and outputs can be identified; an energy budget can be drawn up to describe the exchanges that take place within it.

The Sun

The celestial body that affects us most is our own star, the sun. It is no wonder that many civilizations have made the sun an object of worship, for our existence depends upon it, no less today than in earlier ages. Without the sun's great gravitational force to keep the earth in its orbit, our planet would drift forever through the dark cold reaches of interstellar space. Not only does the sun give us the light and warmth that sustain life, but it is the ultimate source of almost every form of energy we use on earth. The sun also directly affects the climate and weather of our globe, and thus, indirectly, its geology.

Solar Energy

Even at the enormous distance of 93 million miles, a square meter of our planet's surface perpendicular to the sun's rays receives nearly 20,000 calories of radiant energy per minute from its parent star. If the sunlight that strikes the roof of a typical house in the temperate latitudes were converted into electricity, it would provide more than enough energy to make that house self-sufficient in its electrical needs. (Unfortunately the technology to do this efficiently still seems a decade or two in the future.) From these figures, scientists have calculated that the sun's total energy output is an astounding 3.8×10^{33} ergs/sec, or 5×10^{23} horsepower. Explaining the mechanism that produces this immense outpouring of energy was one of the great achievements of modern physics.

Each second, the sun produces more energy than the human race could use in 5 million years at its present rate of consumption.

It was natural for the first scientists who considered this problem to suppose that the sun burned some sort of fuel, like a gigantic bonfire. But chemical combustion is much too unproductive to be a potential source of the sun's energy; even if the entire solar mass were composed of the most energy-rich chemical fuels, all would be consumed in a few thousand years. Yet we know that the sun and the earth are several billion years old, and geologists have evidence that the sun's output of radiation has not varied appreciably for much of that time.

Another theory, popular in the nineteenth century, proposed that the sun was contracting slowly under the influence of its enormous gravity. This process would heat the solar mass, evolving enough energy to account for the observed luminosity of the sun. An advantage of this theory was that it did not require the expenditure of any matter to produce the sun's energy, merely the conversion of gravitational potential energy to heat. But it had a serious defect as well. Calculations indicated that this process of contraction could not have been taking place for more than 100 million years, which could thus be considered the maximum age for the sun. Furthermore, tracing this contraction backward in time, it could be shown that the sun must have been large enough to encompass the earth's orbit a mere 20 million years ago. Thus the earth itself could not be more than 20 million years old. This was incompatible with the findings of geologists and paleontologists, who by the middle of the nineteenth century knew that the earth was far older.

NUCLEAR FUSION In the twentieth century physicists discovered a much more efficient mechanism than chemical combustion for converting mass into energy: nuclear reactions. Such reactions do not involve changing one molecule into another, as in chemical reactions, but transforming one element into another. There are two kinds of nuclear reactions. **Fission**, which occurs in the atomic bomb and in nuclear power reactors, is the splitting of a heavy atomic nucleus to create two or more lighter atoms. **Fusion** is the combination of light atomic nuclei to form a heavier element. Heavy elements like uranium and plutonium, the fuels used in fission reactions, are scarce on the sun. But the sun is mostly hydrogen, the fuel from which the hydrogen bomb derives its awesome power. Could the sun be fusing hydrogen to produce its energy?

Scientists studying this problem in the 1920s and 1930s knew that fusion could not take place in the outer layers of the sun. The surface temperature of the sun is only about 6000°K, and hydrogen fusion requires temperatures on the order of 10 million degrees K. Calculations revealed that conditions near the sun's center, however, where the temperature rises to 15 million degrees K and the pressure to a billion atmospheres, would make fusion possible, and indeed inevitable.

The sun's surface temperature is about 6000°K, but at its center the temperature may reach 15,000,000°K.

Figure 22-1
The proton-proton chain, a series of nuclear fusion reactions responsible for most of the sun's energy. The positron produced in step one collides with free electrons in the solar plasma; the resulting mutual annihilation releases more gamma radiation. In hotter stars a different series of reactions, in which carbon serves as a catalyst, also occurs.

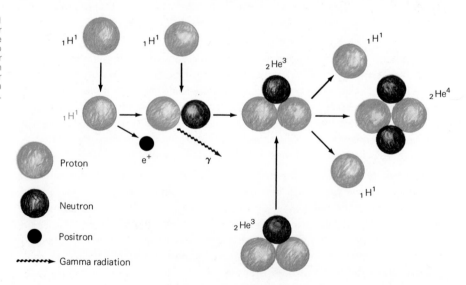

Proton

Neutron

Positron

Gamma radiation

Most of the sun's energy is produced by the fusion of hydrogen into helium through the three step process shown in Figure 22-1. This series of reactions, worked out in 1939 by the German–American physicist Hans Bethe, is known as the proton-proton chain. The four hydrogen nuclei that enter the reactions have a combined mass of 4.0325 atomic mass units, and the resulting helium nucleus has a mass of 4.0039. The mass lost, slightly more than 0.7 percent, is released as energy.

Nuclear fusion is almost ten million times more productive than chemical combustion. Is this enough to account for the tremendous energy production of the sun over billions of years? Using the formula $E = mc^2$, it can be shown that to produce its 3.8×10^{33} ergs/sec, the sun must "burn" 600 million tons of hydrogen each second, of which over 4 million tons will be lost in the form of energy. Though these figures seem huge at first glance, they are really very small compared with the sun's total mass of 2×10^{30} kg. The sun could continue to consume hydrogen at its present rate for nearly 10 billion years before running low on fuel. And the seemingly prodigious loss of over 4 million tons of mass per second is quite insignificant, less than a ten-trillionth of its entire mass per year—not enough to have made any measurable difference in the sun's gravitational force, and hence in the earth's orbit, since the birth of the solar system some 4.6 billion years ago.

SOLAR RADIATION Most of the energy released by hydrogen fusion in the sun's interior takes the form of x-rays and gamma rays—electromagnetic radiation of extremely short wavelength, and thus of high energy. If the sun actually radiated its energy into space in this form, life as it has evolved on earth would be impossible, since these wavelengths are lethal in large quantities. But as the high-energy photons travel outward from the inner regions of the sun, they undergo a series of transformations. Again and again they are absorbed by atoms in the sun and re-emitted, often as two or more photons of lower energy and longer wavelength. This process occurs so many times that energy from the sun's core takes about a million years to reach the solar surface.

By the time it is sent out into space from the sun's surface, much of the original short-wavelength radiation has been transformed into visible light. A somewhat larger part is radiated at the even longer wavelengths of the infrared, which we experience as heat when they strike our skin; a small amount falls within the ultraviolet region of the spectrum, with wavelengths slightly shorter than those of visible light. By spectroscopic analysis of the sun's light, scientists have identified some 70 elements in the sun. In addition, some of the stabler molecules and fragments of molecules have been found in the cooler regions of the solar surface. Though elements other than hydrogen and helium constitute only one or two percent of the sun's mass, they seem to be present in roughly the same proportions to each other as we find in the earth, which suggests that the sun and earth had a common ancestry.

Table 22-1
Properties of the Sun

Quantity	Sun	Earth	Sun:Earth
Mean distance from earth	9.3×10^7 mi		
Diameter	8.64×10^5 mi	7.9×10^3 mi	109:1
Volume	3.4×10^{17} mi³	2.6×10^{11} mi³	1,300,000:1
Mass	2×10^{30} kg	6×10^{24} kg	333,000:1
Acceleration of gravity (at surface)	900 ft/sec²	32.2 ft/sec²	28:1
Density	~150 g/cm³ (core)		
	~10^{-7} g/cm³ (photosphere)		
	1.4 g/cm³ (average)	5.5 g/cm³ (average)	.26:1
Pressure	~10^9 atm (core)		
	~.05 atm (photosphere)		
Temperature	~1.5×10^7 deg K (core)	3500°K	4,300:1
	~6000°K (surface)	286°K	21:1
Solar constant (energy received by earth)	1.36×10^6 ergs/sec-cm²		
Luminosity	3.8×10^{33} ergs/sec		
	5×10^{23} hp		
Period of rotation	~25 days (equator)	24 hr	25:1
	~30–35 days (poles)		

Physical Properties and Structure

Table 22-1 lists some of the sun's vital statistics. Translating its size and mass into earth units may enable us to grasp its immensity more clearly. The diameter of the sun is about 109 times that of the earth. Since the volume of a sphere is proportional to the cube of its radius, it would take 1.3 million earths to equal the sun's volume. The sun's average density is only about one-quarter that of the earth, so its total mass is about a third of a million times the earth's. How much would a person weigh if he could somehow survive on the sun's surface? Since the force of gravity is directly proportional to the mass of the body (333,000 times that of the earth) and inversely proportional to the square of the distance from the center of the body (109 times the radius of the earth), we can easily calculate that his weight would be increased by a factor of $333,000/109^2 = 28.5$. A 150 lb man would weigh 4200 lbs on the sun.

The sun is generally divided into four principal regions, with no distinct boundaries between them. When we look at the sun's disc in the sky, we see the **photosphere**—the brilliant outer skin of the sun. Surrounding it is the sun's atmosphere, a much more diffuse and less luminous envelope of gas extending far out into space. The inner layer of this atmosphere is known as the **chromosphere** and the rest constitutes the **corona**. These three regions comprise only a very tiny fraction of the sun's mass. All the rest lies within the layer of the photosphere, in the sun's **interior**.

THE INTERIOR The solar interior is itself far from uniform. At the center of the sun, the enormous weight of the overlying layers compresses the solar gases until they reach densities on the order of 150 g/cm³—nearly 14 times denser than lead. But the temperature, pressure, and density fall off rapidly outside the interior. In the photosphere, the temperature has dropped to about 6000°K, pressure has decreased to a fraction of an atmosphere, and the density is only a ten-thousandth of the density of air at sea level on earth. By terrestrial standards, the surface of the sun is almost a vacuum!

Because enormously high temperatures are required for nuclear fusion, almost all the sun's energy production takes place in a small central core with a radius roughly 20 percent that of the photosphere, and containing about one-third of the solar mass. This region, where the conversion of hydrogen to helium has been in progress for some 4½ billion years, is by now much richer in helium and poorer in hydrogen than the rest of the sun. Although the sun as a whole is roughly 74 percent hydrogen and 24 percent helium, in the core the proportions may be closer to 38 percent hydrogen and 60 percent helium.

The sun is roughly three-quarters hydrogen, but in its core, where nuclear fusion takes place, much of the hydrogen has been transformed into helium.

THE PHOTOSPHERE   The photosphere is a mere shell, perhaps 200 miles deep, yet it is this region, less than 0.4 percent of the solar radius, that is the immediate source of all the sun's radiation into space. This is due chiefly to the presence of *negative* hydrogen ions, which have temporarily captured an extra electron in the relatively cool region of the photosphere. Though their concentration is low, these ions are exceptionally efficient absorbers and radiators of energy, capturing virtually all light from the deeper regions of the sun and reradiating it out into space. Thus the long process of absorption and reradiation of photons from the sun's nuclear furnace ends in the photosphere.

High-energy radiation from the sun's core is largely transformed into visible and infrared light by the time it finally reaches the photosphere, the sun's thin surface layer.

THE CHROMOSPHERE   In the chromosphere, a cooler, thinner region of gas extending roughly two thousand miles above the photosphere, the temperature drops from 6000° to about 4000°. The relatively cool gases of the upper photosphere and chromosphere give rise to the dark absorption lines in the sun's spectrum. At 4000° the gases of the chromosphere are still hot enough to be luminous, but under ordinary circumstances this luminosity is lost in the far brighter glare of the solar photosphere. During an eclipse, when the photosphere is obstructed by the moon's shadow, the chromosphere briefly becomes visible at the sun's edge, emitting the beautiful red color characteristic of hydrogen.

Figure 22-2
The solar corona, photographed during an eclipse.The corona appears largest at times of maximum sunspot activity.

THE CORONA   During a total eclipse, the corona also becomes visible as an irregular halo extending out several solar radii. The gas of the corona is extremely diffuse but very hot, perhaps a million degrees. This temperature is great enough to produce a high degree of ionization of the constituent atoms. Why the corona should be so much hotter than the photosphere and chromosphere has not been fully explained. Astronomers think that shock waves—mechanical waves similar to the "sonic boom" of jet planes—from the turbulent surface of the sun may be a factor.

In an extremely attenuated form the corona may even be said to extend out beyond the orbit of the earth, for the density of gas in the space between the planets is about 100 to 1000 atoms per in^3, compared with 10 atoms per in^3 in interstellar space. In fact, the corona is not static, but can more accurately be thought of as a continuous flow of gas from the sun. This **solar wind** blows past the earth, and can be detected by space probes. The moon, unprotected by an atmosphere, shows the effect of such bombardment, which has darkened the rocks on the lunar surface.

Solar Phenomena When Galileo first turned his telescope toward the sun, he noticed that the solar disc at times exhibited dark spots. These spots appeared to move around the sun, which prompted Galileo to speculate, correctly, that the entire sun rotates. We now know that the surface of the sun, at the equator, has a rotation period of about 25 days. This figure changes, however, both with latitude (regions closer to the sun's poles take more than 30 days) and with depth (different layers of the atmosphere and interior seem to vary in their velocity of rotation). Thus we know that the sun cannot be a rigid body. The spots observed by Galileo were only the first of many intriguing phenomena of the solar surface and atmosphere discovered by scientists.

SUNSPOTS Sunspots have been intensively studied since Galileo's time. Their apparent darkness is due to their relatively low temperature, about 1500°K cooler than the surrounding regions. Seen in isolation, a sunspot would shine brilliantly, but the contrast with the photosphere, five times brighter still, is enough to make them seem dark. Sunspots vary greatly in lifespan; some appear and disappear in a matter of hours or days, whereas others persist for a month or more. They average several thousand miles in diameter, with occasional giants reaching 100,000 miles.

Sunspots do not appear in constant numbers from year to year, but vary instead in a puzzling but regular pattern. Some years the sun exhibits a mere handful of spots, and may be entirely free of them for months at a time, whereas in other years the total reaches 50, 100, or even 150. (A record number, nearly 200, were

Figure 22-3
A large sunspot group.

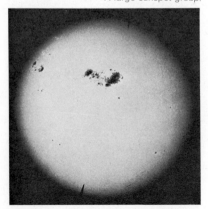
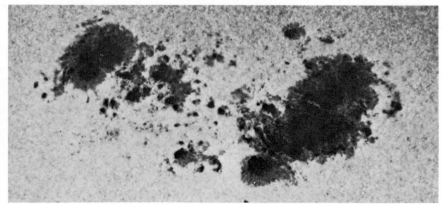

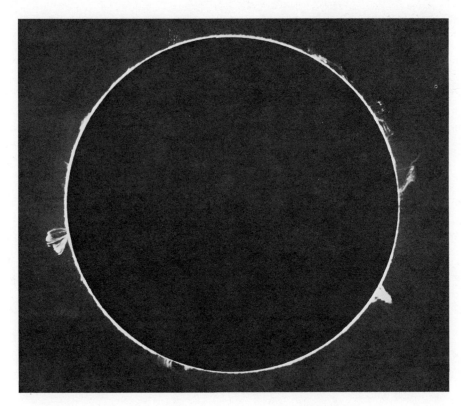

Figure 22-4
Prominences, great tongues of luminous gas
in the sun's atmosphere.

observed in both 1957 and 1958.) These fluctuations follow an unexplained 11-year cycle, with periods of maximum and minimum sunspot activity alternating at intervals of roughly 5½ years. (The next peak can be expected about 1979–1980.)

Sunspots often appear in pairs of opposite magnetic polarity, aligned in an east—west direction on the sun's face, with the lines of force seeming to emerge at one spot and re-enter the sun at the other. This fact is a major clue to the nature of sunspots, which are thought to originate as local disturbances of the solar magnetic field. It is likely that the sun's complex pattern of rotation, with different layers turning at different speeds, contributes to these disruptions.

Sunspots seem to be magnetic disturbances on the solar surface; their number rise and fall in an 11-year cycle.

SOLAR FLARES Sunspots are closely associated with several other unusual solar phenomena. Among these are **prominences**, great tongues or arches of luminous gas that form in the corona, tens of thousands of miles above the photosphere, and **solar flares**, violent explosions on the sun's surface. Both occur much more frequently when the number of sunspots is great, and can often be found near conspicuous sunspot groups.

Flares, which may last for over an hour, are often remarkably bright, emitting unusually large amounts of radiation in the visible, ultraviolet, and x-ray portions of the spectrum. Frequently they hurl material out from the sun as well. Some of this matter may condense in the corona, forming a prominence. More energetic particles, however, are blown out into space, creating an unusually powerful gust or even a gale in the solar wind. Such particles, along with high-energy photons of x-ray wavelength, often reach the earth, affecting the magnetic field and radiation belts that surround our planet. They also disturb the ionosphere, interfering with short-wave transmission and causing spectacular auroral displays.

AURORAS Probably the most beautiful atmospheric phenomenon is the spectacular light show called the aurora. The dark stillness of the night sky is dramatically interrupted by the intrusion of rapidly changing rivers of color, red and green, that appear and disappear, advance and retreat, in intricate, shifting patterns. These lovely displays are usually seen only at very high latitudes; they are common in northern Canada but are relatively rare in the United States. In the northern hemisphere they are called **aurora borealis**, in the south **aurora australis**.

Auroras are caused by high-speed protons and electrons ejected during periods of intense solar activity. As they approach the earth they are deflected by the planet's magnetic field, which extends far out into space. Most particles penetrate through to the earth's atmosphere only in regions near the north and south magnetic poles, where the field is relatively weak. This accounts for the characteristic location of auroral displays.

The interaction of these particles with atoms of the air is similar to the light-producing mechanism in neon signs. The oxygen and nitrogen atoms in the ionosphere, 50 to 100 miles above the earth, are excited to higher energy levels or even ionized by collisions with the fast incoming solar particles. When they return to their normal states they release their extra energy as light of certain characteristic wavelengths, red and blue for nitrogen, red and green for oxygen.

Protons and electrons in the solar wind excite atoms of gas in the earth's upper atmosphere, causing auroras.

The Moon The moon's impact on the consciousness of mankind can be measured by its importance in mythology, where it plays a role second only to the sun. The moon affects many earthly phenomena: the tides, the life rhythms of animals, perhaps even human behavior. Its closeness has often prompted speculation about its size, distance, surface features, and even its possible inhabitability; stories of voyages to the moon abound in the literature of many peoples and eras. Today, thanks to the manned lunar space flights of recent years, we at last have first-hand information about the nature of the moon.

Natural Cycles and Biological Clocks

The temporal cycles that most affect the lives of organisms on earth are clearly defined by astronomical events. The rising and setting of the sun creates a regular pattern of light and dark. The sun's annual journey through the zodiac gives rise to seasons of warmth and cold, rain and drought, growth and decay in various parts of the world. The close relation between our notions of time and astronomical cycles is reflected in the very units we employ to measure time: a day is the 24-hour period during which the earth completes one rotation on its axis; a month is about the time from one new moon to the next; and a year is the time it takes for the earth to complete one full revolution around the sun.

The vital rhythms of living creatures—waking and sleeping, hunting and hiding, mating and nesting—are geared in a variety of ways to these astronomical cycles. Recently it has been discovered that organisms seem to have built-in "clocks," senses of time that are independent of, or only partially dependent on, external stimuli. These biological clocks regulate the early growth, maturation, reproduction, decay, and death of the organism. In addition to the life cycle of their species, however, many organisms exhibit cyclical behavior in their life functions, indicating that biological clocks may divide a lifetime into much smaller intervals.

The most common cycle found in nature is the 24-hour cycle known as the circadian rhythm. Animals such as butterflies, honeybees, songbirds, lions, and humans are diurnal; that is, they have their period of greatest activity during the day. Cockroaches, mice, owls, racoons, and coyotes are examples of nocturnal animals; they are most active at night. Green plants alternate between a daytime rhythm of food production by photosynthesis and a nighttime rhythm of growth and assimilation. Almost every organism studied has been found to exhibit various 24-hour rhythms.

It has been found that these cycles are not activated simply by changing conditions of light and temperature, however. Organisms seem to have their own built-in rhythms that are maintained even under laboratory conditions of steady illumination and constant temperature. There is considerable evidence to indicate that this timekeeping function is built into our very cells. This makes the accuracy of the clock in a plant like the bamboo all the more astonishing: this remarkable organism has a seed to seed cycle of thirty years, accurate to the month.

Human beings isolated in an environment without windows or clocks exhibit a number of biological rhythms that are roughly geared to the 24-hour period. Most people, for example, tend to fall into a regular waking and sleeping cycle of about 25 hours. Students who study all night, medical interns who work 36-hour shifts, and diplomats who fly across the world into radically different time zones all experience the dysfunction that results when their internal timing is thrown out of syncronization with the time in the outside environment. They need time and rest to adjust their biological clock to the local time.

The distance from the earth to the moon was first calculated centuries ago by the method of parallax (Figure 1-11). The average value is about 238,000 miles; since the orbit is an ellipse, the closest approach is about 15,000 miles less, and the furthest separation about 12,000 miles more. The gravitational effect of other planets is also noticeable, and can make a difference of 5000 miles at times.

The moon makes one complete revolution about its parent planet in 27⅓ days, but since in that time the earth moves a considerable distance around the sun, it takes the moon a total of 29½ days to return to its former phase. Like the earth, the moon also rotates on its axis. Eons ago, however, the force of the earth's gravity slowed the moon's rotation until its period came to equal that of its revolution about the earth. As a result, the moon always shows the same face to observers on earth, so mankind did not get to view the far side of the moon until the first space vehicles were placed in orbit about it. The moon shines only by reflected sunlight, and the **phases** of the moon that we see are the result of its changing position with respect to the sun as it circles the earth.

Figure 22-5
Eclipses of the sun and moon. (A) Note that the earth's shadow is much larger than the moon's; as a result, lunar eclipses last longer than solar eclipses and can be seen from the entire night side of the planet. (B) Because of the inclination of its orbit, the moon does not always pass precisely between the sun and the earth. If it did, there would be an eclipse of the sun every month. (Not to scale.)

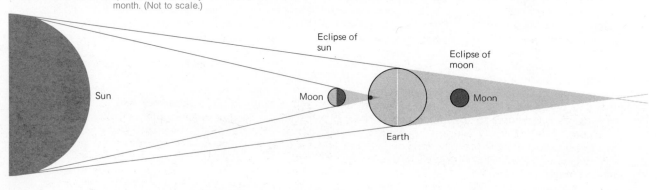

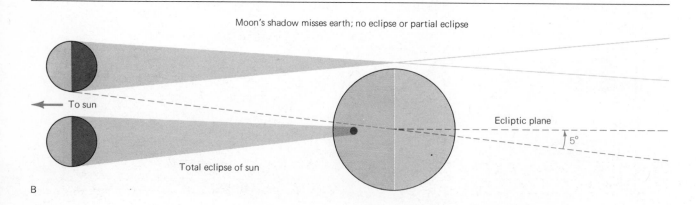

The following description by John Paraskevopoulos (Page and Page, eds., *The Origin of the Solar System,* 1966, pp. 43–4) as he witnessed the eclipse of October 1, 1940, in South Africa, gives us a sense of the emotional impact of a total solar eclipse:

> Only a small slice of the sun remained visible — an extremely thin meniscus (crescent shape). Barely a few seconds before totality, the light of the landscape was very dim, ashy, and queer . . .

> Darkness fell upon us and the stars appeared in the middle of the day. Birds, bewildered by the unexpected darkness, went home to roost. Suddenly, the brilliant corona made its appearance, shining with an uncanny pearly color, and with streamers extending over three million miles beyond the surface of the sun. Two especially fine red-flaming prominences could be seen jutting from the surface of the sun to heights of tens of thousands of miles. . . .

> After three and a half minutes, light came from the western horizon, a dawn in the west. Slowly the sun emerged, light was gradually restored, birds twittered as at dawn, and all animal life started a new day . . .

If the moon passes between earth and sun every month, why don't we see a total solar eclipse each time? The reason is that the moon's orbit is not in exactly the same plane as that of the earth. The two are inclined at an angle of about 5° to each other, which means that the moon will usually pass several thousand miles above or below a line joining the earth and sun. Several times each year the moon is in position to cast its shadow on the earth, but most of these will be **partial eclipses**, with the moon obstructing only a portion of the solar disc. About twice in three years, on the average, the moon will be perfectly positioned to obscure the whole sun.

Then why have you probably never seen a total eclipse? The reason is the size of the moon's shadow. The moon, about $1/400$ the size of the sun, is just about 400 times closer, so that it has the same apparent size in the sky. If it were any smaller or further away, we would never see a total eclipse, for the moon would be too small to cover the sun. The shadow of the moon on the surface of the earth is never more than 165 miles wide. As the earth turns under it, this shadow may touch points along a path up to 12,000 miles long, but no one spot is ever in the full shadow for more than 7 minutes. For any given place on earth, the average frequency of total solar eclipses is once in 360 years.

Total solar eclipses are not rare, but each one is visible from only a small part of the earth's surface.

The lunar eclipse, though less dramatic, is a more familiar sight. Actually there are fewer lunar eclipses than solar each year, but because the earth's shadow is so large, a lunar eclipse may take six hours, with the phase of totality lasting up to an hour and a half. Moreover, such an eclipse can be viewed from more than half the earth. Thus your chance of witnessing an eclipse of the moon is quite good — better than even, in fact, for any given year.

Much of what we know of the moon's nature has come directly from, or been confirmed by, the explorations of astronauts in recent years. Table 22-2 lists some of the moon's basic statistics. Though the mass of the moon is only about $1/81$ that of the earth, it is by far the most massive of the planetary satellites relative to its mother planet. Its density of 3.34 g/cm^3 is comparable to that of rocks in the earth's crust, though considerably less than the overall density of the earth. The lunar surface gravity is about $1/6$ that of earth. This low gravity is insufficient to retain any gases, so the moon has no atmosphere.

The moon's gravity is too weak to retain an atmosphere.

With no atmosphere, there are no clouds and no weather on the moon, nor any of the scattering of sunlight that makes our sky blue; the lunar sky is as black in the daytime (except where the sun glares with undiffused brilliance) as at night. Both day and night skies are filled with a myriad of stars, many too faint to be seen through earth's murky atmosphere. These stars do not twinkle, for twinkling is a result of atmospheric turbulence. Since sound cannot propagate through a vacuum, the moon is also a silent place. Without an atmosphere to serve as insulation, the extremes of temperature on the lunar surface are very great, reaching about 110°C during the two-week long lunar day and −173° at night.

Table 22-2
Properties of the Moon

Property	Value	Moon/Earth
Mean distance	239,000 mi	
Diameter	2160 mi	.272
Mass	7.4 × 10²² kg	.0123
Density	3.34 g/cm³	.605
Surface gravity	5.31 ft/sec²	.165
Period of revolution	27.3 days	.0366
Synodic month*	29.5 days	

* Period from full moon to full moon.

THE LUNAR SURFACE Many of the prominent features of the moon's surface are visible to the naked eye, and astronomers using large telescopes have mapped the hemisphere facing earth in great detail. The **seas**, so named by Galileo because he thought they might be oceans like those on earth, are in reality large, smooth, dry plains, relatively dark in color. There are 14 such regions, comprising about half the lunar surface. Much of the rest is a harsh landscape of jagged, unweathered mountain ranges and rocky, boulder-strewn plateaus. The entire surface, but especially the highlands, is marked with millions of **craters**, ranging in diameter from a few feet to nearly 150 miles—the result of collisions with massive meteorites, unimpeded by an atmosphere. About 2 inches of dust cover the lunar surface, under which there is a layer quite similar to earth's soil—both products of the pulverization of moon rocks by micrometeorites and thermal erosion.

The lunar surface also exhibits great cracks, possibly formed by moonquakes, and domes of seemingly volcanic origin. Such features suggest a history of geological activity, but until the manned landings scientists were uncertain about when this activity took place and what may have caused it. Careful study of rocks brought back by Apollo astronauts now indicates that the moon must have been at least partly molten twice in its history—once as a result of intense meteoric bombardment during or just after its birth, and again some billion years later as a result of heat gradually generated by radioactive rocks in its interior. The radioactive heating apparently initiated a period of intense volcanic activity, lasting more than half a billion years, during which lava from the interior flowed forth onto the moon's surface, filling the great meteor-carved valleys and craters to form smooth basalt seas. All this activity, however, ended

About 3½ billion years ago, the moon underwent a period of intense volcanic activity.

Figure 22-6
The smaller photograph shows the far side of the moon, not visible from earth, photographed by Lunar Orbiter III. The larger photograph shows the lunar landscape, photographed by an Apollo astronaut.

some 3 billion years ago, to judge by radioisotope dating of lunar rocks. The moon, too small to retain very much heat, cooled rapidly, and today is devoid of volcanic activity.

Origin of the Moon

The moon rocks themselves are somewhat similar to those of the earth's crust, though their composition differs slightly in certain respects. One of the chief differences is their absolute dryness; they contain neither free water nor minerals with water molecules in their crystal structures. The oldest rocks found so far are more than 4 billion years old, a figure comparable to the age of the earth. The similarity in age and composition of the earth and moon suggests a common origin for both bodies. Despite these recent discoveries, however, the question of the moon's birth cannot yet be answered with certainty. Four major theories have been proposed to account for it, but none of them has been proven conclusively.

One theory holds that the moon is a fragment of the earth, spun off by centrifugal force at a time when the earth was rotating much faster than it does now.

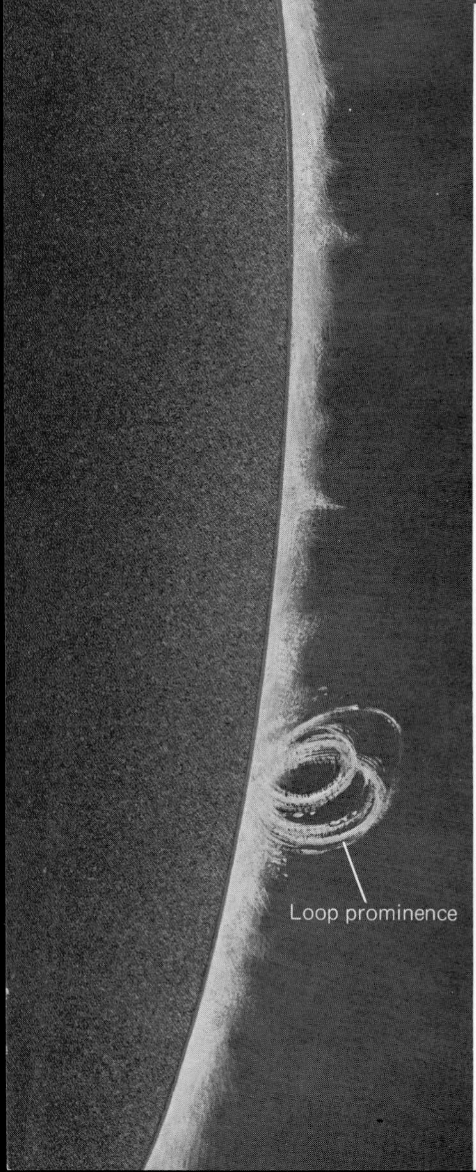

Figure 22-7
The relative sizes of the sun and the planets. (Distances not to scale.)

According to this hypothesis, first proposed in 1898, the Pacific Ocean basin is the scar marking the original location of the lunar mass. It has been calculated, however, that the frictional forces arising from such a process would slow the earth's rotation so much that the moon could never acquire enough angular velocity to break free. An alternative idea suggested more recently is that the earth spun off a ring of gaseous material which later condensed into a solid body to form the moon. But such material would have come from the equator, producing a satellite with an equatorial orbit, whereas our moon's orbit is inclined at a considerable angle to the earth's equatorial plane. And both versions of the spin-off theory are open to objections based on the law of conservation of momentum. Adding the angular momentum of the earth's rotation on its axis and the moon's revolution about the earth, we can find the total angular momentum of the earth–moon system. If the combined mass of earth and moon had once possessed all this angular momentum, it must have been spinning at a rate of one complete rotation every 5 hours. This is only about half the speed necessary to spin off the lunar mass.

A third possibility is that the moon originated as a planetoid elsewhere in the solar system, and was subsequently captured by the earth's gravitational pull. The odds against such a chance encounter and capture are very great, however. The fourth hypothesis is that the earth and moon were born simultaneously, condensing out of neighboring masses of material when the solar system was formed. This theory is very tempting, but it also raises a difficult question: why should the moon's density be so different from the earth's? In the absence of an answer to questions such as this, the problem of the moon's origin must be considered still unsolved.

Table 22-3 provides a brief summary of the data about the nine planets circling our sun. The system also contains at least 32 moons (satellites of the planets); tens of thousands of small planetlike bodies called asteroids; innumerable tiny rocky or metallic fragments known as meteoroids; and an undetermined number of comets, numbering perhaps in the millions or billions. Although the number of bodies is thus very large, most of the system's mass is concentrated in the sun, which has 99.86 percent of it.

All these bodies move in elliptical orbits about the sun. Seven of the planetary orbits are nearly circular, but Mercury and Pluto, the innermost and outermost planets, move in ellipses of considerable eccentricity. Perhaps the most striking aspect of the planetary orbits is that they are all nearly coplanar. (Pluto, with an orbit tilted 17° to the plane of the earth's orbit, is the chief exception.) Because of this, all the apparent motions of the sun and planets that we observe take place in a narrow band of the sky on either side of the **ecliptic**, a line that marks the projection of the orbital plane of the earth onto the heavens.

The 12 constellations along this band, through which the sun seems to move as the earth revolves around it during the year, are called the **zodiac**. (When someone tells you that you were born under the sign of Scorpio, it means that the sun was in that region of the sky on your birthday.) The sun and all the planets also rotate about axes which are in most cases not exactly perpendicular to the plane of the ecliptic. The earth's axis of rotation, for example, is inclined at an angle of 23½°. Here Venus and Uranus are atypical: Venus rotates in a direction opposite to that of the sun, seven of the planets, and most of the satellites; Uranus spins on its side about an axis almost in the ecliptic plane.

The Solar System

Table 22-3
Properties of the Planets

Name	Distance from sun (10⁶ miles)	Distance from sun relative to earth's distance (AU)	Revolution period (earth-years)	Rotation period
Mercury	36	0.39	0.24	59 days
Venus	67	0.72	0.62	243 days
Earth	93	1.00	1.00	23 hr 56 min
Mars	142	1.52	1.9	24 hr 37 min
Jupiter	483	5.20	11.8	9 hr 50 min
Saturn	886	9.54	29.5	10 hr 14 min
Uranus	1780	19.18	84.0	10 hr 49 min
Neptune	2790	30.07	165.0	15 hr
Pluto	3670	39.44	248.4	6.39 days

Mass (earth = 1)	Diameter (miles)	Density (water = 1)	Solar energy (earth = 1)	Number of satellites	Surface gravity (earth = 1)
0.05	3,000	5.4	6.7	0	0.39
0.82	7,600	5.1	1.9	0	0.91
1.00	7,930	5.52	1.00	1	1.00
0.12	4,270	3.97	0.43	2	0.38
317.8	89,000	1.33	0.04	13	2.64
95.2	75,000	0.68	0.01	10	1.13
14.5	30,000	1.48	0.003	5	1.07
17.2	28,000	2.15	0.001	2	1.41
0.1?	4,000?	4?	0.0006	0	?

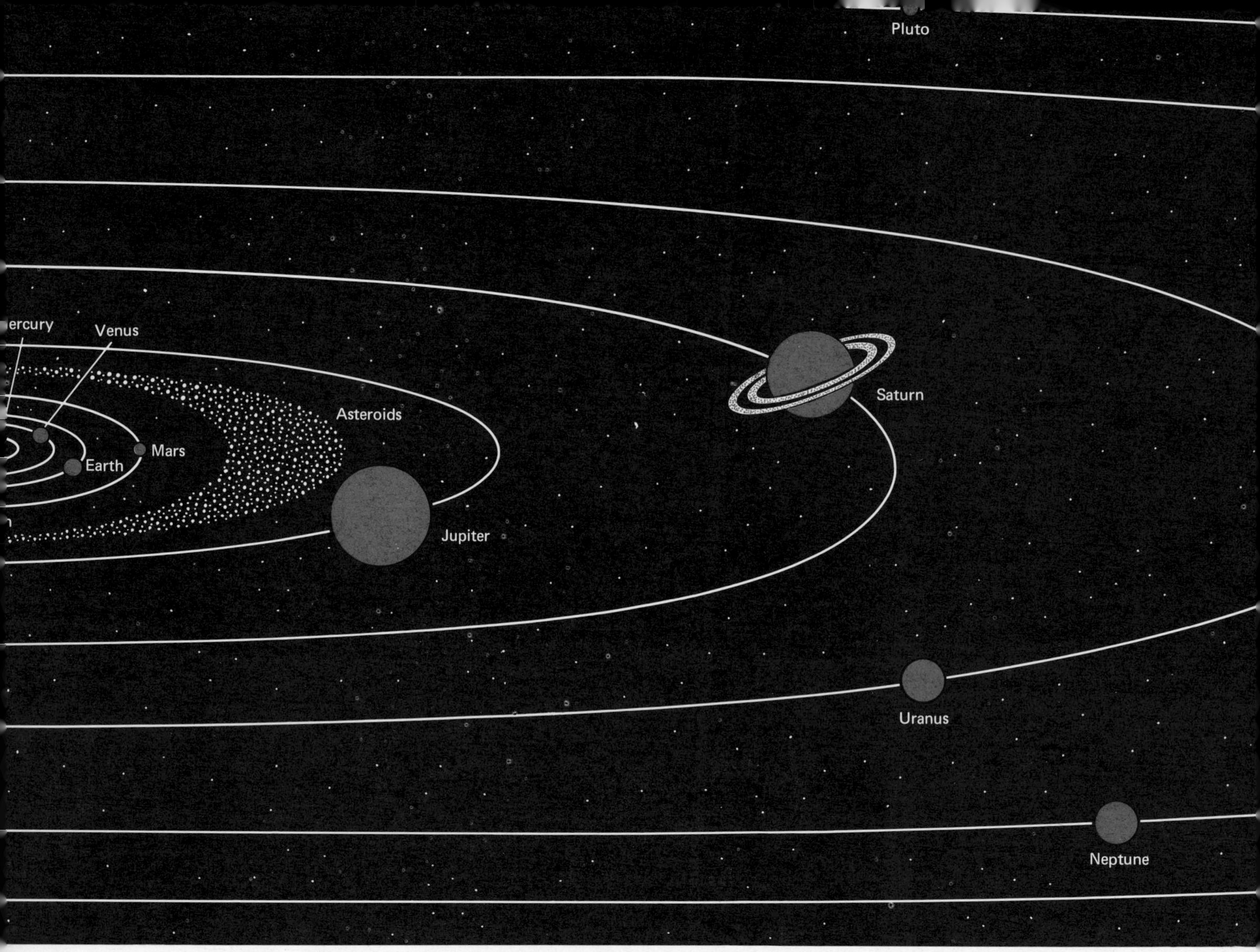

Figure 22-8
The solar system. The orbits of the planets are shown approximately to scale; however, the sizes of the bodies are not to scale.

Since the era of space exploration, our knowledge of the planets has increased dramatically. Spacecraft loaded with instruments have passed close to several of them, radioing back data and photographs to scientists on earth, and one such probe has made a successful soft landing on Venus. These unmanned flights have become our chief source of information about the planets.

THE HOT PLANETS: MERCURY AND VENUS Mercury, the closest planet to the sun, is also the smallest in the solar system, with a diameter of only 3100 miles. It follows a very eccentric orbit, with a nearest approach to the sun of 29 million miles and a furthest distance of 43 million. Because Mercury is so close to the sun, it receives about ten times as much solar radiation as the earth. During the planet's 88-day long "day" the temperature of the sunlit side rises above 400°C (higher than the melting point of lead), while the dark face experiences temperatures of −125°C. The absence of an atmosphere contributes to this extreme temperature range. Pictures radioed back to earth from Mariner 10's close approach to the planet in March, 1974, reveal a barren, rocky surface, pocked with craters, quite similar to the moon's.

Because its orbit lies inside that of earth, Venus, like Mercury, is visible only as a morning star, rising shortly before dawn, or an evening star, setting a few hours after the sun. When it is far enough away from the sun to be seen, it is the brightest of all the heavenly bodies except the moon and sun. This brilliance is the result not only of the planet's closeness to the earth, but also of its very high reflectivity. Seventy-seven percent of the sunlight striking Venus is reflected back into space by its dense mantle of clouds.

Venus's atmosphere traps the sun's heat like a greenhouse.

With a diameter of 7800 miles, nearly the same as that of earth, Venus has often been called our twin planet. The term is misleading, for there are few other similarities. Venus rotates very slowly on its axis; recent radio data indicate the Venusian day is 243 earth days. Moreover, space probes have found that Venus' surface temperature is about 425°C—even hotter than that of Mercury. This is hard to understand at first, since Venus receives only about twice as much solar radiation as earth, and much of it is reflected away by its clouds without ever reaching the planet's surface. But the Venusian atmosphere, almost all carbon dioxide, is an exceptionally efficient insulator. Thus very little heat escapes into space, and the temperature rises to a level that makes the existence of life extremely unlikely. Somewhat surprisingly, Venus seems to have little water, for its atmosphere contains hardly any water vapor (the form in which water would exist at Venusian temperatures).

Figure 22-9
Mercury, seen from a Mariner spacecraft, resembles our moon.

THE RED PLANET: MARS Ever since the announcement by the Italian astronomer Giovanni Schiaparelli (1835–1910) in 1877 that he had observed canal-like markings on the surface of Mars, speculation about possible life elsewhere in the solar system has focused principally on our red neighbor. Though subsequent improvements in observational technique have shown the "canals" to have been a combination of optical illusion and wishful thinking, other evidence has given us reason to suspect that Mars is the one other planet where life is possible, or at least may have been possible in past eras.

Seen through a telescope, Mars exhibits dark surface markings and clearly defined ice caps, similar to those of earth, that change dramatically with the seasons. (The inclination of Mars' axis to the ecliptic is almost identical to the earth's, creating seasons like our own.) In the Martian summer the dark areas expand noticeably, while the polar ice cap shrinks from roughly 2000 miles in extent to only 200. The dark areas were once thought to be vegetation that flourished in the relative warmth of summer, nourished perhaps by water from the melting ice cap. But in recent years, space probes have found that the

amount of water on Mars is very small. The polar ice caps consist at least in part of frozen carbon dioxide (dry ice), and the percentage of water vapor in the Martian atmosphere is extremely low. This atmosphere, composed almost entirely of carbon dioxide, with hardly any oxygen and no nitrogen, is extremely thin; its pressure at the planet's surface is only about $1/300$ atm, about what we would find at an altitude of 120,000 feet above the earth. This thin atmosphere can trap little heat, for though the daytime temperature at the Martian equator may reach a balmy 70°F, it drops at night to a hundred degrees below zero. Nor does such an atmosphere provide any protection against ultraviolet radiation.

These factors make Mars a less inviting habitat for life as we know it than had once been hoped. But it is possible that Mars goes through vast climatological cycles lasting 50,000 years and is currently in a dry Ice Age. Life may be sleeping through this long winter era in the protective form of spores which will

Figure 22-10
A photomosaic of Mars (right), composed of more than 1500 television pictures relayed to earth by Mariner 9 in 1971 and 1972. The north polar ice cap is visible at the top. In the bottom center is the volcano Nix Olympica, 82,000 ft high and 375 mi across. Recent photographs of Mar's surface taken by Mariner spacecraft reveal several meandering channels (left), which strongly resemble riverbeds on earth. Such evidence strongly suggests that running water may once have been plentiful on the Martian surface.

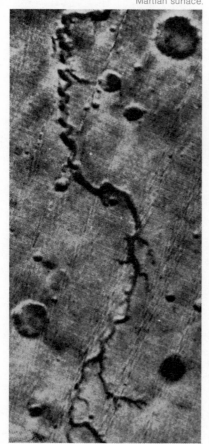
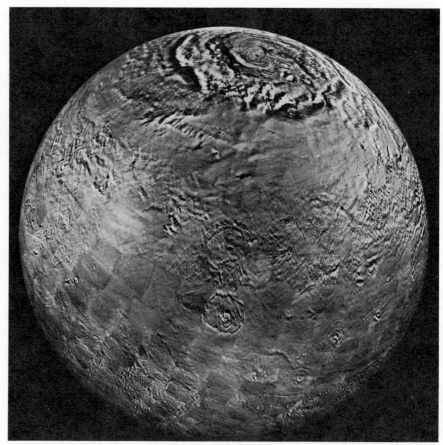

germinate when the spring epoch begins, melting the ice caps and increasing the atmospheric pressure to the point where liquid water could flow on the planet.

THE GIANT PLANETS: JUPITER AND SATURN   Jupiter is the largest of the sun's family, outweighing all the other planets combined. With a diameter 11 times that of earth, its volume could hold 1300 earths. Its mass, however, is equal to only 318 earth masses, which indicates that its density must be considerably less than that of earth. This low density, like that of the other large planets (Saturn, Uranus, and Neptune), suggests that these giants of the solar system consist mostly of hydrogen and helium. The most recent evidence, gathered from the Pioneer 10 fly-by of December 1973, indicates that Jupiter's core is liquid hydrogen, at temperatures of 30,000°C and pressures of millions of atmospheres. In fact, if Jupiter were some 30 times more massive, its gravitational compression would raise its internal temperature to the point where hydrogen fusion begins, and Jupiter would be a small star.

Moving outward from Jupiter's center, the temperature drops steadily, until it reaches −120°C in the upper layers of its atmosphere. This consists of hydrogen, helium, and methane, in which float dense clouds of water, ice, and frozen ammonia. There is probably no clear transition between Jupiter's liquid interior and its thick atmosphere. Jupiter's clouds exhibit patterns of some stability. The most noticeable is the Great Red Spot; first observed in 1877, it has persisted continuously since then, occasionally fading somewhat and then returning to its original distinctness. The Pioneer findings suggest that the Red Spot is the vortex of an enormous tornadolike storm that has been raging in Jupiter's atmosphere for centuries. A strong planetary magnetic field and intense radiation belts have also been detected.

Jupiter rotates very rapidly; its day is only 10 hours long. This fast spin has produced a marked equatorial bulge and flattening at the planet's poles. Jupiter also has the most extensive satellite system of all the planets. Four large inner moons (two of them bigger than Mercury) were discovered by Galileo. There are nine smaller ones, most of them quite distant from the planet, which are probably asteroids captured by Jupiter's great gravitational field. Jupiter takes almost 12 years to complete its journey around the sun.

Saturn is by far the most beautiful object in the solar system. In most respects, it resembles Jupiter: large, flattened by a rapid rate of rotation, and possessing a dense, cloud-filled atmosphere of methane and hydrogen. Since Saturn is further from the sun, it is colder than Jupiter. It is also even lower in density—less dense, in fact, than water. Saturn would float, if you could find an ocean large enough to immerse it in. The magnificent rings, 90,000 to 170,000 miles in diameter, consist of three distinctly visible bands (and a much fainter one discovered in 1971), all lying in the same plane. The rings are inclined considerably to the plane of Saturn's orbit, so the angle at which we view them changes

Figure 22-11
This photograph of Jupiter, taken from a distance of nearly half a million miles by Pioneer 11 spacecraft in December, 1974, shows complex bands of clouds in the giant planet's atmosphere. At the upper right, about 670,000 mi from Jupiter, is the planet's largest satellite, Ganymede (diameter 3500 mi).

Jupiter consists largely of liquid hydrogen; its atmosphere is stormy and filled with clouds of water, ice, and frozen ammonia.

as Saturn moves around the sun. Once every 15 years they are presented to us edge-on, and disappear—an indication that the rings are extremely thin.

Saturn's rings consist of tiny particles of ice and rock in orbit about the planet.

Observing the rings face-on, we find that stars can be seen through them, so they cannot be solid. Evidently they are composed of tiny orbiting particles of meteoric rock, dust, and ice. The explanation for their existence is to be found in the concept of Roche's Limit, the closest distance a satellite can safely approach its parent body. Theoretical calculations show that if a moon comes within a certain critical distance of the planet it circles, the planet's gravitational force, pulling more strongly on the near side of the satellite than on the far side, will tear it apart. Saturn has ten moons circling safely outside Roche's Limit (2.44 times its diameter), but the rings lie inside it. Perhaps a moon somehow drifted inside the limit long ago and was shattered into fragments. But it is more probable that the rings represent material that ended up inside the limit when Saturn's moons were forming, and was thus prevented by Saturn's gravity from coalescing into a single body.

THE REMOTE PLANETS: URANUS, NEPTUNE, PLUTO Uranus and Neptune, quite similar in many respects, are intermediate in size between the earth and Jupiter. Both have densities almost as low as Jupiter's, indicating that they too are made up largely of light elements, presumably hydrogen. Their atmospheres, like those of Jupiter and Saturn, appear to consist largely of methane and hydrogen; these planets are so cold that any ammonia present would crystallize out and fall to the surface like snow. Of Pluto—small, remote, and dimly illuminated—very little is known, except that its density is high, indicating a rocky or metallic composition like the earth's. Pluto takes 248 years for its trip around the sun, travelling in an orbit so eccentric that it actually passes within the orbit of Neptune at times. More may be learned about these outer members of the solar system in the late 1970s, when their positions will be unusually favorable for fly-bys by space probes.

Asteroids, Meteoroids, Comets

Between the orbits of Mars and Jupiter there is a region known as the **asteroid belt,** where thousands of small planetoids circle the sun. The four largest so far discovered have diameters between 200 and 500 miles, but there are over 10,000 others of all sizes, some no bigger than boulders. Most asteroids move in roughly circular orbits, like the planets. Some, however, have been deflected by Jupiter's great gravitational field into highly eccentric orbits that occasionally bring them close to earth. In 1937 one approached within 400,000 miles of this planet—a near miss by astronomical standards. The combined mass of all the asteroids is much less than that of any of the planets. Astronomers think they may represent material left over from the formation of the solar system that did not have enough mass to coalesce into planetary form.

The vast spaces of the solar system contain considerable solid debris, ranging

Figure 22-12
This enormous crater in Arizona,
three-quarters of a mile across and 600 ft
deep, was blasted by a great meteorite
thousands of years ago.

from particles no larger than a grain of sand to chunks of rock and metal the size of a skyscraper, not confined to the asteroid belt. When these **meteoroids** strike the atmosphere at high speeds, friction heats them to incandescence, creating the bright streaks of light across our night skies that we know as **meteors** or shooting stars. Although this fiery passage through our atmosphere vaporizes most of the smaller particles, an occasional larger fragment will survive to reach the earth's surface. These are known as **meteorites,** and are valued by scientists for the clues they can provide to the origin of our solar system; some have been found with ages up to 4.6 billion years, which would make them as old as the earth itself. Most meteorites are stony in character, composed largely of silicates, but others are found which contain high percentages of iron and nickel. On the average, a patient observer on a dark night will see about 10 meteors per hour.

When the earth crosses the orbit of a swarm of meteorites, a meteor shower occurs.

Comets are thought to be loose aggregations of frozen methane, ammonia, and water, intermixed with particles of meteoric dust; they have been likened

Figure 22-13
Halley's comet in 1066, depicted on a medieval tapestry, and on its most recent swing around the sun in 1910. The orbit of this great comet, which has a period of roughly 76 years, was calculated by Edmund Halley in 1705 from the records of earlier sightings in 1531, 1607, and 1682. He predicted that it would appear again in 1758, and it arrived right on schedule.

Millions of comets circle the sun far out in space; some of them pass through the solar system only once in a thousand years.

to dirty snowballs. The head of a comet is generally not more than a few miles in diameter. As it nears the sun, solar radiation vaporizes some of the frozen gases, which stream out behind the head to form a beautiful luminous tail which may be millions of miles long. Spectacular comets are rare, but the total number of comets that circle our sun may run into the millions or billions. The orbits of comets are generally so elongated that they spend most of their time far away from the sun. Many pass through our neighborhood only once in thousands of years. Some comets, indeed, travel in hyperbolic or parabolic paths, looping around the sun once on their journey through space and never returning.

Origin of the Solar System

One of the great unsolved problems in astronomy is the origin of the solar system. An adequate theory would have to explain a great deal of curious data: not only the very diverse sizes and compositions of the planets, but also the peculiarities in the orbits of Mercury and Pluto, the anomalous spins of Venus and Uranus, the presence of the asteroid belt where we might expect to find another planet, and so forth. Some of the regularities of the solar system are as difficult to account for as its irregularities. The simple relationship known as Bode's law, for example, seems to predict the spacings of the planets far too accurately to be coincidental, but though it was first noticed some 200 years ago, its significance is still unexplained. Taking the progression 0, 3, 6, 12, . . . , adding 4 to each number, and dividing the result by 10 gives an excellent approximation, in astronomical units (the earth-sun distance) of the radii of the planetary orbits, including the asteroids but excluding Neptune, which does not fit the pattern (Table 22-4). There are two chief theories of the origin of the solar system: the **nebular hypothesis** and the **encounter hypothesis**.

THE ENCOUNTER HYPOTHESIS Suppose, many eons ago, another star had passed close to our sun. Its gravitational field might have pulled some of the sun's matter out into space, probably in the form of a great gaseous stream stretching out towards the intruder. As the stars passed each other, this temporary bridge of solar material would be swung around like a whip, giving it enough angular momentum to keep some of it in orbit about the sun after the other star had disappeared on its way. The orbiting matter from the sun would then cool and condense, forming planetary fragments. In time, random collisions and gravitational attraction would cause the larger fragments to sweep up the smaller ones, and the planets would be born. This scenario is known as the encounter or near-collision hypothesis of the solar system's formation. Detailed theoretical calculations, however, have shown that many details of this model are open to serious doubt, and the encounter hypothesis, once extremely popular, is not now widely accepted among scientists.

THE NEBULAR HYPOTHESIS The alternative theory, known as the nebular hypothesis, suggests that the sun could have given birth to the planets spontaneously during its own process of formation, without the intervention of any outside force. Modern versions of the nebular hypothesis are variations of an idea first proposed in the eighteenth century by Immanuel Kant (1724–1804), and by the French astronomer Pierre Simon de Laplace (1749–1827). According to this theory, the sun began as a large, diffuse, slowly rotating nebula, or cloud of gas. Under the influence of its own gravitational field, this nebula gradually contracted, spinning faster in the process, just as a skater does when he draws his arms in closer to his sides. As a result of this process, a series of rings formed around the equator of the shrinking cloud of gas (Figure 22–14). Gravitational contraction slowly heated the central mass of gas until it reached the temperature at which nuclear fusion was possible and became a star. Meanwhile, the rings of material left behind by the shrinking sun-to-be coalesced to form the planets. (For some reason, this process was never completed for the material which now forms the asteroid belt.)

This theory, while still uncertain in many of its details, now seems more plau-

Table 22-4
Bode's Law

Planet	Predicted distance	Actual distance (AU)
Mercury	0 + .4 = .4	0.39
Venus	.3 + .4 = .7	0.72
Earth	.6 + .4 = 1.0	1.00
Mars	1.2 + .4 = 1.6	1.52
Asteroids	2.4 + .4 = 2.8	2.8 (avg)
Jupiter	4.8 + .4 = 5.2	5.20
Saturn	9.6 + .4 = 10.0	9.54
Uranus	19.2 + .4 = 19.6	19.18
Neptune		30.07
Pluto	38.4 + .4 = 38.8	39.44

Figure 22-14
The nebular hypothesis of the origin of the solar system. The planets form from rings of material left behind by the contracting cloud of gas that will become our sun.

sible than any alternative yet proposed. One interesting corollary of the nebular theory is that the likelihood of other solar systems circling other stars is very great. For if this explanation of the formation of planets is correct, there is no reason to doubt that almost every star could produce them as a natural byproduct of its own evolution. The encounter theory, by contrast, implies that planets are extremely rare, for the odds against the required near-collision of two stars are immense. Our solar system, in this view, is a freak accident, perhaps unique in the cosmos. But if we accept some version of the nebular hypothesis, the existence of millions or billions of other solar systems must be considered quite probable.

It seems probable that millions of other stars, as well as the sun, have planets circling them.

Glossary

Asteroids are irregular rocky bodies, numbering in the tens of thousands, which range up to 480 miles in diameter. Most of them circle the sun in a belt lying between the orbits of Mars and Jupiter.

The **aurora** is a shimmering, multicolored glow in the night sky seen mainly in high latitudes, the result of atomic particles from the sun striking the earth's ionosphere.

The **chromosphere** is the relatively cool inner layer of the sun's atmosphere.

Comets are small, loose aggregations of ice and meteoric fragments, typically moving in very elongated elliptical orbits which only infrequently bring them close to the sun.

The **corona** is the outer layer of the sun's atmosphere, a region of thin, hot, ionized gases.

A solar **eclipse** takes place when the sun's light is obstructed by the moon. A lunar eclipse takes place when the moon enters the earth's shadow.

The **ecliptic** is the projection of the earth's orbital plane on the heavens.

The **encounter hypothesis** of the formation of the solar system holds that the material of the planets was pulled from the sun by the gravitational attraction of a passing star.

Flares are violent explosive outbursts of energy on the sun's surface.

Meteoroids are small rocky or metallic fragments that often collide with the earth's atmosphere. Most of them are vaporized by the heat of friction as they pass through the air, creating a streak of light in the night sky known as a shooting star or **meteor**. A meteoroid that reaches the earth's surface is called a **meteorite**.

The **nebular hypothesis** of the origin of the solar system holds that the planets were formed from rings of matter left behind by the contracting cloud of gas that ultimately became the sun.

The **photosphere** is the thin outer shell of the sun from which all of its energy is radiated into space.

Prominences are great loops of cool, luminous gas that form in the corona above the sun's surface.

The **proton-proton chain** is a three-step process in which hydrogen is converted to helium by nuclear fusion. It is the main source of the sun's energy.

The **solar wind** is a flow of charged particles from the sun out into the solar system.

Sunspots are relatively cool, dark regions of the solar surface characterized by strong local magnetic disturbances.

The **zodiac** consists of the 12 constellations lying along the plane of the ecliptic, through which the sun and planets appear to move.

Exercises

1 Is it possible that the source of the sun's energy is combustion similar to that which takes place in a household furnace?

2 What is believed to be the source of the energy of the sun? What arguments have led astronomers to this conclusion? Why is energy production in the sun occurring at the center and not at the surface?

3 How do we know which elements are present in the sun? Which element is the most abundant? Why is it possible that the earth and the sun had the same ancestry?

4 How do we know that the sun rotates about its axis? Why are we certain that the sun is not a rigid body?

5 What are sunspots? Are they completely dark? Do they appear at random or is there a pattern in their appearance?

6 The orbit of the moon has a period of approximately one month. Why is there not an eclipse of the sun or the moon every month? What is the moon's phase at the time of a solar eclipse? A lunar eclipse?

7 Why does an observer usually see more lunar eclipses than solar eclipses every year, though solar eclipses are actually more frequent?

8 Why are there such extremes of temperature on the moon?

9 Show with a diagram how planets that are closer to the sun than the earth exhibit phases just like the moon. Why is Venus brightest in the sky when it is less than half full?

10 Why is Venus so much hotter than we would expect for a planet 67 million miles from the sun?

11 Mars and Mercury have about the same surface gravity, but Mars has retained an atmosphere while Mercury has not. Why?

12 Suppose you are speaking by radio with a friend on Jupiter, at a time when it is at its nearest approach to Earth. How long after you finish a sentence will you have to wait to hear his answer?

13 Why are we certain that the rings of Saturn are not solid objects? How are they believed to have formed?

14 What are the asteroids? Where are they found?

15 Is life more likely on Venus or on Mars? Why?

16 Why are moon rocks and meteorites thought to provide information about the formation of the solar system?

17 What are the theories of solar system formation? What key features should a successful theory explain?

One of the oldest human dreams has been to escape from the prison of gravity on the earth's surface, to soar high enough to see beyond the horizon. In our day a handful of men and women have finally achieved this ancient ambition. But even before astronauts traveled beyond the earth's atmosphere, science had placed eyes in space—satellites carrying telescopes, film cameras, TV cameras, infared and ultraviolet scanners, x-ray and cosmic-ray detectors.

Some of these eyes turn outward towards the sun, the planets, and the distant stars and galaxies. Other eyes are directed back towards the earth. Satellites now monitor world-wide weather conditions, inventory vital resources, survey ecological conditions, keep track of pollution, and provide data for geologists and cartographers (map makers). Infrared cameras, which register even slight differences in temperature, can distinguish among different types of vegetation

People have been fascinated by the
possibility of flying since ancient times. In
the story from Greek mythology, Icarus was
imprisoned with his father, Daedalus, on the
island of Crete. Daedalus constructed wings
from feathers and wax so that the two could
escape. But Icarus flew too near the sun,
which melted the wax on his wings, and he
plunged into the Aegean Sea and drowned.

from a height of hundreds of miles. These cameras provide information for early
detection of blight and insect infestation in crops and forests, prediction of
earthquakes and volcanic activity, and location of large schools of surface-
feeding fish. Satellite instruments also provide microwave measurements for
estimating soil moisture content and detecting ore and oil deposits. Cameras in
weather satellites take thousands of pictures of the world's cloud cover every
day, supplying data about the patterns of climate that span the globe. Much of
the information provided by satellite is simply unavailable from a ground-level
perspective; other data could be accumulated only by years of work on the
earth's surface.

These images of earth from space can lead us to see our planet and ourselves
from a new perspective. They help us realize how small the earth is and how
much the destinies of all the living things that share it are intertwined. Ultimately our new vision of the earth from space may inspire us to take better care
of our small world—and of each other.

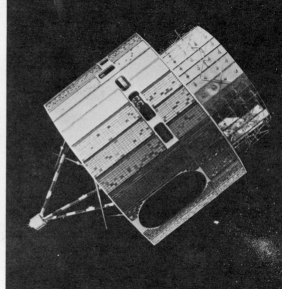

Above right: The Synchronous Meteorological Satellite occupies an orbit 22,300 miles above the earth's equator. At this altitude, its orbital velocity exactly matches the earth's rate of rotation, so that the satellite remains poised over the same point on the earth's surface, monitoring global weather conditions with cameras sensitive to both visible light and infared (heat radiation). Above left: This view of nearly half of the world's weather was taken by SMS-1 from its vantage point over the Atlantic Ocean in July, 1974. Africa is at the right in the photograph; the United States is at the upper left, and South America is the land mass below and to the left of center. Signs of the extreme drought affecting much of the African continent are clearly visible. Hot, dry, dust-laden air has been blown by winds sweeping south and west from the Sahara desert. This air, a light grey mass in the photograph, covers much of the western bulge of the continent and balloons out over a thousand miles into the Atlantic. The mass of dusty air continued westward and reached the coast of Florida a week later. Satellites have proven invaluable in tracking the movements of potentially dangerous meteorological phenomena such as hurricanes and typhoons. When the deadly hurricane Camille, the most powerful storm to strike the North American coast in recorded history, swept ashore from the Gulf of Mexico in August, 1969, she was preceded by warnings from weather satellites that had been photographing her development for almost a week. It is estimated that this advance warning saved 50,000 lives. Middle left: This photo of Hurricane Ginger was taken by the Nimbus-4 weather satellite from an altitude of 690 miles in September, 1971.

Lower left: Despite their brief history, satellite photographs have already led to many new discoveries in a number of fields. Often they show things not visible at all from the ground. This photograph of the Great Salt Lake, Utah, is an example. The line across the lake (upper left) is a railroad causeway that impedes water circulation. South of the causeway the lake water is diluted by inflow from nearby rivers, making it considerably less salty than the water north of the causeway. This difference in salinity has ecological consequences: different forms of algae grow in the two halves of the lake. The species inhabiting the southern part of the lake appears much darker when photographed from space. Until such pictures became available, these differences had not been detected by earthbound scientists.

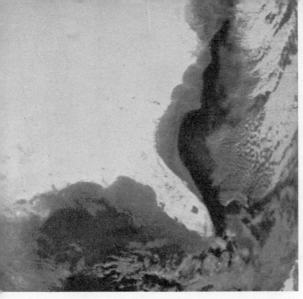
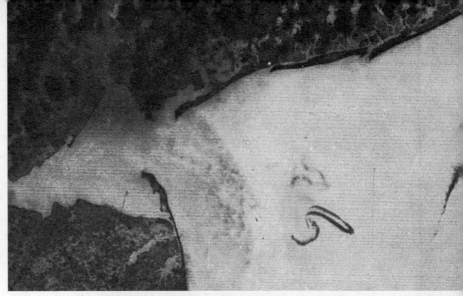

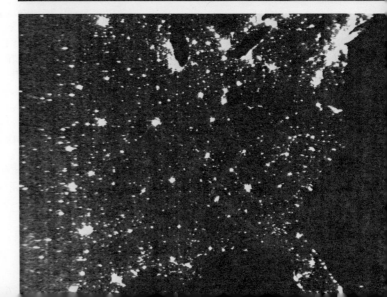

Above left: This photograph of the southern Atlantic coast of the United States was taken by infared photography, which records temperature differences. Warm areas register dark; relatively cooler areas appear lighter. The very dark ribbon extending upward from the tip of the Florida peninsula is the Gulf Stream, a warm ocean current that flows north parallel to the U. S. coastline and then eastward across the Atlantic, giving the British Isles their mild, moist climate. This photograph was taken from an altitude of 900 miles by the NOAA-1 satellite, which circles the earth in a pole-to-pole orbit every 114 minutes.

Above right: Space satellites have proven invaluable in surveying the earth's resources, studying global and regional ecology, and monitoring pollution. This negative print shows the Atlantic Ocean just outside New York Harbor. (The dark land mass at the top is Long Island; Staten Island is to the left, and the New Jersey coast is at the lower left.) The black trail in the lower right-hand portion of the picture is industrial acid wastes, being dumped into the ocean by a barge.

Middle right: This satellite photograph shows the west coast of Nicaragua in Central America. The triangle of water in the lower left of the picture is the Pacific Ocean; above the ocean at the top left is the Gulf of Fonseca with Cosiguina Point forming the southern gulf shore; across from the gulf in the lower right is lake Managua. Parallel to the western coast of Nicaragua are a string of about 40 volcanoes, which run from the northwest to the southeast. This view shows a portion of that string, running roughly from the promontory of Cosiguina Point to the lake; they appear in the photograph as a string of small black circles. The volcanoes mark the location of a geological fault. Shifts along this fault are probably responsible for the earthquake activity in the region. This photograph was taken one day after a disastrous earthquake struck Nicaragua in December, 1972.

Lower right: This nighttime portrait of the eastern half of the United States was taken by an Air Force weather satellite. The lights of the major population centers are clearly visible.

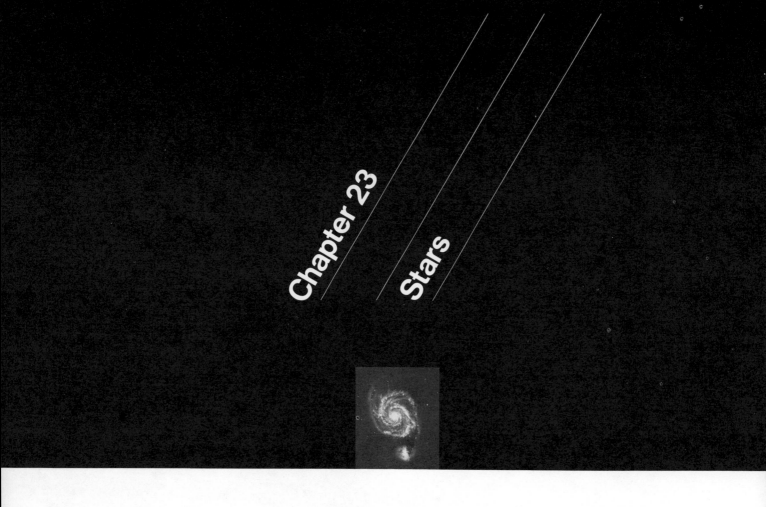

Chapter 23 Stars

It may seem a curious fact that we know much more about the remote stars than about the planets of our own solar system, for the nearest star is about a million times further from us than the nearest planet at its closest approach, and some stars are so distant that their light, traveling at 186,000 mi/sec, takes millions of years to reach us. There are two explanations for this paradox. First, because there are so many stars (some hundred billion in our own stellar system, the Milky Way Galaxy, alone), we see them in all phases of evolution, and can deduce their complete life cycles; by contrast, we have only eight planetary neighbors, all more or less the same age. Second, the planets—relatively cool, solid bodies—shine only by reflected sunlight, which can tell us little about their internal makeup; the stars generate their own radiation in enormous quantities. But while the brilliance of the stars enables us to see them across the vast reaches of space, we could learn little about their nature without some means of analyzing starlight and deciphering its message. The chief instrument for doing this is the spectroscope. Spectroscopy, only slightly more than a century old, has made possible our surprisingly detailed knowledge of the origins, lives, and destinies of the stars.

To a casual eye, the night sky offers a bewildering example of disorder. Countless stars seem scattered at random across the heavens. They range from dim, barely visible specks to bright points of light, some a pure white, some noticeably tinged with color. At first it is hard to find any intelligible pattern in this splendid abundance. But a persistent watcher of the skies soon notices many orderly features. Even in ancient times, thousands of years before the invention of the telescope, men had already learned much about the heavens by patient observation.

The Stars in Space

The most obvious regularity in the celestial realm is the fact that the stars appear never to change position relative to each other; hence the term **fixed stars**, in contrast to the "wandering stars," the planets. Nor (with a very few excep-

Naked Eye Astronomy

Figure 23-1
It is easy to make star trails by putting a camera on a tripod, pointing it towards the night sky, and leaving the shutter open for several hours. As the earth turns, the paths of the stars are recorded on the film. (Stars near the North Pole of the sky never rise or set, but trace circular arcs around the pole star, the last star in the handle of the Little Dipper.)

Figure 23-2
The constellation Orion, represented on an old star map.

The stars are not really fixed, but they are so distant that their motion can only be detected over a period of years with a large telescope.

Telescopic Astronomy

tions) do they ever seem to change in luminosity—again unlike the planets, which vary markedly in brightness as they pursue their intricate paths among the fixed stars.

THE CONSTELLATIONS Since the stars form seemingly unchanging patterns in the heavens, it is not surprising that people have imagined among these patterns various familiar shapes, populating the skies with animals, tools, and mythological heroes. These designs, known as **constellations**, have been handed down through the centuries as part of the traditional lore of various peoples. Thus a familiar grouping of stars in the northern sky was known as the bier to the Arabs, the plow to the early Britons, the bear to the American Indians, and is the Big Dipper to us. But the constellations should not be regarded as the quaint products of idle fancy or primitive superstition. They represent an attempt by men to impose some order on the apparent chaos of the night sky—to name and, in a rough way, to map the various regions of the heavens.

Early astronomy, the product of this same impulse toward order, was necessarily confined to naked-eye observation—that is, charting the movements of the planets and the positions of the stars. Within these limitations, diligent and painstaking observation was practiced by the people of many cultures, including the Greeks, Egyptians, Babylonians, Indians, and Chinese. About 134 BC the Greek astronomer Hipparchus compiled the first systematic star map, carefully marking the position and relative brightness of over 800 stars. Some 250 years later the Egyptian Ptolemy (famous as the codifier of the geocentric or Ptolomaic model of the solar system) expanded Hipparchus' map to more than 1000.

THE MOVEMENT OF THE STARS The full importance of Ptolemy's star chart did not become apparent for almost 1600 years. In 1718 the English astronomer Edmund Halley (1656–1742) noticed that the bright star Arcturus seemed to have moved about 1°—twice the diameter of the full moon—from its location indicated on the ancient star maps. Until then only a few thinkers had conjectured that the stars themselves moved. Halley was the first to show that the "fixed" stars were not truly immobile after all. The motion of a star across our line of sight is called its **proper motion**. In general proper motions are so small that they can be detected only for relatively nearby stars, and even then cannot be measured in a single lifetime without a powerful telescope.

Fortunately, by Halley's day good telescopes were increasingly available. The century from 1750 to 1850 was the golden age of observational astronomy, when man and telescope together explored the universe. In the middle of the nineteenth century, two new instruments, the camera and the spectroscope, gave astronomers a powerful technology, but substantially reduced the role of the human eye.

This period coincides roughly with the lifetime of William Herschel (1738–1822), the greatest of all telescopic observers. Herschel built larger and better telescopes than any astronomer before him; he was a pioneer in the development of the reflecting telescope, invented a century earlier by Newton, and constructed one with a mirror 48 inches in diameter. He spent a lifetime mapping the heavens, cataloging the stars, and observing their motions. Herschel realized that a telescopic survey of the entire sky was beyond the capacity of any man. There are only about 6000 stars bright enough to be visible to the naked eye, and fewer than half of them can be seen from any one place on earth at a time. But even a small telescope vastly increases this number, and Herschel's instrument brought hundreds of thousands of stars into view. (It is estimated that the number within the range of the 200-inch Mt. Palomar telescope is over a billion.)

On a clear night an unaided observer can see about 2500 stars, but the largest telescopes can detect more than a billion.

As an alternative Herschel devised a statistical technique to determine the distribution of the stars: he counted the stars in a large number of small representative areas scattered throughout the skies of the northern hemisphere. (His son, John Herschel, a pioneer in the field of astronomical photography, extended this work to include the southern skies.) As a result of these laborious and systematic observations, Herschel was able to propose a model for the structure of our Galaxy—the great stellar system to which all the stars we see belong—that has been shown to be substantially correct (Chapter 24).

BINARY STARS Another of Herschel's important discoveries was the existence of **binary and multiple star systems**—two or more stars revolving about their common center of gravity. The middle star in the handle of the Big Dipper, for example, is a naked-eye double, called the horse and rider by the Arabs, who of course did not know of the gravitational bond between them. Most binary stars are telescopic objects, however, and sometimes one member of the system is too faint to be detected even telescopically. Sirius, the brightest star in our skies, has a dim companion which revolves about it with a period of 50 years; its presence was inferred, even before anyone observed it in a telescope, because of its gravitational influence on Sirius.

Figure 23-3
Sir John Herschel, son of the great astronomer Sir William Herschel, and himself a pioneer in the use of photography for mapping the heavens.

To Herschel, the discovery of double stars was important because it offered confirmation that Newton's laws were valid in the distant realm of the stars as well as in our solar system. To the modern astronomer double stars are still useful objects of study, for if the orbits of the stars can be determined, their masses can be computed using the laws of Kepler and Newton. Moreover, if they move in an orbit that we happen to view edge-on, they will periodically eclipse each other, permitting determination of their sizes. But neither of these calculations is possible unless the distance to the star system is known. Unfortunately, in Herschel's day, not a single stellar distance had been determined.

If we know the distance to a binary star system, we can compute the masses of the stars from their orbits about each other.

PARALLAX AND STELLAR DISTANCE The discovery of proper motion, however, had fanned the hopes of astronomers interested in solving this age-old problem. If at least some of the stars were close enough for us to measure their proper motion over the years, perhaps they were close enough for us to determine their distance by measuring their **parallax** (Chapter 1). But the parallax of a star is much harder to measure than its proper motion. Proper motions are cumulative; small as they are, they may add up through the centuries to a substantial displacement. But even making observations at six-month intervals, so that a star could be seen from locations on opposite sides of the earth's orbit (186 million miles apart), astronomers had long failed to detect any stellar parallax at all. Ever since the time of the Greeks, this fact had been used to support the geocentric theory of the universe.

Nevertheless, such parallaxes do exist. They are merely much too small to be seen with the naked eye, and can only be perceived by very painstaking observation with a good telescope. It was not until the late 1830s that astronomers succeeded in measuring the first stellar parallaxes. The largest, belonging to Proxima Centauri (a faint star in the southern hemisphere), turned out to be a mere 3/4 of a second of arc, or 1/4800 of a degree. This is the angle subtended by a silver dollar at a distance of about ten miles! Proxima Centauri is thus our nearest star (except for the sun); its distance, computed by trigonometry from its parallax, is slightly more than 25 trillion miles. Since the mile is obviously an inconveniently small unit for expressing such vast distances, astronomers generally employ either the **light-year**, which is the distance travelled by light through a vacuum in one year, or the **parsec**, defined as the distance at which a star would appear to have a parallax of exactly 1 second of arc. The light-year is 5.88 trillion miles, the parsec 19.2 trillion miles, or 3.26 light-years. Proxima Centauri is about 4.3 light-years away; if that star suddenly winked out, we would not know about the catastrophe for over four years. (By comparison, the light we receive from the sun takes just eight minutes to reach us.)

The nearest star, Proxima Centauri, is about 25 trillion miles away; its light takes more than four years to reach us.

Figure 23-4
Star B is further from earth than star A, and its parallax (angle b) is smaller than A's (angle a). The distance to a star of known parallax can be determined by trigonometry. The tangent of half the parallax angle ($\frac{1}{2}$a) is the radius of the earth's orbit, r (93×10^6 mi), divided by the star's distance, d. Thus $d = 93 \times 10^6$ mi/tan ($\frac{1}{2}$a).

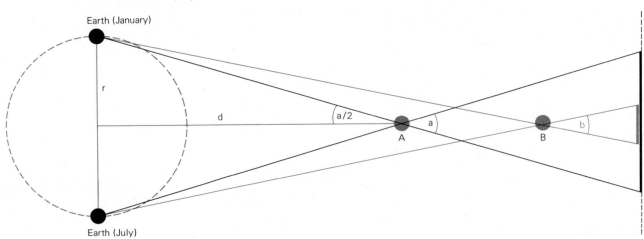

Earth (January)

r

d

a/2

a

A

b

B

Earth (July)

CEPHEID VARIABLES Because the angles of stellar parallax are so small, they are hard to measure precisely even with large modern telescopes, and parallax is a reliable indicator of stellar distances only out to a radius of about 100 light-years. For more remote parts of the universe other methods of estimating distance have been devised. The most important depends on a very fortunate property of certain stars known as **Cepheid variables**. These stars change in luminosity with great regularity, dimming and brightening again in cycles that range from 2 to 45 days. Astronomers have found that there is a very precise relationship between the period of a Cepheid and its intrinsic luminosity. If we know the true brightness of such a star from its period and compare its apparent brightness, it is easy to determine its distance by applying the inverse square law.

Cepheids are often found in star clusters—large aggregations of stars that travel through space together. The discovery of a single Cepheid in a cluster can therefore tell us the distance of thousands of stars at once. This method of determining distances is reliable out to several million light-years. Conversely, once we know the distance of a star, the inverse square law will tell us its intrinsic brightness. Thus the Cepheid yardstick enables us to find the true luminosity of millions of distant stars.

The relationship between the period and brightness of Cepheid variables enables astronomers to find the distance and luminosity of millions of remote stars.

When you mention astronomy, the average person immediately thinks of the telescope, and with some justice. Certainly the invention of the telescope, and Galileo's use of it for astronomical observation in 1609, represent the first great breakthrough that opened up the universe to the eyes of mankind. The progress of astronomy in the last 350 years owes much to the development of larger and better telescopes—from Galileo's little one-inch spyglass to the precision-machined, motor-driven, computer-guided giants of our day. While the camera and spectroscope have enormously increased our knowledge of the universe, they must still be used with a telescope to tell us anything about the stars.

The Astronomer's Tools

THE TELESCOPE The two basic types of telescope, refracting and reflecting, allow us to scrutinize the stars. The reflector, which is considerably less expensive to construct and easier to perfect optically, is the design now used for all large modern telescopes. There is some confusion in many people's minds about the purpose of these instruments. It is not high magnification that is the primary function of large telescopes (though that is sometimes useful), but rather great light-gathering power. This permits astronomers to study objects too remote and faint to be visible to the unaided eye, and to achieve far greater resolving power (the ability to distinguish fine detail) than the unaided eye is capable of. The maximum diameter of the eye's pupil is $\frac{1}{3}$ of an inch. The

200-inch mirror of the Mt. Palomar telescope has a light-gathering area several hundred thousand times that of the naked eye.

ASTROPHOTOGRAPHY   Photography allows the telescope to capture even fainter detail. Astronomers often use very long exposures, letting their plates accumulate light for hours. In this way they can capture the image of objects too dim for visual observation. Consequently, an important feature of any sophisticated telescope is its mounting and clock-drive mechanism, which enable the instrument to follow a heavenly object across the skies as it rises and sets. The object thus appears motionless in the instrument's field of view, making long exposures possible. Further, astrophotography gives us a permanent record of the heavens at a particular time, making it easier to measure the position and brightness of celestial objects, and to detect any change or movement with the passage of time.

The Spectroscope   The telescope and the camera are the basic tools of astronomy. Yet these instruments take us beyond naked-eye astronomy only in degree. They enable us to see a vastly greater number of stars, but even with the largest telescopes and most advanced photographic techniques, the stars remain mere pinpoints of light. We can do little more than count and catalog them, recording their position and wherever possible calculating their distance and intrinsic brightness. For a small number of stars that are part of binary systems we can sometimes

Figure 23-5
The large reflecting telescope of the Lick Observatory at the University of California at Santa Cruz and its parabolic mirror, 10 ft in diameter.

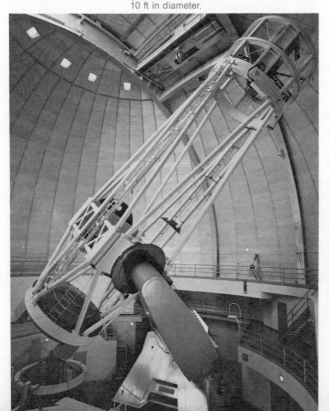

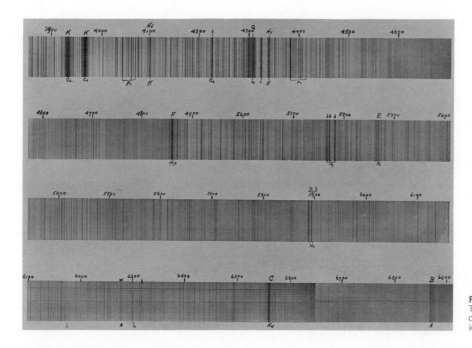

Figure 23-6
The solar spectrum, showing hundreds of dark Fraunhofer (absorption) lines, which identify elements in the sun's atmosphere.

deduce masses and perhaps sizes. But of the nature of the stars—their composition, their structure, their internal conditions, their energy processes, their evolution—observational astronomers from Galileo's time to the latter half of the nineteenth century knew virtually nothing. It was spectroscopy that opened the way to the remarkably intimate knowledge of the stars that we now enjoy.

THE LAWS OF SPECTROSCOPY   In 1814 the German physicist Joseph von Fraunhofer (1787–1826) found that when a ray of sunlight was passed through a narrow slit and diffracted by a prism or grating, the expected continuous spectrum of colors was crossed by hundreds of thin dark lines. These **Fraunhofer lines** represent wavelengths which are missing in the solar spectrum. When the spectra of various stars were obtained, they were found to be similar in nature, with many dark lines in exactly the same places as in the sun's. The significance of these dark-line spectra was not understood at the time, but in 1859 another German scientist, Gustav Kirchhoff (1824–1887), discovered how to interpret Fraunhofer's observations. Kirchhoff's findings may be summarized as three **laws of spectroscopic analysis:**

1   Heated solids, liquids, and many gases at high densities emit light of all wavelengths, forming a **continuous spectrum** (like that of a rainbow).
2   A thin luminous gas emits light only at certain characteristic wavelengths, forming a discontinuous spectrum consisting solely of isolated **bright-lines**. The positions of these lines identify the element emitting them.

3 If light from a hotter, more intense source is passed through a gas, the gas will absorb certain wavelengths, creating **dark lines** in the spectrum, which are exactly those the gas itself emits.

SPECTROSCOPIC ANALYSIS From these laws alone we can immediately learn several important things about the sun and the other stars. Since these bodies emit continuous spectra, they must consist of hot dense matter—almost certainly gases at high pressure. Furthermore, since their continuous spectra exhibit absorption lines, they must be surrounded by atmospheres of thinner, somewhat cooler gas. And from the positions of these lines in the spectrum, astronomers have been able to determine the composition of the stellar atmospheres. Most consist mainly of hydrogen (50 to 80 percent), with helium the next most abundant element; together, the two comprise between 96 and 99.9 percent of the mass of almost all known stars. Absorption lines characteristic of dozens of other elements have also been detected. In addition, careful study of a star's spectral absorption lines can furnish an enormous amount of detailed information about conditions inside the star. Pressure, temperature, degree of ionization, and intensity of stellar magnetic fields all leave their traces in the stellar spectra.

Much of what we know about the composition, temperature, internal conditions, and rotation of stars is determined from their spectra.

THE DOPPLER SHIFT Even the rotation of a star about its axis can usually be detected and measured from spectroscopic evidence. When a luminous object approaches us, the wavelengths of its radiation will be shortened, and thus shifted towards the violet end of the spectrum, by the Doppler effect (Chapter 9). An object receding from us will exhibit a corresponding shift of its spectrum toward the red end. If a star is rotating, its surface on one side will be moving toward us at any given instant while the other side moves away. Thus some of its light will be shifted slightly toward each end of the spectrum, producing a characteristic broadening of the spectral absorption lines.

The Doppler shift also enables us to determine how fast the star itself is moving toward or away from us through space. If the distance of a star with a measurable proper motion is known, we can translate the observed proper motion into its real velocity across our line of sight. But there is no reason to assume that a star is moving *only* perpendicular to our line of sight. More likely it will be moving at an angle, and will have a component of motion toward or away from us that we cannot see in our telescopes. The Doppler shift recorded by the spectroscope enables us to measure this radial component of a star's velocity as well, so that both its actual velocity and direction of motion can be determined.

SPECTRAL TYPES One of the most important things astronomers can learn from the spectrum of a star is its temperature. We know from everyday experience

Figure 23-7
Rotational broadening of spectral absorption lines by the Doppler shift.

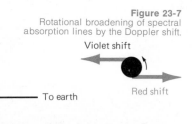

Violet shift

◄——— To earth

Red shift

No rotation spectrum

Rotational broadening of spectral lines

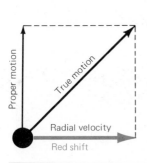

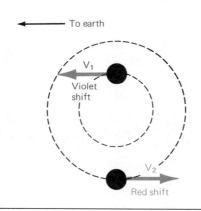

1. Star 1: violet shift
2. Unshifted reference spectrum
3. Star 2: red shift

that hot objects emit light—the coils of an electric broiler, for instance, are usually hot enough to glow a dull red or orange. The more heat a body has, the more radiation it emits at the short wavelength, or blue, end of the spectrum. A body that is white-hot is hot enough to be emitting radiation through the entire visible range, while one that is glowing blue-white is hotter still, radiating primarily at the shorter wavelengths: blue, violet, and the invisible ultraviolet.

Thus the temperature of a star can be estimated from its color. Red stars like Antares, the bright star in the heart of the scorpion (the constellation Scorpius), are relatively cool; yellow stars like the sun are somewhat hotter; white stars are hotter still; and blue stars, like Rigel, which marks the right foot of Orion the hunter, are the hottest. By analyzing a star's light with the spectroscope and determining at what wavelength it emits the most radiation, scientists can convert this approximation into a very precise determination of the star's surface temperature. On the basis of these spectroscopic determinations of temperature, astronomers divide the stars into seven main spectral classes, designated O, B, A, F, G, K, M (generations of students have used the sentence "Oh, Be A Fine Girl, Kiss Me" to help remember these letters). The temperatures and colors corresponding to these classes are shown in Figure 23-10.

For astronomers, unraveling the mysteries of the stars has resembled putting together the pieces of a jigsaw puzzle. Each bit of information has led to further discoveries and made it slightly easier to fit the next into place. We have already seen how knowing the period-luminosity relationship for Cepheid variables gave astronomers a distance scale, and how knowing the distance of a star in turn enables us to determine its luminosity. Similarly, other bits of data have been combined to give a very complete picture of the stellar universe.

Figure 23-8
The absorption lines of a star moving towards the earth are shifted towards the violet end of the spectrum; stars moving away from earth exhibit a red spectral shift.

The color of a star reflects its temperature: red stars are relatively cool, blue or white ones very hot.

The Story of the Stars

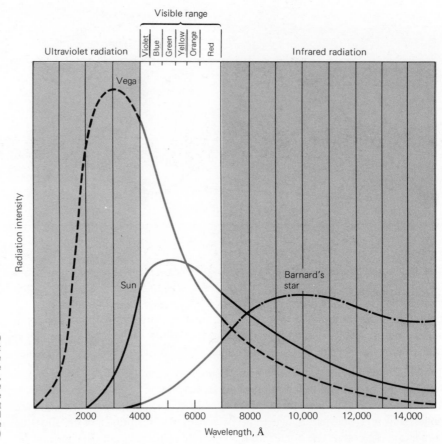

Ultraviolet radiation   Visible range   Infrared radiation

Violet | Blue | Green | Yellow | Orange | Red

Vega

Sun

Barnard's star

Radiation intensity

Wavelength, Å

Figure 23-9
The output of radiation at different wavelengths of the sun (surface temperature ~ 6000°K), a hotter star, Vega (surface temperature ~ 10,000°K), and a cooler star, Barnard's star (surface temperature ~ 3000°K). Hot stars radiate mostly at the blue end of the spectrum, cooler ones at the red end. (The curves are idealized, and the intensities of radiation from the three stars are not to the same scale.)

The H-R diagram

The pieces started to fall into place around 1913, when the astronomers E. Hertzsprung (1873–1967) and H. N. Russell (1877–1957) independently began to compare the intrinsic luminosities of a number of stars with their spectral class. The easiest way to do this is to make a graph like that shown in Figure 23-10. In this graph, intrinsic brightness is plotted on the vertical axis, while spectral type (with the corresponding temperature and color) is recorded along the horizontal axis. When such a Hertzsprung-Russell diagram (or more commonly H-R diagram) is prepared for a large number of stars, a definite pattern emerges. Most of the stars are found to lie along a distinct band running from the lower right (cool, dim stars) to the upper left (hot, bright stars). Some 90 percent of all known stars fall within this **main sequence**.

THE MASS-LUMINOSITY RELATIONSHIP This result agrees with our common-sense expectations, for we know that the hotter a body, the more brightly it will glow. But why should some stars be so much hotter than others? The answer was provided by the English astronomer Arthur Eddington (1882–1944), a pioneer in the study of stellar structure and energy production. Eddington realized that in any stable star there must be an equilibrium between the inward pressure of its enormous weight and the outward pressure in its central regions, which derives from two sources—gas pressure and the pressure of radiation produced by the star's fusion reactions (Chapter 22). Both of these increase with temperature, which in turn rises with increasing compression. Thus we can say that there is an equilibrium between the star's gravitational weight and its counterbalancing temperature. The more massive a star, the more heavily its central regions are compressed by its own great gravitational force; the greater the gravitational compression, the hotter (and thus the brighter) the star must become to maintain equilibrium.

The sun lies near the middle of the main sequence. If the mass-luminosity rule holds, stars above and to the left along the main sequence should be more massive than the sun, and those below and to the right, less massive than the sun. This has been confirmed by observation. The range of stellar masses is actually not very great; the heaviest stars known are only about 50 times more

Figure 23-10
The H-R diagram for a typical population of stars, showing the main sequence, white dwarf, and red giant regions. The evolutionary track of a star such as the sun is superimposed.

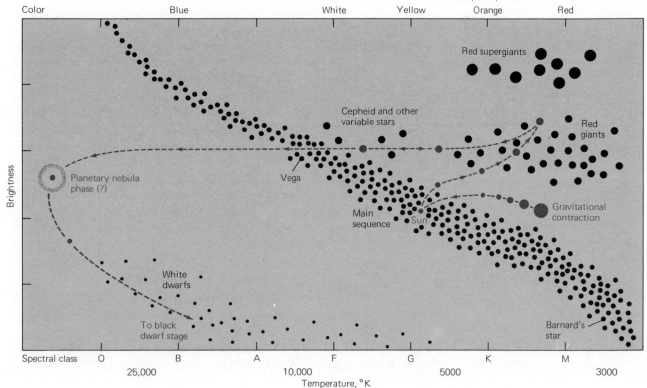

massive than the sun, while the lightest have approximately a twenty-fifth of the solar mass. But since the energy production of a star is roughly proportional to the cube of its mass, the range of stellar luminosities is enormous. The brightest known star is some 600,000 times more luminous than the sun; the faintest is about a millionth as bright as the sun.

RED GIANTS AND WHITE DWARFS   A minority of stars, however, do not fall within the main sequence band on the H-R diagram. In the lower left there is a sprinkling of stars that, though very hot, are nevertheless very faint, and in the upper right there is a region of very cool stars that are extremely luminous. Astronomers studying these stars realized that a very hot blue or white star could radiate little light only if it were very small, while a cool, red star could be very luminous only if it were extremely large. These small, hot stars are called **white dwarfs**, and the large, cool stars are called **red giants** (certain of the largest are even classified as **supergiants**). Many red giants and supergiants are so cool that they emit hardly any visible light at all, radiating almost entirely in the infrared.

The range of sizes represented by the white dwarfs and red giants is very great. Some red giants are large enough to hold almost the entire solar system. But since this great volume is occupied by a mass only a few dozen times that of the sun, the density of such a star is roughly a ten-millionth of the sun's—thin enough to pass for a very good vacuum in an earthly laboratory. A typical white dwarf, by contrast, has a diameter of only 10 or 20 thousand miles, and one has been discovered that is just half the size of the moon. Yet these tiny stars have masses not much smaller than the sun's. This means that their densities are almost unimaginably high. The material of such a star must be 10,000 times denser than lead; a chunk the size of a golf ball would weigh several tons.

Normal atoms cannot pack closely enough to achieve such densities, so we must assume that most of the matter in these stars is **degenerate**. In degenerate matter the structure of the atom is broken down; all the electrons are torn from their orbital paths about the atomic nuclei. The resulting mixture of free electrons and bare nuclei behaves much like a gas, but it is vastly more compressible, since it lacks the electron orbitals that account for most of the volume of a normal atom. The surface gravity of such small but massive stars is very great; the gravitational red-shift predicted by the theory of relativity is large enough to be detected in the spectra of certain white dwarfs.

Red giants and white dwarfs differ from main sequence stars such as our sun chiefly in density; the outer regions of a red giant are almost a vacuum, but a teaspoon of matter from a white dwarf would weigh tons.

Stellar Evolution   Even as the details of the H-R diagram were being filled in, astronomers were searching for evolutionary relationships among the different classes of stars. Did stars evolve up or down the main sequence? Were red giants very old or very young stars? As our knowledge of stellar energy processes increased, the modern theory of stellar evolution began to take shape. We can now describe, in broad outline at least, how a star comes into being, evolves, and dies.

BIRTH OF A STAR   The formation of a star is very much a matter of chance. The density of gas and dust in interstellar space is subject to random variations from the average value. Wherever it happens to be relatively high, atoms of gas and particles of dust in that region may begin to condense under the force of their mutual gravitation, attracting still more matter from surrounding space. At first the result is nothing but a cold, diffuse cloud, concentrated only by comparison with the even thinner interstellar gas. But gradually, as more and more material falls together, the **protostar** becomes denser, and begins to warm up—gravitational potential energy is being converted into heat. (A century ago physicists thought this process might be responsible for the sun's energy.) Eventually gravitational compression raises the protostar's temperature to the point where nuclear fusion can take place (Chapter 22), and it becomes a star.

A star's mass is its destiny. Mass determines how long the star will take to be born, what it will be like during its life, and when and how it will die. A very massive protostellar cloud may require only a million years or less to arrive at the nuclear ignition point, while the condensation period of a smaller one may be measured in tens or even hundreds of millions of years. In either case, however, condensation takes only a small fraction of the star's life. Once at the stage of providing its own nuclear energy, it has reached the main sequence, but just where on the main sequence also depends on its mass. More massive stars are hotter and brighter, occupying the upper left corner of the H-R diagram, while lighter stars, cooler and fainter, lie along the lower right portion of the curve.

All stars remain near their original locations on the main sequence, with relatively little change in size, temperature, or luminosity, for about 99 percent of their lifespan. But that lifespan, too, is a function of the star's mass. Though the hotter, brighter stars begin with a greater mass of hydrogen, they burn this nuclear fuel much more rapidly. The most massive blue giants may run through their original supply in a few million years, while a small red star at the tail end of the main sequence may burn dimly for a hundred billion years or so. Our sun, which has been fusing hydrogen for some 4.5 billion years, probably has some 5 to 8 billion more years left on the main sequence.

OLD AGE OF A STAR   Sooner or later, the hydrogen in the central region of any star must be used up, and a core of helium forms. At this point the star leaves the main sequence, and undergoes a number of changes in rapid succession. The details of a star's late evolution are quite complex, and in part still rather hypothetical, having been worked out through computer simulations of stellar interiors. The main outlines, however, are fairly well established. The helium core of an aging star, unable to maintain its equilibrium through the production of nuclear energy, undergoes gravitational contraction, which causes it to get hotter. This heat makes the outer layers of the star expand and in this process they cool. Thus while the core becomes smaller and hotter, the star as a whole

Stars form from clouds of interstellar gas and dust by gravitational condensation.

Massive stars are hotter and expend their hydrogen in nuclear reactions faster than lighter stars.

grows larger and cooler. It moves off the main sequence, upward and towards the right on the H-R diagram, and becomes a red giant.

Eventually the core reaches the temperature at which helium itself undergoes nuclear fusion to form carbon, and the star contracts and heats up again, moving back to the left on the H-R diagram. Helium fusion cannot last nearly as long or produce nearly as much energy as hydrogen fusion, however, and soon a carbon core forms. At this point the star may move back up to the red giant region once more. If the star is massive enough, other fusion reactions may begin to take place in the carbon core, building up still larger nuclei: carbon (atomic number 6), for example, is transformed into magnesium (atomic number 12), and so on up the periodic table. The entire process of alternately ballooning into a red giant, heating up, and shrinking to leave the red giant region may occur many times over. During this time, too, the star may pass through unstable periods, and pulsate regularly. Many of the variable stars that we observe, including the Cepheids, may belong to this phase of stellar evolution. But whether the star becomes a red giant once, twice, or several times, it is now nearing its end.

The Death of Stars

Several different possible fates await the aging star, and here too its mass will be the deciding factor. Most small stars probably leave the red giant region for the last time when their outer layers become so cool and distended that they separate from the star entirely, to form a thin, expanding shell, known as a **planetary nebula**. (Astronomers, influenced by the nebular hypothesis of the origin of the solar system, at first mistook these shells for embryonic planets.) Inside this tenuous, glowing envelope of gas there is now a new star: the small, extremely hot core of the former red giant. The red giant has become a white dwarf. (See photo in the Color Portfolio.)

Red giants are stars entering old age; many of them eventually blow off their outer layers and evolve into hot, dense white dwarfs, which slowly cool to extinction.

Once the white dwarf stage has been reached the story is almost over. With scarcely any nuclear fuel left, the star's temperature and luminosity fall as it gradually radiates its warmth into space; the star moves downward and to the right on the H-R diagram. After billions of years of slow cooling, nothing will remain but a dark and extremely dense chunk of matter, a fragment of stellar ash.

SUPERNOVAS   Not all stars, however, disappear so placidly; some go out with a bang. As we have seen, very massive stars attain ever higher temperatures in their cores, making possible the successive fusion reactions that build up heavier elements from lighter ones. But even in the most massive stars, all fusion reactions must stop when the nucleus of iron (atomic number 26) is reached. The iron nucleus is the most stable of all the elements—it cannot be either fused or split without absorbing energy rather than releasing energy. Because of this, the formation of iron represents the end of the line for stellar energy production. The star's final gravitational collapse begins.

In the past, such compression had caused the core to heat up, slowing the process of collapse. This time, however, that fails to happen. Theoretical astronomers are not sure why. It may be because the nuclear reactions taking place in the heart of such a star produce immense numbers of neutrinos. These neutrinos, passing easily through matter, are radiated out into space much more rapidly than electromagnetic radiation, carrying vast quantities of energy with them. Possibly, too, the loss of energy occurs when the iron core is broken down into helium again at the enormous temperatures of the star's center.

Whatever the source of the energy leak, the result is catastrophic. The temperature of the core, instead of rising to maintain equilibrium, drops, and the star begins an abrupt and drastic collapse. Within a few minutes it has reached the limit of compression of even degenerate matter, and an explosion of unimaginable violence occurs. Compression and shock waves combine to heat the star to a temperature of billions of degrees. Unburned hydrogen in the cooler outer layers is instantly heated to fusion temperatures and explodes like a giant hydrogen bomb, adding the energy of nuclear fusion to the already prodigious energies being generated in the stellar core. A large fraction of the star's mass—perhaps as much as half—is blown off into space.

Such exploding stars are called supernovas, and they are certainly one of the most awesome phenomena in the universe. At its peak, a supernova releases the energy of billions of suns; it can outshine an entire galaxy. Though they are rare—an average galaxy with its billions of stars will have only one every few hundred years—supernovas have an importance far greater than their numbers suggest. During the brief explosion of a supernova, all the heavier elements beyond iron are created and spewed out into space, along with the elements between helium and iron formed during the star's earlier evolution. Without past supernova explosions to enrich the interstellar gas from which stars condense, our sun and our solar system would consist entirely of hydrogen and helium.

Elements heavier than helium, formed by nuclear reactions in the interiors of stars, are spewed out into space during supernova explosions and are incorporated into newly forming stars.

PULSARS AND NEUTRON STARS What is left after a supernova explosion? At first astronomers thought the remnant would be a white dwarf, surrounded by an expanding envelope of gas, rather like a large planetary nebula. Such a shell of gas had already been found in at least one case. In 1054, Chinese astronomers had observed a remarkable "new star" which was bright enough to be visible in the daytime. After many months it faded from view, but modern astronomers have determined its location from the ancient records. It turns out that in the region of the sky where the supernova was seen, there is a large, bright mass of gas known as the Crab Nebula. Moreover, the Crab Nebula is expanding at some 900 miles per second; calculating backwards from its present size, astronomers have determined that the process must have begun some 920 years ago. Almost certainly the Crab Nebula consists of material blown outward by the supernova of 1054.

But what of the star itself? Before long, many astronomers started to doubt

Figure 23-11
The Crab Nebula, an expanding mass of gas blown outward by the supernova of 1054, photographed in red (right) and blue light. At the center of the nebula is a pulsar, the remnant of the star that exploded over 900 years ago.

whether an ordinary white dwarf could result from such a cataclysmic event as a supernova explosion. Instead, some theorists predicted that the pressures inside a supernova might be so great that the atomic nuclei and electrons would be squeezed together to form a **neutron star**—a solid ball of neutrons, no more than about 20 miles in diameter, but containing much of the star's original mass. Compared to such stars, even the enormously dense white dwarfs are mere soufflés; a teaspoon of material from a neutron star would weigh a billion tons. But there was no direct evidence for the existence of such remarkable objects until 1967.

In that year radio astronomers began to detect mysterious brief, regular bursts of radio waves from several sources in the heavens. The signals were so unvarying in their pattern (the intervals between them, ranging from one-thirtieth of a second to a few seconds, did not vary for any one source by more than one part in 10 million) that astronomers thought at first they were receiving radio messages from some distant civilization. The fact that the pulses themselves lasted only about $1/100$ of a second indicated that their sources, which astronomers named **pulsars**, must be very small—perhaps only about 20 miles in diameter. This was precisely the size predicted for a neutron star.

Pulsars are thought to be rapidly spinning neutron stars, remnants of past supernova explosions.

The possibility that pulsars were in fact neutron stars was greatly strengthened when a very rapid pulsar was discovered in the center of the Crab Nebula. This object proved also to be a visual pulsar—its output of visible radiation turned on and off 30 times a second, like a strobe light. Astronomers now think that pulsars are neutron stars that spin very rapidly, flashing radiation into space in the same manner as a rotating lighthouse beacon. Since a spinning object that radiates energy in this manner must gradually slow down, scientists measured the period of the Crab pulsar with great care, and discovered that the interval between its pulses was indeed lengthening—by about one three-millionth of a second per year.

BLACK HOLES It is hard to believe that our universe could contain anything more extraordinary than supernovas and neutron stars. But the death of certain stars may give rise to an even stranger phenomenon. Theoretical calculations indicate that neutron stars do not necessarily represent the upper limit for the density of matter. Indeed, it now appears that there is no upper limit. If a star, after its supernova explosion, retains a mass greater than approximately twice that of our sun, it may continue to contract beyond the neutron star radius. In fact, there is no known repulsive force between particles great enough to halt this collapse. Theoretically, the entire stellar mass could contract indefinitely: to the size of a pumpkin, to the size of a grain of sand—no limit is known. Physicists can only speculate about the ultimate destiny of such matter; some believe it may leave our universe entirely.

A massive star may finally collapse into a black hole, producing a gravitational field so strong that not even light can escape it.

Once such a mass contracts to a radius of about two miles, its gravitational force is so great that even light can no longer leave its surface. At this point the star blinks out, and becomes merely a bottomless hole in the universe, a **black hole**. Nothing—neither matter nor radiation—can ever escape from such a black hole. While much of the speculation about black holes has so far been purely theoretical, there is now some observational evidence that black holes do exist, swallowing up light and matter like some insatiable demon out of mythology. Some astronomers feel that all the matter in the universe may eventually disappear down a black hole.

Glossary

A **black hole** is the remains of a star which has undergone total collapse, resulting in a gravitational field so intense that not even light can escape from it.

A **binary star** system (double star) consists of two stars in orbit about their common center of gravity. Multiple star systems are also known.

Cepheid variables are pulsating stars that fluctuate regularly in luminosity. Because the intrinsic brightness of a Cepheid can be determined from its period of variation, they are useful in finding stellar distances.

The **Hertzsprung-Russell (H-R) diagram** is a graph on which the intrinsic brightness of a number of stars is plotted against their temperature. Most of the stars on the H-R diagram lie along a band called the **main sequence**, running from the region of dim, cool red stars to the region of bright, hot blue stars. Off the main sequence we find **red giants**—large, diffuse, cool stars entering old age, and **white dwarfs**—small, hot, extremely dense stars in the final stage of their evolution.

A **light-year** is the *distance* that a ray of light, moving at 186,000 mi/sec, travels in a year—about 6 trillion miles.

The **mass-luminosity relationship** predicts that the mass of a main sequence star must be roughly proportional to the cube of its luminosity.

The **parallax** of a star is the apparent shift in its position when seen from opposite sides of the earth's orbit about the sun.

A **planetary nebula** is a shell of gas blown off by an aging red giant star. The small, hot core of the star that remains becomes a white dwarf.

Proper motion is the observed movement of a star across our line of sight.

Pulsars are sources of very regular bursts of radio waves. They are thought to be very rapidly spinning **neutron stars**—stars that have collapsed to form super-dense balls of neutrons only about 10 miles in radius.

Radial velocity is the movement of a star along our line of sight, either approaching or receding from us.

The **red shift** is a phenomenon observed in the spectra of objects moving away from us. Emission and absorption lines from such objects are displaced towards the red end of the spectrum by the Doppler effect.

A **stellar spectrum** consists of a **continuous spectrum** of colors—the radiation from the star's hot, dense interior—on which are superimposed many thin dark **Fraunhofer** or **absorption lines**. These represent wavelengths absorbed by the gases of the star's atmosphere. The positions of these lines identify the elements producing them. Stellar spectra also provide information about the temperature, pressure, degree of ionization, and magnetic field of the star, as well as its radial velocity and rate of rotation.

A **supernova** is an exploding star that may emit as much light as billions of ordinary stars. Supernovas eject much of their original mass into space, enriching the interstellar gas from which new stars are formed with heavier elements built up by nuclear fusion in the interior of the exploding star.

Exercises

1 Are the stars of a particular constellation necessarily near each other in space? Why do you suppose the constellations visible from the northern hemisphere are named after mythological creatures, while many in the southern skies are named after scientific instruments?

2 What is proper motion? Why didn't ancient astronomers discover its existence?

3 Is it possible to determine the true velocity (both magnitude and direction) of a star from a knowledge of its proper motion? What other information is needed? How are such data usually obtained?

4 Why are Cepheid variable stars so useful to astronomers? If we know the distance of a star, what other property can be determined very easily?

5 What can astronomers learn from the study of binary star systems?

6 Briefly outline the role of the spectroscope in astronomy. Why are wavelengths characteristic of particular elements in a star *missing* in its continuous spectrum?

7 How does the color of a star indicate its surface temperature? If two stars in a binary system are different colors, one red and the other blue, but have the same apparent magnitude, what can you deduce about their relative sizes?

8 Why is there a relationship between a star's mass and its luminosity for main sequence stars? How much more luminous than the sun is a star of twice the sun's mass?

9 Two stars of spectral type M are found to differ in luminosity by a factor of a million. Explain the difference in their size and structure that is responsible. If these two stars are the same age, how would you expect their masses to differ? If they have roughly the same mass, what can you conclude about their ages?

10 What are the three possible *final* results of a star's evolution? Have all three objects been observed? If not, how can they be detected?

11 Why do we owe the existence of life on earth to the deaths of stars more massive than the sun? If our sun had been formed some 10 billion years ago, when our universe is thought to have been very young, what kind of planets could it have produced?

12 Why is it unlikely that an advanced species such as man could evolve on a planet circling a star many times more massive than the sun? (Hint: The first living things appeared on earth about 4 billion years ago.)

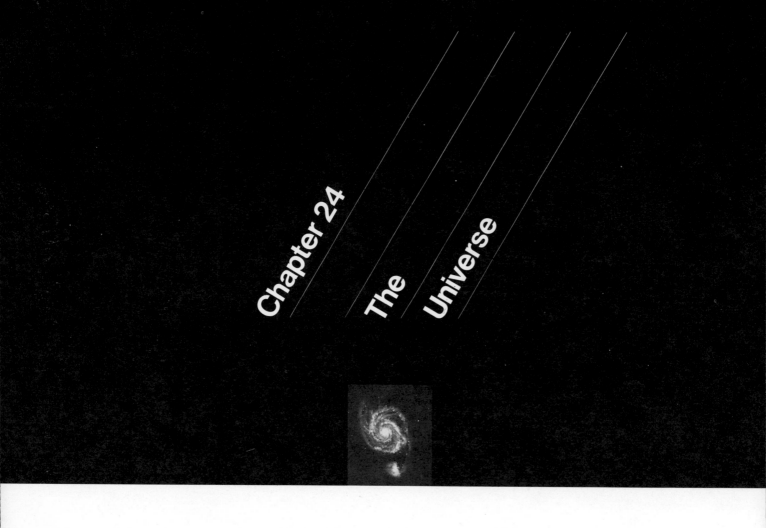

Chapter 24
The Universe

Our Galaxy

On a clear, moonless night, away from the city lights and haze, we can see the sky with its beautiful profusion of stars—brilliant nuclear beacons whose light has traveled trillions of miles over millions of years to reach us. As our eyes become accustomed to the darkness of the sky, we notice a diffuse band of light stretching from horizon to horizon, a tremendous arch across the vault of the heavens. The Greeks named this the **Milky Way**, a poetic metaphor which has stayed with us ever since.

The Greeks speculated that the Milky Way was really a dense band of very faint stars, but this idea was not confirmed until 1610. When Galileo turned his primitive one-inch telescope towards the celestial belt, he found that it indeed consisted chiefly of myriad tiny points of light—stars too dim and closely-spaced to be distinguished by the naked eye. Until then, the distribution of stars in the sky

Figure 24-1
A portion of the Milky Way in the constellation Sagittarius, near the direction of the center of our Galaxy.

had seemed roughly uniform. But the existence of this vast concentration of stars in the Milky Way forced astronomers to alter drastically their picture of the universe. The sun, it seemed, was a member of an immense structure of stars; what we call the Milky Way was merely its distant reaches seen from our position within it. This vast structure came to be known as the **Galaxy** (from the Greek word for "milky way").

When (in the late eighteenth century) Herschel undertook a systematic star count of various regions of the sky, the first quantitative determination of the shape of our Galaxy became possible. Herschel found that the Galaxy seemed round and rather flat, like a double convex lens, or a pair of frisbees joined bottom to bottom. The ribbonlike appearance of the Milky Way is the result of our perspective as we peer out through the Galaxy. When we look in any direc-

Configuration of the Galaxy

The Milky Way consists of hundreds of thousands of distant stars in our own Galaxy.

tion along the central plane, we see the greatest number of stars—so many and so distant that they blend together to form the Milky Way.

Herschel's conjectures about the size and shape of our Galaxy were refined by other astronomers, especially the American Harlow Shapley (1885–1974), who was able to show that Herschel had been mistaken in thinking the sun was near the center of the galactic disc. When Herschel made his star counts, he had found approximately the same star density in all directions along the central plane of the Milky Way. It seemed logical that this would only be true if the stars were observed from the center of the Galaxy. Observing from anywhere else, one would expect to find a higher star density in the direction of the galactic center and lower densities out toward the edge. But Shapley, through studies of groups of stars, was able to show in 1917 that the sun is actually quite far from the center of the Galaxy and that the center lies in the direction of the constellation Sagittarius. More recent investigations have explained how Herschel, despite his correct reasoning, was misled. He did not know that clouds of tiny, interstellar dust particles lie between our solar system and the galactic center, obscuring much of the visible light from that region.

Figure 24-2
Two galaxies that are probably rather similar in structure to our own Milky Way Galaxy. One is seen edge on, the other at about a 45° angle to the central plane.

THE SIZE OF THE GALAXY   Radio frequencies are not obstructed by interstellar dust particles, so astronomers have been able to use them to map our Galaxy with greater precision. Its diameter is now thought to be about 100,000 light-years. Our sun is situated very close to the central plane, some three-fifths of the way out toward the edge of the Galaxy—about 27,000 to 30,000 light-years from the galactic center. The overall shape of the Galaxy is, as Herschel proposed, that of a lens, though a somewhat thinner, flatter lens, in proportion to its diameter, than he imagined.

In the middle of the Galaxy there is a thick bulge, almost spherical in shape, with a radius of 10,000 light-years. This region is known as the **galactic nucleus**, and there the stars are crowded together rather more closely than in the surrounding disc. Outside the nucleus, in the main body of the disc itself, it is now known that the stars are not distributed uniformly, but instead form a series of curved arms, rather like the vanes of an enormous pinwheel, coiling in widening spirals about the galactic nucleus. These spiral arms contain many of the stars of the Galaxy, and our sun appears to be situated in one of them (Figure 24-3). Knowing the volume of the Galaxy and its average star density, and taking into consideration the screening effect of dark galactic dust clouds, it is possible to estimate the number of stars in it—about 100 billion stars.

GALACTIC ROTATION   The flat, circular shape of both the solar system and the Galaxy led the German philosopher Immanuel Kant (1724–1804) to speculate that the stars of the Galaxy are revolving about its center, much as the planets circle the sun. This must be the case, for there is no known repulsive force keeping the stars apart; instead their mutual gravitational attraction can only be

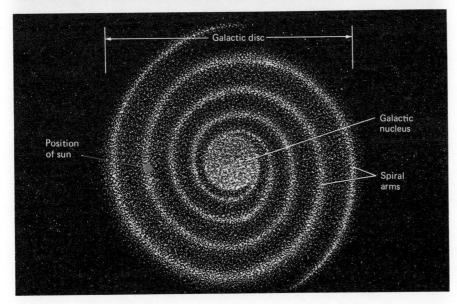

Figure 24-3
A schematic diagram of our Galaxy. The diameter of the galactic disc is about 100,000 light-years. The earth is located in one of the spiral arms, about ⅗ of the way out from the galactic center. Our view of the central regions of the Galaxy is obscured by interstellar dust.

balanced by the centrifugal reactions of orbital motion. If the Galaxy revolved as a rigid body, like a wheel, the velocity of the stars would be proportional to their distance from the galactic center—those on the outer edge would revolve faster than those close to the hub. Instead, the stars seem to follow the pattern of the planets in our solar system, with those nearest the center moving at the greatest velocities. Careful analysis of the Doppler shifts indicates that the sun is moving in a circular orbit around the center of the Galaxy at about 150 mi/sec. At this rate, it completes one revolution in roughly 200 million years. Since most nearby stars are moving in the same direction in very similar orbits, their velocities relative to the sun are usually not too high—on the order of 10 to

Our Milky Way Galaxy is a great pinwheel shaped system 100,000 light-years in diameter; our sun completes one revolution about the galactic center every 200 million years.

20 mi/sec in most cases. A few stars, however, have elongated elliptical orbits similar to those of comets in our solar system. Since such stars are not moving parallel to the sun's path, their velocities relative to the sun are greater—50 mi/sec or more.

In recent years scientists have amassed evidence from a number of sources suggesting that our galaxy has a significant magnetic field. One indication of this is the behavior of cosmic ray particles, which are thought to attain their enormous velocities by interaction with the galactic field. The spiral arms of the galaxy are probably held together at least partly by magnetic forces. The nature of magnetic interactions on a cosmic scale is still largely a mystery, but astronomers are beginning to suspect that magnetic fields may be as important a factor as gravitational fields in the structure of the universe.

Star Clusters

While gravitational attraction keeps the stars revolving about the galactic center, it also can serve as a link between nearby stars, causing them to move together as a group through the Galaxy. Often, large numbers of stars may be united gravitationally in groups called **star clusters**. These clusters can contain anywhere from a dozen or so loosely connected stars to a million or more so closely grouped that even the largest telescopes cannot resolve individual stars in their center. The fact that the stars in clusters move through space together strongly suggests that they were formed together from a single cloud of gas. Thus, the stars of any given cluster are probably about the same age, which makes clusters very useful to astronomers studying stellar evolution. There are two types of star clusters: galactic and globular.

Galactic clusters are relatively loose associations of stars found in the galactic disc. The best-known of these is the Pleiades, called the Seven Sisters by the ancients. The six or seven closely-spaced stars that you can see with the naked eye becomes dozens with binoculars and several hundred with a telescope. Numbers of stars in other galactic clusters vary from 10 or 20 to more than 1000.

The **globular clusters** are much larger, denser, and farther away than the galactic clusters. These spherical clusters range in diameter from 65 to over 350 light-years. Typically, they contain hundreds of thousands of stars—sometimes as many as a million. To someone observing the striking image of a globular cluster in a telescope or on a photographic plate, it might seem that the central regions are so crowded with stars as to leave hardly any space between them. (See photo in the Color Portfolio.) Even in the densest parts of a globular cluster, however, neighboring stars are probably separated by several light-months—that is, by a trillion miles or more.

The 100 billion stars of our Galaxy are widely scattered through space; even in the heart of a globular cluster, neighboring stars are trillions of miles apart.

Unlike most of the stars of the Galaxy, the globular clusters are *not* usually found in or near the plane of the galactic disc. Instead they are distributed in a roughly spherical volume about the galactic center. This spherical region, which also contains a thin scattering of stars not belonging to the globular clusters, is known as the galactic corona or **halo** (Figure 24-3). Its diameter is at least that of the galactic disc itself, 100,000 light-years, and it may extend quite a bit farther in some directions, for certain globular clusters have been observed at distances of 60,000 light-years or more from the galactic plane. Because of their location outside the main disc of the Galaxy, globular clusters are relatively easy for astronomers to observe. Over a hundred are known in our Galaxy, and many have been discovered in photographs of distant galaxies.

Even more difficult to imagine than the immense number of stars in our Galaxy are the vast expanses of nearly empty space through which these stars are scattered. For the Galaxy is far from being crowded; like the Bohr atom, it is mostly empty. The average distance between stars in our Galaxy is over 30 trillion miles. To grasp the implications of this figure, let us imagine a scale model of the solar system, using a ridiculously small scale of 1:10 billion. On this scale, the sun has a diameter of only 5½ inches, about the size of a large grapefruit. The earth shrinks to a dot 1/20 of an inch across—slightly smaller than the "o" in "dot." (The human beings inhabiting an earth on this scale would be a hundred times smaller than the tiniest viruses.) Still, the earth would be about 50 feet from the sun, and Pluto almost 2000 feet, or about 7½ city

Interstellar Matter

Figure 24-4
Left: a galactic star cluster. The presence of large amounts of dust and gas—indicated by the bright nebulosity surrounding the stars—suggests that the cluster is a young one, occupying a region where star formation is still taking place. Right: the Lagoon Nebula in Sagittarius. The dark areas are clouds of interstellar dust not illuminated by starlight.

blocks. The nearest star, about 7000 times as far from the sun as Pluto, would be another grapefruit 2500 miles away—about the distance between New York and San Diego. The rest of the Galaxy would consist of billions of grapefruits, each about 3000 miles from its nearest neighbor. Even on this drastically reduced scale of 1:10 billion, the farthest stars of our Galaxy would still be almost 50 million miles away.

By terrestrial standards these inconceivably vast spaces between the stars contain . . . nothing. It has been estimated that interstellar space contains, on the average, only a single atom of gas per cubic centimeter of its volume, and a mere handful of microscopic solid particles per cubic mile. Such densities represent a vacuum millions of times greater than any that can be achieved in an earthly laboratory. Yet the Galaxy is so huge that these rarefied wisps of gas and dust, spread out through interstellar space, add up to a tremendous amount of matter—enough, perhaps, to constitute several percent of the total mass of the Galaxy, the equivalent of two or three billion stars.

The space between the stars contains scattered atoms of gas (mostly hydrogen) and tiny solid dust particles.

NEBULAS Early astronomers were aware of the presence of matter in "empty" space only in certain regions, where unusually dense, localized concentrations of dust or gas, resembling small luminous clouds, could be seen in telescopes. Many of these bright, hazy areas were discovered in the eighteenth and nineteenth centuries; because they resembled clouds, they were called **nebulas** (the Latin word for "cloud").

Where interstellar dust is relatively concentrated and illuminated by the light of nearby stars, the result is a **reflection nebula**. Since light at the blue end of the spectrum is scattered by such particles more than light of longer wavelengths, reflection nebulas are almost always blue. When these dust clouds are not in the neighborhood of bright stars, they remain dark and may block out the light of the stars lying beyond them. Such regions, known as **dark nebulas**, were once thought to have no stars at all, or even to be "holes in the heavens." They can be seen prominently in any photograph of the Milky Way, and they are often silhouetted against nearby bright nebulas, creating beautiful, intricate patterns of dark and light (Figure 24-4).

There is much more gas than dust in the interstellar spaces, but it has very little effect on our view of distant objects. Compared to the dust particles, the interstellar gas is almost completely transparent. But the atoms in a mass of such gas may be excited by strong magnetic fields or by high energy radiation from hot stars in its vicinity, causing it to shine by fluorescence. Such a body of luminous gas, one of the most beautiful sights in the heavens, is called an **emission nebula**. (See photo in the Color Portfolio.) The hot, ionized gases of an emission nebula radiate light at wavelengths characteristic of the particular atoms present, so the emission and absorption spectra of interstellar gas enable us to determine its composition: 90% H, about 9% He, traces of C, N, O, Fe, Ca, Na.

Figure 24-5
The giant 1000-foot radio-radar telescope "bowl" at the Arecibo Observatory, Puerto Rico.

THE RADIO SPECTRUM In the past two decades, astronomers have discovered through the use of radar that there is considerable radio and microwave radiation from outer space. These wavelengths, longer than those of visible light, pass easily through the earth's atmosphere and the dust of interstellar space. It is not only their penetrating power, however, that makes them so useful. The emission and absorption lines of many atoms, molecules, and molecular groups lie in the radio spectrum.

In the past decade more than 30 compounds of carbon, nitrogen, oxygen, hydrogen, and sulfur have been discovered in space, chiefly through the detection of their characteristic absorption or emission lines by radio telescopes. Some are simple substances: the hydroxyl group, —OH; the cyanogen group, —CN; water, H_2O. More elaborate organic compounds, such as acetaldehyde and methyl alcohol, have also been found. This surprising development has implications for the study of the origins of life, for it suggests that the process of building the organic molecules upon which life depends may have begun in the vast "empty" reaches of galactic space before the creation of the earth.

Cosmic Rays

Various kinds of electromagnetic radiation reach the earth from space: visible light, ultraviolet and infrared, radio waves, x-rays. But the earth is constantly bombarded not only with energy but also with matter—atomic particles traveling at incredibly high velocities from distant reaches of the cosmos. These particles are called **cosmic rays.**

Their discovery was an outgrowth of early studies of ionizing radiation. When an atomic particle with a high velocity, and thus a large kinetic energy, collides with an atom in our atmosphere, it can dislodge an electron from its orbit, creating an ion. Ions can be easily detected because of their electrical charge. Early in this century scientists found that such ionization occurred continuously in our atmosphere. Radioactive elements in the earth's crust were the only natural sources of high-energy particles then known. But the study of ion distribution in the atmosphere produced a surprising result. The higher the test balloons went, the more ionization they detected—just the opposite of what might have been expected if the source of the radiation was in the earth. Evidently the ionizing particles came not from earth but from outer space.

With the advance of rocketry it became possible to send sophisticated equipment to high altitudes to investigate the nature of these energetic particles. About 90 percent of them turned out to be bare protons—that is, hydrogen nuclei. Most of the rest are helium nuclei (alpha particles), with a sprinkling of nuclei of heavier atoms.

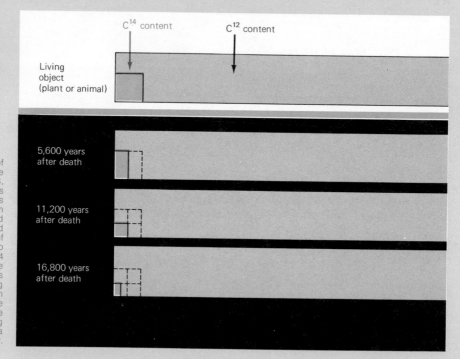

One of the fortunate results of cosmic-ray bombardment is the creation of radioactive carbon, C-14, by the splitting of a nitrogen nucleus in our atmosphere. The C-14 reacts with oxygen in the air to form radioactive CO_2, which is absorbed by living things at a known rate and incorporated into the chemistry of their bodies. After they die, they no longer absorb CO_2, so the C-14 present in their bodies cannot be replenished and slowly disappears by radioactive decay. By measuring the ratio of C-14 to normal carbon (C-12), it is possible to determine the length of time since the specimen died, thus giving archeologists and paleobiologists a calendar of prehistory.

Their proportions seem to reflect the overall abundance of these elements in the universe as a whole, as we know it from the study of stars and of the interstellar gas. All these **primary cosmic ray** particles arrive with speeds close to that of light. They thus possess enormous kinetic energies, typically several billion electron-volts (Bev), with a few ranging as high as 100 billion Bev—10^{20} electron-volts, or about 12 ft-lbs. (To see how remarkable this is, imagine a proton with this energy striking one end of a seesaw, the other end of which supports a 1 lb weight. Though the proton has a rest mass of only 3.7×10^{-27} lbs, its impact would be great enough to raise the weight 12 feet.)

The number of cosmic rays reaching our atmosphere every day lies in the billions of billions. Because they are electrically charged, they are influenced by our planet's magnetic field and tend to spiral in along the lines of magnetic force that emanate from the earth's north and south magnetic poles. The most energetic particles penetrate the field even in the equatorial regions, but most are deflected towards the north or south, so that the intensity of cosmic-ray bombardment increases as one moves towards higher latitudes. Some particles are caught and permanently held by the field, forming the Van Allen radiation belts that girdle the planet.

Primary cosmic rays are so energetic that if they collide with any atomic nucleus in our atmosphere, as they almost invariably do, that nucleus will often be fragmented into several lighter nuclei and subatomic particles—protons, neutrons, mesons, hyperons, electrons, and gamma rays. These **secondary cosmic rays** interact and decay in a variety of complex ways; they may also strike other atoms in the air, generating additional secondary particles. While primary cosmic rays seldom reach the earth's surface, large numbers of secondary particles formed in such "showers" do.

Scientists are not sure where cosmic rays originate or how they attain their immense energies. Since they arrive around the clock in equal numbers from all directions, we can rule out the sun as their principal origin. It has been suggested that supernova explosions could be the initial source of the particles, which then acquire additional energy boosts by their interaction with magnetic fields. But even a magnetic field the size of the Galaxy could not provide anywhere near the known upper limit of cosmic-ray energy. Higher energy particles must come from other galaxies with stronger magnetic fields; the particles travel millions of years through intergalactic space until they reach their final destination in our atmosphere.

A spark chamber consists of layers of metal plates maintained at high electrical potential. A charged particle passing through the chamber causes tiny sparks to leap from plate to plate along its path. The spark track in this photo is that of a meson, produced by a high energy cosmic ray.

Even in cool regions, where gases are not ionized, hydrogen atoms emit a clear, characteristic radio signal at a wavelength of about 21 cm, making it possible to map areas of neutral hydrogen in interstellar space. Since hydrogen is by far the most abundant element in the interstellar gas, and since this gas is concentrated in the spiral arms of the Galaxy, such maps are extremely valuable in determining the size and structure of our Galaxy. Since radio frequencies pass easily through the clouds of interstellar dust that obscure visible light, radio mapping can reveal distant parts of the Galaxy, especially the galactic center, inaccessible to optical observation. Studies of the Doppler shift of the 21-cm line can even tell us about the motion of the interstellar gas and the rotation of the entire Galaxy.

Radiotelescopes enable astronomers to map regions of the galaxy whose visible light cannot reach us and to detect various kinds of atoms and molecules in interstellar space.

The Stellar Populations

We have been discussing our Galaxy as if stars of various size, age, temperature, and luminosity were distributed at random throughout. But in the early 1940s it was discovered that this is not the case. During the World War II blackout of Los Angeles, Walter Baade, using the 100-inch telescope at Mount Wilson Observatory, had the opportunity to take unprecedentedly clear photographs of the Andromeda Galaxy, a galaxy thought to be very similar to our own. These pictures enabled Baade to study the individual stars in the nucleus of the Andromeda Galaxy, the first time such stars had been studied in any galaxy.

The stars in the nucleus and those in the spiral arms of the Andromeda Galaxy seemed to differ considerably. The former were predominantly cooler, reddish stars of moderate size and luminosity; the latter tended to be larger, brighter, hotter, and more varied in their spectral types, including many blue and blue-white giants—a type virtually never found among the stars of the nucleus. Astronomers have named the stars found in the galactic disc **Population I** stars, and those found in the galactic nucleus **Population II** stars. The stars in the globular clusters and the scattered stars of the galactic halo are members of Population II.

OLD AND YOUNG STARS Spectroscopic studies have revealed additional differences between these two stellar classes. Population II stars are composed almost entirely of hydrogen and helium, with hardly any of the heavier elements, whereas Population I stars tend to be much richer in the elements beyond helium. This fact furnishes a clue that helps explain the existence of these two contrasting populations. It is now thought that the first elements, from which the oldest stars were formed, were hydrogen and helium. Elements heavier than helium have been created by nuclear reactions in the interiors of stars, and spewed forth into the Galaxy in the violent explosions that accompany the final phases of stellar evolution. Population II stars would seem, then, to belong to the "first generation" of star formation, dating back to the earliest eons of the

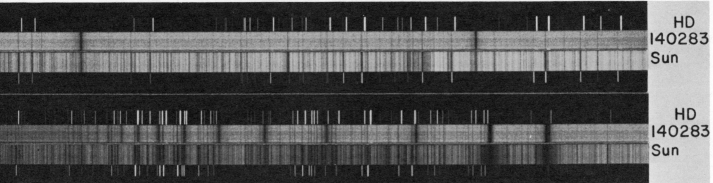

HD
140283
Sun

HD
140283
Sun

Figure 24-6
The solar spectrum (bottom band, labeled "Sun") compared to that of a Population II star (top band, labeled "HD 140283"). Population II stars, formed early in the history of the universe, consist almost entirely of hydrogen and helium. The sun, a relatively young Population I star, contains small amounts of many other elements, as indicated by the greater number of dark absorption lines. The interstellar gas from which stars are formed is gradually becoming enriched with elements heavier than helium, formed in the interior of hot, massive, short-lived stars and spewed forth into space by supernova explosions.

universe. Population I stars, on the other hand, are younger, formed from galactic gas and dust that had already been enriched with substantial amounts of heavier elements from dying stars of earlier generations.

If this is true, it also explains why blue giants are never found in Population II regions. These hot, massive stars, burning their nuclear fuel at a tremendous rate, have short life spans, so any such stars formed early in the existence of our Galaxy must have burned themselves out long ago. Thus, Population II regions contain only the smaller, longer lived, redder stars. The blue giants we see in the spiral arms of the Galaxy—in the vicinity of our own sun, for example—must be relatively young stars, formed quite recently by galactic standards.

This hypothesis receives strong confirmation from another piece of evidence. Astronomers have found that, whereas the Population I regions of the galactic disc are rich in dust and gas, the Population II regions are almost completely free of such interstellar material. The gas and dust in these areas must have been used up in star formation ages ago. In the spiral arms of the Galaxy, there are not only many young stars, but every indication that star formation is still taking place. In fact, photographs of nebulas often show small, dark, spherical shapes, which may be clouds of interstellar material in the process of condensing to form stars.

The Population I stars of the galactic disc are relatively young; the galactic nucleus and globular clusters contain older stars of Population II.

Kant was the first to suggest that the Galaxy was perhaps not the only universe in a vast sea of space. He speculated that other "island universes" might exist, so far from our Galaxy that telescopes could not resolve them into individual stars. Instead, they would appear as hazy patches, rather like the Milky Way as it appears to the naked eye.

Beyond the Milky Way

Figure 24-7
The great galaxy in Andromeda, thought to be very similar to our own. The Andromeda galaxy is a close neighbor, only about 2 million light-years distant. Like our Milky Way Galaxy, it contains about 100 billion stars.

THE ANDROMEDA NEBULA Such sights were in fact known to eighteenth century astronomers. One was the Andromeda Nebula, on a clear night visible to the naked eye as a faint, fuzzy star. Through the telescopes of the period, it resembled a small elliptical cloud not resolved into individual stars. Consequently, most astronomers counted these hazy patches among the gaseous nebulas and were convinced that they were located within our Milky Way.

Near the end of the nineteenth century supernova explosions were observed in the Andromeda Nebula. This offered a method of estimating its distance. Astronomers assumed that the true brightness of these supernovas was about the same as the intrinsic brightness of comparable ones in our own Galaxy whose distance from earth had been determined by other means. If this was so, then the *apparent* brightness of the Andromeda supernovas would serve as an indication of their distance.

Such calculations suggested that the distance of the Andromeda Nebula was at least several hundred thousand light-years away. This definitely placed it far outside the Milky Way Galaxy. Speculation that Andromeda and other objects like it were really distant galaxies like our own was powerfully reinforced by this discovery, but the evidence was not conclusive. Then, in 1917, Hubble focused the new 100-inch Mount Wilson telescope on the Andromeda Nebula and was able for the first time to resolve individual stars on its outskirts. This indisputably confirmed Kant's hypothesis of other "island universes" and in one glance vastly enlarged the size of the known universe.

Other Galaxies

By the turn of the century, over 13,000 nebulas of the Andromeda type were known. Each of them turned out to be another galaxy, far from the Milky Way. It has been estimated that as many as a million galaxies lie within the reach of the 200-inch Mount Palomar telescope, some of them billions of light-years away. Larger telescopes and better photographic methods soon enabled astronomers to observe more detail, and a method of classifying galaxies according to their form was devised. Four major categories are recognized: elliptical, spiral, barred spiral, and irregular. Three of these are shown in Figure 24-8.

GALACTIC EVOLUTION   How are these types of galaxies related? Many astronomers have proposed evolutionary interpretations, in which the different galactic forms represent phases in the life span of a galaxy. One such theory states that the galaxies began as enormous clouds of gas and dust, far greater in diameter than galaxies today. These gradually condensed in spherical form. According to this theory, a galaxy starts its existence as an elliptical galaxy. Its spin slowly causes it to flatten out, and the galaxy evolves from a spherical to a pancake shape. In time, this rotation gives rise to spiral arms, which gradually become looser and more extended. Eventually, the galaxy loses its spiral character entirely, and ends its life as an irregular galaxy.

Figure 24-8
Three of the four principle galactic types. A and B are spirals: A is a tightly coiled spiral; B is a loose, open spiral. C is an elliptical galaxy. D is a barred spiral. A small number of galaxies are classified in a fourth category, irregular.

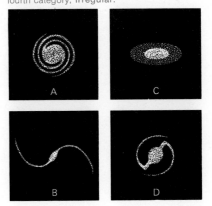

A second theory proposes just the opposite sequence of evolution. It suggests that galaxies develop from a childhood as irregulars, to a middle life as spirals, and finally an old age as ellipticals, after gravitational attraction has pulled the spiral arms in closer and closer to the nucleus.

The discovery of the two stellar populations raised serious objections to both of these hypotheses. The elliptical galaxies were found to consist almost entirely of very old Population II stars, with no dust and gas in the regions between them. Such a galaxy could not possibly evolve into a spiral galaxy containing many young stars and large amounts of interstellar material, as the first theory requires. But spiral and irregular galaxies also contain many Population II stars, so it is questionable whether they could represent recently formed galaxies, as the second theory proposes. The entire question must be considered open. Much more detailed information will have to be amassed before an entirely satisfactory answer can be formulated.

CLUSTERS OF GALAXIES Astronomers were naturally eager to determine the distances of the thousands of known external galaxies. The discovery of Cepheid variable stars in some galaxies provided the first intergalactic yardstick. Using this method of finding distance, astronomers have established that the Andromeda Galaxy, our closest galactic neighbor (except for the two small satellite galaxies known as the Magellanic Clouds, visible only from the southern hemisphere), is between 2 and 2¼ million light-years away. But Cepheids are resolvable only in nearby galaxies; for more remote objects, other techniques had to be devised. Astronomers assumed, for example, that the brightest star in each galaxy had the same intrinsic luminosity as the brightest star of known distance in our own galaxy. Comparison of the apparent brightness of the two stars would then give the ratio of their distances by the inverse square law. Similar comparisons were undertaken using the brightness of supernovas, extending the yardstick still further.

Our local group of galaxies has only about 20 members, but other clusters of galaxies may include many thousands.

Using such methods, astronomers discovered the existence of groups of galaxies. Our own Galaxy seems to belong to a rather small cluster of galaxies, about 3 million light-years in diameter and containing perhaps twenty or so members. Further away from our immediate neighborhood there are many far larger clusters, often containing thousands of galaxies. The giant cluster in the constellation Hercules, for example, is about 300 million light-years distant, and contains roughly 10,000 galaxies. About 3000 such large clusters have been mapped by astronomers, lying at distances of up to 3 billion light-years. There is even evidence that clusters of galaxies are grouped together in superclusters.

Galaxies in Flight

HUBBLE'S LAW In the second decade of this century, astronomers were able to make spectrograms of some of the more prominent spiral nebulas. Surprisingly, the vast majority of spectra from spiral nebulas exhibited strong Doppler red shifts—that is, they were moving away from us, in many cases at startlingly high speeds.

Shortly thereafter, Hubble proved that these objects were external galaxies, and he began to determine their distances by the methods described in the previous section. He discovered that the farther away the galaxy, the faster it was moving. In all cases, the distance of the galaxy was directly proportional to its velocity of recession. This simple but far-reaching principle, known as **Hubble's law**, can be expressed mathematically as:

Velocity $= H \times$ distance

Figure 24-9
Part of the great cluster of galaxies in the
constellation Hercules. Several dozen
galaxies of various types are visible in the
photograph; the cluster may contain as
many as 10,000 in all.

where H is an algebraic constant (Hubble's constant). Recent measurements indicate that H has a value of about 17 km/sec per million light-years. If we assume this simple ratio to be valid throughout the universe, galaxies far distant from us must be receding at enormous velocities. According to the theory of relativity, no object can move at a speed greater than that of light. If this is so, Hubble's law enables us to calculate the size of the universe, for it tells us how far away a galaxy must be to be receding from us with the velocity of light. Setting the velocity in the equation for Hubble's law equal to the velocity of light, c, the maximum observable distance r will be c/H, or about 18 billion light-years. We have already discovered galaxies fleeing us at 90 percent of the speed of light, which implies that we may already be approaching the theoretical limits of the universe.

THE EXPANDING UNIVERSE What does Hubble's law tell us about the makeup of the universe? Since all velocities are recessional (except those of our local group galaxies, which appear to travel through space with us), it would seem to imply that all the galaxies are fleeing our neighborhood. This seems to suggest that we are at the center of the universe—a view that has not been acceptable to scientists and philosophers since the Copernican revolution. If we look at certain analogous situations, we can see that this is not the only conclusion to be drawn from the observable facts, however.

Let us consider a two-dimensional example—the surface of the planet on which we live. Because it is so large, it seems flat to us, though it is actually the surface of a sphere. Suppose the earth itself were expanding at a steady rate of 10 percent a day, like a balloon being inflated. The earth's surface would become larger and larger, and the distance between any two points on it would change accordingly. On Monday morning an observer in New York would find that Washington was 200 miles away, Des Moines 1000, London 3500, and Honolulu 5000. Waking Tuesday morning and repeating his measurements, he would discover that all the cities had moved overnight. Washington would be 220 miles distant, Des Moines 1100, London 3850, and Honolulu 5500. He would be forced to conclude that all the cities he could see were moving away from New York. He might also notice that the more distant the city, the faster it seemed to be fleeing. A city 200 miles away seemed to be receding at 20 miles per day; a city 1000 miles away at 100 miles per day, and one 5000 miles away seemed to be receding at 500 miles per day. This relationship could be formulated into a simple law: the velocity of recession is directly proportional to the distance, or $V = Hd$.

Our hypothetical observer would have noticed that, no matter in which direction he looked, the velocities he observed were always away from New York. He would assume that New York was the center of this strange flight, although he might wonder why it should have been selected to play such a unique role. But his assumption is based on an erroneous interpretation of the situation. New York is not the center, nor are the other cities fleeing it. In fact, there is no center. An observer in any city on earth would perceive exactly the same phenomenon: all other cities receding at a velocity proportional to their distance from him. This is because *all* points are getting further apart, by the same fraction of their initial distance, as the earth's surface gets larger.

All the galaxies are moving further apart as the universe expands; the further a galaxy from us, the greater its velocity of recession.

Astronomers believe that something like this is happening to the observable universe. The galaxies and clusters of galaxies we see are all moving further apart. No galaxy can be said to be the center, and an observer in any galaxy will perceive all other galaxies fleeing his vicinity with speeds that increase with their distance from him. Hubble's law thus describes not a local oddity or peculiar coincidence, but a fundamental characteristic of the cosmos itself: the entire observable universe is expanding. This is the most important, far-reaching, and fascinating discovery of modern astronomy. But its implications are by no means simple or clear, as we shall see later in this chapter.

The development of modern astronomy, particularly radio astronomy, has led in the past two decades to the discovery of many strange objects and cataclysmic events in the distant reaches of the cosmos. These phenomena, not yet fully understood, are of great interest to astronomers. A few of them are discussed in this section.

Mysteries and Catastrophes

COLLIDING GALAXIES Galaxies are, of course, much farther apart than stars within a galaxy. But the distance between neighboring stars is almost always millions of times their diameters, whereas the average separation between neighboring galaxies is only about 10 to 100 times their diameters. So intergalactic space is much more crowded with galaxies, in proportion to their size, than a galaxy is with stars. Because the average separation of stars is so great in comparison with their diameters, stars rarely collide—astronomers have never witnessed such a collision. But galaxies, so close together for their size, collide fairly often. Since it takes hundreds of thousands of years for one galaxy to pass through another, we can observe quite a few such collisions in photographs taken with the larger telescopes.

Figure 24-10
This mysterious object, known as NGC 5128 in astronomical catalogs, is a strong radio source. It may be an exploding galaxy or two galaxies in collision.

These collisions are suprisingly uneventful. This is because a galaxy, as we have seen, is mostly empty space. Its stars are so far apart that, even when two galaxies are passing through each other, stellar collisions remain extremely unlikely. The clouds of gas and dust between the stars, however, may perhaps interact during such collisions. Some astronomers have suggested that as the interstellar material becomes more concentrated and more turbulent due to the collision, the formation of new stars is greatly accelerated. Others think that the collision may sweep the dust and gas right out of one or both of the galaxies into intergalactic space, effectively halting further star formation in those galaxies.

RADIO GALAXIES AND EXPLODING GALAXIES In the 1940s and 50s, astronomers discovered a large number of strong radio sources in the heavens. At first it was believed that most of these were "radio stars," but there were two objections to this theory. For one thing, astronomers knew of no reason why any star should emit substantial amounts of radio waves. Then, too, visible stars were rarely found in the regions from which the radio emissions seemed to emanate. When the most powerful telescopes were focused on these areas, astronomers generally found instead either expanding envelopes of gas—the remains of past supernovas—or distant galaxies, often very peculiar looking ones. Many of these radio galaxies seem to be pairs of galaxies in collision, while others appear to be undergoing some sort of incredibly violent explosion. Whatever the cause of the radio emissions, it must be sufficient to account for a truly staggering release of energy, for some radio galaxies have been found to radiate several times as much energy in the radio spectrum as our own Galaxy puts out at *all* wavelengths combined.

Many distant galaxies have been found to be strong sources of radio emissions.

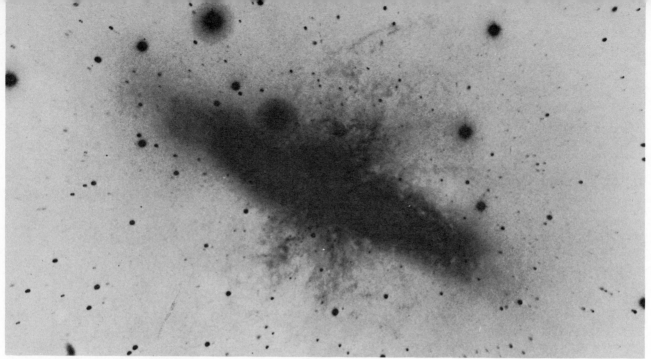

Figure 24-11
M-82, a radio galaxy. This is a negative print, often used by astronomers because it makes fine detail more clearly visible. The galaxy appears to be in the throes of a violent explosion, the nature and cause of which are not yet understood.

When the radio signals from such galaxies are mapped, they are often found to emanate not from the galaxy itself, but more or less symmetrically from two areas on either side of it. Astronomers therefore theorize that the radiation is produced by clouds of interstellar gas blown out of the galaxy into the surrounding space by some extraordinary event, presumably a collision between galaxies or an explosion. This hypothesis is supported by telescopic observation, which reveals visible jets of matter shooting forth from a few of the radio galaxies. We cannot be certain it is adequate for all cases, however. There are some galaxies which appear completely normal, without any sign of an unusual or cataclysmic occurrence, but nevertheless are strong radio sources. Still other radio galaxies emit their radiation from the galactic nucleus, a region usually devoid of interstellar gas.

Another problem is the nature of the postulated explosion. Although there are a number of galaxies that certainly look as if they are exploding (Figure 24-11), neither the cause nor the mechanism of such an explosion is known. A number of interesting theories have been proposed to account for the tremendous quantities of energy evidently released in these upheavals. Some astronomers have suggested a chain-reaction of supernova explosions among the closely spaced stars of the galactic nucleus; others, the gravitational collapse of an overcrowded nucleus.

QUASARS Of all the unusual objects recently discovered in our cosmos, quasars are the most puzzling. They have consequently excited the most interest, speculation, and controversy. When astronomers found that most of the strong radio sources in the heavens were either galaxies or nebulas within our own Galaxy, the idea that there might be "radio stars" was more or less abandoned. But further investigation showed that there were indeed some radio signals that seemed to come from tiny areas of the sky—too small, seemingly, to be anything but stars. And careful telescopic search of these areas did, in most cases, turn up faint starlike objects. At first, these objects seemed to be merely faint stars of our own Galaxy, but their appearance was slightly peculiar. Much more peculiar were their spectra, which contained emission lines at wavelengths corresponding to no known element or ion.

In 1963, astronomers identified the mysterious lines as those of common elements that were shifted much farther toward the red end of the spectrum by the Doppler effect than anyone had anticipated. The shifts indicated that these objects were traveling away from us at enormous velocities, ranging up to about 91 percent the speed of light. Applying Hubble's law, then, we can see that quasars must be very distant—billions of light-years away. Evidently, they cannot be stars, for no star could possibly be luminous enough to see at such a distance. Astronomers named these strange objects "quasi-stellar sources," which was soon shortened to **quasars**.

All quasars radiate quite incredible amounts of energy in the visible and infrared region; the visible radiation of some quasars is on the order of ten times that of a galaxy like our own. But even this remarkable luminosity is dwarfed by their infrared output, which is a thousand times greater. Nothing else in the universe—not even a supernova or the largest known galaxy—even approaches a quasar in energy production.

This outpouring of energy would perhaps be less astonishing if quasars were large objects, but we have evidence that they are not. Many quasars fluctuate irregularly in luminosity from year to year, from month to month, and in some instances from week to week. Whatever internal change is responsible for such fluctuations, it cannot propagate itself through the quasar faster than the speed of light. If a quasar were the size of a galaxy, such a change would take thousands of years to spread through the entire body, and any resultant change in luminosity would occur gradually over such a period. If a quasar can vary in brightness over a period of a week, we must conclude that it is less than a light-week in diameter—far, far smaller than any galaxy.

Though they are relatively small objects, quasars emit more radiant energy than even the brightest galaxies.

ARE QUASARS GALAXIES? We simply do not know what sort of object could exhibit the combination of properties shown by quasars. The great distances, the small size, and the immense energy output of quasars all puzzle astronomers. Perhaps the most promising clue to the riddle is the resemblance between quasars and a very peculiar class of galaxy known as Seyfert galaxies. Seyferts are similar to ordinary spiral galaxies in their general appearance, but their nuclei are tiny and extremely brilliant, radiating perhaps a hundred times as much energy as an ordinary galaxy, mostly in the infrared region of the spec-

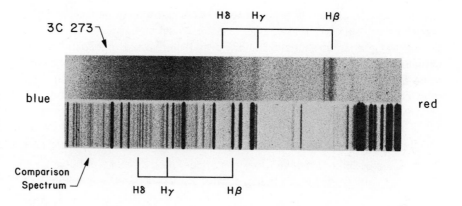

Figure 24-12
The spectrum of quasar 3C273, showing the shift of spectral lines towards the red end of the spectrum. This great Doppler shift indicates that 3C273 is receding from us at about 45,000 km/sec. If so, Hubble's law tells us that it must be more than 2 billion light-years away. Over a thousand quasars have now been discovered, including one that is receding from us with a velocity greater than 90 percent the speed of light, corresponding to a distance of 10 billion light-years.

trum. They are often unusually strong emitters at both ultraviolet and radio wavelengths, and their spectra exhibit strong emission lines, indicative of hot gases in their nuclei. In all these respects they resemble quasars. These similarities between quasars and Seyferts suggest that quasars may be simply the extreme stage of certain processes or accidents that occur in many galaxies, and that we can observe in less extreme form when we study Seyferts or exploding galaxies.

QUASAR ENERGY The central problem concerning quasars is the source of their enormous energy output. The theories advanced on this point are for the most part similar to those invoked to explain the exploding galaxies. Because quasars are so small, some astronomers have speculated that their stars, necessarily very crowded, must often collide. This process would release considerable energy, and if the stars frequently flared up into supernovas as a result, it would account for still more. Other theorists have speculated that a quasar may be a kind of galactic equivalent of a neutron star (Chapter 23)—a galaxy that has collapsed in upon itself. This process might also liberate enough energy to explain the quasar luminosity. At present, though, we just do not know.

Cosmology

How did the universe begin, and how will it end? Such questions, which have always engaged the imagination of mankind, belong to the realm of cosmology, the branch of astronomy that deals with the origin of the universe and its evolution through time.

Our difficulty in understanding the history of the universe is due partly to the fact that we see it only from one point in space; even more limiting is the fact that we exist in only one instant of cosmological time. In a universe in which significant changes occur on a time span of billions of years, our brief century of technologically sophisticated observation is not even the blink of an eye. We are aided by the fact that our large telescopes see far back in time as well as out into space: when we observe a galaxy 2 billion light-years distant, we are seeing it as it was two billion years ago. Even so, we can never watch any one place in the universe as it changes through the eons, and all our cosmological theories are subject to experimental verification at only one point in time. Thus, cosmology cannot today boast of definitive answers to the large questions it proposes, and it may never be able to supply such answers. But the advances of astronomical knowledge in recent decades at least enable us to substitute informed speculation and plausible hypotheses for the blind wonder and mythology of past ages.

All modern cosmological thought must take into account two fundamental facts. The first is the crucial role played by hydrogen in the life of the universe. At the present time, the universe is about three-fourths hydrogen, by mass. This figure must have been higher in the past, for the billions upon billions of stars in the cosmos derive most of their energy from converting hydrogen first into helium and then into heavier elements. Such transformations are irreversible; once the hydrogen has been "burned" in this way, it can never be recovered. This implies that some day all the hydrogen in the universe may be exhausted. There would then be nothing from which new stars could be formed, or at least nothing that they could burn for energy. The cosmos would be cold, void, and dead.

Cosmological Theories

But if the story of hydrogen points to a possible end for the universe, it also implies a possible beginning. If all the other elements have been built up from hydrogen, there must have been a time when there was nothing but hydrogen in the universe. The Population II stars, formed early in the history of the universe, are evidence for this view, for they consist almost entirely of hydrogen and helium (the product of hydrogen fusion), with very few of the heavier elements.

The second basic fact of cosmological significance is the observed recession of the galaxies, as described by Hubble's law. This also implies a possible beginning of the universe. For if we trace the movement of the galaxies backwards in time, we arrive at a moment when they must all have occupied a single small region of space. Could that moment have marked the birth of the universe as we know it?

There are three principal cosmological theories now in favor among scientists, and each answers this question in a different way. One of them, generally known as the big bang model of the universe, says yes. A related theory, known as the pulsating universe model, gives a qualified yes. The third theory, the steady state model, denies that there could ever have been such a beginning. The big bang and pulsating universe models can be classified as evolutionary cosmologies, because they picture a universe that changes with the passage of time. The steady state model, on the other hand, attempts to avoid any implication of large-scale change in the universe. It proposes a universe without beginning or end, one that remains essentially as we now see it from eon to eon. The pulsating universe model, like the big bang, can also be considered an evolutionary cosmology, since it accepts the fact of change. Like the steady state theory, however, it side-steps the notion of a beginning or end of the universe. It postulates instead that the universe repeats endlessly the same cyclical pattern of evolution.

THE BIG BANG MODEL The big bang theory, in its earliest version, was propounded in 1927 by the Belgian astronomer, Abbé Georges Lemaître (1894–1966). It was further developed and championed by two eminent physicists who were also writers and popularizers, Arthur Eddington and George Gamow. The big bang theory proposes that at one time all the matter and energy in the universe was concentrated in a small, inconceivably hot and dense ball, about the size of the solar system; Lemaître called this the primeval atom or cosmic egg. Under such conditions, even protons and electrons can no longer maintain separate existences, but are squeezed together to form neutrons. This mass of compressed neutrons exploded in the "big bang" or beginning of the universe. (Where the original matter of the cosmic egg came from, how long it may have existed before the big bang, and why it exploded, not even the proponents of this theory attempt to say.) The expanding mass of neutrons cooled rapidly, breaking down into protons and electrons, the particles that make up the hydrogen atom. According to the early versions of this theory, other elements were created by nuclear collisions during the early stages of expansion. Modern calculations suggest, however, that the creation of elements other than hydrogen and helium could not have taken place under such conditions, and,

The big bang theory proposes that the expansion of the universe that we now observe began with an enormous explosion between 10 and 20 billion years ago.

Figure 24-13
A schematic representation of the pulsating universe model. If gravitational forces are not strong enough, the universe will continue to expand indefinitely, as the big bang model originally proposed (first three pictures).

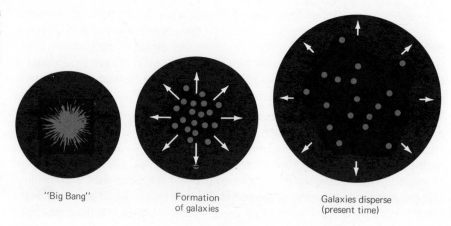

"Big Bang"   Formation   Galaxies disperse
                  of galaxies   (present time)

as we have seen, physicists today favor the theory that such elements were not formed until much later, in the interior of hot stars.

The expanding cloud of matter, whatever its composition, was fragmented into many smaller clouds by the turbulence of the explosion. These clouds, moving at varying velocities, eventually condensed to form the galaxies and clusters of galaxies. The expansion of the universe we observe today is the result of the different velocities imparted to the original matter by the big bang. Although the universe is still expanding, gravity is at work between the galaxies, acting as a rein on their velocities of recession. As time passes, the big bang theory predicts, the rate of expansion must decrease.

And what of the ultimate fate of the universe? That depends on the total mass of the universe. If most of the mass is present in the galaxies known today, the gravitational drag will be insufficient ever to halt the expansion, and the universe will become progressively colder, dimmer, and emptier as the galaxies proceed on their infinite recession.

THE OSCILLATING UNIVERSE Some theorists suggest that cold death is not to be the fate of the universe. If the density of matter in intergalactic space is sufficiently high, the gravitational forces will be great enough to stop the expansion of the universe. Once this happens, gravitational collapse will set in, and the entire process will be reversed. The galaxies will start sailing toward each other, eventually reaching the same spot and compressing down to another mass of primordial neutrons. At that point, presumably, the entire process could begin over again. This second scenario is known as the **pulsating** or **oscillating universe model,** for it predicts an infinite series of expansions and contractions of the universe. This model implies that the world never began and

Gravitational forces may eventually halt the expansion of the universe and cause it to contract; the result might be a pulsating universe, expanding and collapsing in an endless cycle.

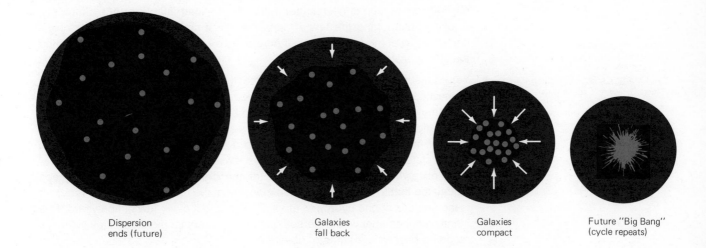

Dispersion
ends (future)

Galaxies
fall back

Galaxies
compact

Future "Big Bang"
(cycle repeats)

will never end, and that the matter of the universe existed indefinitely. Allan Sandage, working from observations of distant objects that indicate greater velocities of recession in past eons than today, has calculated a possible pulsation period of 80 billion years for the universe. He estimates that the universe started on its current expansion some 15 billion years ago.

All of these models depend on certain assumptions about the conditions that prevail and the physical laws that apply elsewhere in the cosmos. Usually, astronomers start by assuming that, for all time and in all regions of the universe, the laws of physics are the same as we know them here on earth and in our solar system. Evolutionary cosmological theories also assume that space is homogeneous, or pretty much the same, in all directions and throughout all its volume. This assumption, that over large regions of the universe the density of matter and radiation are roughly constant, is known as the **cosmological principle**. Evolutionary cosmologies do not extend the cosmological principle to include different periods of time. Such theories call for changes in the density of matter in the universe during the span of cosmic time since the big bang.

THE STEADY STATE THEORY In 1948, Thomas Gold, Hermann Bondi, and Fred Hoyle introduced a cosmological model based on what they felt to be a philosophically more satisfactory set of assumptions. They proposed the **perfect cosmological principle**: the universe is uniform not only in space, but also in time. The resulting cosmology is generally known as the steady state model of the universe. It assumes that there was no time when the universe was, on the average, any different from what we see today, and that there will be no such time. The universe had no beginning, and will have no end. It has always existed and will always exist, without any large-scale change.

In order to reconcile the perfect cosmological principle with the observed recession of the galaxies, some way must be found to keep the density of matter constant as the universe expands. The steady state theory postulates the continuous creation of new matter in the void of intergalactic space. Since this involves abandoning the law of conservation of mass and energy, one of the cornerstones of modern physics, it may seem a rather drastic way of supporting the steady state theory.

The steady state theory postulates that the universe is the same everywhere in space and does not change significantly with time.

Yet, no matter what cosmology is adopted, the problem of the creation of matter is inescapable. Other cosmological theories avoid the question of how and when the matter of the universe came into being, but the implicit assumption of its creation is always there. The steady state theory merely changes the creation of matter from a one-time affair to a continuous process. Although the steady state theory appears to violate the laws of physics, we may simply never have observed the rare occurrence of the creation of matter. It has been calculated that to offset the known expansion of the universe, all that is required is the creation of one hydrogen atom per cubic foot every few billion years. This infinitesimal amount would be sufficient to form replacement galaxies for those receding beyond our limits of observation.

To decide between these rival cosmological theories by observation, we must attempt to test those points on which their predictions differ. The big bang theory, for example, requires the density of galaxies to have been greater in the past than today. The steady state theory holds that this value is forever constant. So if we can determine the distribution of galaxies several billion years ago, it should be possible to distinguish between the two opposing viewpoints. We can theoretically do this simply by observing the distribution of galaxies several billion light-years distant. But at present it is very difficult for us to see many galaxies at such distances, and the uncertainties in measurement are often greater than the differences predicted by the cosmological models being tested. Nevertheless, there are some indications that there really was a greater galactic density in the distant past—support for the big bang theory.

The two cosmologies also predict different ages for the galaxies. According to the evolutionary model, all the galaxies were created at roughly the same time. Thus, when we look far out into space (back in time), we should see younger galaxies than we find in our immediate neighborhood. The steady state model, by contrast, calls for a random distribution of galactic ages, with very distant galaxies no different, on the average, from closer ones. Still another source of evidence is the rate of expansion of the universe, as expressed by Hubble's constant. The steady state theory assumes that the rate of expansion is unvarying, while the evolutionary cosmologies predict that it must be slowing down. If the value of Hubble's constant for distant galaxies, and thus for past eons, could be determined accurately enough, it would allow us to decide between these two cosmologies, and even to choose between a cyclical universe and one that will expand forever. Current evidence on both these points again seems slightly to favor the evolutionary models, but the difficulties in determining ages and distances of galaxies with the needed precision are very great, and the results are far from conclusive.

The discovery of quasars may shed additional light on the situation. If the quasars are really at cosmic distances, as most astronomers believe, the universe must have been quite different 10 billion years ago. This is yet another piece of evidence against the steady state theory. The most recent and perhaps final blow to the steady state theory was the discovery, in 1965, of a weak, diffuse background of microwave emission from space. This radiation is completely isotropic—that is, it is the same from every direction, as if it filled all space rather than emanating from a particular place or object. Unknown to its discoverers, just such a faint glow of microwave radiation had been predicted by Robert Dicke, an American physicist, as one of the results of the big bang. According to Dicke's calculations, at the time of the big bang, the universe must have been flooded with very intense radiation. With the expansion of the universe, the strength of this radiation must have declined, but a residuum of it should still be perceptible as a faint haze of microwaves. The detection of just such radiation was strong support for the big bang theory, and the steady state model, while not decisively defeated, is finding fewer and fewer supporters.

Cosmological Tests

Figure 24-14
A schematic representation of the steady state universe model. In order for the density of matter in the universe to remain constant as the galaxies recede from each other, new matter must be continuously created in interstellar space. As the older galaxies (red) disperse, new galaxies (white) form.

Galaxies disperse

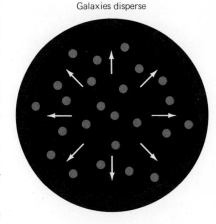

New galaxies form

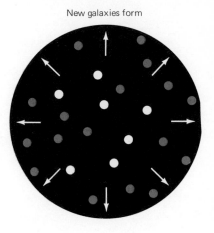

Glossary   The **big bang model** of the universe proposes that the universe began as a small, dense, extremely hot ball of concentrated matter and energy, which exploded, expanded, cooled, and broke up into separate clouds of gas out of which the galaxies eventually condensed.

Star clusters are associations of stars in our Galaxy that travel through space together under the influence of their mutual gravitational attraction. **Galactic clusters** are relatively loose groups of Population I stars found in the galactic disc. **Globular clusters** are more compact, spherical aggregations of Population II stars, usually found far from the plane of the galactic disc.

Cosmology is the branch of astronomy that studies the universe as a whole, especially its origin and its development through time.

A **galaxy** is a vast system of stars, usually numbering in the billions. The majority of galaxies, including our own, have a spiral structure, with a series of pinwheel shaped arms coiling in widening arcs about a denser central nucleus.

Hubble's law, a description of the expansion of the universe, states that the remote galaxies are receding from us with velocities proportional to their distances: $V = Hd$. The constant of proportionality, H (Hubble's constant), is about 17 km/sec per million light-years.

The **interstellar medium** contains molecules of gas, mostly hydrogen, and tiny solid particles (dust) in very low concentrations.

The **Milky Way** is a faintly visible band of light stretching across the night sky. It is formed by myriads of distant stars in the disc of our galaxy.

Nebulas are clouds of gas or tiny dust particles in interstellar space. When near bright stars, they glow by reflection or fluorescence. Dust clouds not illuminated by nearby stars form **dark nebulas**, obscuring the regions beyond them.

Population I stars, found in the spiral arms of the galaxy, are relatively young and include many hot, bright, blue stars. **Population II stars**, found in the galactic nucleus, halo, and globular clusters, tend to be smaller, cooler, and redder. They represent an older generation of stars.

The **pulsating universe model** proposes that the universe undergoes an unending series of expansions and contractions.

Quasars (an abbreviation of "quasi-stellar sources") are objects at the limits of the observable universe which emit enormous amounts of energy, mostly in the infrared spectrum. Quasars may be galaxies undergoing a catastrophic explosion or collapse.

The **steady state model** of the universe proposes that the universe undergoes no large-scale change in time: it has always been and will always be much the same as we now observe it. For this to be possible, matter must be continuously created in interstellar space, so that the density of matter in the universe remains constant despite expansion.

The 21-cm line is a radio wavelength emitted by neutral hydrogen in space. Since radio frequencies can penetrate the interstellar dust that obscures visible light, this radiation enables observers on earth to map our Galaxy in much greater detail than is possible using ordinary optical methods.

Exercises

1 Describe the structure of the galaxy in which we live. Why did the presence of interstellar obscuration lead Herschel into erroneously believing that the sun was at the center of a disc-shaped galaxy? (Consider the problem of trying to locate yourself in a thick fog.)

2 What is the evidence for the existence of interstellar dust and gas?

3 Why are reflection nebulas bluer than the stars that illuminate them?

4 We know that interstellar dust causes greater extinction of blue light than of red light. If you wished to undertake a survey of the nucleus of the Galaxy, obscured beyond hope in the visible region of the spectrum, at what wavelengths would you attempt to do so?

5 If there were any hitherto undiscovered small galaxies in our local group, where do you think they would lie?

6 Explain the various arguments and observations that led astronomers to conclude that there were galaxies similar to our own.

7 What is the 21-cm spectral line of hydrogen? Why did the prediction and discovery of this line lead to a great enlargement of the role of radio astronomy in our quest for knowledge about the universe?

8 What is Hubble's law? Why is there so much uncertainty regarding the exact value of Hubble's constant, H? We know the distance to the galaxy in Andromeda, M-31 quite accurately; why can this not be used to establish an accurate value of H?

9 If you were to make the assumption that the velocities of the galaxies had remained constant throughout the evolution of the universe, what would you estimate the age of the universe to be? (Work back in time till the universe was a point. 1 year $= 3 \times 10^7$ seconds.)

10 How do observations with powerful telescopes tell us about the universe as it appeared billions of years in the past?

11 Observations indicate that galaxies are receding radially away from us at great velocities. This seems to imply that we are at the center of the universe. Are we?

12 What are the observational objections to the steady state model?

13 A spectral line normally found at a wavelength of 5000 Å is discovered in the spectrum of a quasar at a location corresponding to a wavelength of 5500 Å. What is the radial velocity of this quasar?

14 If Hubble's constant is taken to be 17 km/sec per million light-years, determine the distance of the quasar in problem 13.

15 If the quasar of problems 13 and 14 is found to have the same apparent magnitude as a galaxy 70 million light-years from earth, how much brighter is it in reality?

There are millions of stars in the Cygnus region, which is only a portion of our Milky Way. Is it possible that some form of life exists somewhere among these stars?

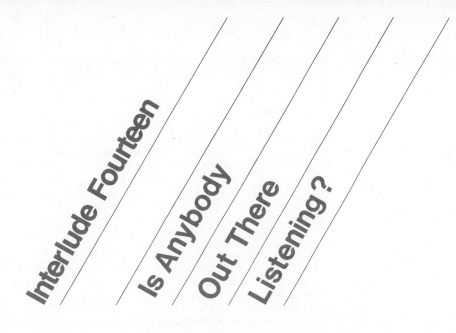

Interlude Fourteen

Is Anybody Out There Listening?

A search is on which — if successful — will alter our perception of ourselves and the entire human species more profoundly than the Copernican sun-centered revolution, which displaced the human home on earth from the center of the universe, or Darwin's theory of evolution, which demoted the human kind to merely the latest among the earthly species of animals and plants. The search is taking place in chemical laboratories here on earth, in orbiting satellites, in unmanned spacecraft traveling through the solar system, on the hostile surfaces of the moon and Mars, in the beautiful swirling rings of Saturn, on meteors and comets, and through radio signal beams that traverse the vast interstellar distances of our galactic space.

The search is for life elsewhere. Not necessarily little green men from Mars or mutant living blobs that engulf everything in their path or even monolith beacons buried in the moon by intelligent beings—although none of these can be totally ruled out. Many scientists from several disciplines, including biology, chemistry, physics, and astronomy, are convinced that somewhere out there life does exist. The study of life beyond the earth is called **exobiology**. Its first task is to find out whether it has a genuine subject to study.

Strong arguments support a positive answer. After all, life exists here on earth, one planet out of nine in our solar system. Within the Milky Way galaxy, of which our solar system is a part, there are some 100 billion stars, some of which undoubtedly contain planetary systems of their own. And the Milky Way is only one of some 10 billion such galaxies so far known by telescopic evidence, each with about 100 billion stars, and each with perhaps an equal number of planets. If only a small proportion of these planets were like our earth in chemical and geological properties, there would be millions of planets like earth in the Milky Way alone. Can this small speck of matter called earth be unique, the only place where life has originated?

Consider that the chemical components of living matter as we know it here on earth are structures based on atoms we know to exist throughout the universe of stars and galaxies. Further, we have reason to believe that the same laws of physics and chemistry govern atoms throughout the universe, to the most distant galaxies even at the beginning of the universe. The chemicals of life—such essential compounds as water, methane, ammonia, and hydrogen—have been observed spectroscopically, scattered throughout interplanetary and interstellar space, and within the surfaces of other planets here and there.

What would prevent biochemical chains of causation similar to those on earth from bringing life into existence elsewhere too? Life on earth probably began some two billion years after the planet formed, or so it seems from the earliest fossil records. Within that great epoch, most likely within the last hundreds of millions of years, it is believed that chance combinations of hydrogen, nitrogen, oxygen and carbon atoms occurred under the energetic catalytic action of lightning bolts and cosmic rays striking into the primeval nutrient soup of the oceans. Eventually the fundamental molecules of nature were formed: the *amino acids* for construction into proteins, which either build the bodily structures or form the biochemical control mechanisms, and the *nucleotides*, which then assembled into the nucleic acids, especially DNA, which provides the mechanism for reproduction of life forms. Once molecular structures had been formed which could reproduce themselves, then the same geochemical

Pioneer 11 was launched in April, 1973, to take photographs of Jupiter and Saturn. This portrait of Jupiter (below left) looks down on its north pole and was taken in December, 1974, when the spacecraft was 1.3 million miles from Jupiter's surface. Pioneer 11 will fly 100 million miles above the plane containing the sun and most of the planets in 1977 on its way to Saturn; at this height, it will be able to view parts of our solar system from 17° above the sun's equator, about twice as high as any previous observation. In September, 1979, the spacecraft will fly past Saturn; preliminary plans call for it to pass between the inner ring and the planet's visible surface and then to fly outbound for a close look at the Saturn moon Titan. This artist's rendering from NASA shows the Saturn fly-by. The spacecraft will then continue away from earth and out of the solar system.

environment which made that possible provided a home for the evolution of life.

How unusual was the earth experience that produced life out of these building blocks? How likely what Jacob Bronowski has called the "evolution of complexity" among the chemical structures? Maybe we will find bits of life on Mars. If we do, then it will no longer be so reasonable to think of life as an accident in our unique solar system and our unique planet. If there were two such places where life evolved this would suggest that life may be a regular occurrence on earthlike planets. Or so the exobiologist argues.

The search for life on other planets of our solar system is underway. By the end of the 1970s, Mercury, Venus, Mars, Jupiter, Saturn, and several of their moons will have been rather closely examined by American or Soviet robot space missions. Our solar system may not be the most likely place to look for extraterrestrial life, for all we know, but it offers important advantages: first, life is known to exist on one of the planets of our sun's system, the earth, and so the neighboring planetary environments seem possibly suitable for at least the earlier stages of biological development; second, the planets after all are *here*, close by when compared with the enormous distances to other suns and their solar systems.

The space probes of the late 1960s and the early 1970s have provided exciting new information about the physical conditions of neighboring planets. Some of this indicates that suitable life-sustaining environments do exist elsewhere within our solar system. Given our present knowledge, Mars is most promising of all.

Mars may now be in the grip of an ice age, and if so it may once have contained running water, which leads some scientists to speculate that primitive forms of life have already developed on that planet. These life forms may have died out during the coming of the ice, or they may have adapted to that hostile environment, waiting until times and conditions are more suited for the flourishing of life. Two American Viking spacecraft will land on Mars in mid-1976, with automated robot biological laboratories to test the Martian soil and atmosphere for life.

The striking similarities between Jupiter's present surface and atmosphere and those of the earth millions of years ago have led some exobiologists to conjecture that simple forms of life may right now be developing on that huge planet. If life has already evolved on Jupiter, it is probably located somewhere between the boiling surface and frozen cloud tops.

The planets are not the only places to look. From knowledge of the chemical evolution of life here on earth and of the existence of the required molecules in interstellar matter, we can expect that meteors, comets, and interplanetary dust clouds have environments to sustain life at its most elementary stage.

While space probes provide some hope for the discovery of extraterrestrial life within our solar system, it is more probable that life—in both its primitive and its intelligent forms—exists in other solar systems of the Milky Way and in other galaxies. Exobiologists face the question of how to test such a hypothesis. The rest of us may wonder what to do if the hypothesis is correct.

The vast almost endless distances of intragalactic space make it impossible at present—on a practical as well as theoretical level—to travel there and back. The distances to other galaxies seem utterly beyond speculative travel. But within the local galaxy, we need not travel at all. *Why not talk to them? Why not listen to them?* Scientists are trying to communicate with any intelligent far-off civilizations. Their method is to use powerful radio telescopes designed to listen for messages from outer space and to transmit our own messages to other stars and their solar systems.

Since 1960, more than 300 astro-chemically promising stars have been

Two unmanned Viking spacecraft are scheduled to land on Mars in 1976 to conduct various chemical and biological experiments on the planet's surface. This device, called a surface sampler, will be deployed from the spacecraft on an extendable boom that will unroll like a metal tapemeasure to permit the scoop to gather soil and bring it into the spacecraft to be tested.

scanned for possible radio signals directed toward our sun and its planets. In some instances, messages translated into what was hoped is a universal code have been transmitted outward to space targets from the antennas of radio-telescopes on earth. One such message from the Arecibo telescope in Puerto Rico in 1974 was directed toward two neighboring stars. Unfortunately, even at the speed of light, it will take about 24,000 years for the message to arrive there, and of course another 24,000 years to receive an answer if there is anyone listening at that time who also wishes to answer.

The search for extraterrestrial life, and especially for intelligent civilizations, has stimulated much argument in the scientific community. Many believe the task to be hopeless. Others prefer, whatever their estimate of the task of exobiology, that scientists should focus their imagination, intelligence, and socially-financed resources on human problems here on earth. Wait, they say, until our earthly home has been put in order, for there is profound crisis for the human species which demands all the competence we can find among ourselves. But there are those who favor a continued, even an expanded search for life out there. In his book *Other Worlds*, the American astronomer Carl Sagan argued for the search this way:

> I cannot say I believe that there is life out there. But it is possible. Some of us think it is probable. Our first halting steps into space have shown us other worlds far stranger and more interesting than imagined by authors of the most exotic fiction.

> For the first time in the history of our species, we have devised tools—unmanned space vehicles and large radio telescopes—to search for extraterrestrial life. I would be very ashamed of my civilization if, with these tools at hand, we turned away from the cosmos.

If you want to talk to an alien from another solar system, what language do you use? And what do you say? This is a visualization of the Arecibo message that is being transmitted by radio in the direction of the star cluster Messier 13. It is the latest earthling attempt to contact intelligent life on whatever planets may be orbiting the estimated 300,000 stars in Messier 13. The message is encoded in a binary number system—one which can be written using only two symbols—and has a total of 1,679 characters. It is designed to inform extraterrestrial beings about earth and earthlings. At the top of the message is a description of the number system used in the code, a listing of numbers 1–10 in binary notation. In the second row are the atomic numbers of hydrogen, carbon, nitrogen, oxygen, and phosphorus. The next group of lines show the structures of the phosphate group, deoxyribose, thymine, adenine, guanine, and cytosine—the building blocks of DNA. Below these is a crude sketch of a double helix molecule along with a figure indicating the number of atoms in the components of DNA molecules. Underneath the double helix is a picture of an earthling; to the right are figures showing the height of the earthling and to the left are numbers indicating earth's population. The next row is a schematic of the solar system, with the sun on the right and the earth displaced to show that it is the source of the message. Beneath the solar system is a picture of the telescope used to transmit the message; the last row gives the diameter of the telescope. The Arecibo message was first transmitted on November 16, 1974, and is repeated automatically whenever the telescope is not in use for other work. It takes 169 seconds to transmit the entire message.

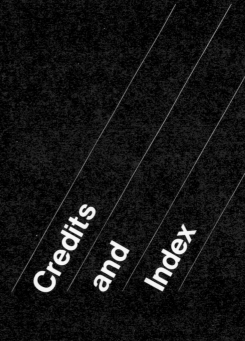

Credits
and
Index

Interlude *Cygnus stars* Lick Observatory
Motion of Pluto Hale Observatories

Chapter One *Figure 1-1* N. Y. C. Public Library Picture Collection
Figure 1-2 Gerry Cranham/Rapho-Photo Researchers, Inc.
Figure 1-4 Dennis Stock/Magnum Photos
Figure 1-5 U. P. I. Photo
Figure 1-7 From *PSSC Physics*, D. C. Heath, Lexington, Mass., 1965
Figure 1-8 The Bettmann Archive, Inc.

Interlude *Stonehenge* Freelance Photography Guild
Stonehenge British Tourist Authority
Hand dial Reprinted by permission of the publisher Horizon Press, New York, from *Clockwork Man* by Lawrence Wright, copyright 1968.
Cannon sundial Adler Planetarium, Chicago
Sundial sculpture The *Times*, London
Hourglass Smithsonian Institute, number 59716
Waterclock Crown Copyright, The Science Museum, London
Grandfather clock Erich Hartmann/Magnum Photos
Galileo's pendulum G. Sansoni, Florence, Italy
Big Ben British Tourist Authority
Quartz crystal clock Lent to Science Museum, London by G. P. O.
Atomic clock U. S. Dept. of Commerce

Chapter Two *Figure 2-1* U. P. I. Photo
Figure 2-2 NASA
Figure 2-3 Harold E. Edgerton, M. I. T., Cambridge, Massachusetts
Figure 2-8 From *PSSC Physics*, D. C. Heath, Lexington, Mass., 1965.
Figure 2-12 Gerry Cranham/Rapho-Photo Researchers, Inc.
Box The Bettmann Archive, Inc.

Chapter Three *Figure 3-1* © Rube Goldberg, permission granted by King Features Syndicate
Figure 3-3 Joseph Needham, *Science and Civilization in China*, Cambridge University Press.
Figure 3-4 Bruce Roberts/Rapho-Photo Researchers, Inc.
Figure 3-5 U. P. I. Photo
Figure 3-6 Ray Ellis/Rapho-Photo Researchers, Inc.
Figure 3-7 Gerry Cranham/Rapho-Photo Researchers, Inc.
Figure 3-8 Claude Gazuit/Rapho-Photo Researchers, Inc.
Box Escher Foundation-Haags Gemeentemuseum-The Hague

Interlude *Bob Beamon* U. P. I. Photo
Bob Seagren U. P. I. Photo

Igor Ter-Ovanyesan Phil Bath/Visual Track and Field Techniques

Chapter Four *Figure 4-1* C. Whelan/Rapho-Photo Researchers, Inc.
Figure 4-2 Fritz Goro
Figure 4-3 U. P. I. Photo
Figure 4-4 Fundamental Photographs for the Granger Collection
Box Richard Holt/Photo Researchers

Chapter Five *Figure 5-1* The Burndy Library
Figure 5-10 Rudolf Freund/Photo Researchers, Inc.
Box Inge Morath/Magnum Photos

Chapter Six *Figure 6-1* The Burndy Library
Figure 6-4 Westinghouse, Inc.
Figure 6-6b From *PSSC Physics*, D. C. Heath, Lexington, Mass., 1965
Figure 6-7 N. Y. C. Public Library Picture Collection
Figure 6-11 Sylvania, Inc.
Box From Lennart Nilsson, *Behold Man*, Little, Brown and Company, Boston, 1974

Interlude *Kite* Bruce Roberts/Rapho-Photo Researchers, Inc.
Franklin portrait The Franklin Institute

Color Portfolio *Page 1* Fritz Goro
Page 2 Fritz Goro
Page 3: razor blades Ken Kay
Page 3: willows Douglas Faulkner
Page 4 Bettmann Archive, Inc.
Page 5 Fritz Goro
Page 6 J. Paul Kirouac
Page 7-9 Fritz Goro
Page 10–11 Ken Kay
Page 12: top left Charles Gellis/Photo Researchers, Inc.
Page 12: other three crystals H. L. Garrett/Dow Chemical
Page 13 Fritz Goro
Page 14: butterfly Alan Blank from National Audubon Society
Page 14: moth E. J. Howard from National Audubon Society
Page 15 Douglas Faulkner
Page 16 Fritz Goro
Page 17 Douglas Faulkner
Page 18 Fritz Goro
Page 19–20 Lee Boltin
Page 21: emeralds Carl Frank/Photo Researchers, Inc.
Page 21: diamonds Courtesy of Tiffany and Company
Page 22: iridescent limonite Lee Boltin
Page 22: two photos of moon rock Fritz Goro
Page 23: lumachelle Lee Boltin
Page 23: fossil algae Fritz Goro
Page 24: Aurora Borealis Charlie Ott from National Audubon Society
Page 24: sunset A. L. Goodman/Rapho-Photo Researchers, Inc.

Page 25: *rainbow glory* Fritz Goro
Page 25: *rainbow at Yosemite* Dick Rowan/Photo Researchers, Inc.
Page 26: *hurricane* Bradley Smith/Photo Researchers, Inc.
Page 26: *bridge* Rudolf Freund/Photo Researchers, Inc.
Page 27 Myron Wool/Photo Researchers
Page 28 NASA
Page 29: *planets* Copyright by the California Institute of Technology and Carnegie Institution of Washington
Page 29: *sun in ionized light* Susan Rayfold
Page 30 Copyright by the California Institute of Technology and Carnegie Institution of Washington
Page 31: *global star cluster in Hercules* U. S. Naval Observatory
Page 31: *planetary nebula in Aquarius* Copyright by the California Institute of Technology and Carnegie Institution of Washington
Page 32 Copyright by the California Institute of Technology and Carnegie Institution of Washington

Chapter Seven
Figure 7-1 Wards Natural Science Establishment, Inc.
Figure 7-2 From *Dutton's Navigation and Piloting* by G. D. Dunlap and H. H. Shufeldt. Copyright 1972, United States Naval Institute.
Figure 7-5b Bernice Abbott/Photo Researchers, Inc.
Figure 7-7 Holt, Rinehart and Winston
Figure 7-11, 13b General Electric
Figure 7-14 Paolo Koch/Rapho-Photo Researchers, Inc.

Interlude
Walking truck General Electric
Robot Comics Marvel Comics Group

Chapter Eight
Figure 8-4 Bettmann Archive, Inc.
Figure 8-8 Bernice Abbot/Photo Researchers, Inc.

Interlude
Disk on moon NASA
Laser drill International Laser Systems
Diamond hologram Cartier

Chapter Nine
Figure 9-2a,b Bernice Abbott/Photo Researchers, Inc.
Figure 9-2c Carl Purcell/Photo Researchers, Inc.
Figure 9-3 H. M. Null/Rapho-Photo Researchers, Inc.
Figure 9-4 Bernice Abbott/Photo Researchers, Inc.
Figure 9-8a Fritz Henle/Photo Researchers, Inc.
Figure 9-8b Photo Researchers, Inc.
Figure 9-9 Bernice Abbott/Photo Researchers, Inc.
Figure 9-10 Fundamental Photographs for the Granger Collection

Figure 9-11 J. Paul Kirouac
Box Leonard Lee Rue III/Bruce Coleman

Chapter Ten
Figure 10-7 Albright-Knox Art Gallery, Buffalo, New York; courtesy of George F. Goodyear and The Buffalo Fine Arts Academy
Figure 10-10 Line art: Virginia Kudlak Photo: courtesy of the Archives, California Institute of Technology
Box Illustrations by John Tenniel

Chapter Eleven
Figure 11-5 AIP Neils Bohr Library
Figure 11-7 Gamov
Figure 11-11 AIP Neils Bohr Library

Chapter Twelve
Figure 12-1 AIP Neils Bohr Library
Figure 12-2 Fritz Goro
Figure 12-3 Lawrence Radiation Laboratory, Berkeley, California
Figure 12-4 Brookhaven National Laboratory
Figure 12-6 Cornell University
Figure 12-7 Bibal/Rapho-Photo Researchers, Inc.
Box These cartoons were prepared by Dr. David L. Judd and Ronald G. MacKenzie, Lawrence Berkeley Laboratory, University of California, to accompany Dr. Judd's keynote speech at the International Conference at Gatlinburg, Tennessee, published in *IEEE Transactions on Nuclear Science,* Vol. NS-13, No. 4, 1966.

Interlude
Fermilab Tony Frelo, Fermilab, National Accelerator Lab Photo
Fermilab tunnel Fermilab Photo
CERN Photo CERN
Brookhaven tunnel Brookhaven National Laboratory
Brookhaven cosmotron Brookhaven National Laboratory

Chapter Thirteen
Figure 13-1 The Granger Collection
Figure 13-2 John R. Freeman and Co.
Figure 13-3, 4 The Bettmann Archive
Figure 13-6 American College of Radiology
Figure 13-7 Photo Researchers, Inc.
Figure 13-10 After J. B. Marion, *Physical Science in the Modern World,* Harcourt Brace Jovanovich, 1974, p. 531.

Chapter Fourteen
Figure 14-4 LIFE Photo by Fritz Goro
Figure 14-5 DeBeers Consolidated Mines Ltd.
Figure 14-7, 8 Photo Researchers, Inc.
Figure 14-9 The Bettmann Archive, Inc.
Figure 14-11 Bill Brooks/Bruce Coleman
Box Courtesy of Professor M. H. F. Wilkins, Kings College, London

Interlude
Converter General Motors

Chapter Fifteen
Figure 15-1 NASA
Figure 15-4a L. L. Smith/Photo Researchers, Inc.
Figure 15-4b The Bettmann Archive, Inc.

Index